Advanced Structured Materials

Volume 141

Common engineering materials reach in many applications their limits and new developments are required to fulfil increasing demands on engineering materials. The performance of materials can be increased by combining different materials to achieve better properties than a single constituent or by shaping the material or constituents in a specific structure. The interaction between material and structure may arise on different length scales, such as micro-, meso- or macroscale, and offers possible applications in quite diverse fields.

This book series addresses the fundamental relationship between materials and their structure on the overall properties (e.g. mechanical, thermal, chemical or magnetic etc.) and applications.

The topics of *Advanced Structured Materials* include but are not limited to

- classical fibre-reinforced composites (e.g. glass, carbon or Aramid reinforced plastics)
- metal matrix composites (MMCs)
- micro porous composites
- micro channel materials
- multilayered materials
- cellular materials (e.g., metallic or polymer foams, sponges, hollow sphere structures)
- porous materials
- truss structures
- nanocomposite materials
- biomaterials
- nanoporous metals
- concrete
- coated materials
- smart materials

Advanced Structured Materials is indexed in Google Scholar and Scopus.

More information about this series at http://www.springer.com/series/8611

Holm Altenbach · Victor A. Eremeyev · Leonid A. Igumnov
Editors

Multiscale Solid Mechanics

Strength, Durability, and Dynamics

Springer

Editors
Holm Altenbach
Faculty of Mechanical Engineering
Otto von Guericke University Magdeburg
Magdeburg, Sachsen-Anhalt, Germany

Leonid A. Igumnov
Research Institute for Mechanics
State University of Nizhny Novgorod
Nizhny Novgorod, Russia

Victor A. Eremeyev
Faculty of Civil and Environmental Engineering
Gdańsk University of Technology
Gdańsk, Poland

ISSN 1869-8433 ISSN 1869-8441 (electronic)
Advanced Structured Materials
ISBN 978-3-030-54927-5 ISBN 978-3-030-54928-2 (eBook)
https://doi.org/10.1007/978-3-030-54928-2

This Springer imprint is published by the registered company Springer Nature Switzerland AG
The registered company address is: Gewerbestrasse 11, 6330 Cham, Switzerland

Preface

The present book is a collection of papers of leading Russian scientists devoted to the description and simulation of dynamics and strength of materials and durability of systems. The problems arising from the operation of machines and systems under multifactorial influences are considered. The design and experimental assessment of the processes of deformation, the development of damage, the destruction of materials and structures are studied. For given loading history, the appearance of macroscopic material discontinuities are determined. The fundamental dependences on the strain rates of the ultimate characteristics of strength, dynamics, etc., are formulated.

When modeling, the significant influence on the rate of the processes of deformation, accumulation of damage, fracture, and such factors as the type of deformation trajectory should be taken into account. In addition, the temperature change, the stress state, the history of the stress state changes, etc., are considered. In the framework of damage mechanics, the mechanisms of degradation of the initial strength properties of structural materials under various conditions of thermo-force loading are studied. The kinetics of the stress–strain state in the corresponding places of structural elements is introduced in the theoretical models and simulations. The numerical experiments are based on variational-difference and finite element technologies. The correctness of the simulations are proofed by experiments.

The results of numerical modeling of the processes of visco-plastic deformation of structural elements are presented for several applied problems. Together with elements of the analysis of the current state of the experimental, theoretical, and numerical studies of the behavior of materials under quasi-static, cyclic, and dynamic loading, the development and implementation of new simulation methods are presented. The identification and verification of mathematical models of deformation, damage and fracture of structural materials is carried out including experimental methods of high-speed deformation based on the Kolsky method.

The problems of basic experiments related to identification of the material parameters and functions used material models are discussed. The study of the processes of deformation and fracture of structural materials is carried out from the perspective of an experimental-theoretical approach. Finally, applications are presented including not only mechanical loading, but also the influence of the temperature regimes, etc. Considering the scale factor multiscale approaches are applied.

The editors hope that the readers of this book can get an excellent insight into actual theoretical, numerical and experimental research results of Russian scientists working in the field of mechanics.

Magdeburg, Germany — Holm Altenbach
Gdańsk, Poland — Victor A. Eremeyev
Nizhny Novgorod, Russia — Leonid A. Igumnov
June 2020

Contents

Contributors

Dmitry R. Abashev Department of Technical Mechanics, Moscow Polytechnic University, Moscow, Russian Federation

Levon V. Agamirov Moscow Power Engineering Institute, National Research University, Moscow, Russia

Vladimir L. Agamirov "Volga" Ltd, Sevastopolsky Prospect, Moscow, Russia

Sergei M. Aizikovich Don State Technical University, Rostov-on-Don, Russia

Olga M. Babeshko Kuban State University, Krasnodar, Russia

Vladimir A. Babeshko Federal Research Centre the Southern Scientific Centre of the Russian Academy of Sciences, Rostov-on-Don, Russia

Dmitry V. Balandin National Research Lobachevsky State University of Nizhny Novgorod, Mathematical Center, Nizhny Novgorod, Russian Federation

Vladimir V. Balandin Research Institute for Mechanics, National Research Lobachevsky State University of Nizhny Novgorod, Nizhny Novgorod, Russian Federation

Valentin G. Bazhenov Research Institute for Mechanics, National Research Lobachevsky State University of Nizhny Novgorod, Nizhny Novgorod, Russian Federation

Valentin S. Bondar Department of Technical Mechanics, Moscow Polytechnic University, Moscow, Russian Federation

Anatoly M. Bragov Research Institute for Mechanics, National Research Lobachevsky State University of Nizhny Novgorod, Nizhny Novgorod, Russian Federation

Igor Demin National Research Lobachevsky State University of Nizhny Novgorod, Nizhny Novgorod, Russian Federation

Dmitry Dudin Perm National Research Polytechnic University, Perm, Russia

Nadezhda S. Dyukina Research Institute for Mechanics, National Research Lobachevsky State University of Nizhny Novgorod, Nizhny Novgorod, Russian Federation

Vladimir I. Erofeev Mechanical Engineering Research Institute of Russian Academy of Sciences, Nizhny Novgorod, Russia

Olga V. Evdokimova Federal Research Centre the Southern Scientific Centre of the Russian Academy of Sciences, Rostov-on-Don, Russia

Vyacheslav A. Firsov Kazan National Research Technical University named after A.N. Tupolev, Kazan, Russian Federation

Leonid V. Fomin Research Institute of Mechanics, Lomonosov Moscow State University, Moscow, Russia

Mikhail M. Gadenin Mechanical Engineering Research Institute of the RAS, Moscow, Russia;
Joint Stock Company Afrikantov OKB Mechanical Engineering, Nizhny Novgorod, Russia

Ruslan K. Gazizullin Kazan National Research Technical University named after A.N. Tupolev, Kazan, Russian Federation

Dimitri Georgievskii Mechanical and Mathematical Department, Moscow State University, Moscow, Russia

Sergey I. Gerasimov Russian Federal Nuclear Center–All-Russian Research Institute of Experimental Physics, Sarov, Nizhny Novgorod Region, Russian Federation

Irina G. Goryacheva Ishlinsky Institute for Problems in Mechanics RAS, Moscow, Russian Federation

Sergey Gurbatov National Research Lobachevsky State University of Nizhny Novgorod, Nizhny Novgorod, Russian Federation

Leonid A. Igumnov Research Institute for Mechanics, National Research Lobachevsky State University of Nizhny Novgorod, Nizhny Novgorod, Russian Federation

Rashit A. Kayumov Kazan State University of Architecture and Engeneering, Kazan, Russia;
Kazan National Research Technical University named after A.N. Tupolev, Kazan, Russia

Dmitry A. Kazakov Research Institute for Mechanics, National Research Lobachevsky State University of Nizhny Novgorod, Nizhny Novgorod, Russian Federation

Ilya Keller Institute of Continuous Media Mechanics of the UB RAS, Perm, Russia

Sergey A. Kholmogorov Kazan National Research Technical University named after A.N. Tupolev, Kazan, Russian Federation

Irina V. Kirillova Saratov State University, Saratov, Russian Federation

Ekaterina A. Korovaytseva Institute of Mechanics, Lomonosov Moscow State University, Moscow, Russia

Leonid Yu. Kossovich Saratov State University, Saratov, Russian Federation

Vasily L. Kotov National Research Lobachevsky State University of Nizhny Novgorod, Nizhny Novgorod, Russian Federation

Alexander E. Kraus Khristianovich Institute of Theoretical and Applied Mechanics SB RAS, Novosibirsk, Russia

Evgeny I. Kraus Khristianovich Institute of Theoretical and Applied Mechanics SB RAS, Novosibirsk, Russia

Anna V. Leonteva Mechanical Engineering Research Institute of Russian Academy of Sciences, Nizhny Novgorod, Russia

Igor I. Lipatov Central Aerohydrodynamic Institute (TsAGI), Moscow, Russian Federation

Elena E. Lisenkova Mechanical Engineering Research Institute of RAS, Nizhny Novgorod, Russian Federation

Artem Lisin National Research Lobachevsky State University of Nizhny Novgorod, Nizhny Novgorod, Russian Federation

Alexandr N. Litvinenko South Federal University, Rostov-on-Don, Russia

Andrey K. Lomunov Research Institute for Mechanics, National Research Lobachevsky State University of Nizhny Novgorod, Nizhny Novgorod, Russian Federation

Maksim V. Makarov Kazan National Research Technical University named after A.N. Tupolev, Kazan, Russian Federation;
Kazan (Volga region) Federal University, Kazan, Russian Federation

Yulia Yu. Makhovskaya Ishlinsky Institute for Problems in Mechanics RAS, Moscow, Russian Federation

Nikolay A. Makhutov Mechanical Engineering Research Institute of the RAS, Moscow, Russia;
Joint Stock Company Afrikantov OKB Mechanical Engineering, Nizhny Novgorod, Russia

Alexey O. Malkhanov Mechanical Engineering Research Institute of RAS, Nizhny Novgorod, Russian Federation

Sergey V. Maslov Mechanical Engineering Research Institute of the RAS, Moscow, Russia;
Joint Stock Company Afrikantov OKB Mechanical Engineering, Nizhny Novgorod, Russia

Evgeny E. Maslov Russian Federal Nuclear Center VNIIEF, Sarov, Nizhny Novgorod, Russia

Vladimir S. Metrikin Research Institute for Mechanics, National Research Lobachevsky State University of Nizhny Novgorod, Nizhny Novgorod, Russian Federation

Andrei V. Metrikine Faculty of Civil Engineering and Geosciences, Delft University of Technology, Delft, The Netherlands

Ivan A. Modin Research Institute for Mechanics, National Research Lobachevsky State University of Nizhny Novgorod, Nizhny Novgorod, Russian Federation

Julia A. Morozova Institute of Mathematics, Mechanics and Computer Science named after I.I. Vorovich, Southern Federal University, Rostov-on-Don, Russia

Anatoly S. Okonechnikov Moscow Aviation Institute, Moscow, Russia

Maxim Yu. Orlov Tomsk State University, Tomsk, Russia

Yulia N. Orlova Tomsk Polytechnic University, Tomsk, Russia

Vitaly N. Paimushin Kazan National Research Technical University named after A.N. Tupolev, Kazan, Russian Federation;
Kazan (Volga region) Federal University, Kazan, Russian Federation

Irina L. Pankratieva Institute of Mechanics, Lomonosov Moscow State University, Moscow, Russia

Vladimir A. Panov Mechanical Engineering Research Institute of the RAS, Moscow, Russia

Igor S. Pavlov National Research Lobachevsky State University of Nizhny Novgorod, Nizhny Novgorod, Russian Federation

Egor V. Petrakov Research Institute for Mechanics, National Research Lobachevsky State University of Nizhny Novgorod, Nizhny Novgorod, Russian Federation

Van Khiem Pham Moscow Institute of Physics and Technology, Dolgoprudny, Moscow, Russian Federation

Sergey N. Pichkov Mechanical Engineering Research Institute of the RAS, Moscow, Russia

Dmitrii K. Plotnikov Southern Mathematical Institute of the Vladikavkaz Scientific Center of the Russian Academy of Sciences, Vladikavkaz, Russia

Natalya V. Polyakova Kazan National Research Technical University named after A.N. Tupolev, Kazan, Russian Federation

Vitaly A. Polyansky Institute of Mechanics, Lomonosov Moscow State University, Moscow, Russia

Nikolay Pronchatov-Rubtsov National Research Lobachevsky State University of Nizhny Novgorod, Nizhny Novgorod, Russian Federation

Sergey G. Pshenichnov Institute of Mechanics, Lomonosov Moscow State University, Moscow, Russia

Dmitry O. Reznikov Mechanical Engineering Research Institute of the RAS, Moscow, Russia;
Joint Stock Company Afrikantov OKB Mechanical Engineering, Nizhny Novgorod, Russia

Vladimir I. Romanov Russian Federal Nuclear Center VNIIEF, Sarov, Nizhny Novgorod, Russia

Alexander A. Ryabov Russian Federal Nuclear Center VNIIEF, Sarov, Nizhny Novgorod, Russia

Marina Yu. Ryazantseva Institute of Mechanics Lomonosov Moscow State University, Moscow, Russia

Vladimir M. Sandalov National Research Lobachevsky State University of Nizhny Novgorod, Nizhny Novgorod, Russian Federation

Fedor G. Sarafanov National Research Lobachevsky State University of Nizhny Novgorod, Nizhny Novgorod, Russian Federation

Georgy F. Sarafanov Mechanical Engineering Research Institute of Russian Academy of Sciences, Nizhny Novgorod, Russian Federation

Ivan I. Shabalin Khristianovich Institute of Theoretical and Applied Mechanics SB RAS, Novosibirsk, Russia

Ravil Shabaykin Mechanical and Mathematical Department, Moscow State University, Moscow, Russia

Farid R. Shakirzyanov Kazan State University of Architecture and Engeneering, Kazan, Russia;
Kazan National Research Technical University named after A.N. Tupolev, Kazan, Russia

Ashot V. Shekoyan Institute of Mechanics of the National Academy of Sciences of the Republic of Armenia, Yerevan, Armenia

Maksim A. Shishov Kazan National Research Technical University named after A.N. Tupolev, Kazan, Russian Federation;
Kazan (Volga region) Federal University, Kazan, Russian Federation

Alexander S. Skaliukh Southern Federal University, Rostov-on-Don, Russia

Vladimir A. Skripnyak National Research Tomsk State University, Tomsk, Russia

Vladimir V. Skripnyak National Research Tomsk State University, Tomsk, Russia

Aleksey Spivak National Research Lobachevsky State University of Nizhny Novgorod, Nizhny Novgorod, Russian Federation

Natalya S. Starodubrovskaya Nizhny Novgorod State Technical University n.a. R.E. Alekseev, Nizhny Novgorod, Russian Federation

Dmitry V. Tarlakovskii Moscow Aviation Institute, Institute of Mechanics Lomonosov Moscow State University, Moscow, Russia

Andrey S. Vasiliev Don State Technical University, Rostov-on-Don, Russia

Alexander O. Vatulyan Institute of Mathematics, Mechanics and Computer Science named after I.I. Vorovich, Southern Federal University, Rostov-on-Don, Russia

Vladimir A. Vestyak Moscow Aviation Institute, National Research University, Moscow, Russia

Sergey Ye. Vlasov Scientific Research Institute for System Analysis, Russian Academy of Sciences (SRISA), Moscow, Russian Federation

Ivan A. Volkov Research Institute for Mechanics, National Research Lobachevsky State University of Nizhny Novgorod, Nizhny Novgorod, Russian Federation

Sergey S. Volkov Don State Technical University, Rostov-on-Don, Russia

Tatiana N. Yuzhina Research Institute for Mechanics, National Research Lobachevsky State University of Nizhny Novgorod, Nizhny Novgorod, Russian Federation

Andrei V. Zemskov Moscow Aviation Institute, Institute of Mechanics Lomonosov Moscow State University, Moscow, Russia

Dmitry V. Zhegalov Research Institute for Mechanics, National Research Lobachevsky State University of Nizhny Novgorod, Nizhny Novgorod, Russian Federation

Chapter 1
Practical Methods for Fatigue Characteristics Assessing Based on the Monotonic Diffusion Distribution Under Random Censorship Conditions

Levon V. Agamirov, Vladimir L. Agamirov, and Vladimir A. Vestyak

Abstract The article offers practical methods for processing the results of fatigue testing of materials and failure of structural elements under conditions of random censorship. A monotonic diffusion distribution is considered as the basic probability distribution of a random value of durability until a critical state is reached, a feature of the physical model of which is the justification of the process of irreversible destruction of mechanical systems. The methods under consideration are aimed at obtaining quick and effective estimates, eliminating the need to use complex equations of maximum likelihood for repeatedly censored sampling sets. It is shown that in the field of high durability and low fracture probabilities, the use of the traditional log-normal distribution leads to significant errors in the estimation of quantiles. In these cases, it is recommended to use the proposed models. The adequacy of the proposed approaches was verified by the Monte Carlo method, which showed good agreement between the experimental and calculated parameters.

Keywords Fatigue testing · Monotonic diffusion · Probabilistic model · The Monte Carlo method · Bootstrap modeling · Quantile method

---

L. V. Agamirov
Moscow Power Engineering Institute, National Research University,
Krasnokazarmennaya Street, 14, Moscow 111250, Russia
e-mail: itno_agamirov@mail.ru

V. L. Agamirov
"Volga" Ltd, Sevastopolsky Prospect, 56/40, Build. 1, Moscow 117342, Russia
e-mail: avhere@yandex.ru

V. A. Vestyak (✉)
Moscow Aviation Institute, National Research University,
Volokolamskoe Highway, 4, Moscow 125993, Russia
e-mail: kaf311@yandex.ru

H. Altenbach et al. (eds.), *Multiscale Solid Mechanics*,
Advanced Structured Materials 141,
https://doi.org/10.1007/978-3-030-54928-2_1

1.1 Introduction

Studies on the structural materials resistance to fatigue failure show that the use of normal or log-normal distributions in the fatigue characteristics dispersion models often leads to errors in the distribution parameters estimation [1, 5–7, 9, 14, 19, 20, 22, 23, 26, 29] especially, when evaluating quantile (gamma-percent) values of durability and endurance limits. These errors are significantly noticeable when testing at low levels of variable stresses amplitudes due to the manifestation of threshold values [26], which require the use of three-parameter distributions. From a mathematical point of view, their use is extremely undesirable due to the need for a significant increase in the volume of the sample population. One way to solve this problem is to use probabilistic models based on the physical processes of damage accumulation in structural materials, as well as leading to the failure of machine parts and structural elements. To this end, to solve the problems, in this paper, we consider the monotonic diffusion distribution model (inverse normal or DM distribution). It is known that the mechanical system degradation process due to the fracture processes irreversibility (mechanical wear, low-cycle and high-cycle fatigue) is considered to be a continuous diffusion type Markov wear process with independent increments, whose implementations are monotonous [17, 18, 27, 28].

In [8, 13, 18], it is noted that the diffusion distribution better describes the experimental data compared with two-parameter models. It is important that the shape parameter of this distribution is the variation coefficient. This is a stable characteristic of durability and endurance, often determined a priori based on mass fatigue tests [26]. For example, in [28] a table of variation coefficient values for various technical means is given in Table 1.1.

The results analysis of an operational damage to aircraft full-scale structural elements is complicated by the formation of repeatedly censored samples [2, 3, 8, 13, 14], which are formed because of determining the critical structural element operating time that has reached or not a critical state at the time of technical inspection. Randomly censored samples can also be formed during fatigue tests of materials, components of engines, aggregates and airframes of aircraft when the object

Table 1.1 Variation coefficients values of the basic destruction processes

Type of destruction	Variation coefficients
Static destruction	0.05–0.15
Fatigue	
Low cycle	0.15–0.40
Multi-cycle	0.40–1.00
Contact	0.40–1.20
Wear	
Mechanochemical	0.20–0.50
Abrasive	0.40–1.00
Aging	0.40–1.00

due to the time limit does not reach a critical state and is removed from testing. Maximum likelihood estimates (MLEs) in case of a repeatedly censored sample are generally determined from a rather complex nonlinear equation system [3, 4]. As practice shows, the solution of such systems causes significant mathematical difficulties associated with their bulkiness, as well as a number of local extrema in minimized functions, etc. This leads to the fact that stable solutions are obtained only for large volumes and small degrees of sample censorship. They also significantly depend on the initial approximations. Hence, it is not possible to use the least squares method in the classical formulation [15, 16], because the mathematical expectations and the order statistics covariance matrix for the DM distribution are uncertain. In this regard, this paper proposes some alternative estimation methods applicable in a small and censored sample. The adequacy of the solutions is checked by the Monte Carlo method.

1.2 Statistical Estimation Methods of a Randomly Censored Sample Parameters Based on the Monotonic Diffusion Distribution

A statistical properties' complete list and estimates of the DM distribution are described in [8, 12, 13, 18, 27, 28]. Here, we give only the simplest and most necessary formulas for further exposition.

The density of the DM distribution is determined by the equation:

$$f(x) = \frac{a + x}{2\gamma x\sqrt{2\pi ax}} \cdot e^{-\frac{(a-x)^2}{2\gamma^2 ax}}, \tag{1}$$

where a and γ are distribution parameters. The distribution function is determined by the equation:

$$F(x) = \Phi\left(\frac{x - a}{\gamma\sqrt{ax}}\right), \tag{2}$$

$\Phi(z)$ is the Laplace integral. The P-level distribution quantile is given by the equation:

$$x_p = a \cdot \left(1 + \frac{\gamma^2 z_p^2}{2} + \gamma z_p\sqrt{1 + \frac{\gamma^2 z_p^2}{4}}\right), \tag{3}$$

where z_p is the P-level quantile of the normalized normal distribution.

The expected value and variance are determined from the equations:

$$M(x) = a \cdot \left(1 + \gamma^2/2\right), \tag{4}$$

$$D(x) = a^2\gamma^2 \cdot \left(1 + 5\gamma^2/4\right). \tag{5}$$

DM distribution moment estimates are obtained from Eqs. (4), (5) by the formulas:

$$\hat{a} = \frac{5\overline{x}^2 - s^2}{4\overline{x} + \sqrt{\overline{x}^2 + 3s^2}} \tag{6}$$

$$\hat{\gamma} = \left[\frac{2 \cdot \left(\overline{x}\sqrt{\overline{x}^2 + 3s^2} + s^2 - \overline{x}^2\right)}{5\overline{x}^2 - s^2}\right]^{0.5}, \tag{7}$$

$$\overline{x} = \frac{\sum_{i=1}^{n} x_i}{n}, \ s^2 = \frac{\sum_{i=1}^{n} (x_i - \overline{x})^2}{n-1}, \tag{8}$$

where n is sample size; $\overline{x}$,s^2 are sample mean and variance. If the shape parameter a priori value is known, the estimate of which can be the variation coefficient, then the estimate of the scale parameter is determined by the formula:

$$\hat{a} = \frac{1}{\sum_{i=1}^{n} i} \cdot \sum_{i=1}^{n} i \cdot x_i \cdot \left(1 + \frac{\gamma^2 z_{pi}^2}{2} + \gamma z_{pi}\sqrt{1 + \frac{\gamma^2 z_{pi}^2}{4}}\right), \tag{9}$$

where z_{pi} is quantile of the normalized normal distribution corresponding to the estimate probability of a ith failure $\hat{P}_i = i/(n+1)$.

In [21], the following equation was substantiated for the lower confidence estimate x_{low} (with confidence probability β) of the kth order statistic in a sample with volume n and distribution function $F(x)$

$$F(x_{\text{low}}) = \frac{k}{k + (n - k + 1) \cdot f_{2(n-k+1),2k;\beta}}, \tag{10}$$

where $f_{2(n-k+1),2k;\beta}$ is quantile of β-level F-Fisher distributions with degrees of freedom $2(n - \mathrm{k} + 1)$ and $2k$. Upper confidence estimate x_{up} of the kth order statistic:

$$F(x_{\text{up}}) = \frac{k}{k + (n - k + 1)/f_{2k,2(n-k+1);\beta}}. \tag{11}$$

1.2.1 Bootstrap Method

In this paper, to apply the discussed above mathematical models in randomly censored samples, we propose the formation of an equivalent ("quasi-complete") sample based on the initial one censored by using random number bootstrap modeling [10] in a range limited by observable ordinal statistics. To this end, the Monte Carlo method was used to simulate a complete sample (based on the diffusion distribution), in which objects that did not reach a critical state were formed randomly by modeling evenly distributed random numbers in a given range. The proportion of such objects ranged from 0 to 50% with sample sizes of 10, 15, 20, and 40. The censored observations were then replaced by randomly selected observations from the same sample that reached a critical state. Subsequently, the sample is sorted, and an equivalent "quasi-complete" sample is formed, the parameters of which were determined using moment estimates (6)–(8). Confidence intervals for the quantiles of the distribution were determined by two methods. The first method consisted in assessing confidence limits x_{low} and x_{up} for ordinal statistics (10), (11). However, this approach has a significant limitation associated with the probability level dependence on the observation volume (as noted above, the probability estimate for the ith order statistics was determined as $\hat{P}_i = k/(n+1)$). This method allows us to estimate the confidence region for the empirical distribution function, but it makes prediction in the low-probability region difficult when assigning a resource to critical parts and structural elements of aircraft. In this regard, an approximate estimate of confidence limits is proposed, obtained on the basis of the well-known equation for the quantile variance:

$$D(\hat{x}_p) = \frac{\hat{P} \cdot (1 - \hat{P})}{(n - r) \cdot f^2(\hat{x}_p)}, \tag{12}$$

$$x_{\text{up,low}} = \hat{x}_p + z_{\beta,1-\beta} \cdot \sqrt{D(\hat{x}_p)}, \tag{13}$$

where $D(\hat{x}_p)$ is the quantile estimate variance; $\hat{x}_p$, $f^2(\hat{x}_p)$ are determined by Eqs. (1), (3) by replacing the parameters a, γ with their moment estimates; $z_{\beta,1-\beta}$ is the quantile of the normalized normal distribution for a given confidence probability; β, r are the numbers of censored observations.

1.2.2 Quantile Method

The quantile method is to approximate the empirical distribution function $\hat{P}_i$ by a theoretical law, the parameter estimates of which minimize the objective function:

$$\left[\Phi\left(\frac{x_i - \hat{a}}{\hat{\gamma}\sqrt{\hat{a}x_i}}\right) - \hat{P}_i\right]^2 \to \min. \tag{14}$$

Given random censorship, in (14), x_i are the observations values that have reached the object's critical state. In this case, the empirical distribution function P is estimated in accordance with a nonparametric technique for multiple-censored sample, the calculation algorithm of which is considered in detail in [2, 3, 11, 24]. Confidence boundaries for the distribution quantiles are determined by Eqs. (10)–(13) from the obtained parameter estimates. Along with simplicity, this method has an obvious computational drawback, consisting in the need to apply a particular minimization procedure. In this work, the deformable polyhedron method was used as such a procedure. The initial approximations were moment estimates (6)–(8), calculated from observations without taking into account censored values. In this method, the calculation duration is increased many times, which is especially critical for multiple simulations. However, for estimating the parameters, the method is quite applicable. Especially, taking into account the obvious range limitation of possible functional values (14), the probabilities lie in the range from 0 to 1. This allows us to quickly find the optimal estimates $\hat{a}$, $\hat{\gamma}$ by means of "fines," whose value range also has limitations like $x_1 < \hat{a} < x_n$, $0 < \hat{\gamma} < \gamma_{\max}$.

1.2.3 Monte Carlo Model Testing

To test the proposed models, statistical modeling was done with multiple (up to 2000 times) test repetitions. In each of the tests, the parameters of equivalent samples were estimated using Eqs. (6), (7) and the confidence level probability using Eqs. (10), (11). The P-level distribution quantile corresponded to the first-order statistics depending on the test volume from 10 to 40. Some simulation results are presented in Table 1.2. The calculation results showed good convergence for the shape and scale parameters up to 5% relative errors. A slightly larger error of up to 7% was observed in the estimation of confidence probabilities for distribution quantiles. Approximate formulas (12), (13) for confidence boundaries give a slightly larger error (up to 10%) for the confidence side (i.e., in the margin along the width of the confidence interval). The quantile method (14) in the simulation showed results close in accuracy, but significantly inferior in terms of calculation time.

1.2.4 Statistical Processing of Test Results

Logarithmically normal distribution, often distorts the real picture of scattering of fatigue properties of materials and structural elements real fatigue properties dispersion. Moreover, it does not correspond to the fatigue physical model at low load levels due to the manifestation of cycle sensitivity threshold. Alternatively, in order to illustrate the use of DM distribution for processing fatigue tests, a fatigue test statistical analysis of samples with stress concentration varying degrees of VT3-1 titanium alloy was done [25] (Table 1.3), characterized by a very large dispersion in

Table 1.2 Statistical modeling results of a randomly censored sample from the monotonic diffusion distribution with following input data: parameter a = 1.0; shape parameter γ = 0.4, two-sided confidence coefficient $\beta = 0.9$; $\hat{a}$, $\hat{\gamma}$ parameter estimates; n is the sample size; $h = r/n$ is the degree of censorship; r is the number of censored sample elements

h/n	10			15		
	$\hat{a}$	$\hat{\gamma}$	$\hat{\beta}$	$\hat{a}$	$\hat{\gamma}$	$\hat{\beta}$
0	0.9959	0.4110	0.9225	0.9991	0.4074	0.9500
0.1	1.0109	0.4008	0.9135	1.0154	0.3998	0.9270
0.2	1.0234	0.3995	0.9025	1.0255	0.3982	0.9155
0.3	1.0288	0.3954	0.8685	1.0226	0.3956	0.8975
0.4	1.0397	0.3880	0.8445	1.0395	0.3862	0.8765
0.5	1.0414	0.3846	0.8430	1.0407	0.3892	0.8700
h/n	20			40		
	$\hat{a}$	$\hat{\gamma}$	$\hat{a}$	$\hat{\gamma}$	$\hat{a}$	$\hat{\gamma}$
0	0.9989	0.4028	0.9505	0.9997	0.4037	0.9650
0.1	1.0142	0.3982	0.9390	1.0125	0.3998	0.9635
0.2	1.0216	0.3972	0.9180	1.0230	0.3950	0.9570
0.3	1.0304	0.3910	0.9105	1.0341	0.3913	0.9540
0.4	1.0391	0.3899	0.9005	1.0435	0.3906	0.9395
0.5	0.9989	0.4028	0.9505	0.9997	0.4037	0.9650

durability. The table provides estimates of average values $\lg \overline{N}$, standard deviations $s_{\lg N}$ of the durability logarithm for different levels of variable stress amplitudes in a symmetric cycle, as well as moment estimates $\hat{a}$, $\hat{\gamma}$ of the monotonic diffusion distribution. Censored observations in this case are absent. The table shows that the shape parameter of the diffusion distribution can reach a value of 1.47, which generally does not contradict the data in Table 1.1. The last two rows of Table 1.3 present the 0.01 level durability quantile estimates calculated by diffusion [Eq. (3)] and logarithmically normal $\hat{N}_{0,01} = 10^{\lg \overline{N} - 2,326 \cdot s_{\lg N}}$ models, respectively. In all cases where the durability exceeds 10^6 cycles and $s_{\lg N}$ is greater than 0.4, there is a significantly underestimated quantile value calculated by the log-normal model. Figure 1.1 in coordinates $\hat{P}$ (empirical probability)—$N/\hat{a}$ (relative durability) shows a generalized distribution function. The calculated curve in Fig. 1.1 is obtained from the combined sample.

1.3 Conclusion

A method is proposed for estimating the parameters of the monotonic diffusion distribution under random censorship conditions, which is based on the "quasi-complete" sample formation using mixed bootstrap modeling. A quantile method

Table 1.3 Durability N (in cycles) of smooth and notched samples of titanium alloy VT3-1, α_σ—theoretical stress concentration coefficient, σ_a—stress amplitude of a symmetrical cycle

α_σ	1.0	1.4	1.9		2.36	
σ_a, MПa	450	350	270	215	250	200
1	233,884	278,612	95,060	1,588,547	44,668	833,681
2	397,192	1,183,042	99,083	1,999,862	154,882	1,213,389
3	414,954	2,466,039	100,000	2,999,163	163,305	1,432,188
4	425,598	2,630,268	107,895	3,303,695	223,357	3,318,945
5	635,331	3,539,973	125,893	3,396,253	268,534	3,357,376
6	674,528	3,899,420	137,088	3,899,420	405,509	11,194,379
7	799,834	4,487,454	144,877	6,295,062	434,510	12,560,300
8	1,196,741	7,961,594	162,181	10,690,549	500,035	23,604,782
9	2,202,926	8,570,378	236,048	13,708,818	1,114,295	25,118,864
10	2,786,121	11,994,993	250,035	16,788,040	1,264,736	50,003,454
11	4,295,364	14,454,398		20,183,664	1,482,518	62,230,029
12	5,888,437	14,689,263				
13	12,676,519	15,922,087				
14	16,255,488	38,018,940				
15	17,060,824					
16	27,605,779					
17	28,444,611					
$\lg \overline{N}$	6.3800	6.7246	5.1386	6.7377	5.5542	6.8747
$s_{\lg N}$	0.7259	0.5558	0.1555	0.3926	0.4674	0.6822
$\hat{a}$	3,440,019	5,824,387	135,597	5,630,365	386,731	9,847,388
$\hat{\gamma}$	1.473819	1.091297	0.388233	0.860298	0.920511	1.264102
$DM - \hat{N}_{0,01}$	251,421	699,621	56,526	965,095	60,143	933,119
$Log - \hat{N}_{0,01}$	49,149	270,272	59,837	667,711	29,302	194,046

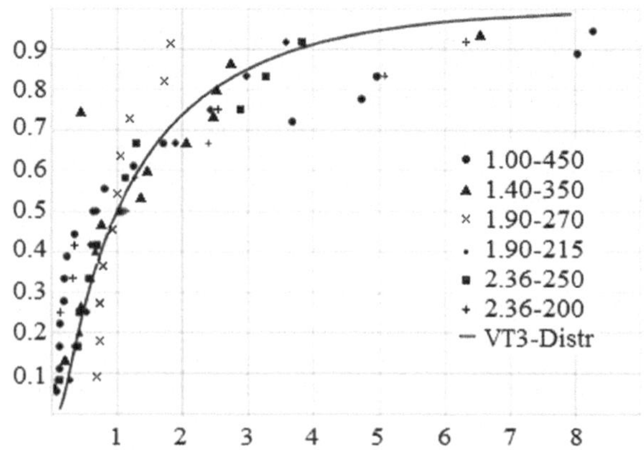

Fig. 1.1 Empirical distribution function of the titanium alloy VT3-1 relative durability with varying degrees of stress concentration and stress amplitudes. $\hat{P}$, empirical probability. $N/\hat{a}$, relative durability

based on the empirical distribution function nonparametric estimation during censorship is also proposed. It is shown that, as a first approximation, scale parameter moment estimates and the diffusion distribution shape in both cases are effective. An approximate method for constructing a confidence region for distribution quantiles is also proposed. The models were tested by the Monte Carlo method, and it showed satisfactory convergence between the experimental and calculated estimates.

In order to illustrate the method, a statistical analysis of the fatigue test results is done for a titanium alloy with significant dispersion of the fatigue properties characteristics. It is shown that for large durability and scattering, a significant error is observed in the estimation of the quantile calculated by the logarithmically normal model. This can lead to unjustified underestimation of the resource during strength calculations. In this regard, we recommend using the monotonic diffusion distribution to predict the fatigue test results in the region of low loads and probabilities.

References

1. ASTM E739 2015: Standard Practice for Statistical Analysis of Linear or Linearized Stress-Life SN and Strain-Life. American Society for Testing and Materials
2. Agamirov, L., Podzhivotov, N.: Damage and equivalent running time of a maneuverable aircraft glider. Aviat. Ind. **1**, 47–54 (2018)
3. Agamirov, L., Agamirov, V., Vestyak, V.: Estimation of statistical characteristic of aeronautical materials and structures with the empirical distribution function under random censoring. IOP Conf. Ser. J. Phys. Conf. Ser. **1158**, 022003 (2019). (IOP Publishing)
4. Agamirov L.V., Agamirov V.L., Vestyak V.A. Statistical analysis of the test results of aircraft products under conditions of random censorship. Software Products and Systems, No. 1, Volume 30, 2017, pp. 124–129
5. Bandara, C., Siriwardane, S., Dissanayake, U., Dissanayake, R.: Developing a full range S-N curve and estimating cumulative fatigue damage of steel elements. Comput. Mater. Sci. **96**, 96–101
6. Bandara, C., et al.: Full range S–N curves for fatigue life evaluation of steels using hardness measurements. Int. J. Fatigue. **82** (2015)
7. Boyer, H.E.: Atlas of Stress-Strain Curves, 2nd Revised edn, 808 p (2002)
8. Chhikara, R., Folks, J.: The Inverse Gaussian Distribution. Theory, Methodology and Application. Marcel Dekker, New York (1989)
9. Correia, J., et al.: Fatigue life prediction of notched details made of puddle iron based on variable fatigue strength reduction factors concept. In: Conference: International Symposium on Notch Fracture (ISNF), At Santander, Cantabria, Spain (2017)
10. Davison, A., Hinkley, D.: Bootstrap Methods and Their Application, p. 592. Cambridge University Press, Cambridge (2006)
11. DSTU 3004-95: Reliability of technology. Methods for assessing reliability indicators from experimental data, 122 p (1996)
12. DSTU 3942-99. Reliability of technology. Test plans for monitoring the mean time between failure (failure)—Part 2. Diffusion distribution, 34 p (2000)
13. Folks, J., Chhikara R.: The inverse Gaussian distribution and its statistical applications. J. R. Stat. Soc. Ser. B **40**, 263–289 (1978)
14. Goedel, F., et al.: Methodology for assessment of statistical planning effects on the S-N curve determination using Monte Carlo simulations. Fatigue Fract. Eng. Mater. Struct. (2018)
15. Kendall, M., Stewart, A.: Statistical conclusions and relationships. M: Science, 899 p (1973)

16. Kendall, M., Stewart, F.: Distribution theory. M: Science (1966)
17. Kordonsky, H.B., Friedman, Y.V.: Some questions of the probabilistic description of fatigue life. Zavod. Lab. **7**, 829–847 (1976)
18. Larin, M.M.: Unbiased estimates of variance and some other characteristics of the inverse normal distribution. Tech. Cybern. **6**, 134–137 (1986)
19. Lyamkin, V., Starke P., Boller, C.: Cyclic indentation as an alternative to classic fatigue evaluation. In: Conference: 7th International Symposium on Aircraft Materials, At Compiègne, France (2018)
20. Mirza, O., et al.: Experimental investigation of retrofitting techniques for steel bridge girders subject to fatigue failure. J. Steel Struct. Constr. **4** (2018)
21. Nicholas, A., et al.: Predictive inferences for future order statistics coming from an inverse Gaussian distribution. In: Proceedings of the World Congress on Engineering 2014, vol. II, WCE 2014, London, UK, 2–4 July 2014
22. Reicher, V.L.: Fatigue damage. LATMES, 238 p (2006)
23. Serensen, S.V., Kogaev V.P., Shneiderovich R.M.: Bearing capacity and strength calculation of machine parts. Mech. Eng., 488 (1975)
24. Skripnik, V.M., Nazin, A.E., Prikhodko, Yu.G.: Reliability analysis of technical systems by censored samples. Radio Commun. (1988)
25. Stepnov, M.N., Agamirov, L.V.: On the statistical laws of fatigue resistance of VT3-1 titanium alloy. Zavod. Lab. **1** (1980)
26. Stepnov, M.N., Hyacinthov, E.V.: Fatigue of light structural alloys. Mech. Eng. , 317 (1973)
27. Strelnikov, V.P.: Parameterization of the DM distribution. Math. Mach. Syst. **2**, 117–124 (2007)
28. Strelnikov, V.P.: Probabilistic-physical methods for studying the reliability of machines and equipment. Reliab. Qual. Control **9**, 3–7 (1989)
29. Strzelecki, P., Sempruch, J.: Experimental method for plotting S-N curve with a small number of specimens. Pol. Marit. Res. **23**(4), 129–137 (2016)

Chapter 2
Earthquakes and Cracks of New Type Complementing the Griffith–Irwin's Crack

Vladimir A. Babeshko, Olga V. Evdokimova, and Olga M. Babeshko

Abstract The paper is devoted to the issue on the existence of a new type of cracks, complementing the Griffith–Irwin cracks. These cracks explain one of the mechanisms of fracture of brittle solids, consisting in the appearance of a singular concentration of contact stresses in converging lithospheric plates. The main difference between the Griffith–Irwin cracks and the cracks of new type is that the Griffith–Irwin cracks have a smooth boundary with an apex angle of 180°. Griffith explained their formation as the result of continuous deformation of an unbounded plate with an elliptical cavity into a cavity with rectilinear lateral boundaries resulting from compression of the elliptical cavity from the sides. There are smooth curves at the apices of the cavity. Cracks of new type are formed as a result of compression from the sides of a rectangle instead of an ellipse. The result is a cavity with a piecewise-smooth boundary and an apex angle equal to zero. The study was conducted under the assumption of the absence of friction in the contact area of the lithospheric plates with the base in the first case and the presence of only tangential contact stresses in the second case. In addition, it has been shown that numerous studies of wedge-shaped cavities turning into cracks when the wedge angle tends to zero give rise to a Griffith crack and cannot generate a cracks of new type.

Keywords Block element method · Lithospheric plates · Griffith–Irwin's cracks · Exterior forms · Block structures · Boundary value problems · Bodies with coverings · Starting earthquakes · Cracks of new type

---

V. A. Babeshko (✉) · O. V. Evdokimova
Federal Research Centre the Southern Scientific Centre of the Russian Academy of Sciences, Chehova St., 41, Rostov-on-Don, Russia 344006
e-mail: babeshko41@mail.ru

O. V. Evdokimova
e-mail: evdokimova.olga@mail.ru

O. M. Babeshko
Kuban State University, Stavropolskaya St., 149, Krasnodar, Russia 350040
e-mail: babeshko49@mail.ru

H. Altenbach et al. (eds.), *Multiscale Solid Mechanics*,
Advanced Structured Materials 141,
https://doi.org/10.1007/978-3-030-54928-2_2

2.1 Introduction

The discovery of the starting earthquakes and the study of their properties under different conditions of contact of lithospheric plates with a deformable base showed the identity of medium fracture processes in the zones of convergence of lithospheric plates on the Conrad discontinuity. The problems of contact of lithospheric plates with a base in the absence of friction, small vertical effects, which can be neglected, and strong horizontal effects were considered earlier. The presence of vector of tangential stresses in the contact zone was taken into account. In all these cases, by the convergence of the lithospheric plates, a singular concentration of contact stresses was detected. Only the case of the full adhesion contact of lithospheric plates with a deformable base in the presence of both vertical and horizontal effects remained unexplored. The results of this study are published in this article. The obtained results also show the occurrence of singular features in contact stresses by full proximity of the ends of lithospheric plates. This result made it possible to put forward the assumption that when the ends of lithospheric plates are close, but there is no interaction between them, a cracks of new type is created, the banks of which are the ends of the converged lithospheric plates. However, there was a doubt that the reason for such behavior of contact stresses in the zone of converged lithospheric plates may be a simplified model of lithospheric plates described by two-dimensional Kirchhoff plates on a three-dimensional base. Authors have investigated the formulated boundary problem for a block structure consisting of three-dimensional lithospheric plates on a three-dimensional base in [1] and the existence of the cracks of new type have been confirmed again. The question of the possibility of emergence of cracks of a new type when considering wedge-shaped cavities when the wedge angle tends to zero remains open. The assumption is justified by the fact that the wedge-shaped cavity has, like the crack of a new type, a piecewise-smooth boundary and may remain in the crack formed. In this article, based on the previously obtained and recognized study results of transition of the wedge-shaped cavity into a crack, made in [2], it is proved that in the limit. there is a Griffith–Irwin crack with a smooth boundary and an apex angle of 180°. This fact fully proves the existence of the cracks of new type in the zone of convergence of lithospheric plates which have not been previously described. The authors analyzed the number of publications on cracks of the various scientists [2–59] including the last of time, but the information about the cracks of new type have not detected.

The wedge-shaped cavities considered in this publication with angles of various sizes down to zero. However, many researchers of the Russian scientists did not get into the review because of the language barrier. Therefore, the fundamental monograph N. F. Morozov built a rigorous mathematical theory of cracks [48] which is absent in the review. The review did not get great articles V. P. Matveenko who developed a numerically analytical method for studying singularities in pyramids and cones [6, 7, 33, 34, 38, 39]. Also the review does not include interesting studies on the selection of singularities in trihedral angles and cracks by E. V. Glushkov and N. V. Glushkova [4, 5, 16–19], as well as a paper by one of the authors of this article who was among the first to investigate singular features in a trihedral wedge and crack [60]. Perhaps, it is for this reason that some mutual misunderstanding in the theory of cracks often occurs.

The analysis of the origin of the Griffith cracks, described by their discoverer, is quite easy to compare with the cracks of new type. Griffith studied cracks by taking a plate with a cavity in the shape of an ellipse, a circle, in particular [20]. Then, he used a smooth continuous mapping and "squeezed" the cavity on the sides until it took the shape of a slit-like crack, Fig. 2.1. At the top of the crack, the boundary remained rounded and smooth, as in an ellipse with an apex angle of 180°. In order to obtain the shape of cracks of the new type we need to follow the algorithm that led to the starting earthquake, that is, we should bring together the ends of the lithospheric plates. We can also follow the technique used by Griffith, taking a rectangular cavity instead of a cavity in the form of an ellipse. The rectangle has a non-smooth boundary due to the corners. Then, we should compress it continuously from the sides until it takes the shape of a slit, Fig. 2.2. The angle at the top of such a slit is zero. As the last of the considered problems showed, fracture of such a crack occurs in the same way as in the case of starting earthquakes, that is, it differs from the process of fracture of the Griffith–Irwin cracks. A more detailed study of the cracks of the new type made it possible to reveal the relation that determines the possibility of their fracture. The study showed that the cracks of the new type are more susceptible to fracture than

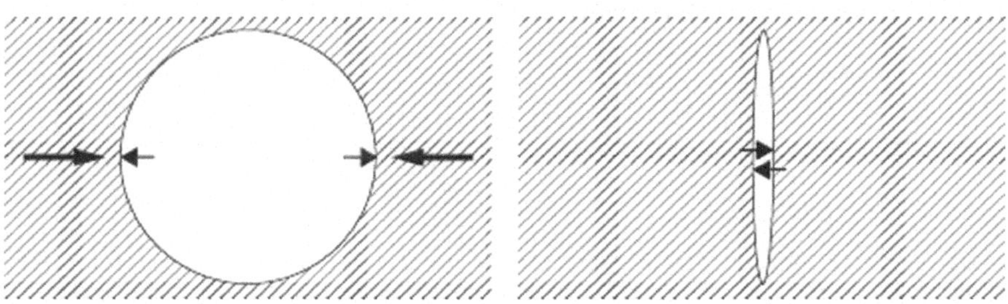

Fig. 2.1 It is the Griffith's model of virtual compressing of the ellipse or the cercal in the direction of the thick arrows till the crack will be appeared. Thin arrows represent the stresses action on the shores of the crack. The Griffith's crack has the smooth boundary

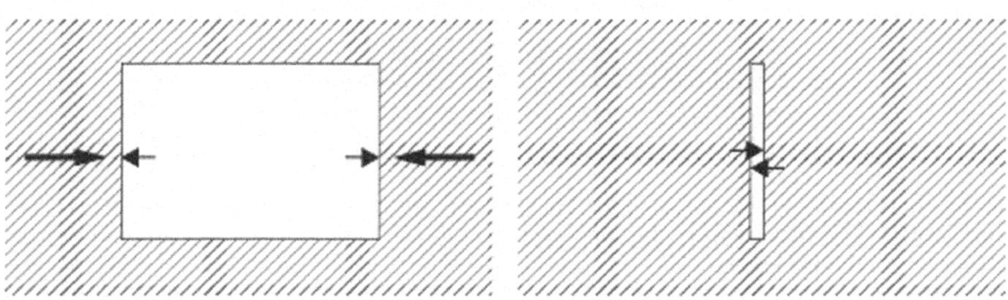

Fig. 2.2 To get the crack of new type, it is enough to replace the ellipse or the cercal by the rectangle in the Griffith's model. The crack of new type has the piece-smooth boundary

those of Griffith–Irwin. It is possible that by the fracture of brittle materials the cracks of new type arise in the first stage after dislocations. Then, these cracks are fractured and either turn into the Griffith–Irwin cracks or completely destroy the material.

2.2 Rigid Contact of the Lithosphere Plates with the Base

The lithospheric plates are assumed to be semi-infinite Kirchhoff plates in the form of half-planes, the boundaries of which are parallel to each other at the distance of 2θ, $\theta \geq 0$. The mechanical properties of each plate are assumed to be different. Let us consider that coordinate axes lie in the plane of the plates, and axis is directed along the external normal to the base, [61–63]. Let us consider the statics of plates, which are horizontal in the plane and rigidly coupled to the base, Fig. 2.3. Then, the equations of the boundary value problem for plates can be written in the form

$$R_b(\partial x_1, \partial x_2)u_b - s_b(x_1, x_2) = 0, \quad b = l, r \tag{2.1}$$

Each plate is considered as a manifold with an end face, where $u_b = \{u_{1b}, u_{2b}, u_{3b}\}$ is the displacement vector of the plate points along the horizontal u_{1b}, u_{2b} and vertical u_{3b} directions of the median surface; $b = l$ for the left plate and $b = r$ for the right one. The following notation is used:

$$\mathbf{s}_b(x_1, x_2) = \left\| \begin{matrix} -\varepsilon_{5b}s_{1b}(x_1, x_2) & 0 & 0 \\ 0 & -\varepsilon_{5b}s_{2b}(x_1, x_2) & 0 \\ 0 & 0 & \varepsilon_{53b}s_{3b}(x_1, x_2) \end{matrix} \right\|, \quad s_{nb}(x_1, x_2)$$

$$= (t_{nb} + g_{nb})$$

$$R_b(\partial x_1, \partial x_2)u_b$$

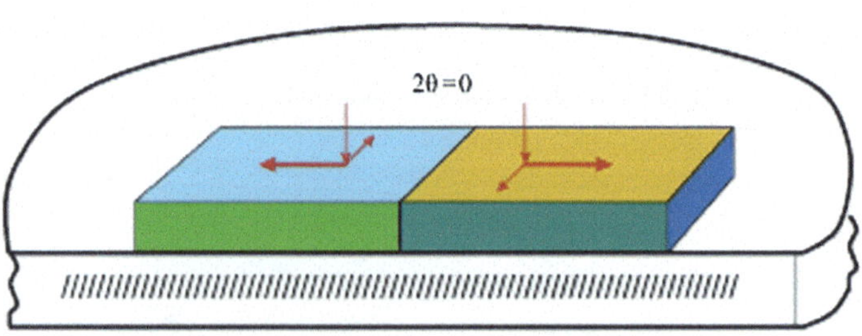

Fig. 2.3 Scheme of toward motion of the lithospheres plates which have infinite width and semi-infinite length. The thicknesses of the upper plates are h_b and lower is H. The red arrows mean the external actions on the plates

$$= \left\| \begin{matrix} \left(\frac{\partial^2}{\partial x_1^2} + \varepsilon_{1b}\frac{\partial^2}{\partial x_2^2}\right)u_{1b} & \left(\varepsilon_{2b}\frac{\partial^2}{\partial x_1 \partial x_2}\right)u_{2b} & 0 \\ \left(\varepsilon_{2b}\frac{\partial^2}{\partial x_1 \partial x_2}\right)u_{1b} & \left(\frac{\partial^2}{\partial x_2^2} + \varepsilon_{1b}\frac{\partial^2}{\partial x_1^2}\right)u_{2b} & 0 \\ 0 & 0 & \left(\frac{\partial^4}{\partial x_1^4} + 2\frac{\partial^2}{\partial x_1^2}\frac{\partial^2}{\partial x_2^2} + \frac{\partial^4}{\partial x_2^4}\right)u_{3b} \end{matrix} \right\| \quad (2.2)$$

The Fourier transform of the differential part of system of Eq. (2.1) has the form

$$R_b(-i\alpha_1 - i\alpha_2)U_b = \left\| \begin{matrix} (\alpha_1^2 + \varepsilon_{1b}\alpha_2^2)U_{1b} & (\varepsilon_{2b}\alpha_1\alpha_2)U_{2b} & 0 \\ (\varepsilon_{2b}\alpha_1\alpha_2)U_{1b} & (\alpha_2^2 + \varepsilon_{1b}\alpha_1^2)U_{2b} & 0 \\ 0 & 0 & -(\alpha_1^2 + \alpha_2^2)U_{3b} \end{matrix} \right\|$$

$$U_b = Fu_b, \quad G_b = Fg_b, \quad T_b = Ft_b$$

$$u_b = \{u_{1b}, u_{2b}, u_{3b}\}, \quad g_b = \{g_{1b}, g_{2b}, g_{3b}\}, \quad t_b = \{t_{1b}, t_{2b}, t_{3b}\}$$

Here, the normal stresses t_{3b} and g_{3b} act on the plate from above and below, respectively. Similarly, the stresses g_{1b}, g_{2b} and t_{1b}, t_{2b} have the tangential direction, g_{2b} and t_{2b} denote the unit normal vectors to the end faces of the lithospheric plates. The following notations are used, cf. [61–63]:

$$U_b = F_2 u_b, \quad G_b = F_2 g_b, T_b = F_2 t_b, \quad b = l, r$$

$$M_b = -D_{b1}\left(\frac{\partial^2 u_{3b}}{\partial x_2^2} + v_b\frac{\partial^2 u_{3b}}{\partial x_1^2}\right), \quad D_{b1} = \frac{D_b}{H^2}, \quad D_{b2} = \frac{D_b}{H^3}$$

$$Q_b = -D_{b2}\left(\frac{\partial^3 u_{3b}}{\partial x_2^3} + (2 - v_b)\frac{\partial^3 u_{3b}}{\partial x_1^2 \partial x_2}\right), \quad u_{3b}, \quad \frac{\partial u_{3b}}{H\partial x_2}$$

$$D_b = \frac{E_b h_b^3}{12(1 - v_b^2)}; \quad \varepsilon_{53b} = \frac{12(1 - v_b^2)H^4}{E_b h_b^3}; \quad \varepsilon_6^{-1} = \frac{(1 - v)H}{\mu};$$

$$\varepsilon_{1b} = 0.5(1 - v_b); \quad \varepsilon_{2b} = 0.5(1 + v_b); \quad \varepsilon_{5b} = \frac{1 - v_b^2}{E_b h_b};$$

$$g_{1b} = \mu_{0b}\left(\frac{\partial u_{1b}}{\partial x_3} + \frac{\partial u_{3b}}{\partial x_1}\right); \quad g_{2b} = \mu_{0b}\left(\frac{\partial u_{2b}}{\partial x_3} + \frac{\partial u_{3b}}{\partial x_2}\right);$$

$$\mu_{0b} = \frac{\mu_b}{H}; \quad x_3 = 0; \quad g = \{g_{1b}, g_{2b}\}$$

Here, denote $\frac{\partial u_{3b}}{H\partial x_2}$ the angle of rotation of lithospheric plates on the borders around the axis ox_1, M_b the bending moments of the ends of lithospheric plates; Q_b the crosscutting forces of the end of lithospheric plates; μ_b is the shear modulus; v_b is the Poisson ratio; E_b is the Young modulus, h_b is the thickness of lithosphere plate; H is the thickness of basal crust of the Earth; g_b are the contact stress, horizontal, g_{1b}, g_{2b} and vertical g_{3b} vectors, acting tangentially and normally to the boundary of the base in the domains Ω_b; t_b they are the external stress, horizontal, t_{1b}, t_{2b} and vertical t_{3b} vectors, acting tangentially and normally to the plates in the domains

Ω_b; Here, denote $F_2 \equiv F_2(\alpha_1, \alpha_2)$, $F_1 \equiv F_1(\alpha_1)$ are the two-dimensional and one-dimensional operators of the Fourier transform, respectively. The expressions for the normal N_{x_2} and tangential $T_{x_1 x_2}$ stress components to the median plane at the end faces of the plates are given by the relations

$$T_{x_1x_2} = \varepsilon_7\left(\frac{\partial u_2}{\partial x_1} + \frac{\partial u_1}{\partial x_2}\right), \quad N_{x_2} = \varepsilon_8\left(\frac{\partial u_2}{\partial x_2} + \nu\frac{\partial u_1}{\partial x_1}\right),$$
$$\varepsilon_7 = \frac{E}{2(1+\nu)H}, \quad \varepsilon_8 = \frac{E}{(1-\nu^2)H}$$

The deformable base is describe by the following model

$$\mathbf{u}(x_1, x_2) = \varepsilon_6^{-1}\frac{1}{4\pi^2}\int_{-\infty}^{\infty}\int \mathbf{K}(\alpha_1, \alpha_2)\mathbf{G}(\alpha_1, \alpha_2)\mathrm{e}^{-i\langle\alpha,x\rangle}d\alpha_1 d\alpha_2,$$
$$x \in \Omega_l, \quad x \in \Omega_r, \quad x \in \Omega_\theta, \quad \langle\alpha, x\rangle = \alpha_1 x_1 + \alpha_2 x_2$$

$$\Omega_l(|x_1| \le \infty; \quad x_2 \le -\theta), \quad \Omega_r(|x_1| \le \infty; \quad \theta \le x_2), \quad \Omega_\theta(|x_1| \le \infty;$$
$$-\theta \le x_2 \le \theta)$$

$$\mathbf{K} = \|K_{mn}\|, \quad m, n = 1, 2, 3, \quad \mathbf{K}(\alpha_1, \alpha_2) = O(A^{-1}), \quad A = \sqrt{\alpha_1^2 + \alpha_2^2} \to \infty$$
$$\varepsilon_6^{-1} = \frac{(1-\nu)H}{\mu}, \quad \mathbf{G}(\alpha_1, \alpha_2) = \mathbf{F}_2(\alpha_1, \alpha_2)\mathbf{g}$$

Here, **g** is the vector of tangential and normal stresses at the plate boundary. The matrix functions $K(\alpha_1, \alpha_2)$ are referred to as the symbols of the system of integral equations, and some type of them are given in [55]. For example, for an elastic layer with a fixed lower face in the static case, this system has the form

$$\mathbf{g}_l(x_1, x_2) = \sigma_{1l}(x_1, x_2)(-x_2 - \theta)^{-0,5+i\gamma} + \sigma_{2l}(x_1, x_2)(-x_2 - \theta)^{-0,5-i\gamma}, \quad x_2 < -\theta,$$
$$\mathbf{g}_r(x_1, x_2) = \sigma_{1r}(x_1, x_2)(x_2 - \theta)^{-0,5+i\gamma}$$
$$+\sigma_{2r}(x_1, x_2)(x_2 - \theta)^{-0,5-i\gamma}, \quad x_2 > \theta \quad \gamma > 0$$

$$M(u) = \frac{(1-\nu)(3-4\nu)(sinh 4u + 4u)}{u^2\Delta}, \quad N(u) = \frac{2 sinh 2u}{u^3 cosh 2u},$$
$$P(u) = -\frac{(1-2\nu)(3-4\nu)sinh^2 2u - 4u^2}{u\,\Delta(u)},$$
$$K(u) = \frac{(1-\nu)(3-4\nu)(sinh 4u - 4u)}{\Delta(u)},$$
$$\Delta(u) = u\left[(3-4\nu)sinh^2 2u + 4u^2 + 4(1-\nu)^2\right], \quad u = \sqrt{\alpha_1^2 + \alpha_2^2}$$

Matrix (2.2) of the boundary value problem is block diagonal. It consists of the second-order matrix on the diagonal, representing a matrix operator or a vector operator, as well as a non-diagonal scalar operator. Since the operators are independent, this makes it much easier to study the boundary value problem at the stage of external analysis by applying results of [61–63].

2.3 External Analysis of the Boundary Value Problem

The boundary value problems for each block are set in a topological space induced by a three-dimensional euclidean space. After that, they are reduced to functional equations by applying the Stokes formula in the topological space. Let us present the functional equations corresponding to the above-listed operators of the boundary value problem. The functional equations of the scalar operator have the form [61–63]

$$R_{3b}(-i\alpha_1, -i\alpha_2)U_{3b} \equiv (\alpha_1^2 + \alpha_2^2)^2 U_{3b} = -\int\limits_{\partial\Omega_b} \omega_{3b} + S_{3b}(\alpha_1, \alpha_2)$$

$$S_{3b}(\alpha_1, \alpha_2) = \varepsilon_{53b}\mathbf{F}_2(\alpha_1, \alpha_2)(t_{3b} + g_{3b}), \quad b = l, r$$

ω_{3b} are the external forms and have the following expressions for the left (l) and right (r) lithospheric plates

$$\omega_{3l} = e^{i\langle\alpha,x\rangle}\left\{-\left[\frac{\partial^3 u_{3l}}{\partial x_2^3} - i\alpha_2\frac{\partial^2 u_{3l}}{\partial x_2^2} - \alpha_2^2\frac{\partial u_{3l}}{\partial x_2} + i\alpha_2^3 u_{3l} + 2\frac{\partial^3 u_{3l}}{\partial x_1^2\partial x_2} - 2i\alpha_2\frac{\partial^2 u_{3l}}{\partial x_1^2}\right]dx_1 + \left[\frac{\partial^3 u_{3l}}{\partial x_1^3} - i\alpha_1\frac{\partial^2 u_{3l}}{\partial x_1^2} - \alpha_1^2\frac{\partial u_{3l}}{\partial x_1} + i\alpha_1^3 u_{3l}\right]dx_2\right\},$$

$$\omega_{3r} = -e^{i\langle\alpha,x\rangle}\left\{-\left[\frac{\partial^3 u_{3r}}{\partial x_2^3} - i\alpha_2\frac{\partial^2 u_{3r}}{\partial x_2^2} - \alpha_2^2\frac{\partial u_{3r}}{\partial x_2} + i\alpha_2^3 u_{3r} + 2\frac{\partial^3 u_{3r}}{\partial x_1^2\partial x_2} - 2i\alpha_2\frac{\partial^2 u_{3r}}{\partial x_1^2}\right]dx_1 + \left[\frac{\partial^3 u_{3r}}{\partial x_1^3} - i\alpha_1\frac{\partial^2 u_{3r}}{\partial x_1^2} - \alpha_1^2\frac{\partial u_{3r}}{\partial x_1} + i\alpha_1^3 u_{3r}\right]dx_2\right\}$$

The matrix functional equations of the boundary value problem for the vector case have the following form for each plate:

$$\mathbf{R}_{12b}(-i\alpha_{1b}, -i\alpha_{2b})\mathbf{U}_{12b} = -\int\limits_{\partial\Omega_b} \boldsymbol{\omega}_{12b} + \mathbf{S}_{12b}(\alpha_{1b}, \alpha_{2b}), \quad \mathbf{U}_{12b} = \{U_{1b}, U_{2b}\},$$

$$\boldsymbol{\omega}_{12b} = \{\omega_{1b}, \omega_{2b}\}, \quad \mathbf{S}_{12b}(\alpha_{1b}, \alpha_{2b}) = -\varepsilon_{5b}\mathbf{F}_2(\alpha_{1b}, \alpha_{2b})(\mathbf{g}_b + \mathbf{t}_b), \quad b = l, r.$$

$$\mathbf{S}_{12b}(\alpha_{1b}, \alpha_{2b}) = \{S_{1b}, S_{2b}\}$$

$$\mathbf{R}_{12b}(-i\alpha_{1b}, -i\alpha_{2b}) = -\left\| \begin{matrix} (\alpha_{1b}^2 + \varepsilon_{1b}\alpha_{2b}^2) & \varepsilon_{2b}\alpha_{1b}\alpha_{2b} \\ \varepsilon_{2b}\alpha_{1b}\alpha_{2b} & (\alpha_{2b}^2 + \varepsilon_{1b}\alpha_{1b}^2) \end{matrix} \right\|$$

Here, $\boldsymbol{\omega}_b$ is the external representation of a vector

$$\omega_{1l} = -e^{i\langle\alpha,x\rangle}\left\{-\left(\varepsilon_{1l}\frac{\partial u_{1l}}{\partial x_2} + \varepsilon_{2l}\frac{\partial u_{2l}}{\partial x_1} - i\varepsilon_{1l}\alpha_2 u_{1l}\right)dx_1 + \left(\frac{\partial u_{1l}}{\partial x_1} - i\alpha_1 u_{1l} - i\varepsilon_{2l}\alpha_2 u_{2l}\right)dx_2\right\},$$

$$\omega_{2l} = -e^{i\langle\alpha,x\rangle}\left\{-\left(\varepsilon_{2l}\frac{\partial u_{1l}}{\partial x_1} + \frac{\partial u_{2l}}{\partial x_2} - i\alpha_2 u_{2l}\right)dx_1 + \left(\varepsilon_{1l}\frac{\partial u_{2l}}{\partial x_1} - i\varepsilon_{1l}\alpha_1 u_{2l} - i\varepsilon_{2l}\alpha_2 u_{1l}\right)dx_2\right\},$$

$$\omega_{1r} = e^{i\langle\alpha,x\rangle}\left\{-\left(\varepsilon_{1r}\frac{\partial u_{1r}}{\partial x_2} + \varepsilon_{2r}\frac{\partial u_{2r}}{\partial x_1} - i\varepsilon_{1r}\alpha_2 u_{1r}\right)dx_1 + \left(\frac{\partial u_{1r}}{\partial x_1} - i\alpha_1 u_{1r} - i\varepsilon_{2r}\alpha_2 u_{2r}\right)dx_2\right\},$$

$$\omega_{2r} = e^{i\langle\alpha,x\rangle}\left\{-\left(\varepsilon_{2r}\frac{\partial u_{1r}}{\partial x_1} + \frac{\partial u_{2r}}{\partial x_2} - i\alpha_2 u_{2r}\right)dx_1 + \left(\varepsilon_{1r}\frac{\partial u_{2r}}{\partial x_1} - i\varepsilon_{1r}\alpha_1 u_{2r} - i\varepsilon_{2r}\alpha_2 u_{1r}\right)dx_2\right\}.$$

Let us apply an external analysis [61–63] for these functional equations, including the factorization of the coefficient of the functional equation (matrix-function or function), the calculation of Leray residue forms, the assembly of pseudo-differential equations, the extraction of the necessary equations from the boundary equations, and the solving of the latter. The obtained solutions are introduced into the external forms of the functional equations of each plate. Then, they are matched with a base, forming a new topological space, called the factor-topological one. Applying the approach described in detail in the above studies, we arrive at the Wiener–Hopf matrix functional equation

$$G_l(\alpha_1, \alpha_2) = G_-(\alpha_1, \alpha_2), \quad G_r(\alpha_1, \alpha_2) = G_+(\alpha_1, \alpha_2)$$

$$MG_+ = G_- + V + K_1^{-1}U_\theta, \quad M = K_1^{-1}K_2,$$
$$K_1 = \varepsilon_6^{-1}K - \varepsilon_l R_l^{-1}, \quad K_2 = \varepsilon_r R_r^{-1} - \varepsilon_6^{-1}K$$

$$V = K_1^{-1}\left(R_l^{-1}\int\limits_{\partial\Omega_l}\omega_l + R_r^{-1}\int\limits_{\partial\Omega_r}\omega_r - \varepsilon_l R_l^{-1}T_l - \varepsilon_r R_r^{-1}T_r\right), \quad U_\theta = F_2 P_\theta u(x_1, x_2)$$

which, apart from the presence of unknowns $\mathbf{G}_{\pm}(\alpha_1, \alpha_2)$ also contains their functionals as unknowns of the form $\mathbf{G}_{\pm}(\alpha_1, \alpha_{2\pm})$. The latter have to be subsequently determined from a certain system of algebraic equations [61–63]. These articles report methods of determining the functionals entering into external forms. In order to study the regularities of solving the functional equation, the factorization approach, presented in [55], was used. The study of the properties of the solutions of this matrix functional equation made it possible to obtain new results, along with those described in [61–63].

In the case when the end faces of the plates are at the distance 2θ, $(\theta > 0)$, the contact stresses at the end faces of the plates have the form [55]

$$\mathbf{g}_l(x_1, x_2) = \boldsymbol{\sigma}_{1l}(x_1, x_2)(-x_2 - \theta)^{-0,5+i\gamma} + \boldsymbol{\sigma}_{2l}(x_1, x_2)(-x_2 - \theta)^{-0,5-i\gamma}, \ x_2 < -\theta,$$
$$\mathbf{g}_r(x_1, x_2) = \boldsymbol{\sigma}_{1r}(x_1, x_2)(x_2 - \theta)^{-0,5+i\gamma} + \boldsymbol{\sigma}_{2r}(x_1, x_2)(x_2 - \theta)^{-0,5-i\gamma}, \ x_2 > \theta \ \gamma > 0$$

Vectors $\boldsymbol{\sigma}_{1l}, \boldsymbol{\sigma}_{1r}$ are continuous in both parameters. The parameter γ is determined by the mechanical characteristics of the base. For example, for case under consideration, we have $\gamma = \operatorname{arcth}\frac{1-2\nu}{2(1-\nu)}$ where ν is the Poisson ratio of the base material [55].

When the end faces of the plates are at a zero spacing that $\theta = 0$, the contact stresses have the form

$$\begin{aligned} \mathbf{g}_l(x_1, x_2) &\to \boldsymbol{\sigma}_{4l}(x_1, x_2)x_2^{-1} \\ \mathbf{g}_r(x_1, x_2) &\to \boldsymbol{\sigma}_{4r}(x_1, x_2)x_2^{-1} \end{aligned} \tag{2.3}$$

The scheme of behavior of Earth's surface in the area of epicenter is shown in Fig. 2.4.

On the presented engraving are the consequences of the possible earthquake in 1884 when the Earth's crust movements occur in all three directions.

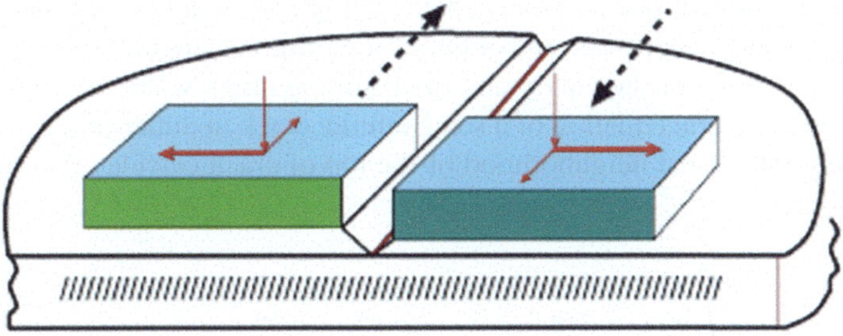

Fig. 2.4 External actions on the plates can cause such displacements of the surface of Earth in the epicenter of earthquake

Fig. 2.5 Such movements of the Earth's surface may have occurred during the earthquake of December 22, 1884, in Andalusia. The crack has length of 3 kilometers

2.4 On the Properties of Wedge-Shaped Cavities and Griffith' Cracks

In review of papers on singularities in wedge-shaped domains [53, 54], the cases of wedge-shaped cavities degenerating into cracks when the wedge angle tends to zero are considered. The opinion is expressed that in the process of such transformation cracks of a new type are also formed. Below it is shown that this opinion is erroneous. In all cases of transformation of the wedge-shaped cavity into a crack when the angle tends to zero only, the Griffith crack is obtained.

We use the results of one of these papers [2] Fig. 2.6. It applied an unquestioned rigorous mathematical approach and obtained the values of singularities in the neighborhood of the vertex of the cavity and the resulting crack when the angle is zero. Let us write the integral equation of a semi-infinite crack in dimensionless form and use the singularity in the neighborhood of the top of crack calculated in this paper. Thus, we obtain

$$\int_{-\infty}^{0} k(x-\xi)f(\xi)d\xi = q(x), \quad -\infty < x \leq 0$$

$$k(x) = \frac{1}{2\pi}\int_{-\infty}^{\infty} K(u)\mathrm{e}^{-iux}du,$$

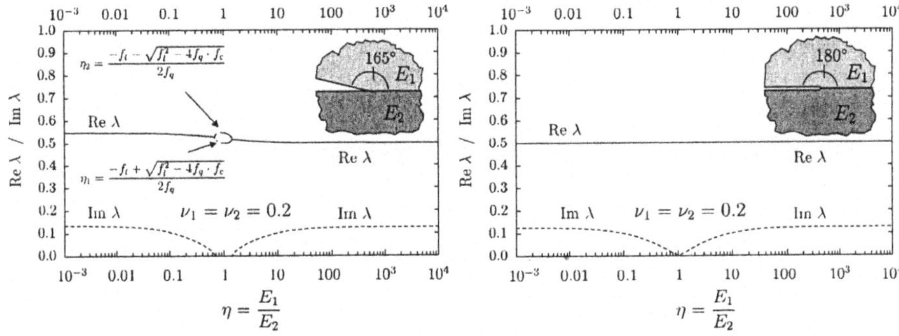

Fig. 2.6 Calculated in [2], the indices λ of the singularities of the stresses $r^{\lambda-1}$ in the material arising from the approach from the right to the top of the wedge-shaped cavity, the figure on the left, and to the crack, the figure on the right. It is $\lambda = 0.5$ for the crack when equal modules of Young $E_1 = E_2$

Here, *q(x)* are the stresses acting on the shores of the cracks; *f(x)* are the behavior of the shores of the crack which determines the shape of the crack; *K(u)*—Fourier transform of the kernel of an integral equation. We assume that the crack shores are symmetric and the same stresses act on both shores. The integral equations have the same appearance in the case of crack Griffith and of a crack of new type. For the sake of simplicity, consider the case of the same material for each semi-plane. In the present form, the kernel of an integral equation is a generalized function since the analytic function *K(u)* is growing

$$K(u) = |u|[(1 + O(u^{-1})], \quad |u| \to \infty$$

In the static case, we assume that the function *K(u)* has no zeros and poles on the real axis [55]. Considering the result of article [2], shown in Fig. 2.6, we find that the stresses approaching the top of crack on the right that is, for, $x \to 0, \ x > 0$ described by function

$$\psi(x) \to c_1 \frac{1}{\sqrt{x}}, \quad 0 < x < \infty, \quad c_1 = \text{const.}$$

Its Fourier transform has the form

$$\int_0^\infty \psi(x) e^{iux} dx \to c_1 \sqrt{\frac{\pi i}{u}}, \quad -\infty < u < \infty$$

We assume that it is given continuously differentiable function $q(x) \in L(-\infty, 0)$ By the Lebesgue–Riemann theorem, its Fourier transforms decrease at infinity. Then

$$Q(u) = \int_{-\infty}^{0} q(x)e^{iux}dx = \frac{1}{iu}q(0) - \frac{1}{iu}\int_{-\infty}^{0} q'(x)e^{iux}dx \to O\left(u^{-1}\right), \quad |u| \to \infty$$

Let us continue an integral equation on the entire axis, we have

$$\int_{-\infty}^{0} k(x-\xi)f(\xi)d\xi = q(x) + \psi(x), \quad -\infty < x < \infty$$

Applying the Fourier transform to integral equation on the whole axis, we obtain a functional equation of the form

$$K(u)F(u) = Q(u) + \Psi(u), \quad F(u) = \int_{-\infty}^{0} f(x)e^{iux}dx, \quad \Psi(u) = \int_{0}^{\infty} \psi(x)e^{iux}dx$$

From here, we find

$$F(u) = \frac{Q(u) + \Psi(u)}{K(u)}$$

or

$$f(x) = \frac{1}{2\pi}\int_{-\infty}^{\infty} \frac{Q(u) + \Psi(u)}{K(u)}e^{-iux}du$$

The derivative has an expression

$$f'(x) = -\frac{i}{2\pi}\int_{-\infty}^{\infty} u\frac{Q(u) + \Psi(u)}{K(u)}e^{-iux}du$$

A simple analysis of the integral under such conditions for given functions allows us to prove the continuity of the derivative everywhere, except for the point $x = 0$ which has the property of

$$f'(x) = c_2((-x)^{-1/2}), \quad x < 0, \quad x \to 0, \quad c_2 = \text{const}$$

This means that the tangent at the apex of the horizontal crack occupies a vertical position. Integrating this relation into the neighborhood of zero, taking into account the continuity of the derivative, on the interval $-\infty < x < 0$, we conclude that the function $f(x)$ is smooth, without top of crack, and behaves at zero as

$$f(x) \to c_3\sqrt{-x}, \quad x < 0, \quad x \to 0 \quad c_3 = \text{const}$$

This means that the shores of the crack are continuous converge, having a smooth-shaped crack of Griffith.

Thus, a Griffith crack is obtained, despite the fact that it originated from a cavity with a piecewise-smooth boundary. The latter means that in the considered papers [53, 54], cracks of a new type did not appear and were not described by anyone.

The assumption that cracks of a new type arise in the case of development of a wedge-shaped cavity in such a way that it penetrates a multi-layered elastic medium, perpendicularly running into layers, is also erroneous. This follows from the results of the study of a similar problem in the remarkable monograph by Morozov [48]. Based on the results obtained in this monograph, we again come to the above integral equation with a slight correction.

2.5 Conclusions

Thus, studying the method block element earthquake prediction as well as hidden defects in the bodies with coatings, the authors suddenly discovered the existence of the cracks of a new type. They complement the Griffith–Irwin's cracks. The existence of cracks of new type expends the understanding of the processes of destruction of media and materials. In seismology, it allows to give a new explanation to foreshocks and aftershocks. Considering the hierarchy of Griffiths–Irwin cracks and the cracks of new type, it can be assumed that after dislocations, the first cracks of the new type appear, which have a piecewise-smooth boundary and are destroyed earlier then the Griffith–Irwin cracks with a smooth boundary. Currently, the study of cracks of new type is continuing, and possibly, their new properties will be revealed.

Acknowledgements This work was supported by the Russian Foundation for Basic Research, projects nos (19-41-230003), (19-41-230004), (19-48-230014), (18-08-00465), (18-01-00384), (18-05-80008), the GZ UNC RAS, project №. 01201354241, (00-18-04), and Ministry of the science Russian Federation, project FZEN-2020-0020.

References

1. Babeshko, V.A., Evdokimova, O.V., Babeshko, O.M.: A new type of cracks adding to griffith-irwin cracks. Dokl. Phys. **64**(2), 102–105 (2019). https://doi.org/10.1134/S10283358191030042
2. Sator, C., Becker, W.: Closed-form solutions for stress singularities at plane bi- and trimaterial junctions. Arch. Appl. Mech. **82**, 643–658 (2012)
3. Balars, J., Sladek, J., Sladek, V.: Stress analysis by boundary element methods. Elsevier, Amsterdam (1989)

4. Barsukov, S.A., Glushkov, E.V., Glushkova, N.V.: The stress singularity at the corner points of the crack front located at the interface between two media. Izv. RAS. Fur. Solid Body. **2**, 77–85 (2002)
5. Barsukov, S.A., Glushkov, E.V., Glushkova, N.V.: Stress singularity at front corner points of an interfacial crack between two media. Izvestia RAN. Mekh. Tverdogo Tela. **2**, 77–85 (2002)
6. Borzenkov, S.M., Matveenko, V.P.: Semianalytic singular elements for plane and spatial problems in the theory of elasticity. Izv. RAS. MTT **6**, 48–61 (1995)
7. Borzenkov, S.M., Matveenko, V.P.: Optimization of elastic bodies in the vicinity of singular points. Izv. RAS. MTT **2**, 93–100 (1996)
8. Cherepanov, G.P.: Brittle fracture mechanics. Nauka, Moscow (1974)
9. Comninou, M.: The interface crack. Trans. ASME J. Appl. Mech. **44**, 631–636 (1977)
10. Comninou, M.: The interface crack in a shear field. Trans. ASME J. Appl. Mech. **45**, 287–290 (1978)
11. Comninou, M., Schmuser, D.: The interface crack in a combined tension-compression and sh. ear field. ASME J. Appl. Mech. **46**, 345–348 (1979)
12. Broutman, L.J., Krock, R.K. (eds.): Composite Materials. Academic Press, New York and London (1973–1974)
13. Dugdale, D.S.: Yielding of steel sheets containing slits. J. Mech. Phys. Solids **8**(2), 100–108 (1960)
14. Sih, G.C (ed.): Elastodynamic Crack Problems. Noordhoff, Leyden (1977)
15. Gautesen, A.K., Dunders, J.: The interface crack in a tension field. Trans. ASME J. Appl. Mech. **54**, 93–98 (1987)
16. Glushkov, E.V., Glushkova, N.V., Munz, D., Yang, Y.Y.: Analytical solution for bonded wedges under thermal stresses. Int. J. Fract. **106**, 321–339 (2000)
17. Glushkov, E., Glushkova, N., Lapina, O.: 3D Elastic stress singularity at polyhedral corner points. Int. J. Solids Struct. **36**, 1105–1128 (1999)
18. Glushkov, E.V., Glushkova, N.V., Lapina, O.N.: Indicators of singularity of elastic stresses at the point where the crack reaches the surface, Izv. RAS. Fur. Solid Body. **5**, 146–153 (1998)
19. Glushkov, E.V., Glushkova, N.V., Lapina, O.N.: Singularity exponents of elastic stresses at the point where the crack reaches the surface. Mech. Solids **33**(5), 117–122 (1998)
20. Griffith, A.: The phenomena of rupture in solids. Trans. Roy. Soc., London 221A, 163–197 (1920)
21. Guz, A.N., Guz, I.A., Men'shikov, A.V., Men'shikov, V.A.: Penny-shaped crack at the interface between classic half-spaces under the action of share wave, Int. Appl. Mech. **45**(5), 534–539 (2009)
22. Guz, A.N., Guz, I.A., Men'shikov, A.V., Men'shikov, V.A.: Stress intensity factors for materials with interface crack under harmonic loading, Int. Appl. Mech. **46**(10), 1093–1100 (2011)
23. Guz, A.N., Zozulya V.V.: Investigation of the effect of friction contact in III Mode crack under action of the SH-wave harmonic load, CMES (Comput. Model Eng. Sci.). **22**(2), 119–128 (2007)
24. Guz, I.A., Menshikov, O.V., Menshikov, V.A.: Application of boundary integral equations to elastodynamics of an interface crack. Int. J. of Fracture **140**, 277–284 (2006)
25. Huang Y., Gao, H.: Intersonic crack propagation—part II: suddenly stopping crack. J. Appl. Mech. **69**, 76–80 (2002)
26. Hutchinson, J.W., Mear, M., Rice, J.R.: Crack paralleling an interface between dissimilar materials. Trans. ASME. J. Appl. Mech. **54**(4), 828–832 (1987)
27. Irwin, G.R.: Analysis of stresses and stain near the end of a crack traversing a plate. Trans. ASME. J. Appl. Mech. **24**(N3), 361–364 (1957)
28. Irwin, G.: Fracture dynamics, Fracture of metals, ASM, Cleveland. 147–166 (1948)
29. Kaminskii, A.A., Usikova, G.L., Dmitrieva, E.A.: Experimental study of the distribution of plastic strains near a crack tip during static loading. Int. Appl. Mech. **30**(11), 892–897 (1994)
30. Kaminskii, A.A., Kipnis, L.A., Kolmakova, V.A.: On the dugdail model for a crack at the interface of different media. Int. Appl. Mech. **35**(6), 58–63 (1999)

31. Kirugulige, M.S., Tippur, H.V.: Mixed-mode dynamic crack growth in functionally graded glass-filled epoxy. Exp. Mech. **46**(2), 269–281 (2006)
32. Kit, H.S., Khaj M.V., Mikhas'kiv, V.V.: Analisis of dynamic stress concentration in an infinite body with parallel penny-shaped cracks by BIEM. Eng. Fract. Mech. **55**(2), 191–207 (1996)
33. Korepanova, T.O., Matveyenko, V.P., Sevodina, N.V.: Numerical analysis of the stress singularity at the apex of a cone with a non-smooth side surface. Comput. Continuum. Mech. **3**(3), 68–76 (2010). https://doi.org/10.7242/1999-6691/2010.3.3.28
34. Korepanova, T.O., Matveyenko, V.P., Sevodina, N.V.: Numerical analysis of the stress singularity at the vertex of spatial intersecting cracks. Comput. Continuum. Mech. **4**(3), 68–73 (2011). https://doi.org/10.7242/1999-6691/2011.4.3.28
35. Leblond, J.B., Frelat, J.: Crack kinking from an interface crack with initial contact between the cracks lips. Europ. J. Mech. A. Solids. **20**, 937–951 (2001)
36. Loboda, V.V., Sheveleva, A.E.: Determing prefracture zones at a crack tip between two elastic orthotropic bodies. Int. Appl. Mech. **39**(5), 566–572 (2003)
37. Loeber, J.F., Sih, G.C.: Transmission of anti-plane shear waves past an interface crack in dissimilar media. Eng. Fract. Mech. **146**, 699–725 (1973)
38. Matveenko, V.P.: The method of numerical analysis of the stress singularity at the corner points of three-dimensional bodies. Izv. RAS. MTT **5**, 71–77 (1995)
39. Matveenko, V.P., Nakaryakova, T.O., Sevodina, N.V., Shardakov, I.N.: The singularity of stresses at the vertex of homogeneous and compound cones under different boundary conditions. PMM **72**(3), 477–484 (2008)
40. Menshykov, O.V., Menshykov, M.V., Guz, I.A.: Elastodynamics contact problem for an interface crack under harmonic loading. Eng. Fract. Mech. **80**, 52–59 (2012)
41. Menshykov, O.V., Menshykov, M.V., Guz I.A., Mickucka, V.: 2-D and 3-D contact problems for interface cracks under harmonic loading. In: Constanda, C., Harris, P.J. (ed) Integral Methods in Science and Engineering: Computational and Analystic Aspects, pp. 241–252. Springer, New York, Dordrecht (2011)
42. Menshykov, O.V., Menshykov, M.V., Guz, I.A.: Linear interface crack under plane share wave. CMES (Comput. Model Eng. Sci.). **48**(2), 107–120 (2009)
43. Menshykov, O.V., Menshykov, M.V., Guz, I.A.: Modelling crack closure for an interface crack under harmonic loading. Int. J. Fract. **165**, 127–134 (2010)
44. Menshykov, O.V., Menshykov, M.V., Guz, I.A.: An iterative BEM for the dynamic analysis of interface crack contact problems. Eng. Anal. Bound. Elem. **35**(5), 735–749 (2011)
45. Mikhas'kiv, V.V.: Opening-function simulation of the three-dimensional interaction of cracks in an elastic body. Int. Appl. Mech. **37**(1), 75–84 (2001)
46. Mikhas'kiv, V.V., Butrak I.O.: Stress concentration around a spheroidal crack coused by a harmonic wave incident at an arbitrary angle. Int. Appl. Mech. **42**(1), 61–66 (2006)
47. Mikhas'kiv, V.V., Sladek J., Sladek V., Stepanyuk, A.L.: Stress concentration near the elliptical crack in the interface between elastic bodies under steady-state vibrations, Int. Appl. Mech. **40**(6), 664–671(2004)
48. Morozov, N.F.: Mathematical Questions of the Theory of Cracks. Nauka, Moscow (1984)
49. Qu, J.: Interface crack loaded by a time-harmonic plane wave. Int. J. Solids Struct. **31**(3), 329–345 (1994)
50. Rangarajan, R., Chiaramonte, M.M., Hunsweck, M.J., Shen, Y., Lew, A.J.: Simulating curvilinear crack propagation in two dimensions with universal meshes, Int. J. Numer. Meth. Eng. **102**(3–4), 632–670
51. Rice, J.R., Sih, G.C.: Plane Problems of cracks in dissimilar media. Trans. ASME. J. Appl. Mech. **32**, 218–224 (1965)
52. Rice, J.R.: Elastic fracture mechanics concepts for interface cracks. Trans. ASME. J. Appl. Mech. **55**, 98–103 (1988)
53. Sinclair, G.B.: Stress singularities in classical elasticity I. Appl. Mech. Rev. **57**, 251–298 (2004)
54. Sinclair, G.B.: Stress singularities in classical elasticity II. Appl. Mech. Rev. **57**, 385–439 (2004)

55. Vorovich, I.I., Babeshko V., A.: Dynamic mixed problems of the theory of elasticity for nonclassical regions, Moscow (1979)
56. Williams, M.L.: Stress singularities resulting from various boundary conditions in angular corners of plates in extension. Trans. ASME. J. Appl. Mech. **19**(4), 526–535 (1952)
57. Williams, M.L.: The stresses around a fault or crack in dissimilar media. Bull. Seismol. Soc. Amer. **49**(2), 199–204 (1959)
58. Zhang, C., Gross, D.: On wave propagation in elastic solids with cracks. Comp. Mech. Publ. South-ampton, UK, Boston, USA, (1998)
59. Zozulya, V.V.: Investigation of the contact of edges of cracks interactions with a plane longitudinal harmonies wave. Sov. Appl. Mech. **27**, 1191–1195 (1991)
60. Babeshko, V.A., Glushkov, E.V., Glushkova, N.V., Lapina, O.N.: The peculiarity of stresses in the vicinity of the top of the elastic trihedron. Rep. Russ. Acc. Sci. **318**(5), 1113–1116 (1991)
61. Babeshko, V.A., Evdokimova, O.V., Babeshko, O.M., Zaretskaya, M.V., Gorshkova, E.M., Mukhin, A.S., Gladskoi, I.B.: Origin of starting earthquakes under complete coupling of litosphere plates and a base. Dokl. Phys. **63**(2), 70–75 (2018). https://doi.org/10.1134/s10283358 18020015
62. Babeshko, V.A., Evdokimova, O.V., Babeshko, O.M.: On the possibility of predicting some types of earthquake by a mechanical approach. Acta Mech. **229**(5), 2163–2175 (2018). https:// doi.org/10.1007/s00707-0172092-0
63. Babeshko, V.A., Evdokimova, O.V., Babeshko, O.M.: On a mechanical approach to the prediction of earthquakes during horizontal motion of litospheric plates. Acta Mech. **229**(11), 4727–4739 (2018). https://doi.org/10.1007/s00707-018-2255-7

Chapter 3
Optimal Attenuation of Transverse Vibrations for a Cantilever Beam

Dmitry V. Balandin and Egor V. Petrakov

Abstract The multi-criteria problem of transverse vibrations attenuation of a cantilever beam is considered by the active and passive damping methods. The mathematical model of beam is provided by Bernoulli–Euler's hypotheses with the linear viscosity. Perturbation acting on the beam belongs to a class L_2 of functions. The beam mode is described by Krylov functions. The normal form method is used to convert to the main coordinates. A model of active vibration isolation applied along the entire length of the cantilever beam and a model connected to a vertical base at one point were constructed. The beam model connected to a viscoelastic base within the passive vibration isolation is considered. The task of transverse vibrations attenuation is a state feedback control problem with two controlled outputs. Two criteria are introduced, namely the maximum beam deflection and the control force. The generalized H_2 norm is used as a measure of functional evaluation in the synthesis of optimal regulators. The search for optimal feedback is based on the use of linear matrix inequalities and efficient algorithms for solving, implemented in the MATLAB package. Synthesis of Pareto-optimal control is implemented on the basis of Germeyer convolution. The optimal values of the functional under distributed and concentrated forces are given with respect to two criteria for active and passive damping methods. The paper includes a comparison of vibration isolation for different attenuation methods.

Keywords Console beam · Transverse vibrations · Bernoulli–Euler model · Vibrations damping · Control · Generalized H_2 norm · Pareto set · Linear matrix inequalities

---

D. V. Balandin (✉)
National Research Lobachevsky State University of Nizhny Novgorod, Mathematical Center, 23 Gagarin Avenue, Building 2, Nizhny Novgorod 603950, Russian Federation
e-mail: dbalandin@yandex.ru

E. V. Petrakov
Research Institute for Mechanics, National Research Lobachevsky State University of Nizhny Novgorod, 23 Gagarin Avenue, Building 6, Nizhny Novgorod 603950, Russian Federation
e-mail: pet.e@mail.ru

H. Altenbach et al. (eds.), *Multiscale Solid Mechanics*,
Advanced Structured Materials 141,
https://doi.org/10.1007/978-3-030-54928-2_3

3.1 Introduction

The problem of reducing vibration affects many areas of engineering, instrumentation and construction. The urgency of the problem increases due to the increase in the size of structures, increasing influence of external disturbance. In modern structures inexpediently dampen vibrations, changing the geometry and materials of the object, as the design is often becomes economically inefficient. New effective methods of vibration protection involve changing the mass or stiffness of vibration isolation. The principle of vibration isolation is widely used in the protection of buildings and structures from seismic effects. Often, the vibration isolating application is located between the foundation and the building.

Vibration isolation systems are represented by various structures of supports, which have a given value of stiffness and viscosity. Vibration isolation is described by passive and active types of damping [1].

There are connected methods of vibration protection for nearby buildings. As space in big cities is limited, buildings located close to each other are connected by damping devices [2–10], thereby reducing the response of buildings to seismic impacts. A striking example of such buildings is the Triple Towers in Tokyo [11, 12]. Damping devices can be located along the entire height of the building or in separate areas or points.

There is no complete understanding in choosing the best type of attenuation, at moment. The authors [13] suggest using semi-active or passive damping for strong resonant vibrations instead of active damping, because the problem of stability is very important for active damping methods. If the active damping method is used incorrectly, the active forces control can amplify vibration. The authors [14] consider that passive vibration systems of attenuation are unacceptable due to low efficiency or exceeding the permissible weight, and active vibration damping implemented by actuators is more effective.

3.2 Mathematical Model

Consider a cantilever beam lying on elastic foundation subject to some perturbation. Active (Fig. 3.1a, b) and passive (Fig. 3.1c) attenuation of vibrations of a cantilever beam lying on elastic foundation with uniformly distributed control along the length (Fig. 3.1a, c) and control at one point (Fig. 3.1b) is investigated.

Suppose the mass and dimension of the base are so much greater than the mass and dimensions of the beam that the effect of the beam on the base is negligible. The values of two criteria providing optimal vibration attenuation (the level of the control force and the maximum deflection of the beam) are estimated. Suppose the mass and dimensions of the base are so much greater than the mass and dimensions of the beam that the effect of the beam on the base is negligible.

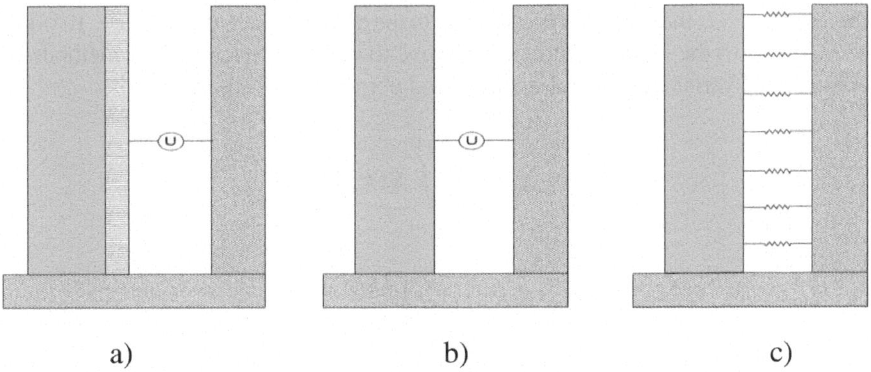

Fig. 3.1 Attenuation of a cantilever beam: **a** active uniformly distributed control along the length, **b** active damping at one point; **c** passive uniformly distributed control along the length

Bernoulli–Euler model is adopted to describe the transverse vibrations of the beam with internal friction (linear model of Kelvin–Voigt viscosity). Thus, the equation of transverse vibrations of the beam is described:

$$\frac{\partial^2 \omega}{\partial t^2} + a^2\left(\frac{\partial^4 \omega}{\partial x^4} + \vartheta_0 \frac{\partial^5 \omega}{\partial x^4 \partial t}\right) = -\frac{u_0}{\rho A} + v_0(t) \tag{3.1}$$

where $\omega = \omega(x, t)$ is deflection of the beam, $a^2 = EI/\rho A$, E is Young's modulus of elasticity, I is the second moment of area, ρ is the density, and A is the cross-sectional area of the beam. $\vartheta_0 = \frac{\vartheta}{a^2}$ is coefficient of internal viscosity, u_0 is the control of system, and $v_0(t)$ is the acceleration acting on the base (function belonging to class L_2). The dynamic system (3.1) contains six constant parameters E, I, ρ, A, ϑ and l (beam length). After the introduction of dimensionless variables

$$\omega' = \frac{\omega}{l}, x' = \frac{x}{l}, t' = \frac{a}{l^2}t, u = \frac{l^3}{EI}u_0, v(t') = \frac{l^3}{a^2}v_0(t)$$

dynamic system (3.1) is converted (hereafter the strokes are omitted) into

$$\ddot{\omega} + \beta K \dot{\omega} + K\omega = -u + v(t') \tag{3.2}$$

where $K = \frac{\partial^4}{\partial x^4}$ is the differential operator of the fourth degree, $\beta = a\vartheta_0/l^2$ is dimensionless damping coefficient of the system. If the control uses a sensor installed at x_1, the control is written as:

$$u = \theta^{(1)}\omega(x_1') + \theta^{(2)}\dot{\omega}(x_1')$$

where $\theta^{(1)}, \theta^{(2)}$ is the feedback coefficients in the displacement and velocity. To bring the Eq. (3.2) to the main coordinates (relative to the time function), the methods of separation of variables (Fourier method) and normal forms are used:

$$\omega(x,t) = \sum_{i=1}^{\infty} X_i(x)T_i(t), \int_0^1 X_i X_j \mathrm{d}x = \begin{cases} 1, i = j \\ 0, i \neq j \end{cases},$$

$$\int_0^1 K X_i X_j dx = \begin{cases} \lambda_i^4, i = j \\ 0, i \neq j \end{cases} \tag{3.3}$$

where X_i is shape function and λ_i is eigenvalue. The shape functions and eigenvalues are found from the boundary conditions for a cantilever beam: $X(0) = 0$, $X'(0) = 0$, $X''(1) = 0$, $X'''(1) = 0$.

Shape functions are described using Krylov functions. A detailed study of the eigenvalues λ and the shape functions X for a cantilever beam is given in [15]. Conversion of Eq. (3.2) to the main coordinates leads to the ratio:

$$\ddot{T}_i + \beta\lambda_i^4\dot{T}_i + \lambda_i^4 T_i = -\int_0^1 X_i(u - v(t))\mathrm{d}x \tag{3.4}$$

3.3 Two-Criterion Problem

To solve the control problem, the first m-eigenmodes of the beam are used and the equation of motion in matrix form is written for them:

$$\dot{T} = \begin{pmatrix} 0_{mxm} & I_{mxm} \\ -\Lambda^4 & -\beta\Lambda^4 \end{pmatrix} T + B_u\theta^{(0)}T + B_v v(t) \tag{3.5}$$

where $T = \left(T_1 \ldots T_m, \dot{T}_1 \ldots \dot{T}_m\right)^T$ is state vector, $\Lambda = \mathrm{diag}(\lambda_1 \ldots \lambda_m)$ is diagonal matrix of eigenvalues, and B_u, B_v are the vectors of the control and perturbation. To control a distributed system: $B_u = B_v$. This is the problem of control theory, i.e. searching for linear state feedback:

$$\begin{aligned} \dot{T} &= AT + B_u u + B_v v \\ z_1 &= CT \\ z_2 &= Du \end{aligned}$$

$$u = \theta^{(0)} T \tag{3.6}$$

where z_1, z_2, C and D are measurable outputs as to both the criteria and the corresponding matrices of outputs. In practical implementation, feedback regulators are subject to a variety of technological requirements (e.g., relative stability, damping decrement, stiffness and damping of the actuator controller). The most important practical requirement is the limitation of the resource of control and its implementation [16].

The problem of damping the vibrations of the cantilever beam on an elastic base is transformed into a two-criterion problem. At the output of the system, the level of the control force and the maximum deflection of the beam are considered. Both values are inversely dependent. It is necessary to minimize both criteria:

$$(J_1(\theta), J_2(\theta)) \to \min$$

3.4 Solution of Two-Criterion Problem

Often, the search for linear feedbacks in the state space is associated with the canonical representation of the controlled object and the construction of a modal control that provides the given eigenvalues (modes) of the matrix of a closed system. The construction of the modal control is reduced to finding the characteristic polynomial of matrix A, choosing the canonical basis and solving a system of linear equations.

At the same time, an alternative way of synthesis of stabilizing regulators is possible, based on the application of the theory of linear matrix inequalities and effective algorithms for their solution implemented in the MATLAB package [17].

For linear matrix inequalities, Pareto-optimal control laws are synthesized using the Germeyer convolution [18, 19]:

$$J_\alpha(\theta) = \max\left\{\frac{J_1(\theta)}{\alpha}; \frac{J_2(\theta)}{1-\alpha}\right\}, \quad \alpha \in (0, 1) \tag{3.7}$$

The output vectors of the dynamic system (3.7) are converted to the form: $C_\alpha = C/\alpha$, $D_\alpha = D/1 - \alpha$.

The generalized H_2 norm is used to minimize the functional of J_α, and linear matrix inequalities for the formulation of the problem (3.7) have the form:

$$\begin{pmatrix} AY + YA^T + B_u Z + Z^T B_u^T & B_v \\ B_v^T & -I \end{pmatrix} < 0, \begin{pmatrix} Y & C_\alpha^T \\ C_\alpha & \alpha^2\gamma^2 \end{pmatrix} \geq 0,$$
$$\begin{pmatrix} Y & Z^T D_\alpha^T \\ D_\alpha Z & (1-\alpha)^2\gamma^2 \end{pmatrix} \geq 0 \tag{3.8}$$

where $Z = \theta Y$ and $Y = X^{-1}$ is inverse Lyapunov matrix. The values of the criteria minimizing the functional (3.8) are found from the expressions:

$$J_1 = \left\| H_{z_1 v} \right\|_{g_2} = \sup_{v \in L_2} \frac{|z_1|_{g_\infty}}{\|v\|_2} = \lambda_{\max}^{\frac{1}{2}}\left(\mathrm{CYC}^T\right)$$

$$J_2 = \left\| H_{z_2 v} \right\|_{g_2} = \sup_{v \in L_2} \frac{|z_2|_{g_\infty}}{\|v_2\|} = \lambda_{\max}^{\frac{1}{2}}\left(D\theta Y \theta^T D^T\right)$$

3.5 Numerical Results

Analysis of the solution of the system of Eqs. (3.5) using the MATLAB software package shows that the curves for 4–10 modes coincide with the curves for 2 and 3. Curves for the first, second, and third modes are shown in Fig. 3.2.

Since the first three modes of vibration are sufficient to describe the set of solutions, the functional for only the first three modes of vibration is considered in further studies.

In the case of a concentrated control force, the mathematical model (3.2) is described by the equation:

$$\ddot{\omega} + \beta K \dot{\omega} + K\omega = -u\delta(x - x_0) + v(t)$$

where x_0 is the point with the concentrated control force applied. This paper deals with the case where the control force is applied to the middle of the beam. The problem with a concentrated control force is of practical interest, and it is easier to use a point dynamic active control for attenuation oscillations [20]. The features of the choice of the point of application of the concentrated control force are not studied here.

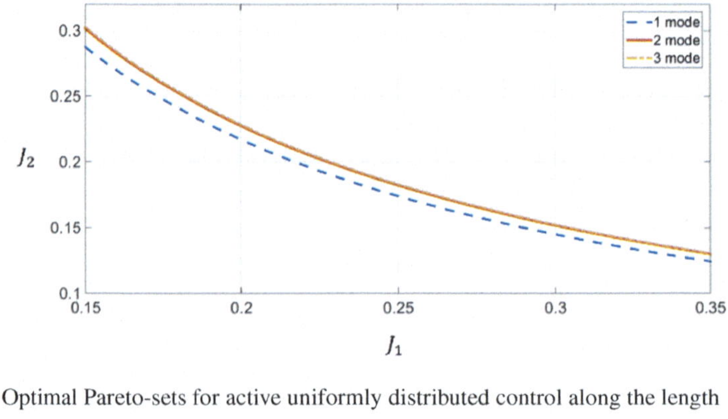

Fig. 3.2 Optimal Pareto-sets for active uniformly distributed control along the length

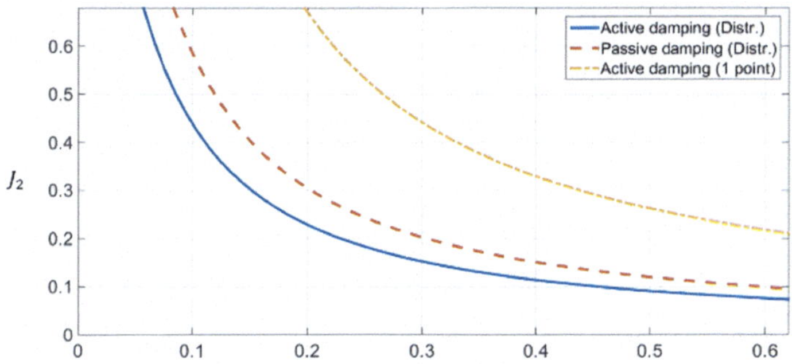

Fig. 3.3 Optimal Pareto-sets for the different methods of attenuation

The optimal passive and active vibration attenuation for a cantilever beam with a viscoelastic coupling with a vertical fixed platform, described by the equation, is found by the method of linear matrix inequalities,

$$\ddot{\omega} + C\dot{\omega} + K\omega = -K_s'\omega - D_s'\dot{\omega}$$

Here

$$K_s' = \frac{l^4 K_s}{EI}, \quad D_s' = \frac{l^2 D_s}{\rho A}$$

are dimensionless stiffness and damping coefficients. The relations between J_1 and J_2 criteria minimizing the $J_\alpha(\theta)$ functional for different attenuation methods are shown in Fig. 3.3. The solution was found using the MATLAB package.

The analysis of Fig. 3.3 shows that the best result is obtained by using distributed attenuation. The difference between the result of the use of passive and active damping is negligible. The active concentrated force in the middle of the beam is the simplest method of applying vibration protection, but the least effective compared to distributed vibration attenuation. The following is an assessment of the efficiency of damping the beam vibrations at some external perturbation, belonging to the class L_2, and described by the equation:

$$\ddot{\omega} + C\dot{\omega} + K\omega = v(t)$$

Vibration record of violent vibrations and the results of active and passive damping are shown below in Fig. 3.4.

Analysis of vibration time history shows that the effect of the use of both passive and active vibration damping. It should be noted that the difference between the result of the use of passive and active damping is negligible.

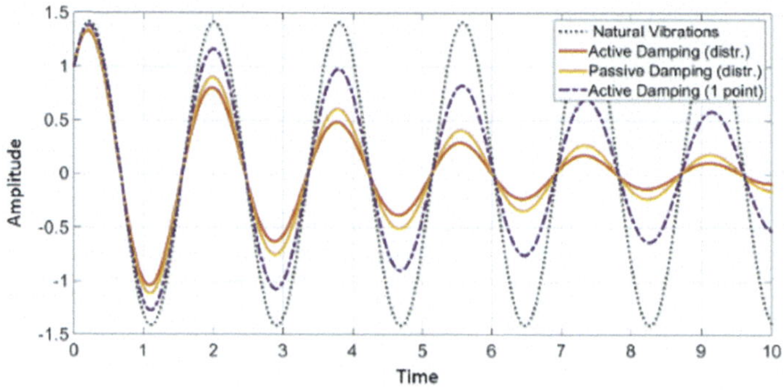

Fig. 3.4 Vibration time history for the different methods of attenuation

3.6 Conclusion

The system of active and passive damping of vibrations of cantilever beam lying on an elastic foundation is considered. To solve the problem, two connected criteria are introduced: the level of the control force and the maximum deflection of the beam. The results of passive and active control of vibration damping of the cantilever beam using a distributed vibration damper and an active concentrated control force are analyzed. Analysis of the results showed that the active concentrated force in the middle of the beam is the simplest method of applying vibration protection, but the least effective, compared with distributed damping, and the difference between the result of the use of passive and active distributed damping is negligible. It should be noted that taking into account the difference in technological, energy and financial costs for active and passive systems, the choice of one or another of them, taking into account efficiency, remains open.

Acknowledgements The work is financially supported by the Federal Targeted Program for Research and Development in Priority Areas of Development of the Russian Scientific and Technological Complex for 2014–20 under the contract No. 075-15-2019-1702 (unique identifier RFMEFI60519X0183).

References

1. Khlebnikov, M.V., Sherbakov, P.S.: Ogranichennoe lineinoe upravlenie, optimal'noe po kvadratichnomu kriteriu special'nogo vida [Limited linear control optimal by a quadratic criterion of special type]. Proceedings of ISA RAN **63**(2), 85–89 (2013)
2. Balandin, D.V., Ezhov, E.N., Petrakov, E.V., Fedotov, I.A.: Multi-criteria problems for optimal protection of elastic constructions from vibrations. Mater. Phys. Mech. **40**, 239–245 (2018)
3. Bhaskararao, A., Jangid, R.: Seismic analysis of structures connected with friction dampers. Eng. Struct. **28**(5), 690–703 (2006)

4. Bian, Y., Gao, Z.: Nonlinear vibration absorption for a flexible arm via a virtual vibration absorber. J. Sound Vib. **399**, 197–215 (2017)
5. Christenson, R.E., Spencer Jr., B., Johnson, E.A.: Semiactive connected control method for adjacent multidegree-of-freedom buildings. J. Eng. Mech. **133**(3), 290–298 (2007)
6. Korenev, B.G., Reznikov, D.M.: Dinamicheskie gasiteli kolebanii [Mass Damper]. Nauka Publ, Moscow (1988)
7. Szmidt, T., Pisarskin, D., Bajer, C., Dyniewicz, B.: Double-beam cantilever structure with embedded intelligent damping block: Dynamics and control. J. Sound Vib. **401**, 127–138 (2017)
8. Xu, Y.L., Ng, C.: Seismic protection of a building complex using variable friction damper: experimental investigation. J. Eng. Mech. **134**(8), 637–649 (2008)
9. Xu, Y.L., Zhan, S., Ko, J., Zhang, W.: Experimental investigation of adjacent buildings connected by fluid damper. Earthquake Eng. Struct. Dynam. **28**(6), 609–631 (1999)
10. Zhang, Y.F., Iwan, W.D.: Statistical performance analysis of seismic excited structures with active interaction control. Earthquake Eng. Struct. Dynam. **32**(7), 1039–1054 (2003)
11. Asano, M., Yamano, Y., Yoshie, K., Koike, Y., Nakagawa, K., Murata, T.: Development of active-damping bridges and its application to triple high-rise buildings. JSME Int. J. Ser. C. **46**(3), 854–860 (2003)
12. Xu, Y.L., He, Q., Ko, J.: Dynamic response of damper-connected adjacent buildings under earthquake excitation. Eng. Struct. **21**(2), 135–148 (1999)
13. Zhang, W.S., Xu, Y.L.: Vibration analysis of two buildings linked by maxwell model-defined fluid dampers. J. Sound Vib. **233**(5), 775–796 (2000)
14. http://www.obayashi.co.jp/english/services/technologies/#anc02
15. Kasagi, M., Fujita, K., Tsuji, M., Takewaki, I.: Automatic generation of smart earthquake-resistant building system: hybrid system of base isolation and building-connection. Heliyon. **2** (2016). https://doi.org/10.1016/j.heliyon.2016.e00069
16. Balandin, D.V., Kogan, M.M.: Pareto optimal generalized H_2-control and vibroprotection problems. Autom. Remote Control. **78**(8), 1417–1429 (2017)
17. Christenson, R.E., Spencer Jr., B., Hori, N., Seto, K.: Coupled building control using acceleration feedback. Comput. Aided Civil Infrastruct. Eng. **18**(1), 4–18 (2003)
18. Balandin, D.V., Kogan, M.M.: Sintez zakonov upravlenia na osnove lineinyh matrichnyh neravenstv [Control Law Synthesis Based on Linear Matrix Inequalities]. Fizmatlit Publ, Moscow (2007)
19. Hazanov, H.S.: Mechanicheskie kolebaniya sistem s raspredelennymi parametrami [Mechanical Vibrations of Systems with Distributed Parametrs]. SSAU Publ, Samara (2002)
20. Bharti, S., Dumne, S., Shrimali, M.: Seismic response analysis of adjacent buildings connected with MR dampers. Eng. Struct. **32**(8), 2122–2133 (2010)

Chapter 4
Numerical Study of the Mutual Influence of Nearby Buried Structures Under Seismic Influences

Valentin G. Bazhenov and Nadezhda S. Dyukina

Abstract Models, methods, algorithms, software tools have been developed to describe the basic laws of the behavior of buried structures under seismic influences, taking into account gravity and nonlinear effects of contact interaction with the ground and reducing computational costs to an acceptable level. The proposed methods of modeling non-reflective boundary conditions and restoration of the vibration effect on the surface of the experimental accelerogram can significantly reduce computational costs. The gravity field necessary for adequate description of seismic oscillations of massive structures is taken into account after damping procedure of the kinetic energy of the computational domain. The developed computational methods are used to study the mutual influence of two different nearby buried structures under seismic influences. A method of reducing the mutual influence of nearby structures by sand dumping is considered. The numerical simulation of seismic isolation is carried out taking into account the granularity and elastic–plastic properties of sand. A complex nonlinear relationship between the vibrations of structures, the distance between them and the presence or absence of sandy seismic isolation is shown.

Keywords Numerical simulation · Contact interaction · Mutual influence · Nearby buried structures · Seismic vibrations · Seismic isolation

4.1 Introduction

Numerical simulation of seismic vibrations of large-sized buried structures is carried out at the design stage and is a necessary element for assessing the seismic resistance of nuclear power plants and other large industries [7, 14, 25, 27]. The complexity of such calculations is related to the consideration of the soil adjacent to the buried

---

V. G. Bazhenov (✉) · N. S. Dyukina
Research Institute for Mechanics, National Research Lobachevsky State University of Nizhny Novgorod, 23 Gagarin Avenue, Building 6, Nizhny Novgorod 603950, Russian Federation
e-mail: bazhenov@mech.unn.ru

N. S. Dyukina
e-mail: ndyukina@inbox.ru

H. Altenbach et al. (eds.), *Multiscale Solid Mechanics*,
Advanced Structured Materials 141,
https://doi.org/10.1007/978-3-030-54928-2_4

structure: assumption the soil as a system of visco-elastic constraints [6, 8, 17, 20, 26] requires theoretical justification; the introduction an array of adjacent soil described by finite elements into the calculations significantly increases the complexity of the task [15, 18, 19, 22]. Earlier, the authors proposed a mathematical model and numerical method for solving problems of interaction of a buried structure with the ground under seismic influences, taking into account the nonlinear effects of contact interaction with the ground, the field of gravity, and reducing computational costs to an acceptable level [9, 11]. Due to the high rate of propagation of seismic waves in the ground, the results of the research can be influenced even by large objects that are significantly remote [10, 12, 13, 21, 28]. The developed computational methods are used to analyze the mutual influence of two buried structures; the series of computational experiments shows how the seismic vibrations of structures depend on the distance between them and their sizes [4]. If the project does not allow placing structures at a considerable distance from each other, eliminating unwanted mutual influence, the device between the structures of the damping layer can solve the problem [16, 24]. A series of computational experiments on seismic vibrations of two nearby structures separated by a seismic insulating layer was carried out. A complex relationship between the behavior of structures, the fact of the presence or absence of seismic isolation between them and the distance is established.

4.2 Mathematical Model and Numerical Technique

The variational approach [1, 2, 23] is used to describe deformations of bodies within the framework of the continuum mechanics hypothesis. To describe the motion of continuums in Lagrangian variables in a fixed Cartesian coordinate system, the equations following from the variational principle of power balance are applied:

$$\iint\limits_{\Omega} \left(\frac{1}{2}\sigma_{i,j}\left(\delta\dot{u}_{i,j} + \delta\dot{u}_{j,i}\right) + \rho\ddot{u}_i\delta\dot{u}_i - \rho f_i\delta\dot{u}_i \right) \mathrm{d}\Omega - \int\limits_{G} p_i\,\delta\dot{u}_i \mathrm{d}S - \int\limits_{G} q_i\,\delta\dot{u}_i \mathrm{d}S = 0$$

Here $\sigma_{i,j}$ are the stress tensor components, $\dot{u}_i$ are the velocities, p_i, q_i are the components of the surface load and contact pressure, f_i are the components of the mass force per unit mass ($i = x, y, z$). Contact interaction with friction is described in the local coordinate basis associated with the contact surface, the conditions of non-penetration along the normal to the contact surface and based on the condition of dry friction-tangentially.

The soil medium massif is modeled by a rectangular parallelepiped, homogeneous in its characteristics or consisting of areas of homogeneity. The authors have developed and used non-reflecting boundary conditions on the lateral surfaces of the array, allowing reducing the horizontal dimensions of the calculated area. At the lower boundary of the soil mass, velocity components are set, recovered from the experimental accelerogram set on the surface by a special technique [9], which allows compensating for multiple re-reflections of seismic waves from the lower boundary—thus reducing the vertical dimensions of the soil mass. The numerical solution of the defining system of Eqs. (1) under the given initial and boundary conditions is based on the variational-difference method of discretization by spatial coordinates and the explicit scheme of integration by time [2]:

$$(\dot{u}_\alpha)_j^{k+1/2} = (\dot{u}_\alpha)_j^{k-1/2} + (F_\alpha)_j^k \frac{\Delta t^{k+1/2}}{(M)_j^k},\ (u_\alpha)_j^{k+1} = (u_\alpha)_j^k + (\dot{u}_\alpha)_j^{k+1/2} \Delta t^{k+1}$$

$$\Delta t^{k+1} = t^{k+1} - t^k,\ \Delta t^{k+1/2} = \frac{1}{2}\left(\Delta t^{k+1} + \Delta t^k\right); \quad \alpha = x, y, z$$

Here, F_j is the generalized forces acting on the node j, M is the mass in the node j.

The described methods of the solution, algorithms of modeling of contact interaction and a field of gravity are realized in the certified program complex "Dynamica-3." Parallelization of algorithms allowed reducing computational costs and increasing the efficiency of calculations.

The authors substantiate the feasibility of modeling sufficiently hard soils in seismic problems by an ideally elastic continuous medium. For soft soil bases—sands and clays—a transversal isotropic model is proposed, taking into account soil compaction with depth [11].

4.3 Analysis of the Mutual Influence of Seismic Vibrations of Two Nearby Structures

In the numerical study of the mutual influence of nearby structures, the structures of the pumping station and the reactor compartment buried into the soil at 9 m were considered. The variants were considered when the distance between the structures (the pumping station and the reactor compartment) was: (a) $L = \infty$, (b) $L = 48$ m, (c) $L = 80$ m, (d) $L = 120$ m, (e) $L = 200$ m. Mechanical characteristics of structures and soil are given in Table 4.1. The seismic effect was given as the seismogram components corresponding to an earthquake with magnitude 6.

Table 4.1 Characteristics of structures and soil

Geometrical and mechanical characteristics	Dimensions			
Pumping station	24 × 24 × 19	1.4 × e + 10	8400	802.7
Reactor compartment	64 × 64 × 75	1.4 × e + 10	8400	802.7
The top layer of soil	$H = 50$	4.5 × e + 7	9.64	2030
The bottom layer of soil	$H = 250$	3.6 × e + 9	78,000	2030

Analysis of changes in the maximum mutual displacements of the walls of structures and soil [4] allowed to draw the following conclusions:

- Mutual influence of seismic vibrations of structures gives different effects for different buildings;
- For a larger reactor structure is characterized by a decrease in the intensity of the oscillations with increasing distance *L*;
- As the distance *L* increases, the roll of less overall construction of the pumping station changes to the opposite one;
- On horizontal mutual displacement are more affected by ripple effects from shear reflected waves;
- Vertical displacements are more influenced by the sediment of a nearby building and associated with changes in the properties of a certain volume of adjacent soil;
- The distance between structures less than their overall dimensions is critical and leads to a progressive roll of structures toward each other over time.

4.4 Influence of Seismic Isolation Between Structures on Their Mutual Influence During Seismic Impact

The influence of seismic isolation between structures on the results of calculations is analyzed. As a seismic isolation was considered sand backfill, thickness 8 m, equidistant from the structures and buried at the same depth with them. The elastic–plastic model of Grigoryan's [5] soft soil environment with specification of functional dependences according to features of the considered problem is applied for modeling of sand dumping. According to this model, compression to p* and discharge of the soil medium are described by different dependencies. The initial compression is characterized by less stiffness than the subsequent loading after unloading. Since the pressure in the soil does not exceed 10 MPa during the propagation of seismic waves, the deformation processes in the sand are satisfactorily described by the linear dependences [5], different for the processes of compression (above the achieved density) and unloading.

Geometric and mechanical parameters of structures and soil layers assumed the same values as in calculations without backfilling. Mechanical characteristics of sand seismic insulation: elastic constants $K = 1.6 \times 10^9$ N/m, $G = 320$ MPa, density $\rho = 2000$ kg/m^3. The view of the computational domain covered by the finite-difference grid is shown in Fig. 4.1. Non-reflecting boundary conditions were modeled at the lateral boundaries of the soil mass, and vertical displacement velocities were set at the lower boundary of the lower soil layer:

$$\dot{u}_z(x, y, -300, t) = \begin{cases} 0, & t < 4\ \text{s} \\ 1, & t \geq 4\ \text{s} \end{cases}$$

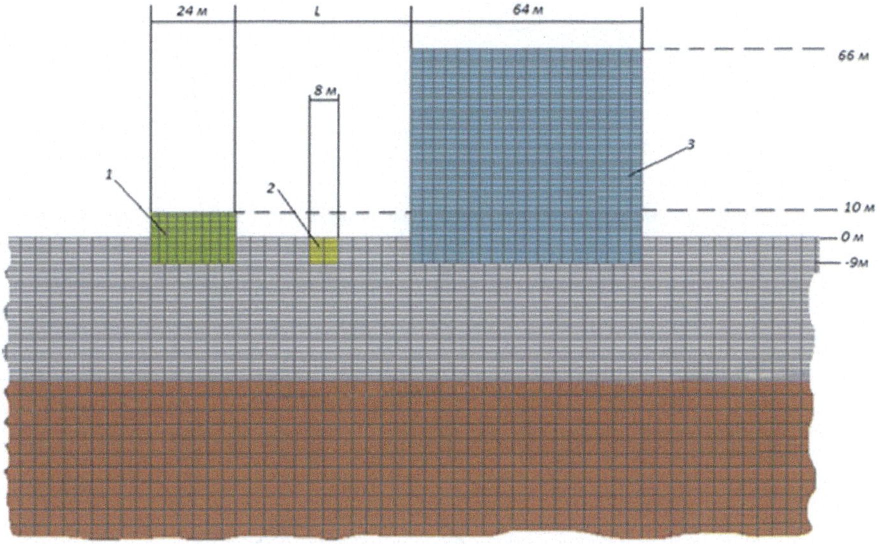

Fig. 4.1 View of the computational domain covered by a finite-difference grid

Mutual horizontal and vertical displacements of walls of constructions and soil for cases of the presence and absence of seismic insulating backfill were analyzed. A complex relationship between the behavior of structures, the fact of the presence of seismic isolation between them and the distance are established. As an example, the mutual horizontal (Fig. 4.2) and vertical (Fig. 4.3) displacements of the right wall of the pumping station and the soil for $L = 48$ m, $L = 80$ m and $L = 129$ m are given. Figures 4.4, 4.5, 4.6 and 4.7 show the maximum mutual displacements of the right and left walls of structures for cases with and without a seismic insulating layer. It can be seen that the right wall of the pumping station and the left wall of the reactor compartment have more significant displacements, which confirms the presence of mutual influence of closely located structures.

Difference of vertical displacements related to the displacements of the mass center (Fig. 4.8) was calculated for the reactor compartment as:

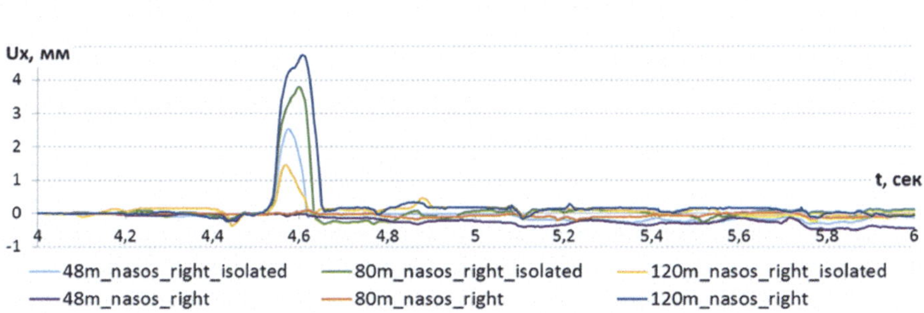

Fig. 4.2 Mutual horizontal displacement of the right wall of the pumping station and the soil

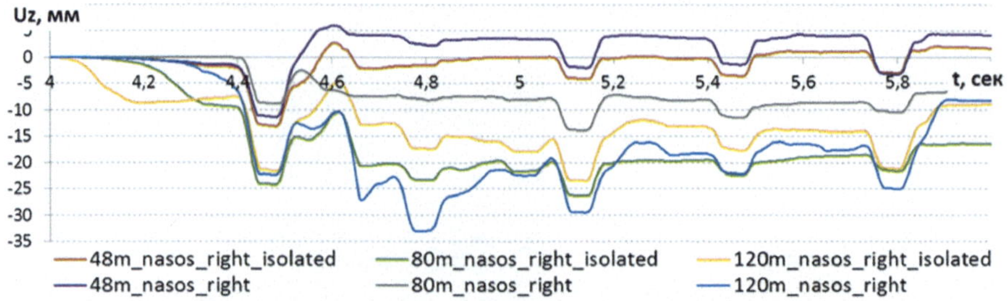

Fig. 4.3 Mutual vertical displacement of the right wall of the pumping station and the soil

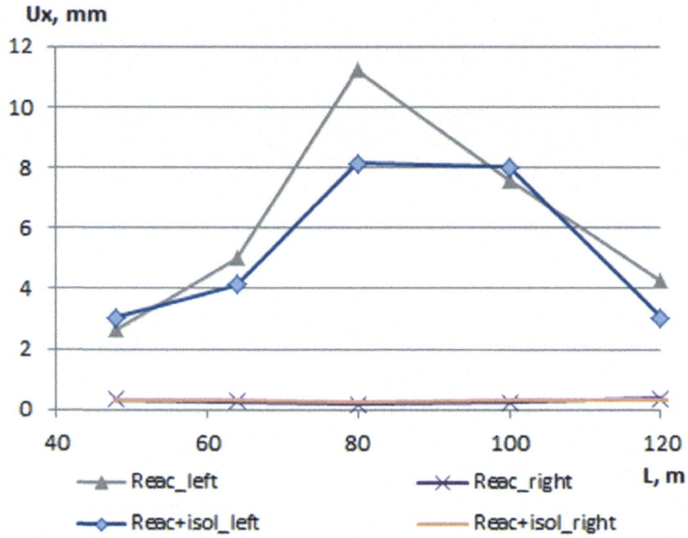

Fig. 4.4 Maximal mutual horizontal displacements of the reactor compartment depending on the distance between the structures

$$\Delta U_z = \frac{u_z^{\text{left}} - u_z^{\text{right}}}{U_z^{\text{mass center}}},$$

for the pumping station as

$$\Delta U_z = \frac{u_z^{\text{right}} - u_z^{\text{left}}}{U_z^{\text{mass center}}}.$$

Since ΔU_z in the Fig. 4.8 take mostly positive values, we can say that a nearby structure increasing the seismic activity. Also, Fig. 4.8 clearly shows that the seismic isolation between the structures reduces the pitching of a smaller overall structure.

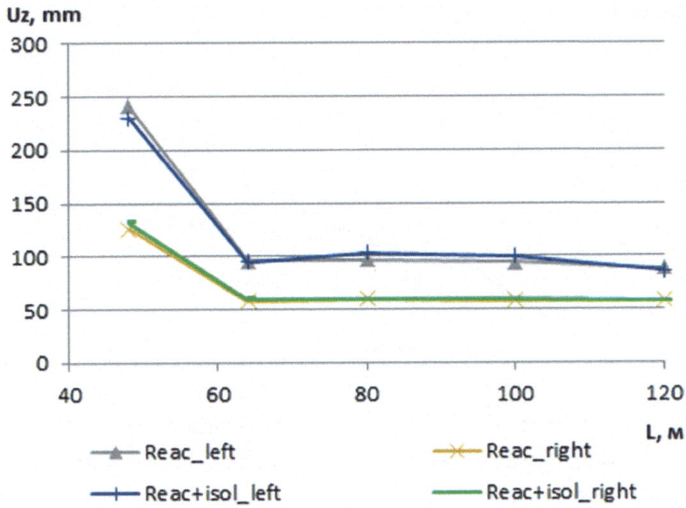

Fig. 4.5 Maximal mutual vertical displacements of the reactor compartment depending on the distance between the structures

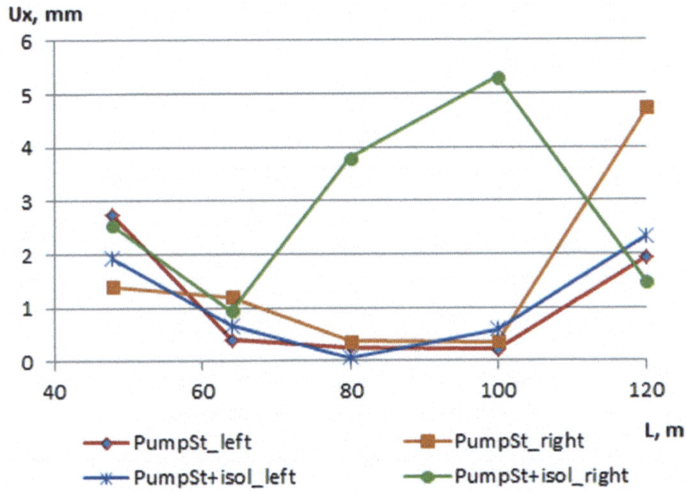

Fig. 4.6 Maximal mutual horizontal displacements of the pumping station depending on the distance between the structures

4.5 Conclusion

For a larger structure, the distance from other structures (L) is more important than the presence of a seismic insulating layer between the structures. If the distance between the structures is larger than the overall dimensions of the building base, the

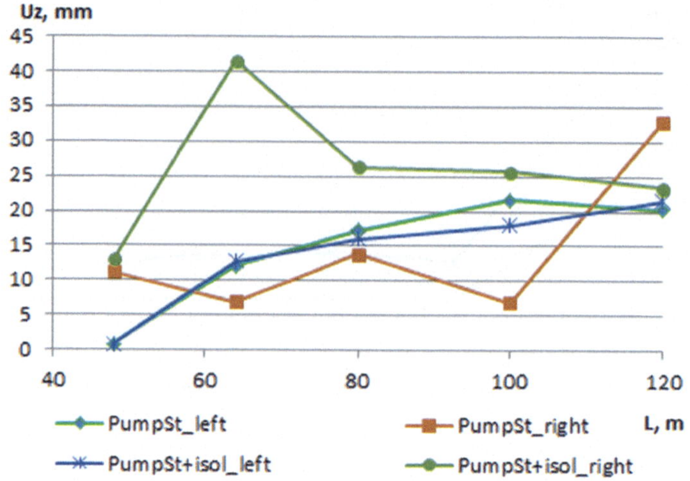

Fig. 4.7 Maximal mutual vertical displacements of the pumping station depending on the distance between the structures

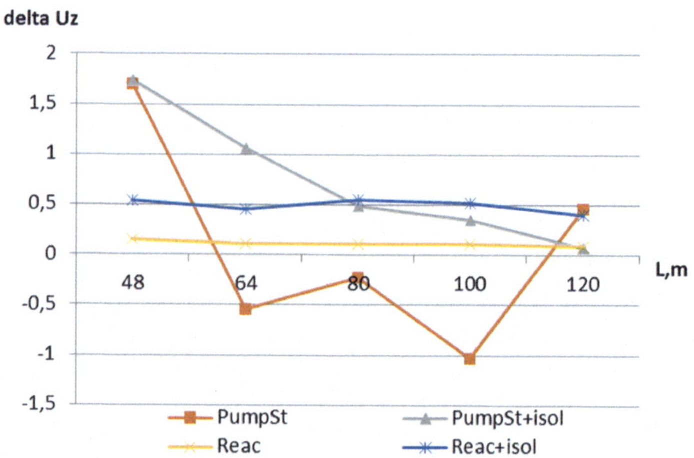

Fig. 4.8 Difference of vertical displacements related to the displacements of the mass center

further increase L, as well as the presence of seismic isolation, does not significantly affect the result of calculations. The described method of seismic protection was ineffective for the less overall structure of the pumping station at $L < 120$. The conducted computational experiments did not reveal a stable seismic insulating effect from sand dumping. Thus, seismic isolation has a negative impact on the less massive structure of the pumping station and has a slight positive impact on the more massive structure of the reactor compartment. Note that passive seismic isolation is often unstable [3, 16]. Therefore, the effectiveness of the applied vibration protection for each case should be selected and justified individually.

Acknowlegements The reported study was funded by RFBR, project number 20-01-00667.

References

1. Abrosimov, N.A., Bazhenov, V.G.: Nonlinear Problems of Dynamics of Composite Structures, p. 391. Publishing House of UNN, Nizhny Novgorod (2000)
2. Bazhenov, V.G., Chekmarev, D.T.: Solving Problems of Unsteady Dynamics of Plates and Shells by the Variational-Difference Method, Publishing House of UNN, Nizhny Novgorod (2000) 118c. (In Russian)
3. Bazhenov, V.G., Dyukina, N.S.: Structure effect of soil medium on the seismic response of buried structures. Comput. Mech. Modern Appl. Softw. Syst. (CMMASS'2019). (In Russian)
4. Bazhenov, V.G., Dyukina, N.S.: Numerical study of seismic vibrations of closely located buried large structures. Lobachevskii J. Math. **40**(11):1915–1921 (2019)
5. Bazhenov, V.G., Kotov, V.L.: Mathematical Modeling of Non-stationary Processes of Impact and Penetration of Axisymmetric Bodies and Identification of Soil Media Properties. Moscow, FIZMATLIT (2011), p. 208. (In Russian)
6. Berezanceva, E.V., Klyachko, M.A., Uzdin, A.M.: Peculiarities of soil structure interaction in construction with artificial bases. In: Proceedings of the Second International Conference on Recent Advances in Geotechnical Earthquake Engineering and Soil Dynamics, St. Louise, Missouri (1991)
7. Construction in seismic areas SNIP II-7-81*, Moscow (2011). (In Russian)
8. Dowrick, D.: Earthquake Resistant Design and Risk Reduction. Wiley, Ltd (2009)
9. Dyukina, N.S., Bazhenov, V.G.: Methods of Numerical Investigation of Seismic Stability of Ruined Structures. LAP LAMBERT Academic, Saarbrücken (2012), p. 141. (In Russian)
10. Dyukina, N.S., Bazhenov, V.G.: Numerically analyzing vibrations of two closely located submerged large-size structures. Prob. Strength Plast. (2018) 10.32326/1814-9146-2018-80-3-349-358. (In Russian)
11. Dyukina, N.S., Bazhenov, V.G.: Simulation of the seismic-stability problem for buried constructions. J. Mach. Manuf. Reliab. **44**(2), 189–193 (2015)
12. Elwood D.E.Y., Martin, D.: Ground response of closely spaced twin tunnels constructed in heavily over consolidated soils. Tunnel. Underground Space Technol. (2016)
13. Favorskaya, A.V., Petrov I.V.: Study of seismic isolation by full-wave numerical modeling. Doklady Earth Sci. **481**(2), 1070–1072
14. Ferraioli, M.M., Mandara, A.: Base Isolation for Seismic Retrofitting of a Multiple Building Structure: Design, Construction, and Assessment. Hindawi Limited in Mathematical Problems in Engineering (2017), pp 1–24
15. Gusev, G.N., Tashkinov, A.A.: Numerical modeling of force interaction between slab-pile foundations and soil massif. J. Appl. Mech. Tech. Phys. (2012). (In Russian)
16. Huang, C., Wang, X., Zhou, H., Liang, Y.: Damping effects of different shock absorbing materials for tunnel under seismic loadings. J. Vibroengineering (2019)
17. Jarernprasert, S., Bazan-Zurita, E., Bielak, J.: Seismic soil-structure interaction response of inelastic structures. Soil Dyn. Earthq. Eng. **47**, 132–143 (2013)
18. Kochetkov, A.V., Poverennov, EuYu., Savikhin, A.O.: Numerically modeling hydroelasto-plastic deformation of a pipeline with a liquid under impact loading. Prob. Strength Plast. **77**(3), 244–252 (2015). (in Russian)
19. Kvasov, I.E., Petrov, I.B.: High-performance computer simulation of wave processes in geological media in seismic exploration. Comput. Math. Math, Phys (2012)
20. Nakhaei, M., Ghannad, M.A.: The effect of soil–structure interaction on damage index of buildings. Eng. Struct. **30**, 1491–1499 (2008)

21. Ng, C.W., Lee, K.M., Tang, D.K.: Three-dimensional numerical investigations of new Austrian tunnelling method (NATM) twin tunnel interactions. Canad. Geotech. J. **41**, 523–539 (2007)
22. Orekhov, V.V., Negahdar, H.: Efficiency of trench barriers used to protect structures from dynamic loads and study of stress-strain state of soil using strain-hardening model of soil behaviour. Vestnik MGSU **4**, 53–60 (2013)
23. Pobedrya, B.E.: Numerical Methods in the Theory of Elasticity and Plasticity. MSU, Moscow (1981), p. 366. (In Russian)
24. Pronin, E.S., Rusinov, A.V.: Patent No. 006553 (1994). (In Russian)
25. Seismic Analysis of Safety-Related Nuclear Structures and Commentary. ASCE4-98. Reston, Virginia, USA (1999)
26. Shustov, V.: Earthquake-protective pneumatic foundation. In: Proceedings of 7th Annual International Symposium on Smart Structures and Materials 3988 (2000), pp. 412–418
27. Uzdin, A.M., Sandovich, T.A., Amin, A.N.M.S.: Fundamentals of the Theory of Seismic Resistance and Earthquake-Resistant Construction of Buildings and Structures. Vedeneyev VNIIG, Saint-Petersburg, Russia (1993), p. 176. (in Russian)
28. Zhang, J., Li, M., Yang, X.: Analysis and optimization of mutual influence of single channel tunnel construction blasting. Vibroengineering PROCEDIA (2017), pp. 79–85

Chapter 5
Dynamic Deformation and Failure Criterion of Cylindrical Shells Subjected to Explosive Loading

Valentin G. Bazhenov, Alexander A. Ryabov, Vladimir I. Romanov, and Evgeny E. Maslov

Abstract The results of numerical and experimental investigations are presented in the paper for impulse loading of cylindrical shell of constant thickness subjected to shock wave of spherical charge explosion of different trinitrotoluol equivalent (TNT). Numerical simulations are conducted with use of LOGOS software, developed at Russian Federal Nuclear Center VNIIEF. Validity of the numerical results is supported by closeness to the results of full-scale tests. Different failure criteria of shell dynamic strength are analyzed in the paper. Limiting deformations obtained in experiments with specimens for two-axial stress state is not an unequivocal criterion of strength of shell subjected to explosion loading. Specific power of deformation can be used as numerically–experimentally confirmed energy criteria of strength of cylindrical shell. Critical value is $N_p^* = 22$ W/mm^3 for investigated shell.

Keywords Cylindrical shell · Explosive loading · Elastic–plastic deformations · Dynamic failure criterion

5.1 Introduction

A problem of dynamic strength of engineering containment objects subjected to shock waves is very important to ensure safety of structures in atomic power engineering, chemical industry, in other ecologically dangerous plants and in scientific explosive experiments. Selection of reliable failure criteria of dynamic strength is one of the most important points in this problem.

V. G. Bazhenov (✉)
Research Institute for Mechanics, National Research Lobachevsky State University of Nizhny Novgorod, 23 Gagarin Avenue, building 6, Nizhny Novgorod, Russian Federation 603950
e-mail: bazhenov@mech.unn.ru

A. A. Ryabov · V. I. Romanov · E. E. Maslov
Russian Federal Nuclear Center VNIIEF, 37 Mira Ave, Sarov, Nizhny Novgorod, Russia 607188
e-mail: aaryabov@cdfirst.vniief.ru

H. Altenbach et al. (eds.), *Multiscale Solid Mechanics*,
Advanced Structured Materials 141,
https://doi.org/10.1007/978-3-030-54928-2_5

Some results of experimental and numerical investigations are presented in [1–17] for a problem of dynamic strength of structure elements: rings, cylindrical and spherical shells under explosive loading. In [5] based on many experiments in explosive loading of spherical and cylindrical vessels, it is suggested to use specific work of surface forces at rupture of material. In the general case, specific work depends on unsoundness of materials, temperature, strain rate and on the other factors, which have not been quite investigated yet. Grady [3] also suggests energy criteria of dynamic strength, and it is supposed that surface destruction of structure is caused by the part of kinetic energy. Criteria of critical plastic strains is developed in [6, 7] for thin wall vessels which are destroyed in the range of large plastic deformations. But experiments with explosive loading of spherical shells [15] with constant internal radius and different thickness show that there is no any regularity with critical failure strains. So, it should be noted that critical strains can be considered only as a low level of dynamic failure strength criteria of spherical shells loaded by internal explosion. Curran [1] notes that under explosive loading of shells during a high strain rate, process of dynamic plastic flow is transformed from isothermal to adiabatic one. In some cases, emitted heat is concentrated in thick zones located along maximal shear stresses volumes, and it flows to high rise of plasticity. But practical application of suggested model, which includes formation of adiabatic shearing volumes, is not possible at present time because of absence of needed experimental data.

Authors of the monograph [16] came to a conclusion that, possibly at present time, promising investigations of dynamic strength criterions of structures subjected to impulse loading are those which are based on the conception of mechanics [8, 18] of damaged continuum. This approach allows to analyze a deformation process including degradation of mechanical properties of a material and collection of damages up to getting by damage function its critical level $\omega = 1$. In dynamic deformations of plastic materials, damage function ω can be used on the basis of scheme of linear summation of damages [4]:

$$\omega(\mu) = \int_{0}^{\mu} \frac{\mathrm{d}\mu}{\mu^*\big(\sigma_{ij}(\mu)\big)},$$

where μ—variable which characterizes a process of elastic–plastic deformation, μ^*—limiting value of this variable corresponding to failure of material with fixed stress state, described by stress tensor σ_{ij}. In a general case, for getting μ^*, one can use some strength criterion of structural materials such as:

$$f\big(\sigma_{ij}, \varepsilon_{ij}\big) = 0.$$

At this, strength criterion should take into account the influence of stress status, strain rate and temperature on to distraction, and it should have the required degree of experimental confirmation. It is noted in [14] that in spite of reached results and collected experience, there are not quite enough scientific data for theoretical general conclusion of dynamic strength of the structures subjected to explosive loading. The purpose of this article is continuation of investigations of explosive deformations of cylindrical shell for getting well-funded numerical–experimental criterions of dynamic strength.

5.2 Statement of Problem and Defining Relations

The problem is considered for dynamic loading of steel cylindrical shell of constant thickness subjected to explosion of spherical charge located in the center of cross section in axis of the shell. Dynamic deformations of the shell are described according to the Lagrange approach [4]. There are two coordinate systems: The first one is a general coordinate system, $X = [X_1 X_2 X_3]$, and the second one is a local coordinate system, $x = [x_1 x_2 x_3]$, with directional cosines n_{ij}:

$$x_i = n_{ij}\, X_j,\ \ i, j = \overline{1, 3} \tag{5.1}$$

Strain rates in the local $\dot{\varepsilon}_{ij}$ and in general basis $\dot{e}_{ij}$ connected by the following relations:

$$\dot{\varepsilon}_{ij} = n_{im} n_{jk} \dot{e}_{mk}, \tag{5.2}$$

Kinematic relations are described in velocities and defined in current state metrics:

$$\dot{e}_{ij} = (\dot{U}_{i,j} + \dot{U}_{j,i})/2,\ (i, j = \overline{1, 3})\quad X_i = X_i|_{t=0} + \int\limits_0^t \dot{U}_i dt, \tag{5.3}$$

where displacements U_i in general coordinate system X.

The equation of motion is written in the form of the variation principle of balance of virtual powers of work [17].

$$\int\limits_{\Omega} \sigma_{ij}\delta\dot{\varepsilon}_{ij} dV + \int\limits_{\Omega} \rho\ddot{U}_i\delta\dot{U}_i dV = \int\limits_{\Gamma p} P_i\delta\dot{U}_i d\gamma + \int\limits_{\Gamma p} P_i^q\delta\dot{U}_i d\gamma,\quad \left(i, j = \overline{1, 3}\right) \tag{5.4}$$

where σ_{ij}—components of stress tensor, ρ—density, P_i^q—contact pressure, P_i—distributed load, Ω—volume of structure, 0413_q—contact surface, Γ_p—surface of external pressure is applied, $\delta\dot{\varepsilon}_{ij}$, $\delta\dot{U}_i$—variations of strain rate $\dot{\varepsilon}_{ij}$, and $\dot{U}_i$ at the surface with given boundary conditions $\delta\dot{U}_i = 0$.

The components of stress rate deviator $\dot{\sigma}'_{ij} = \dot{\sigma}_{ij} + \dot{P}\delta_{ij}$ and elastic parts of strain rate deviator components $\left(\dot{e}^e_{ij}\right)' = \dot{e}_{ij} - \dot{e}\delta_{ij}/3 - \dot{e}^p_{ij}$ are connected based on Hooke's law in the current state metrics.

$$D_J\sigma'_{ij} = 2G\left(\dot{e}^e_{ij}\right)', \quad D_J\sigma'_{ij} = \dot{\sigma}'_{ij} - \dot{\omega}_{ik}\sigma'_{ij} - \dot{\omega}_{jk}\sigma'_{ik},$$
$$\dot{P} = -K\dot{e}^e, \quad \dot{P} = -\dot{\sigma}_{ii}/3, \dot{e} = \dot{e}^e_{ii}, \dot{e}^p_{ii} = 0 \tag{5.5}$$

where K, G—moduli of dilatation and shearing, P—pressure, D_J—Jaumann derivative, describing rotation of continuum particle as rigid body, δ_{ij}—Kronecker's symbol. Rates of plastic strains are defined with associated flow rule:

$$\dot{e}^p_{ij} = \dot{\lambda}\sigma'_{ij}, \sigma'_{ij}\sigma'_{ij} = \tfrac{2}{3}\sigma_i^2(æ), \tag{5.6}$$

where $\sigma_i = \sigma_i(æ)$—dependence of effective stress on Odkwist parameter at monotonic loading. Solving of Eqs. (5.1)–(5.6) is conducted on the basis of space discretization by finite element method and explicit scheme of integration in time realized in LOGOS software [19].

Blasting of high-explosive charge, explosion products flying-off, transmission of shock wave and impact on to the shell are considered as axis-symmetrical task on basis of Euler approach and is simulated in LOGOS software as well. Van Leer numerical method of second order is used for solving gas-dynamic problem. Noted processes are simulated in air volume, restricted by cylindrical shell and opened at both ends.

5.3 Experimental and Numerical Results and Analysis

Deformation properties of low-alloy structural steel 09G2S (GOST standards 19281–73 and 19281–89) have been experimentally investigated [20–23] with several tens of cylindrical tube and some other types of specimens under different (tension, torsion, internal pressure) static and dynamic loads up to a level of strain rates of 800 s^{-1}. Experiments show that engineering, conditional, logarithmic and true stress–strain curve, obtained by method suggested in [23], are close to each other up to the level of effective strain ε_i~ 7%. In the range of large strains, engineering and true stress–strain curve differ from each other and strongly depend on stress state. Dynamic properties of steel's deformation start to appear with strain rates more than 500 s^{-1}. Experimental tests indicate that limiting, failure strains are strongly depended on stress state. Figure 5.1 demonstrates dependences of limiting effective strains $[\varepsilon_i]$ on parameter of stress state Π [24].

Experimental results in engineering strains are marked as circlets in Fig. 5.1 curve «1» corresponds to conditional and curve «2» to true stress–strain curve. Experiments show that in pure shearing ($\Pi = 0$), destruction of material appears with engineering strains of $[\varepsilon_i]$ ~ 60%, in axial tension (Π = 0,6), $[\varepsilon_i]$ ~ 25…30%, and in two-axial tension (Π = 0,8), engineering and true stress–strain curve are close to each other and do not exceed ε_i ~ 4…5%. In two-axial tensions, destruction of the tube specimens is realized as longitudinal crack, Fig. 5.2.

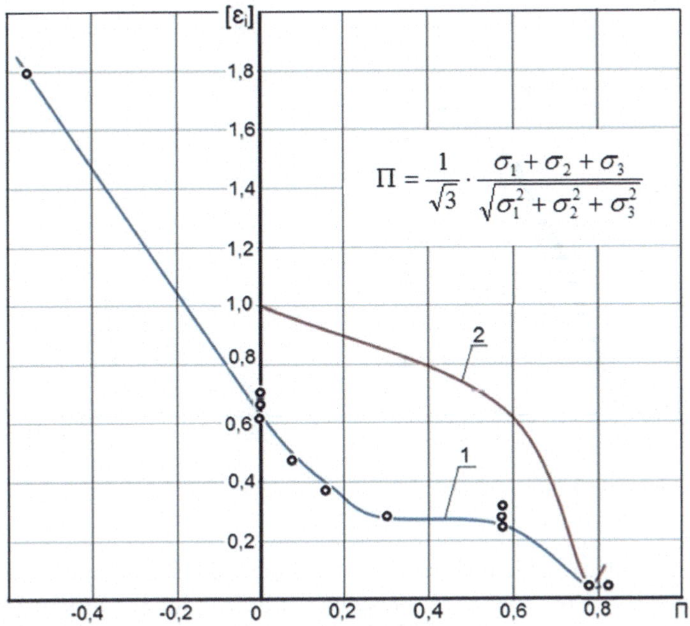

Fig. 5.1 Limiting strains depend on stress state factor Π

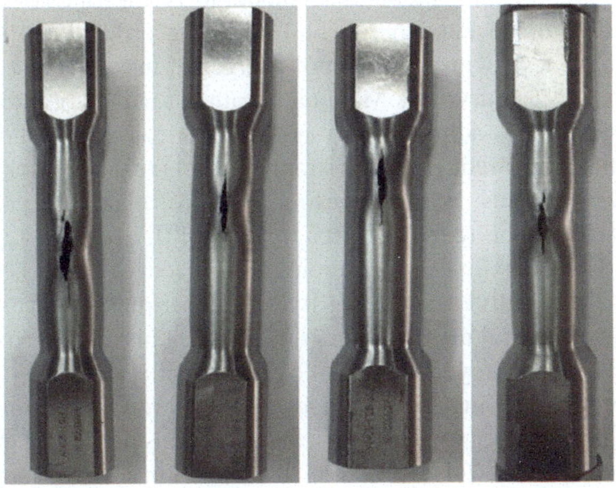

Fig. 5.2 Destroyed tube specimens

The experimental tests also show that local thinning of tube and cylindrical specimens appears before failure, demonstrating instability of deformation process. If we suppose the instability of deformation process is a failure criterion according to Consider criteria [18, 24–26] so experiments demonstrate that in two-axial stress state destruction of tube specimens takes place at effective strain level $[\varepsilon_i]$ ~ 3.6%.

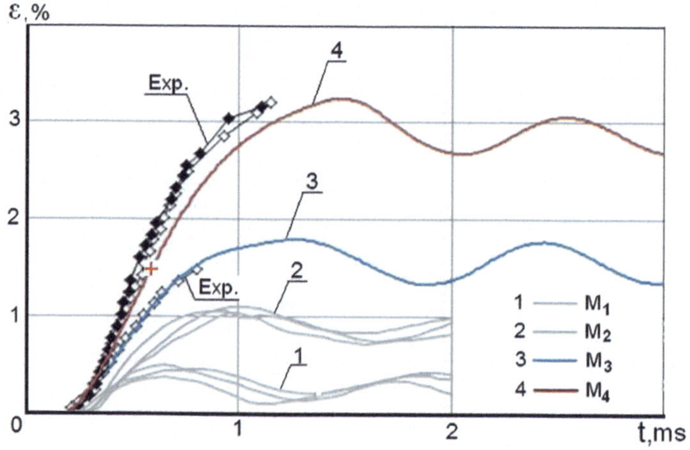

Fig. 5.3 Circumferential strains of shell

Let us consider dynamic deformation of cylindrical shell of constant thickness *h*, *R/h* ~ 20, *L/R* ~ 4 (*R*—radius, *L*—length of the shell) subjected to explosion of spherical charges of different power (from M_1 to $M_4 = 2.6M_1$ kg TNT), located in middle cross section on the shell axis. Dependences of maximal circumferential strains of shells from time above the center of explosion are presented in Fig. 5.3.

Validity of computational modeling is confirmed by closeness of numerical curves (solid lines) and experimental curves, recorded with different measuring equipment in the tests with explosion of charges of masses $M_3= 1.9M_1$ (curve 3) $M_4= 2.6M_1$ kg TNT (curve 4). Maximal strain rates do not exceed 150 s^{-1} in all cases of explosive loadings. Analysis of numerical results shows that in explosion of M_1, maximal circumferential strain (curve 1) reaches level of ε_{max}~ 0.5…0.6% on t_{max}~ 0.5…0.6 ms. With increasing explosive material mass to $M_2 = 1.5M_1$, maximal deformation (curve 2) rises to ε_{max}~ 1.0…1.1% at t_{max}~ 0.9…1.1 ms. With M_3= $1.9M_1$, maximal circumferential strain (curve 3) reaches ε_{max}~ 1.8% at t_{max}~ 1.4 ms. All the shells keep its integrity in all experiments with explosion in the range of explosive material masses from M_1 to M_3. On blast with explosive mass $M_4= 2.6M_1$, the shell is destroyed at level of circumferential strain ε_d~ 1.4% (small cross in curve 4) on t_d~ 0.57 ms and continues to be deformed, reaching maximal level of deformations ε_{max}~ 3.4% in numerical simulations as well as in experiment. With explosion of $M_4= 2.6M_1$, destruction of the shell in form of longitudinal crack with bifurcations on the both ends, Fig. 5.4, happens in conditions of two-axial tension stress state, Π= 0.82 at the stage of active deformation of the shell up to *t* ~ 0.75 ms.

Calculations and experiments demonstrate that with explosion of $M_3= 1.9M_1$, maximal circumferential strains exceed critical level of strain ε_d~ 1.4%, and shell keeps its integrity. Experimentally defined level of critical strain ε_d with cylindrical shell is also significantly below than limiting levels of deformations experimentally defined on specimens with deformational criteria, calculated with engineering strains

Fig. 5.4 Destroyed cylindrical shell

$[\varepsilon_i]$ ~ 4…5% and with Consider criteria, calculated with true strains $[\varepsilon_i]$ ~ 3,6%. So, limiting deformation is not a simple criterion of dynamic strength of cylindrical shell.

Specific power of deformation of shell in the most loaded point on external surface of the shell above the center of explosion is presented in Fig. 5.5. Destruction of the shell is marked by small cross on curve «4» with explosion of $M_4 = 2.6M_1$. Specific power of deformation reaches level of $N_p^* = 22$ W/mm^3 in the moment of destroying of the shell. Many numerical and experimental investigations confirm that steel shell keeps its strength if the level of specific power of deformation does not exceed N_p^*.

Thus, specific power of deformation can be considered as energy criteria of dynamic strength of shell reflecting structural and deformational properties of the material in conditions of two-axial stress state.

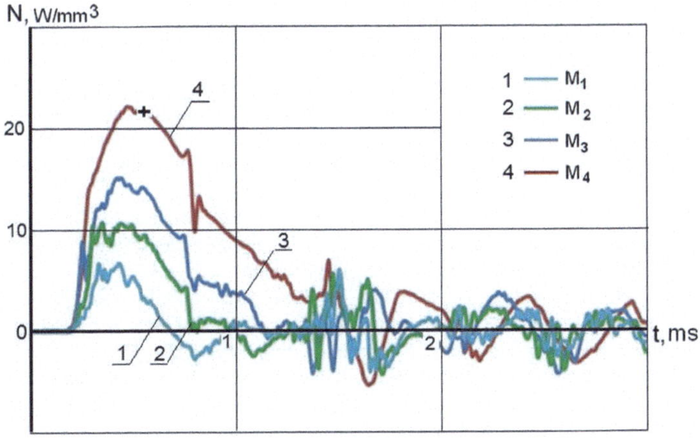

Fig. 5.5 Specific power of deformation

5.4 Conclusions

According to numerical results and experimental data, it is necessary to point out the following:

1. Limiting deformations obtained in experiments with specimens for two-axial stress state is not an unequivocal criterion of strength of shell subjected to explosion loading.
2. Specific power of deformation can be used as numerically–experimentally confirmed energy criteria of strength of cylindrical shell. Critical value is N_p^* $= 22$ W/mm^3 for investigated shell.

Acknowledgements The reported study was funded by RFBR, project number 20-01-00667.

References

1. Curran, D.R.: Dynamical failure. In: *Dynamics of Impact*, pp. 275–293. Mir, Moscow (1985) (in Russian)
2. Destruction of different scale objects under blast. Monography/Edited by Ivanov, A.G., RFNC–VNIIEF, Sarov (2001) (in Russian)
3. Grady, D.E.: Fragmentation of rapidly expanding rings. J. Impact Eng. **5**(6), 272–279 (1987)
4. Glushak, B.L., Novikov, S.A., Rusanov, A.I., Sadyrin, A.I.: Destruction of Deformed Continuums by Impulse Loads. NNSU, Nizhny Novgorod (1992). (in Russian)
5. Ivanov, A.G. Mineev, V.N.: About scale criterion with brittle failure of structures. DAN USSR **220** (3), 552–552 (1975) (in Russian)
6. Ivanov, A.G.: Dynamic destruction of objects in the range of deep plastic strains. AMTP **2**, 146–151 (1986). (in Russian)
7. Odintcov, V.A., Chudov, L.A.: Expansion and destruction of shells subjected to detonation products. Probl. Dyn. Elastic Plast. Mediu. **5**, 85–96 (1975) (in Russian)
8. Odincov, V.A., Gryaznov, E.F., Kolobanova, A.B.: Propagation of damage in the wall of cylindrical shell under impulse loading. Mech. Impuls. Process. **436**, 178–216 (1985). (In Russian)
9. Ryabov, A.A., Romanov, V.I., Tareev, S.A., Vorontcov, S.V., Narozhny, A.T.: Numerical investigations of dynamic deformation of large protective chamber under internal explosive loading. Tr. RFNC VNIIEF Sarov **3**, 468–472 (2002). (In Russian)
10. Ryabov, A.A., Romanov, V.I., Maslov, E.E., Syrunin, M.A., Fedorenko, A.G., Tcoy, A.P.: Numerical and experimental investigations of deformation and destruction steel cylindrical shells under explosive loading. Probl. Strength Plasticity. **65**, 122–127 (2003). (In Russian)
11. Ryabov, A.A., Romanov, V.I., Tareev, S.A., Kukanov, S.S.: Numerical analysis of dynamic deformation of blast–protection chamber and its calculation–experimental failure criterion. Trudy RFNC-VNIIEF. Sarov **6**, 394–405 (2004). (In Russian)
12. Ryabov, A.A., Romanov, V.I., Kukanov, S.S., Maslov, E.E.: Specific features of propagation of shock waves in protective chamber with simultaneous explosion of two charges. In: Proceedings of International conference VII Chariton scientific readings. Extreme states of matter/ Detonation/ Chock waves. Sarov, 14–18 Mar (2005) (in Russian)
13. Ryabov, A.A., Romanov, V.I., Maslov, E.E.: Specific features of dynamic deformation of cylindrical shells under explosive loading and failure criterions. Trudy RFNC–VNIIEF. Sarov, vol. 10, pp. 244–253 (2006) (in Russian)

14. Ryabov, A.A., Romanov, V.I., Kukanov, S.S., Maslov, E.E.: Numerical modeling of dynamic deformation of the model of multi-usable blast protective chamber under explosion loading. In: Youth in Science: Proceedings of V scientific–technical conference. Sarov: FSUE RFNC–VNIIEF, pp. 511–515 (2007) (in Russian)
15. Tcypkin, V.I., Cheverikin, A.M., Ivanov, A.G.: Investigations of steel shells under one time explosion loading. Probl. Strength. **10**, 94–105 (1982). (in Russian)
16. Volkov, I.A., Korotkikh, YuG: Equations of state of viscous–elastic–plastic damaged mediums. FIZMATLIT, Moscow (2008). (in Russian)
17. Washizu, K.: Variational Methods in Elasticity and Plasticity. Pergamon Press, Oxford (1982)
18. Jonson, J.N.: Ductile fracture of rapidly expanding rings. J. Appl. Mech. **50**(3), 668–674 (1983)
19. Spiridonov, V.F., Tcyberev, K.V., Ryabov, A.A. et al.: Software package LOGOS. Modulus for solving static stress analysis tasks and modal analysis. In: Shagaliev, R.M. (ed.) XIII International Seminar Supercomputing and Mathematical Modeling, pp. 493–502. RFNC–VNIIEF, Sarov (2012) (in Russian)
20. Bazhenov, V.G., Lomunov, V.K., Osetrov, S.L., Pavlenkova, E.V.: Experimental and computational method of studying large elastoplastic deformations of cylindrical shells in tension to rupture and constructing strain diagrams for an inhomogeneous stress–strain state. J. Appl. Mech. Tech. Phys. **54**(1), 100–107 (2013)
21. Bazhenov, V.G., Kazakov, D.A., Nagornykh, E.V., Osetrov, D.L., Ryabov, A.A.: Numerical and experimental study of elastoplastic tension-torsion processes of cylindrical specimens from steel 09G2S under large deformations. Fundam. Appl. Probl. Tech. Technol. **4–2**(324), 76–82 (2017). (in Russian)
22. Bazhenov, V.G., Kazakov, D.A., Nagornykh, E.V., Osetrov, D.L., Ryabov, A.A.: The experimental and theoretical study of large deformations of cylindrical samples from steel 09G2S with stress concentrators under tension-torsion loading to failure. PNRPU Mech. Bulletin. **4**, 69–81 (2018). (in Russian)
23. Bazhenov, V.G., Nagornykh, E.V., Osetrov, D.L., Ryabov, A.A.: Numerical and experimental analysis of tension–torsion processes in cylindrical samples made of 09g2s steel under large deformations before destruction. Uchenye Zapiski Kazanskogo Universiteta. Seriya Fiziko–Matematicheskie Nauki. **160**(3), 495–507 (2018) (in Russian)
24. Kazakov, D.A., Kapustin, S.A., Korotkikh, YuG: Modeling of Deformation and Failure Processes of Materials and Structures. NNSU, Nizhny Novgorod (1999). (in Russian)
25. Degtyarev, V.P.: Deformation and Failure In High Stressed Structures. Mashinostroenie, Moscow (1987). (in Russian)
26. Kachanov, L.M.: Foundation of Fracture Mechanics. Nauka, Moscow (1974). (in Russian)

Chapter 6
Plasticity of Materials with Additional Hardening Exposed to Cyclic Loading

Valentin S. Bondar and Dmitry R. Abashev

Abstract This paper presents mathematical models of elastoplastic strain of additionally isotropically hardened materials exposed to simple (proportionate) and complex (disproportionate) cyclic loading. It uses a simplified version of the plasticity theory, which is a particular version of the theory of inelasticity. This version is a single-surface combined-hardening flow theory. The paper presents the results of calculation- and experiment-based analysis of elastoplastic strain and fatigue destruction of materials subjected to a variety of disproportionate cyclic loads. Complex loading processes, effects of additional isotropic hardening, and destruction processes are adequately described by a single, rather simple version of the plasticity theory, which is where the advantage of the herein presented mathematical models lies.

Keywords Plasticity · Non-proportional cyclic loading · Additional hardening · Fatigue destruction · Non-proportionality parameter

6.1 Introduction

Up until the 1980s, plasticity theory and its defining equations were evolving calmly [1]. Surprises were encountered first when experimental research [2–24] into disproportionate (complex) cyclic loading of some stainless steels discovered a considerable hardening that was more than twice that observed in proportional cyclic loading. This phenomenon, later dubbed additional isotropic hardening, could not be described by any theory.

The effect was first described by [2–6]. The approach was taken further by [7–9, 11, 18, 19, 22–27]. However, each paper had a different parameter to respond to the disproportionate loads. Kadashevich [28] gives some thoughts on the parameters of

V. S. Bondar (✉) · D. R. Abashev
Department of Technical Mechanics, Moscow Polytechnic University, Moscow, Russian Federation 107023
e-mail: tm@mospolytech.ru

D. R. Abashev
e-mail: abashev.d@gmail.com

H. Altenbach et al. (eds.), *Multiscale Solid Mechanics*,
Advanced Structured Materials 141,
https://doi.org/10.1007/978-3-030-54928-2_6

disproportionality, while [29] try to justify and substantiate the choice of a single disproportionality parameter. Papers [30, 31] present the best-substantiated choice of a single disproportionality parameter based on the criteria for alignment with the effects of disproportionate loading observed experimentally; the equations for these effects are presented as well.

This paper dwells upon mathematical modeling of these phenomena using a version of single-surface plastic flow theory with combined hardening. The loading surface displacement is described by the Novozhilov–Chaboshe model that implies the total displacement is a total of displacements, each of which has its own evolutionary equation. The first evolutionary equation that describes Type I displacements is the Ishlinky–Prager equation. The Type II displacement equation is the Amstrong–Frederick–Kadashevich equation. The Novozhilov–Chaboshe hypothesis coupled with the Ishlinsky–Prager and Amstrong–Frederick–Kadashevich equations are equivalent to a three-termed evolutionary equation used in the theory of inelasticity [30].

Kinetic equations of damage accumulation are presented herein and use the work of Type II backstresses on the field of plastic strain as the value of energy spend to damage the material. The paper presents the results of calculation- and experiment-based analysis of elastoplastic strain and fatigue destruction of materials subjected to a variety of disproportionate cyclic loads.

6.2 Mathematical Modeling of Elastoplastic Strain

It uses a simplified version of the plasticity theory [31–33], which is a particular version of the theory of inelasticity [30]. This version is a single-surface combined-hardening flow theory. Its applicability is limited to small strains of initially isotropic metals at temperatures that entail no phase transformations, at such strain rates where dynamic and rheological effects are negligible. Consider the materials that showcase additional isotropic hardening when exposed to disproportionate cyclic loading.

Below is a summary of the basic equations for this plasticity theory version.

1. $\dot{\varepsilon}_{ij} = \dot{\varepsilon}^e_{ij} + \dot{\varepsilon}^p_{ij}$
2. $\dot{\varepsilon}^e_{ij} = \frac{1}{E}\left[\dot{\sigma}_{ij} - \nu\left(3\dot{\sigma}_0\delta_{ij} - \dot{\sigma}_{ij}\right)\right]$
3. $f\left(\sigma_{ij}\right) = \frac{3}{2}\left(s_{ij} - a_{ij}\right)\left(s_{ij} - a_{ij}\right) - C^2 = 0$
4. $\dot{C} = q_\varepsilon \dot{\varepsilon}^p_{u*}, \dot{\varepsilon}^p_{u*} = \left(\frac{2}{3}\dot{\varepsilon}^p_{ij}\dot{\varepsilon}^p_{ij}\right)^{1/2}, q_\varepsilon = q_{\varepsilon 0} + q_{\varepsilon A}, q_{\varepsilon 0} = \frac{dC_p}{d\varepsilon^p_{u*}}, q_{\varepsilon A} = \theta_A(C_A - C),$
   $\theta_A = (1 - A)\theta_0 + A\theta_1, C_A = (1 - A)C_0 + AC_1, C_0 = C_p + d_0\left(C_{\max} - C_p\right),$
   $C_1 = d_1 C_p, A = 1 - \left(\frac{\dot{s}_{ij}\dot{e}_{ij}}{\dot{\sigma}_u\dot{\varepsilon}_u}\right)^2, \dot{\sigma}_u = \left(\frac{3}{2}\dot{s}_{ij}\dot{s}_{ij}\right)^{1/2}, \dot{\varepsilon}_u = \left(\frac{2}{3}\dot{e}_{ij}\dot{e}_{ij}\right)^{1/2}$
5. $\dot{a}_{ij} = \frac{2}{3}g\dot{\varepsilon}^p_{ij} + \left(\frac{2}{3}g_\varepsilon\varepsilon^p_{ij} + g_a a_{ij}\right)\dot{\varepsilon}^p_{u*}, g = E_a + \beta\sigma_a, g_\varepsilon = \beta E_a, g_a = -\beta$

6. $\dot{\varepsilon}^p_{ij} = \frac{df}{d\sigma_{ij}}\dot{\lambda} = \frac{3}{2}\frac{s^*_{ij}}{\sigma^*_u}\dot{\varepsilon}^p_{u*}$, $s^*_{ij} = s_{ij} - a_{ij}$, $\sigma^*_u = \left(\frac{3}{2}s^*_{ij}s^*_{ij}\right)^{1/2}$, $\dot{\varepsilon}^p_{u*} = \frac{1}{E_*}\frac{3}{2}\frac{s^*_{ij}\dot{\sigma}_{ij}}{\sigma^*_u}$,
$\dot{\varepsilon}^p_{u*} = \frac{3G}{E_*+3G}\frac{s^*_{ij}\dot{\varepsilon}_{ij}}{\sigma^*_u}$, $E_* = q_\varepsilon + g + g_\varepsilon \varepsilon^{p*}_u + g_a a^*_u$, $\varepsilon^{p*}_u = \frac{s^*_{ij}\varepsilon^p_{ij}}{\sigma^*_u}$, $a^*_u = \frac{3}{2}\frac{s^*_{ij}a_{ij}}{\sigma^*_u}$
7. $\sigma^*_u < C \cup \dot{\varepsilon}^p_{u*} \le 0$ – elastic, $\sigma^*_u = C \cap \dot{\varepsilon}^p_{u*} > 0$ – plastic
8. $\dot{\omega} = \alpha\omega^{\frac{\alpha-1}{\alpha}}\frac{a^{(2)}_{ij}\varepsilon^p_{ij}}{W_A}$, $\alpha = \left(\sigma_a / a^{(2)}_u\right)^{n_a}$, $a^{(2)}_u = \left(\frac{3}{2}a^{(2)}_{ij}a^{(2)}_{ij}\right)^{\frac{1}{2}}$, $a^{(2)}_{ij} = a_{ij} - \frac{2}{3}E_a\varepsilon^p_{ij}$
$W_A = (1 - A)W_0 + AW_1$, $\omega = 1$ – destruction

Here, $\dot{\varepsilon}_{ij}$, $\dot{\varepsilon}^e_{ij}$, $\dot{\varepsilon}^p_{ij}$ are the tensors of elastic, total, and plastic strain rates; e_{ij} is the strain deviator; σ_{ij}, s_{ij}, s^*_{ij}, a_{ij} are the stress tensor, stress deviator, active stress deviator, and backstreses deviator; ε^p_{u*} is the accumulated plastic strain; ω is damage.

This version of the plasticity theory closes with the following material functions to be defined experimentally;

E, ν are the elastic parameters;
E_{ao}, β, σ_a are the parameters of anisotropic hardening;
$C_p(\varepsilon^p_{u*})$ is the function of isotropic hardening;
W_0 is the energy of destruction in case of proportional loading;
n_α is the nonlinearity of the damage accumulation process (equals 1.5 for nearly any structural steel or alloy);
θ_0, θ_1, d_0, d_1 are the absolute values of additional isotropic hardening and softening;
W_1 is the energy of destruction in case of disproportionate loading.

The first ten material functions can be found by testing under the conditions of proportional uniaxial stress state. The basic experiment in this case uses the following data:

- elastic parameters found conventionally;
- tensile plastic strain diagram;
- tense plastic strain diagram after pre-compression;
- data on single-block and dual-block cyclic tension and compression at a constant span of plastic strain: number of pre-destruction cycles and maximum cycle-specific stress as a function of the number of cycles.

The remaining five material functions can be found by testing using disproportionate cyclic loading. The basic experiment in this case uses the following data:

- maximum stress intensity diagram for the cycle $\sigma_u\max$ as a function of accumulated plastic strain ε^p_{u*} in case of proportional cyclic loading;
- diagram $\sigma_u\max(\varepsilon^p_{u*})$ for disproportionate loading along a circular strain path until additional hardening stabilizes, followed by proportional cyclic loading until softening stabilizes;
- data on the fatigue destruction caused by disproportionate cyclic loading along the circular strain trajectories of different radii.

Bondar [30]; Bondar et. al. [32] details upon identifying the material functions from the basic experiment data. The same papers present the material functions for the steels considered herein.

All papers [2–13, 16, 18–24] that seek to describe the effect of additional isotropic hardening use the disproportionality parameter proposed by [2–6] $A = 1 - \left(\frac{\dot{\varepsilon}^p_{ij}\,\dot{s}_{ij}}{\dot{\varepsilon}^p_{u*}\,\dot{\sigma}_u}\right)^2$, which corresponds to the squared cosine of the angle between the vectors of plastic strain and stress rates. Korotkikh and Makovkin [26]; Makovkin [27] proposes a disproportionality parameter $A = 1 - \left(\frac{2}{3}\frac{\dot{\varepsilon}^p_{ij}\,\dot{e}_{ij}}{\dot{\varepsilon}^p_{u*}\,\dot{\varepsilon}_u}\right)^2$ that equals the cosine of the angle between the plastic strain rate and strain rate vectors. As shown in [31], these parameters alone cannot describe the core pattern of additional isotropic hardening, which is that additional hardening increases when the material is exposed to cyclic loading along the circle-shaped strain paths of increasing radii. These parameters are basically not affected by the radii and will return the same value of additional hardening regardless of the radius. Kadashevich and Mosolov [25] offered the parameter $A = 1 - \left(\frac{\dot{e}_{ij}\,\dot{s}_{ij}}{\dot{\varepsilon}_u\,\dot{\sigma}_u}\right)^2$, which equals the cosine of the angle between the strain rate vector and the stress rate vector; this parameter has been [29, 31] proven to match the patterns of additional isotropic hardening and hence its usage for the mathematical modeling of elastoplastic disproportionate cyclic strain.

6.3 Additional Isotropic Hardening in Case of Disproportionate Cyclic Loading

Consider the hardening processes first, i.e., processes where the strain paths follow in the order of increasing loading disproportionality, i.e., cross to square to circle to circles of greater radii. This will correspondingly increase the additional hardening along the paths. Figures 6.1 and 6.2 present the experimental results [2, 5, 6, 20, 21] as circles as well as the calculated results as solid curves for loading along such trajectories. Experimentation used thin-walled tubular specimens of 316 stainless steel subjected to axial force and torque (*P* and *M* are experiments), where the axial and shear strain were measured at room temperature. In Fig. 6.2, Curve 1 corresponds to a calculation adjusted for additional hardening; Curve 2 corresponds to a non-adjusted calculation. Additional isotropic hardening results in more than double hardening. Figure 6.1 shows a cross-hardening effect: additional hardening occurs first, followed by slow softening when the proportional cyclic loading is changed from alternating torsion to tensile–compressive loading. Note that, at greater radii (Fig. 6.2), loading becomes ever more disproportionate, which results in greater additional hardening, although the path of strain has its curvature reduced. Thus, not only the curvature of the strain path, but also the location of the path with respect to the origin (the zero stress–strain state) determines the complexity of the loading process.

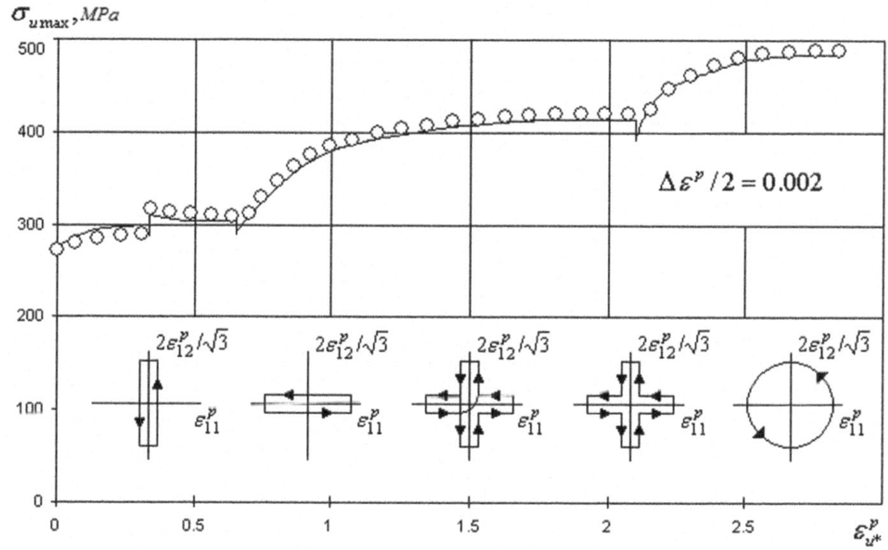

Fig. 6.1 316 stainless steel. Additional isotropic hardening induced by proportional and disproportionate loading along strain paths of increasing disproportionality

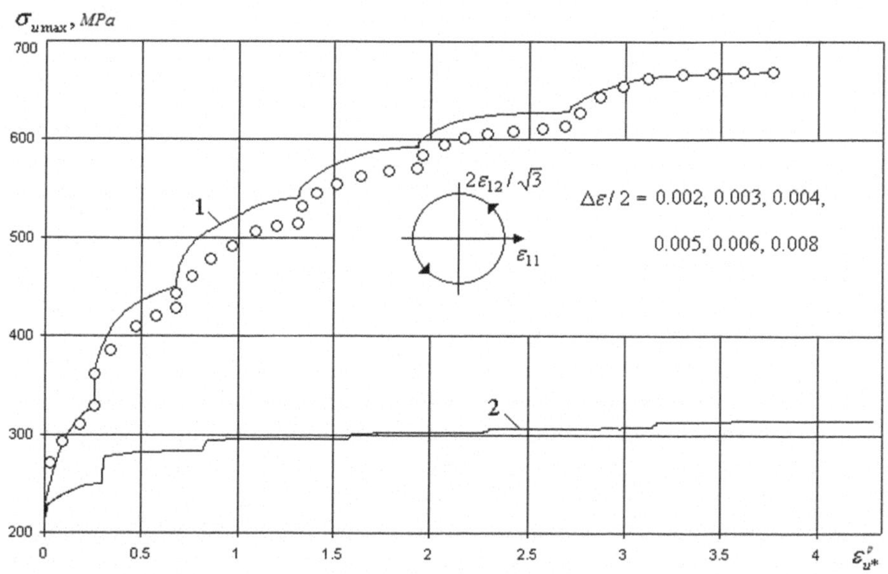

Fig. 6.2 316 stainless steel. Additional isotropic hardening induced by disproportionate cyclic loading along circular paths of increasing radii

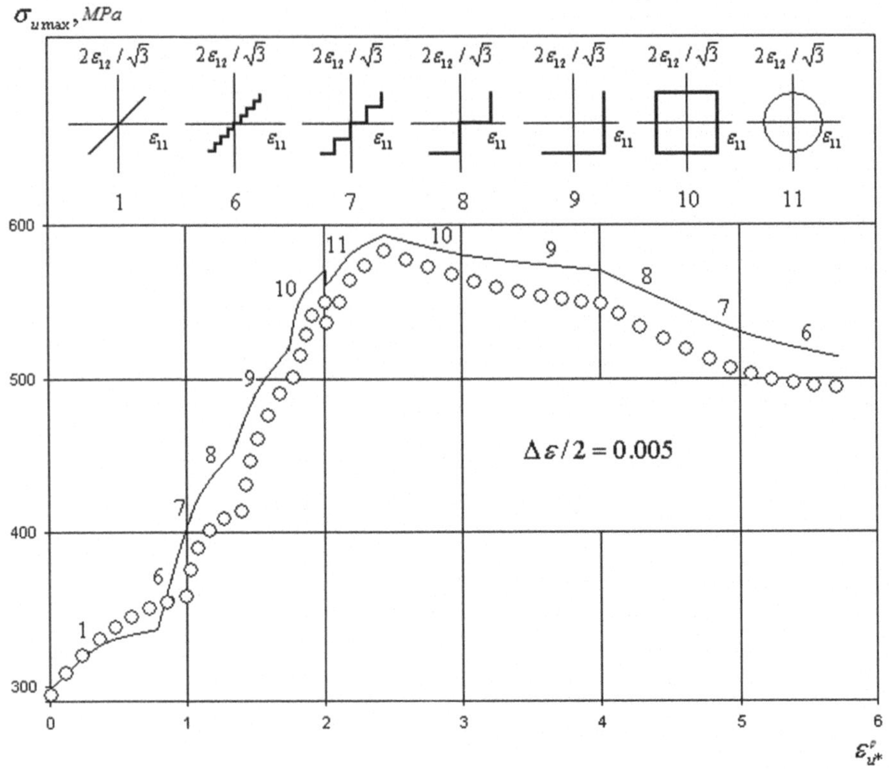

Fig. 6.3 316 stainless steel. Additional isotropic hardening and softening induced by proportional and disproportionate loading along strain paths of increasing and decreasing disproportionality

Next, consider the loading processes, in which softening follows hardening. Figure 6.3 shows the calculated (solid curve) and experimental [2, 5, 6] (circles) results for a very complex cyclic loading process.

Here, additional hardening increases with disproportionality as the material is loaded along the following trajectories: a beam, an eight-stage path, a four-stage path, a two-stage path, a single-stage path, a square, and a circle. Further loading along a squared, a single-stage, a two-stage, a four-stage, and an eight-stage trajectories results in softening, as the disproportionality decreases.

The analytical and experimental research of additional hardening and fatigue destruction induced by disproportionate cyclic loading (tension–compression or alternating torsion) used 304 stainless steel at room temperature. Figure 6.4 shows the calculated results as solid curves and experimental data [34] as checkmarks, dark and bright circles, dark triangles, diamonds, and squares; these data apply to fatigue destruction whether caused by proportionate or disproportionate cyclic loading. Figure 6.4 shows the pathes of cyclic-load strains.

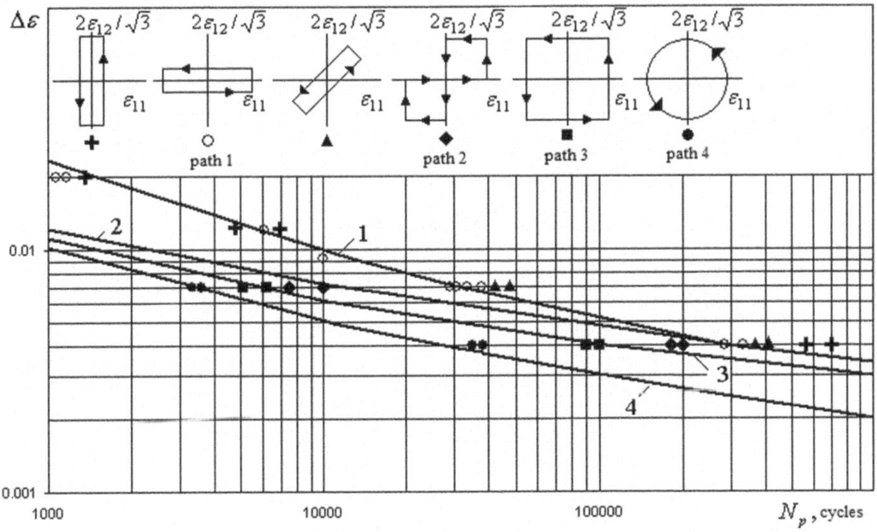

Fig. 6.4 304 stainless steel. Fatigue destruction induced by proportionate and disproportionate cyclic loading

Analysis of experimental and calculation data shows that given the same span of strain, disproportionate cyclic loads inflict greater damage than their cyclic counterparts, and the longevity might be reduced by an order of magnitude. Figure 6.5 shows how additional hardening affects the fatigue destruction in disproportionate cyclic loading. Curve 4 corresponds to a calculation adjusted for additional hardening; Curve 5 corresponds to a non-adjusted calculation. Dark circles show the experiment data per [34]. As shown in Fig. 6.5, non-adjustment for additional hardening and destruction energy as a function of disproportionality may result in overestimating or underestimating the projected longevity compared to the actual longevity of a material that exhibits that property.

Figures 6.6, 6.7 and 6.8 show the adjustment for additional isotropic hardening; they show cyclic diagrams with (thick curves) or without (thin curves) adjustment for additional hardening. Experimental results per [34] are shown as bright circles.

There is significant quantitative difference between the adjusted and non-adjusted cyclic diagrams while they fit well from the qualitative standpoint.

6.4 Conclusions

Complex loading processes, effects of additional isotropic hardening, and destruction processes are adequately described by a single, rather simple version of the plasticity theory, which is where the advantage of the herein presented mathematical models lies. There are far less materials functions (11 parameters and 1 function in

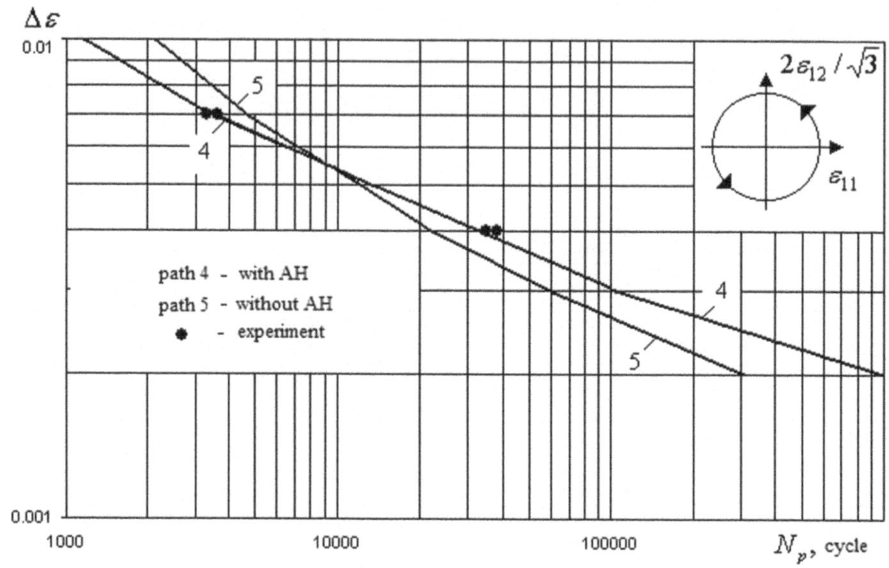

Fig. 6.5 304 stainless steel. How adjustment for additional isotropic hardening affects fatigue destruction

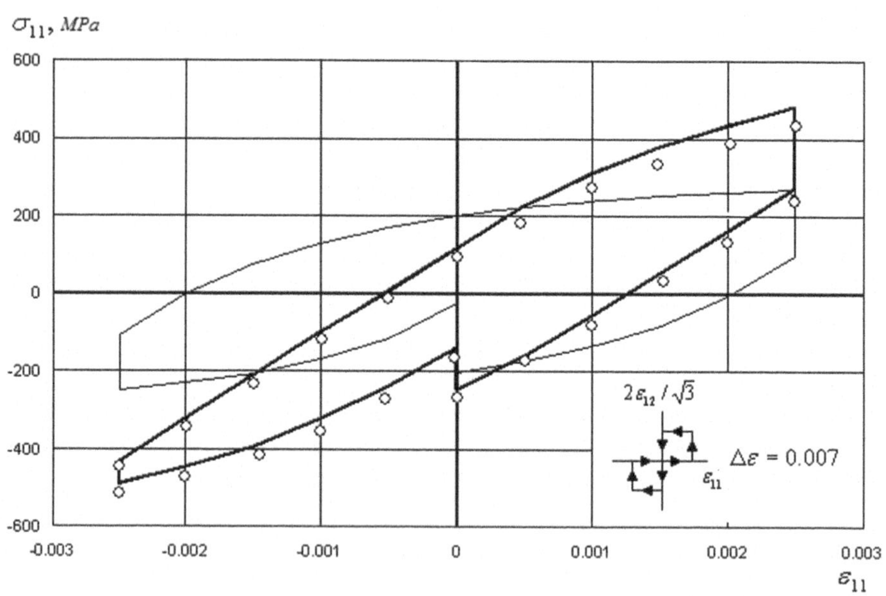

Fig. 6.6 304 stainless steel. How additional isotropic hardening affects the cyclic diagram when straining along a double-square trajectory

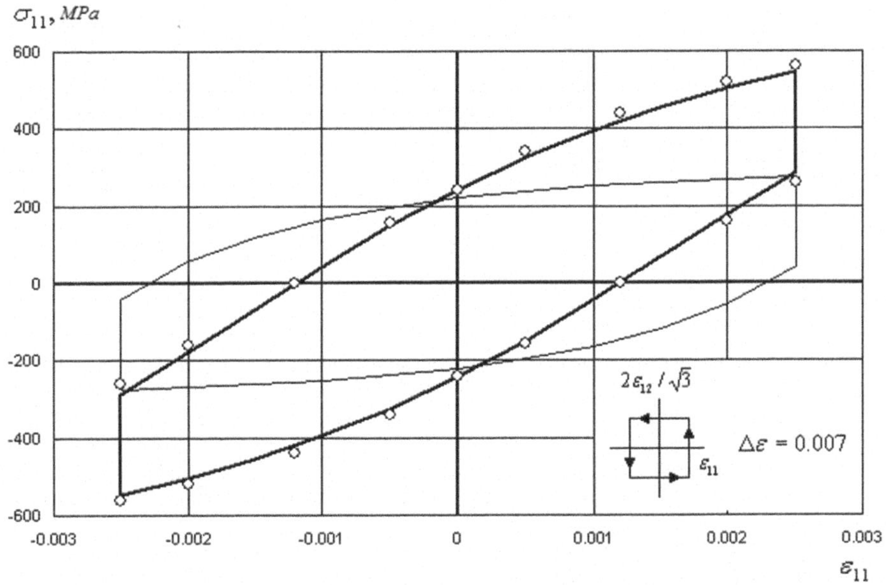

Fig. 6.7 304 stainless steel. How additional isotropic hardening affects the cyclic diagram when straining along a square path

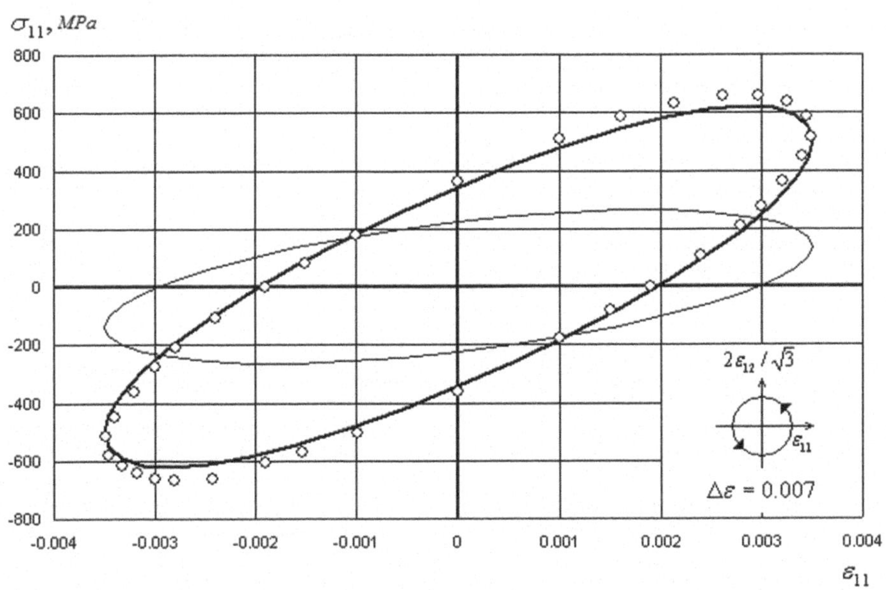

Fig. 6.8 304 stainless steel. How additional isotropic hardening affects the cyclic diagram when straining along a circular path

this case; 9 parameters and 1 function without destruction) than there are material functions and parameters closing the today's theories. Besides, the basic experiment and the material function identification method for the considered version of the plasticity theory are well-defined, relatively simple, and easy to use. Experiment-vs-calculation comparison proves the proposed mathematical modeling approach accurate and reliable.

References

1. Kadashevich, I.I., Pomytkin, S.P.: Ob effekte neproportsional'nosti pri slozhnom tsiklicheskom nagruzhenii. *Prikladnye problemy prochnosti i plastichnosti. Chislennoe modelirovanie fiziko-mekhanicheskikh sistem. Mezhvuz. sb.*, Moscow, Tovarishch, nauchn. izd. KMK, pp. 171–175 (1995)
2. Benallal, A., Calloch, S., Laborderie, C., Marquis, D.: Hardening of metals under cyclic nonproportional loadings in tension-torsionand triaxial tension. In: International Seminar on Multiaxial Plasticity, Cachan France, pp. 1–7 (1992)
3. Benallal, A., Lemaitre, J., Marquis, D., Ronsset, M.: Nonproportional loadings in plasticity and viscoplasticity: experimentation and modelling. In: Proceedings the International Conference on Non Linear Mechanics, Shanghai, China (1985)
4. Benallal, A., Le Gallo, P., Marquis, D.: An experimental investigation of cyclic hardening of 316 stainless steel and of 2024 aluminum alloy under multiaxial loading. Nucl. Eng. Des. **114**, 345–353 (1989)
5. Benallal, A., Marquis, D.: Constitutive equations for nonproportional cyclic elasto-viscoplasticity. J. Eng. Mater. Technol. **109**, 326–337 (1987)
6. Benallal, A., Marquis, D.: An experimental investigation of cyclic hardening of 316 SS under complex multiaxial loadings. In: Proceedings of the 9th SMIRT Conferences paper L 10/3, Lausanne (1987)
7. Delobelle, P., Robinet, P., Bocher, L.: Experimental study and phenomenological modelization of ratcheting under uniaxial and biaxial loading on an austenitic stainless steel. Int. J. Plasticity **11**, 295–330 (1995)
8. Hassan, T., Taleb, L., Krishna, S.: Influence of non-proportional loading on ratcheting responses and simulations by two recent cyclic plasticity models. Int. J. Plasticity **24**, 1863–1889 (2008). https://doi.org/10.1016/j.ijplas.2008.04.008
9. Haupt, P., Kamlah, M.: Representation of cyclic hardening and softening properties using continuous variables. Int. J. Plasticity **11**, 267–291 (1995)
10. Kanazawa, K., Miller, K.J., Brown, M.W.: Cyclic deformation of 1% Cr-Mo-V steel under out-of-phase loads. Fat. Eng. Mat. Struc. **2**, 217 (1979)
11. Kang, G.Z., Gao, Q., Yang, X.J., Sun, Y.F.: An experimental study on uniaxial and multiaxial strain cyclic characteristics and ratchetting of 316L stainless steel. J. Mater. Sci. Technol. **17**, 219–223 (2001)
12. Kremple, E., Lu, H.: The hardening and rate dependent behavior of fully annealed AISI type 304 stainless steel under biaxial in phase and out-of-phase strain cycling at room temperature. ASME J. Eng. Mater. Technol. **106**, 376–382 (1984)
13. Krieg, R.D.: A practical two surface plasticity theory. J. Appl. Mech. **42**, 641–646 (1975)
14. Lamba, H.S., Sidebottom, O.M.: Cyclic plasticity for nonproportional paths: parti and 2-comparison with predictions off three incremental plasticity models. ASME J. Eng. Mater. Technol. **100**, 96–111 (1978)
15. McDowell, D.L.: An experimental study of the structure of constitutive equations for nonproportional cyclic plasticity. ASME J. Eng. Mater. Technol. **4**, 307–315 (1985)

16. McDowell, D.L., Socie, D.F.: Transient and stable deformation behavior under cyclic nonproportional loadings. ASTM-STP 853. In: Proceedings of the International Symposium on Biaxial-Multiaxial I atigue, San Francisco, pp. 64–87 (1982)
17. Ohashi, Y., Kawai, M., Kaito, T.: Inelastic behavior of type 316 stainless steel under miiltiaxial nonproportional cyclic stressings at elevated temperature. ASME J. Eng. Mater. Technol. **2**, 6–15 (1985)
18. Olschewski, J., Pilvin, P.: Phenomenological vs. micromechanical approaches for multiaxial loading. In: Preprints of MECAMAT'92. International Seminar on Multiaxial Plstisity, Cachan France (1992)
19. Tasnim, H., Stelios, K.: Ratcheting of cyclically hardening and softening materials: II multiaxial behavior. Inter. J. Plasticity **10**(2), 185–212 (1994)
20. Tanaka, E., Murakami, S., Ooka, M.: Effects of plastic strain amplitudes on non-proportional cyclic plasticity. Acta Mech. **57**, 167–182 (1985)
21. Tanaka, E., Murakami, S., Ooka, M.: Effects of strain path shapes on nonproportional cyclic plasticity. J. Mech. Phys. Solids **33**(6), 559–575 (1985)
22. Tanaka, E.: A non-proportionality parameter and a cyclic viscoplastic constitutive model taking into account amplitude dependences and memory effects of isotropic hardening. Eur. J. Mech. A/Solids **13**, 155–173 (1994)
23. Zhang, J., Jiang, Y.: Constitutive modeling of cyclic plasticity deformation of a pure polycrystalline copper. Int. J. Plasticity **24**, 1890–1915 (2008). https://doi.org/10.1016/j.ijplas.2008.02.008
24. Ziebs, J., Meersmanu, J., Kuhn, H.-J.: Effects of proportional and nonproportional staining sequences on hardening/softening behaviour of IN 738 LC at elevated temperatures. In: Preprints of MECAMAT'92. International Seminar on Multiaxial Plstisity, Cachan France (1992)
25. Kadashevich, I.I., Mosolov, A.B.: O sootnosheniiakh endokhronnoi teorii plastichnosti s "novoi" meroi vnutrennego vremeni pri slozhnom tsiklicheskom nagruzhenii. Tekhnologiia legkikh splavov **3**, 32–36 (1990)
26. Korotkikh, I.G., Makovkin, G.A.: O modelirovanii protsessov neproportsional'nogo uprugoplasticheskogo deformirovaniia na baze uravnenii plastichnosti s kombinirovannym uprochneniem. In: *Prikladnye problemy prochnosti i plastichnosti. Chislennoe modelirovanie fiziko-mekhanicheskikh protsessov: Mezhvuz. sb.*, Moscow, Tovarishch, nauch. izd. KMK, pp. 5–10 (1997)
27. Makovkin, G.A.: Sravnitel'nyi analiz parametrov neproportsional'nosti slozhnogo uprugoplasticheskogo deformirovaniia. *Vestnik Nizhegorodskogo universiteta im. N.I. Lobachevskogo. Seriia mekhanika*, Vyp. 1, N.Novgorod, Izd-vo NNGU, pp. 30–36 (1999)
28. Kadashevich, Iu.I., Pomytkin, S.P.: Ob effekte neproportsional'nosti pri slozhnom tsiklicheskom nagruzhenii. Prikladnye problemy prochnosti i plastichnosti. Chislennoe modelirovanie fiziko-mekhanicheskikh sistem. Mezhvuz. sb., Moscow, Tovarishch, nauchn. izd. KMK, pp. 171–175 (1995)
29. Bondar, V.S., Titarev I.A.: Variant teorii plastichnosti dlia proportsional'nykh i neproportsional'nykh tsiklicheskikh nagruzhenii. *Problemy prochnosti i plastichnosti: Mezhvuz. sbornik*, Vyp. 63, N.Novgorod, Izd-vo NNGU, pp. 5–17 (2001)
30. Bondar, V.S.: Inelasticity: Variants of the Theory, p. 194. Begell House, New York (2013)
31. Bondar, V.S., Danshin, V.V.: Plastichnost'. Proporcional'nye i neproporcional'nye nagruzhenija (Plasticity. Proportional and disproportionate loading), Moscow, FIZMATLIT, pp. 176 (2008)
32. Bondar, V.S., Danshin, V.V., Makarov, D.A.: Mathematical modelling of deformation and damage accumulation under cyclic loading. PNRPU Mech. Bull. **2**, 125–152 (2014)
33. Bondar, V.S., Danshin, V.V., Alkhimov, D.A.: Analysis on cyclic defonnation and low-high-cycle fatigue in uniaxial stress state. PNRPU Mech. Bull. **4**, 52–71 (2016). https://doi.org/10.15593/perm.mech/2016.4.04
34. Socie, D.: Multiaxial fatigud damage models. ASME J. Eng. Mater. Technol. **3**, 9–21 (1988)

Chapter 7
Dynamic Compressibility of Birch Under Various Types of Stress-Strain State

Anatoly M. Bragov, Andrey K. Lomunov, and Tatiana N. Yuzhina

Abstract The results of dynamic compression tests of air-dry birch samples across the fibers are presented. Dynamic tests were carried out on an installation with a split Hopkinson bar according to the Kolsky method at strain rates of ~2000 s^{-1} at normal temperature. Dynamic diagrams of deformation are obtained for uniaxial stressed and uniaxial strained states, as well as for a some intermediate state (deformation in the birch board array). Comparison of deformation diagrams for different types of stress-strain state shows a significant effect of the type of stress-strain state on the behavior of the material under study. Using the obtained diagrams, some mechanical characteristics of the material are determined.

Keywords Wood · Birch · Dynamic compression · Model · Multicycle load

7.1 Introduction

In order to carry out a reliable numerical analysis of the designed designs of transport containers for transporting hazardous substances, using wood as components that dampen shock loads, reliable mathematical models are required that take into account its complex multicomponent structure. To obtain appropriate models, a large-scale study of the anisotropy of the physico-mechanical properties of wood for various

A. M. Bragov (✉) · A. K. Lomunov · T. N. Yuzhina
Research Institute for Mechanics, National Research Lobachevsky State University of Nizhny Novgorod, 23 Gagarin Avenue, Building 6, Nizhny Novgorod, Russian Federation 603950
e-mail: bragov@mech.unn.ru

A. K. Lomunov
e-mail: lomunov@mech.unn.ru

T. N. Yuzhina
e-mail: yuzhina_tatiana@mech.unn.ru

H. Altenbach et al. (eds.), *Multiscale Solid Mechanics*,
Advanced Structured Materials 141,
https://doi.org/10.1007/978-3-030-54928-2_7

types of stress-strain state is necessary. Wood models of deformation and destruction are now actively developing for numerical simulation of the behavior of technically complex structures incorporating wood elements. An analysis of the literature showed that wood is a complex natural polymer composite material with porosity and pronounced anisotropy of mechanical characteristics. A significant effect of the strain rate on the deformation and strength characteristics of wood is also noted.

Another important feature of the behavior of wood is a significant difference in behavior depending on the type of stress state. Thus, the deformation of the southern yellow pine under tension and shear is brittle, while the deformation diagram under compression has a plastic section [1].

Among other light materials, the tree has a high ability to absorb energy, since it can collapse up to 70–80% of its original height under compression [2–5]. For the construction field, a static load is usually assumed, and therefore, only quasistatic data and strength values are required. However, for applications such as impact strength [2, 5], the dependence of the dynamic properties of wood on the strain rate is required. In order to use efficiently wood in these applications, its corresponding characteristic in dynamic load mode is of paramount importance.

The behavior of birch under compression across the fibers is investigated at room temperature for various types of stress-strain states.

7.2 Material and Specimen

To study the dynamic properties of birch, samples were cut out from a plank across the fibers. To do this, tablets with a diameter of 20 mm were cut from an air-dry birch plank of 10 mm thick using a diamond crown, and birch plates 80 × 80 × 10 mm in size were prepared.

Table 7.1 shows the data [6] on the mechanical properties of birch with a density of 0.62 g/cm^3 and humidity of 9%. The designations of the axes and planes of symmetry correspond to the designations adopted in Fig. 7.1.

When testing wood under uniaxial deformation, the samples were placed between the ends of the measuring bars in a jacket restricting the radial deformation of the sample during its longitudinal compression. Like measuring bars, the jacket of the original configuration [7] was made of high-strength aluminum alloy.

Table 7.1 The mechanical properties of birch

E_r, MPa	E_t, MPa	E_a, MPa	G_{rt}, MPa	G_{at}, MPa	G_{ar}, MPa	μ_{tr}/μ_{rt}	μ_{at}/μ_{ta}	μ_{ra}/μ_{ar}
1126	629	16,600	192	1043	1095	$\frac{0.38}{0.78}$	$\frac{0.470}{0.018}$	$\frac{0.034}{0.490}$

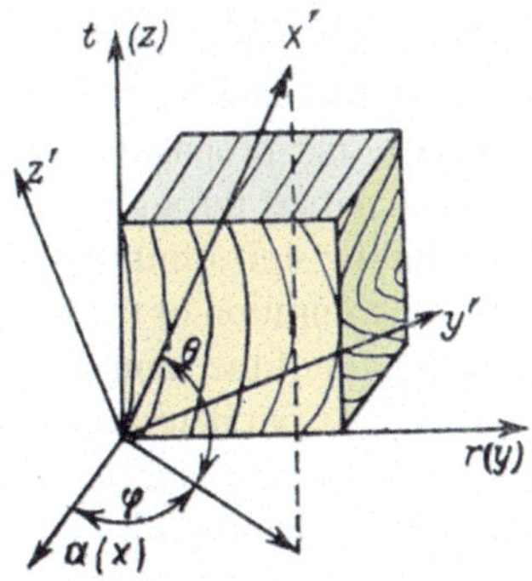

Fig. 7.1 Axes and planes of symmetry of the elementary volume of wood

7.3 Method of Investigation

To test of air-dry birch samples, the installation [8, 9] was used. Its operation is based on the Kolsky method [10] with a split Hopkinson pressure bar (SHPB) as the main measuring tool (Fig. 7.2). This figure also shows the main parametric dependences of the method for determining stress, strain, and strain rate in a sample.

The Kolsky method is based on the one-dimensional theory of the propagation of elastic waves in long thin bars. Traditionally, a system for testing under uniaxial compression consists of two long bars (loading and supporting) with sufficiently high yield strength as well as a thin sample in the form of a tablet located between their ends. Using a compact gas gun, an elastic compression pulse with amplitude proportional to the velocity of the impactor is excited in the loading bar.

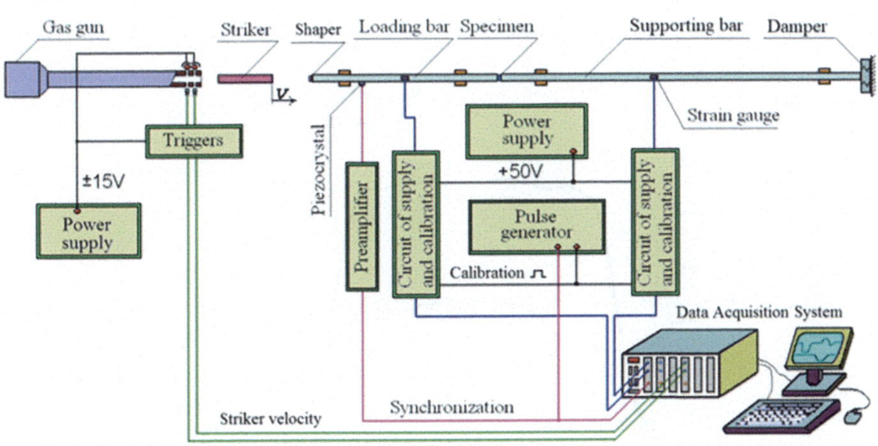

Fig. 7.2 Scheme of the experimental setup for compression tests

When approaching the sample, this pulse, due to the difference in acoustic impedances of the materials of the bars and the sample, is separated: part of the pulse is reflected from the interface and propagates back into the loading bar by a tensile wave, while the other part passes through the sample into the supporting bar by a compression wave. Assuming uniformity of the stress-strain state of the sample along its axis (due to the long duration of the loading pulse compared to the travel time of the wave along the length of the sample), based on these pulses using the formulas of the Kolsky method, we can determine the parametric dependences of development of the stress $\sigma(t)$, strain $\varepsilon(t)$, and the strain rate $\dot{\varepsilon}(t)$ in the sample during the test and further, after synchronization, eliminating the time as a parameter, construct a diagram of the deformation of the sample in the form of dependences $\sigma(\varepsilon)$ and $\dot{\varepsilon}(\varepsilon)$.

During the tests, traditional waveforms were obtained, on which two beams were recorded: from the first (loading) measuring bar (upper beam) and from the second (supporting) measuring bar (lower beam). An example of such an oscillogram is shown in Fig. 7.3a. In this case, both the incident $\varepsilon^I(t)$ and reflected $\varepsilon^R(t)$ pulses are recorded and marked with markers on the upper beam, while the pulse $\varepsilon^T(t)$ passing

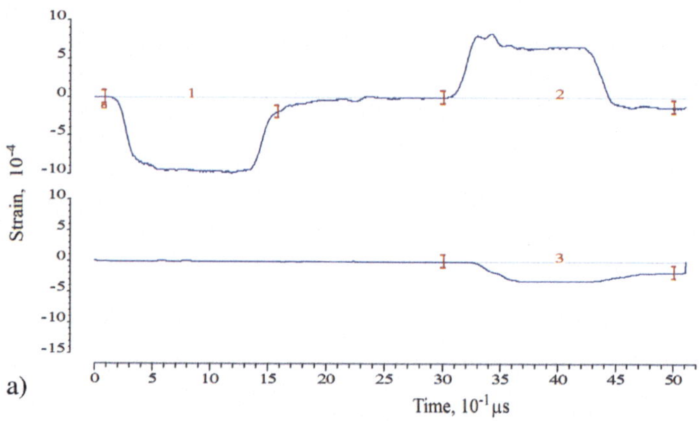

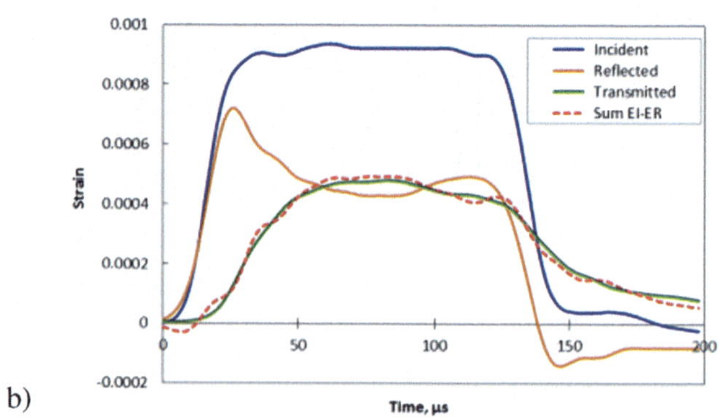

Fig. 7.3 Conventional pulse registration in measuring bars in wood testing

through the sample is recorded on the lower beam. Figure 7.3b shows the same pulses in the synchronization mode with the addition of the total pulse $\varepsilon^R + \varepsilon^T$. It is clearly seen that the reflected and transmitted pulses do not return to the "zero" line after the end of the loading process, i.e., the process of unloading a wood sample has a very long duration.

In addition, it is clearly seen that the amplitude of the reflected pulse is significant (up to 80% of the amplitude of the incident wave). The reason for this fact is the very large difference in the acoustic impedances ρC of the measuring bars and the sample. This reflected pulse propagates to the end face of the first measured bar and is reflected from this free end (since there is no contact with the striker) again as a compression wave. This secondary wave loads the sample again, and its significant part is reflected again, etc. [9]. Thus, the sample undergoes many load-unload cycles with a gradually decreasing amplitude. The pause between cycles is equal to the propagation time of the pulse along the first bar back and forth. The oscillogram gives some idea of this process (Fig. 7.4), obtained with a slower scan of the oscilloscope.

It is clearly seen that the sample during one test is loaded many times, getting a certain deformation in each cycle. When a traditional recording, only the main first loading cycle is measured, whereas subsequent cycles are ignored. This is the reason that, when testing materials with a low acoustic impedance, the calculated residual deformation obtained as a result of processing pulses of only one first load cycle (during traditional registration) does not coincide with the actual residual deformation measured with relatively intact samples.

As noted above, the study of the behavior of materials under cyclic loading is a very urgent problem, since in a real situation with high-speed exposure onto elements of various structures it is quite possible a repeatedly cycle loadings due to reflections and wave interference. Therefore, it is of interest to record such additional cycles of loading the sample and construct the corresponding dynamic diagrams.

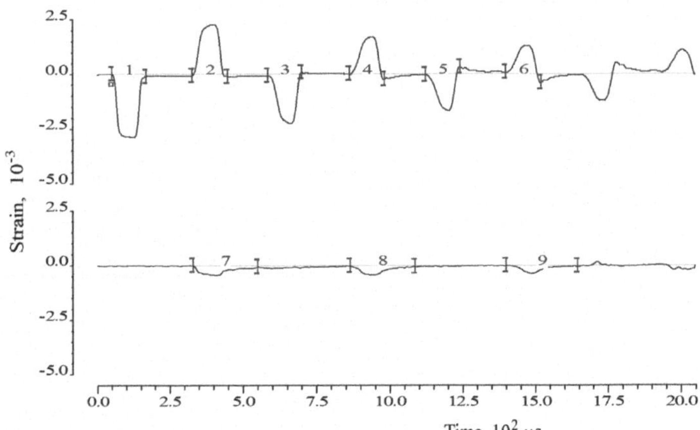

Fig. 7.4 Initial strain pulses during registration of additional loading cycles of the sample

In order to be able to carry out tests during cyclic loading of samples and to record repeated loading cycles in the course of one experiment, it is necessary to exclude the return of the compression wave εt transmitted through it from the rear end of the support bar to the sample, which can affect the loading process in subsequent cycles and distort the recording of subsequent loading cycles. For this, the length of the support bar should be increased in comparison with the length of the loading bar as many times as the loading cycles are supposed to be recorded. The registration time of the test process should be accordingly increased. Figure 7.5 shows the wave pattern of pulses in the measuring bars when registering three loading cycles [9].

To implement tests of samples under the condition of volumetric stress state and uniaxial deformation, the tested sample was located into a rigid jacket [11, 12]. It should be noted that the sample was placed in a jacket with a small gap between its lateral surface and the inner surface of the jacket equal to ~0.1 mm; therefore, a stress-strain state close to volume occurred only after eliminating this gap. Prior to deletion of the gap, as will be shown below, the behavior of the sample in the jacket was similar to the behavior of the sample during testing without restrictions on its radial deformation, i.e., under uniaxial stress state.

In addition to the two indicated types of tests under uniaxial stress state and uniaxial deformation, a test cycle was conducted at some intermediate stress-strain

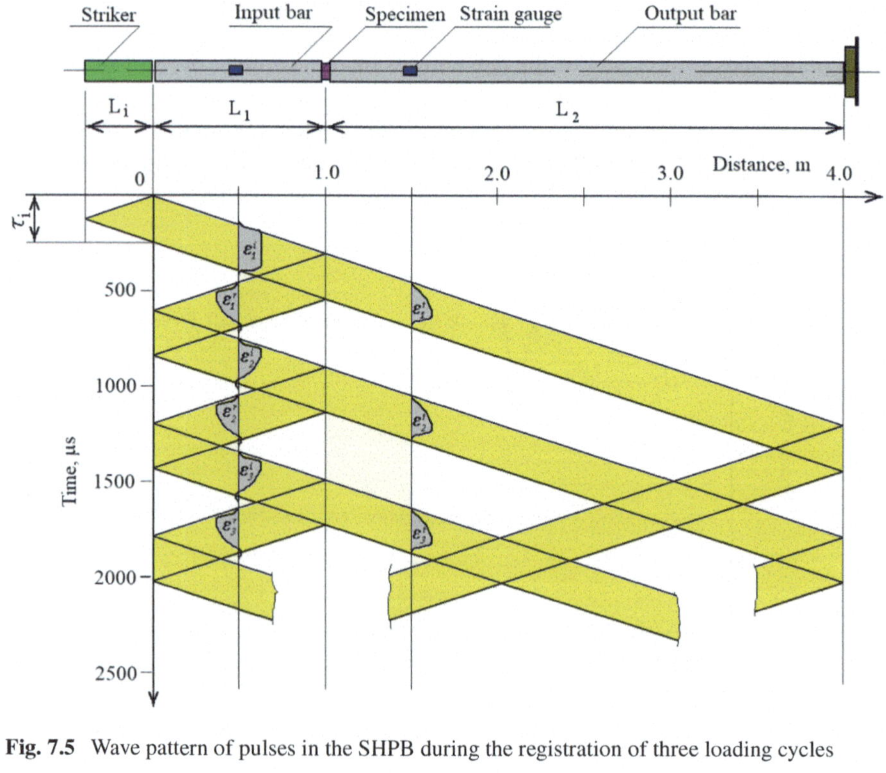

Fig. 7.5 Wave pattern of pulses in the SHPB during the registration of three loading cycles

state, when the material itself played the role of the confining jacket. In this case, a wooden plate 80 × 80 × 10 mm in size was placed between the ends of the measuring bars with a diameter of 20 mm and was subjected to dynamic compression. Such local loading of the wood massive is similar to its loading in a real design; however, the stiffness of the material restricting the radial widening is small, and at large load amplitudes, the material surrounding the compressible zone is destroyed, and the restriction effect is lost. But the confining effect is manifested, in contrast to the test mode in the jacket, from the very beginning of loading.

7.4 Results and Discussion

A cycle of dynamic testing of birch samples was carried out under compression across the fibers at normal temperature under various types of stress-strain state. In each mode, 3–5 tests were carried out, the results of which were averaged. The amplitude of the loading wave, which is directly proportional to the velocity of the impactor, determined the strain rate at which the samples were deformed.

The samples in the form of tablets tested under uniaxial stress state are shown in Fig. 7.6a. The destruction of such samples was accompanied by spallation on the lateral surface and splitting along the fibers.

Samples tested in a rigid confining jacket were compressed under uniaxial deformation. Samples that retained visible integrity under loading, when removed from the jacket, disintegrated into fragments.

In addition to samples in the form of tablets, tests were carried out on rectangular plates cut from a board. The characteristic appearance of such fragments after the test is shown in Fig. 7.6b. Each fragment was tested twice. Deformation of a fragment of the board was characterized by crushing of fibers in the transverse direction in

Fig. 7.6 Birch samples after the test. **a** In the form of tablets. **b** As a fragment of the board

the contact zone of the fragment and the measuring bars. The fragment destruction occurred in the unloading waves in the direction perpendicular to the fiber direction.

As mentioned earlier, due to the small acoustic impedance ρC of the wood, the sample is subjected to a large number of loading cycles with gradually decreasing amplitude, and its plastic deformation reaches a large value. The use of measuring bars with a length of 1.5 and 3 m made it possible to register three cycles of the sample load in one experiment and accurately determine in which loading cycle the sample fracture occurred.

As a result of tests for each mode, parametric dependences $\sigma(t)$, $\varepsilon(t)$ and $\dot{\varepsilon}(t)$ were obtained, and as well as stress-strain curves $\sigma(\varepsilon)$ with corresponding histories of changes in the strain rate $\dot{\varepsilon}(\varepsilon)$.

Figure 7.7 shows a set of obtained averaged diagrams as a result of tests without a jacket in the condition of uniaxial stress state, whereas Fig. 7.8 shows a similar set during compression in the rigid jacket under the condition of uniaxial deformation. Next, a similar set of averaged diagrams obtained by loading a sample in the form of a fragment of a board (combined stress-strain state) is shown in Fig. 7.9.

Each figure shows the obtained parametric curves for three-cycle loading and the resulting diagrams $\sigma(\varepsilon)$ and $\dot{\varepsilon}(\varepsilon)$. Histories of changes in the strain rate are shown in the lower part of the graphs by dashed lines, and the corresponding axis is located on the right. As indicated above, the process of unloading the sample has a long duration, and it is not possible to register it completely with the existing length of the measuring bars.

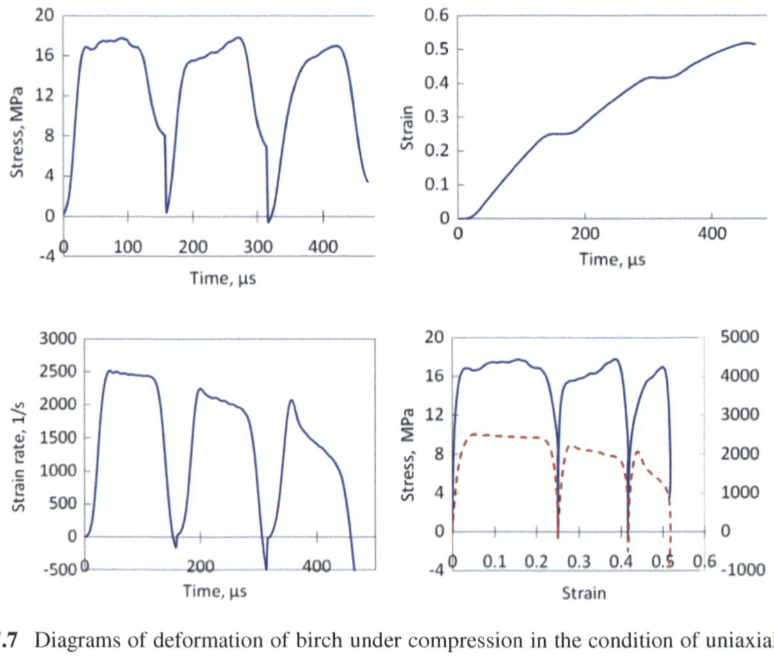

Fig. 7.7 Diagrams of deformation of birch under compression in the condition of uniaxial stress state (without jacket)

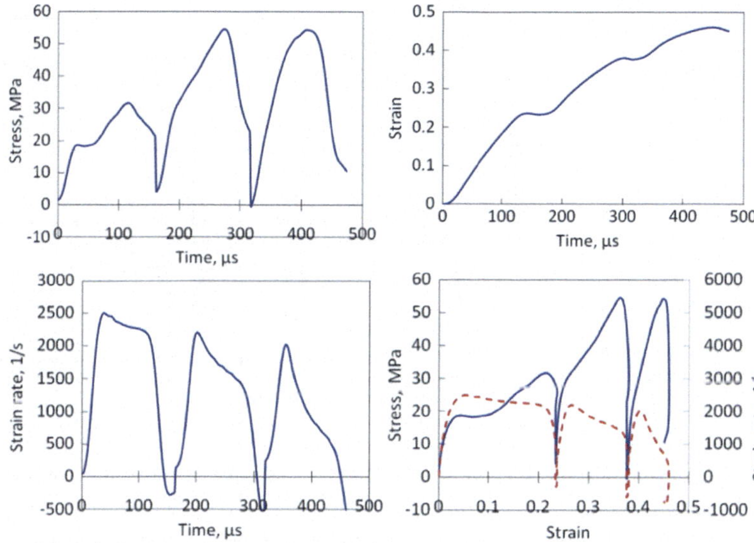

Fig. 7.8 Diagrams of deformation of birch under compression in the condition of a uniaxial deformed state (in a jacket)

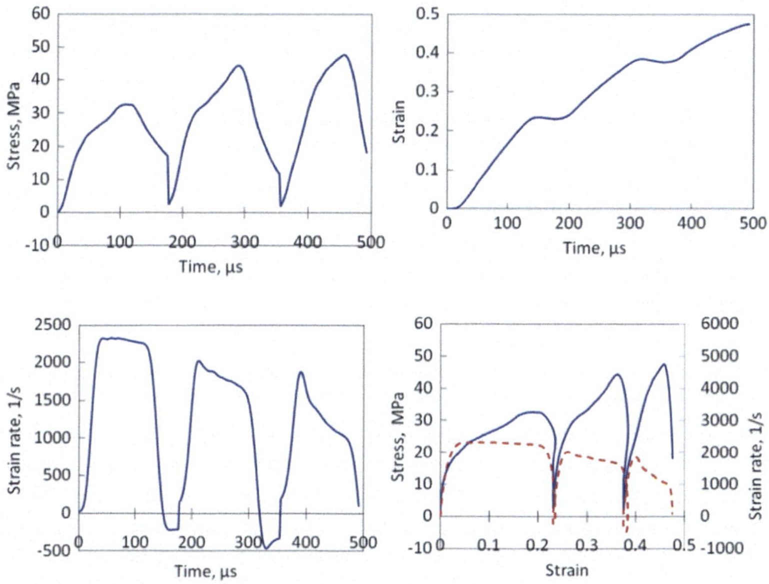

Fig. 7.9 Diagrams of deformation of birch under compression in the condition of a combined stress-strain state (fragment of a board)

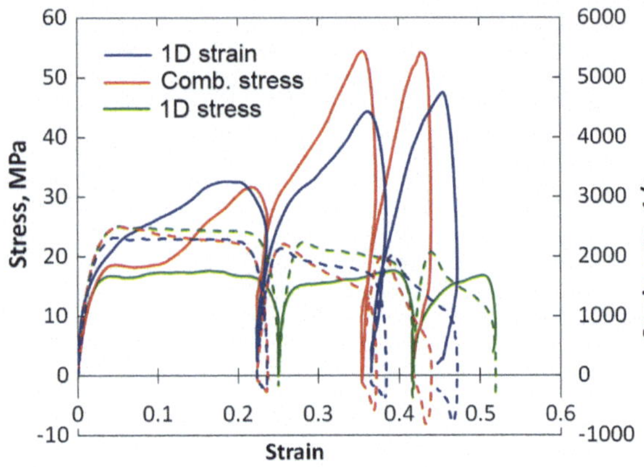

Fig. 7.10 Comparison of the dynamic compressibility of birch under different types of stress-strain state

As can be seen, when birch is compressed across the fibers under the condition of a uniaxial stress state, the stress amplitude during repeated cycles decreases slightly, moreover further compression of the material occurs with a gradually decreasing value of the obtained deformation in each subsequent cycle. When the wood is compressed in the board massive, on the contrary, the stress amplitude during repeated cycles increases, but the degree of deformation obtained in each subsequent cycle also decreases. When a sample is compressed in a jacket (under the condition of uniaxial deformation), the stress amplitude in the second loading cycle increases, but it decreases in subsequent cycles. The degree of deformation achieved with three registered loading cycles for any type of stress-strain state is 40–45%. In further cycles, the samples receive additional deformation, but it is impossible to register these cycles.

An analysis of the obtained three-cycle pulses and in tests with any type of stress-strain state allowed us to conclude that during the three registered load cycles, the sample did not fail, and therefore, it was destroyed in subsequent load cycles, which cannot be registered.

Figure 7.10 shows a comparison of the strain diagrams for the three studied modes of stress-strain state. The dashed lines at the bottom of the graphs show the corresponding curves of the strain rate.

It is clearly seen that during compression across the fibers, the behavior of birch under the condition of a uniaxial stress state is similar to ideal plasticity: the material receives a certain degree of deformation during each compression cycle (25, 17 and 10%) at almost constant stress amplitude. When testing samples in a rigid confining jacket, due to the presence of a small gap (~0.1 mm) between the side surface of the specimen and the inner surface of the jacket, the state of uniaxial deformation occurs only when the deformation reaches ~10%, and before this deformation, the material behaves similarly to the uniaxial stress condition. After eliminating this gap, a significant increase in the stress developed in the sample is observed, due to the limitation of its lateral deformation. In the third loading cycle, a stress increase is not observed due to insufficient amplitude of the loading wave in this cycle. Under the condition of a combined stress-strain state (compression of a board fragment), the influence of a wood massive that limits the lateral widening of the sample causes an intense increase of stress from the very beginning of loading in the first load cycle, whereas in subsequent cycles the amplitude of the achieved stress is less than under uniaxial deformation.

Thus, the behavior of birch during loading of samples with the transverse direction of the fibers strongly depends on the type of stress-strain state, which should be taken into account when identifying the model of wood used in a layered protective structure subject to dynamic loads.

7.5 Conclusions

Using the Kolsky method, samples of air-humidity birch were tested at strain rates of ~2000 s^{-1}. To assess the influence of the type of stress-strain state on the behavior of the material, in addition to samples in the form of tablets, rectangular fragments of the board were tested, as well as samples in a rigid jacket that impeded the radial widening of the sample. To assess the effect of the loading history on the behavior of birch during high-speed deformation, tests were carried out with registration of additional loading cycles. According to the test results, the modules of the load and hardening branches, the conditional yield strength and the energy absorption of the material are determined.

In the case of free widening of the sample in the radial direction, there is a lack of strain hardening in the first loading cycle. In subsequent cycles, hardening is negligible. The stress-strain curves of birch when tested both in the form of tablets in a jacket and fragments of a board are characterized by a noticeable increase in the modulus of the hardening branch with increasing strain. It can be noted that the behavior of the material in the case of testing a fragment of the board is intermediate between the case of free widening of the sample in the radial direction and the case under volumetric stress state and uniaxial deformation.

In the case of testing samples in the form of tablets, the effect of preliminary deformation on the modulus of the load branch is negligible. At the same time, a decrease in the value of the conditional yield strength in subsequent cycles was noted, which is apparently caused by the accumulation of damage.

When testing samples in a rigid jacket, an increase in the modules of the load and hardening branches was noted in subsequent loading cycles, which is apparently due to the limitation of the radial widening of the sample. In this case, the highest value of the conditional yield strength was observed in the second loading cycle.

Tests of board fragments were characterized by a decrease in the values of the modules of the load branches in subsequent loading cycles, which may be caused by the accumulation of damage in the load-unloading waves in the entire fragment massive. As in the case of testing specimens in a rigid jacket, the highest value of the conditional yield strength was observed in the second loading cycle.

The obtained experimental results will serve as the basis for the subsequent identification of the model of deformation and destruction of wood.

Acknowledgements Experimental investigations under uniaxial stress state were supported by a grant of the Russian Science Foundation (16-19-10237-P). Analysis of behavior of birch using multicycle loading methodology was supported by the RFBR (grant 18-08-00808).

References

1. Murray, Y.D., Reid, J.D., Faller, R.K., Bielenberg, B.W., Paulsen, T.J.: Evaluation of LS-DYNA wood material model 143. Report No. FHWA-HRT-04-096 (2005)
2. Adalian, C., Morlier, P.: "WOOD MODEL" for the dynamic behaviour of wood in multiaxial compression. Holz als Roh- und Werkstoff **60**(6), 433–439 (2002)
3. Gibson, L.J., Ashby, M.F.: Cellular solids: structure and properties. Cambridge University Press, Cambridge (1997)
4. Johnson, W.: Historical and present-day references concerning impact on wood. Int. J. Impact Eng. **4**(3), 161–174 (1986)
5. Neumann, M., Herter, J., Droste, B.O., Hartwig, S.: Compressive behaviour of axially loaded spruce wood under large deformations at different strain rates. Eur. J. Wood Wood Prod. **69**(3), 345–357 (2011)
6. Ashkenazi, E.K., Gonov, E.V.: Anisotropy of structural materials. Handbook. Mashinostroenie, Leningrad (1980). (in Russian)
7. Bragov, A.M., Kotov, V.L., Lomunov, A.K., Sergeichev, I.V.: Measurement of the dynamic characteristics of soft soils using the Kolsky method. J. Appl. Mech. Tech. Phys. **45**(4), 580–585 (2004)
8. Bragov, A.M., Lomunov, A.K.: Methodological aspects of studying dynamic material properties using the Kolsky method. Int. J. Impact Eng. **16**(2), 321–330 (1995)
9. Bragov, A.M., Lomunov, A.K., Sergeichev, I.V.: Modification of the Kolsky method for studying properties of low-density materials under high-velocity cyclic strain. J. Appl. Mech. Tech. Phys. **42**(6), 1090–1094 (2001)
10. Kolsky, H.: An investigation of the mechanical properties of material at very high rates of loading. Proc. Phys. Soc. (Lond.) **62B**, 676–700 (1949)
11. Bazhenov, V.G., Bragov, A.M., Kotov, V.L., Zefirov, S.V., Kochetkov, A.V., Krylov, S.V., Lomunov, A.K.: Analysis of the applicability of a modified Kolsky's method for dynamic tests of soils in a deformable casing. J. Appl. Mech. Tech. Phys. **41**(3), 519–525 (2000)
12. Konstantinov, A.Y., Lomunov, A.K., Iuzhina, T.N., Gray III, G.T.: Investigation of wood anisotropy under dynamic loading. Probl. Strength Plast. **80**(4), 555–565 (2018)

Chapter 8
On Description of Fast Diffusion in a Coupled Multicomponent System with Microstructure Within the Framework of the Thermodynamics of Irreversible Processes

Dmitry Dudin and Ilya Keller

Abstract We consider a model of quasi-static coupled processes of diffusion, viscoelastic deformation, chemical reactions, and microstructure evolution at constant temperatures for which the physical and geometric relations are linear. A one-dimensional model problem is posed in which the equilibrium equations are satisfied automatically. The introduction of small spatial perturbations in the nonlinear system of equations of the model leads to an eigenvalue problem, the solution of which gives the spectrum of relaxation times and the corresponding eigenvectors. The introduction of the microstructure evolution and elastic properties leads to two corresponding branches of relaxation times. Asymptotic expressions for the interdiffusion coefficients are obtained for infinitely small and large wavelengths. Depending on the ratio of the characteristic elastic, thermal energy, and microstructure energy, fast or slow diffusion occurs in the system. These types of diffusion accompany the usual thermal diffusion and viscous relaxation. Non-classical diffusion processes are controlled by the microstructure gradient and (or) the mean stress gradient value.

Keywords Fast diffusion · Interdiffusion · Coupled processes · Rheological processes · Microstructure · Metal alloys · Perturbation method

8.1 Introduction

To ensure reliability indicators of modern machines and mechanisms, it is required to study the coupled processes of diffusion mass transfer, chemical reactions, and microstructure evolutions with accompanying viscoelastic deformations in multi-

D. Dudin
Perm National Research Polytechnic University, Perm, Russia
e-mail: dmitryovj@yandex.ru

I. Keller (✉)
Institute of Continuous Media Mechanics of the UB RAS, Perm, Russia
e-mail: kie@icmm.ru

H. Altenbach et al. (eds.), *Multiscale Solid Mechanics*,
Advanced Structured Materials 141,
https://doi.org/10.1007/978-3-030-54928-2_8

component metal alloys. Such processes occur, for example, when the surface of a machine part is treated with carbonation or nitriding to protect against wear, fatigue, and cracking. The problem of predicting the durability of turbine parts under the combined influence of aggressive media, high temperatures, and mechanical loads remains relevant [4]. Authors of the paper [10] have experimentally studied the decomposition of an Al-12%Mg solid solution under developed plastic deformations. The application of a large shear action in Bridgman anvils for 300 seconds leads to complete decomposition (homogenization) of the solid solution. The possibility of describing this decomposition as an abnormally fast diffusion (9 orders of magnitude faster than the self-diffusion rate) in a coupled mechanodiffusion system with a microstructure within the framework of thermodynamics of irreversible processes is a non-trivial open question that requires some research.

One of the models of a multicomponent structurally in homogeneous medium is proposed by Knyazeva [5–7]. Wilmanski builds a model of mixing in pore-viscoelastic bodies taking into account the evolution of porosity [12]. To study coupled diffusion and rheological processes, Stephenson [11] and later Brassart et al. [1] used a one-dimensional problem that allows a homogeneous stationary solution. Further by perturbing unknown variables, they obtained dependences of relaxation times on the perturbation wavelength that allowed to exhaustively study the physics of the process in asymptotic cases.

The model presented in this paper is based on the phenomenological model Brassart et al. [1] which describes the associated quasi-static diffusion and rheological processes in an isotropic isothermal two-component medium. We generalize this model to the case of a three-component medium with reversible chemical reactions, elastic properties, and microstructure evolution. In order to study the relaxation of small spatial perturbations, a one-dimensional model problem is posed. The physical and geometric relations of the model are represented as linear dependency that is sufficient within the analysis used. We study the asymptotics of the eigenvalue problem corresponding to the relaxation of perturbations by a diffusion or viscous mechanism. The spectrum of interdiffusion coefficients is qualitatively investigated depending on the thermodynamic parameters.

8.2 Extended Brassart Model

We consider a three-component isotropic continuous medium that consists of the particles A, B, AB characterized by atomic concentrations $C_A,\ C_B,\ C_{AB}$ per unit of material volume in the reference configuration. A reversible chemical reaction is permitted according to the chemical equation

$$\nu_A A + \nu_B B \rightleftharpoons \nu_{AB} AB,$$

where $\nu_A,\ \nu_B < 0,\ \nu_{AB} > 0$ are the stoichiometric coefficients of reactants and product of a chemical reaction; A, B, AB are the chemical formulas of the components.

In the medium, microstructure may occur over time and it is determined by the scalar microstructural variable H per unit of material volume in the reference configuration.

8.2.1 Strain Tensor and Its Components

It is assumed that geometrically linear processes occur in the medium. They are described by a small strain tensor $\boldsymbol{\varepsilon}$ consisting of elastic $\boldsymbol{\varepsilon}^e$ and viscous $\boldsymbol{\varepsilon}^v$ parts

$$\boldsymbol{\varepsilon} = \boldsymbol{\varepsilon}^e + \boldsymbol{\varepsilon}^v. \tag{8.1}$$

Each of these tensors are represented by the sum of the mean and deviatoric components

$$\boldsymbol{\varepsilon} = \frac{1}{3}\varepsilon_m \mathbf{I} + \mathbf{e},\ \boldsymbol{\varepsilon}^e = \frac{1}{3}\varepsilon_m^e \mathbf{I} + \mathbf{e}^e,\ \boldsymbol{\varepsilon}^v = \frac{1}{3}\varepsilon_m^v \mathbf{I} + \mathbf{e}^v. \tag{8.2}$$

Denote the swelling ratio Ω

$$\Omega\left(C_A, C_B, C_{AB}, \varepsilon_m^e\right) = 1 + \varepsilon_m \tag{8.3}$$

dependent on the concentration of atoms and volume elastic deformation. It is assumed that this variable is homogeneous function of degree one of concentrations

$$\Omega = V_A\left(\xi_A, \xi_B\right) C_A + V_B\left(\xi_A, \xi_B\right) C_B + V_{AB}\left(\xi_A, \xi_B\right) C_{AB} + \varepsilon_m^e, \tag{8.4}$$

where V_A, V_B, V_{AB} are the partial volumes; $\xi_A = C_A/(C_A + C_B + C_{AB})$, $\xi_B = C_B/(C_A + C_B + C_{AB})$ are the composition variables. Equations (8.3), (8.4) determine the swelling ratio of the material that occurs due to changes concentrations and elastic deformations. Differentiating (8.3) and (8.4) in time leads to

$$\begin{aligned}
\dot{\Omega} &= \frac{\partial \Omega}{\partial C_A}\dot{C}_A + \frac{\partial \Omega}{\partial C_B}\dot{C}_B + \frac{\partial \Omega}{\partial C_{AB}}\dot{C}_{AB} + \dot{\varepsilon}_m^e,\\
\dot{\Omega} &= \dot{V}_A C_A + \dot{V}_B C_B + \dot{V}_{AB} C_{AB} + V_A \dot{C}_A + V_B \dot{C}_B + V_{AB}\dot{C}_{AB} + \dot{\varepsilon}_m^e.
\end{aligned} \tag{8.5}$$

The condition of compatibility of these equations gives

$$\begin{aligned}
&V_A = \frac{\partial \Omega}{\partial C_A},\ V_B = \frac{\partial \Omega}{\partial C_B},\ V_{AB} = \frac{\partial \Omega}{\partial C_{AB}},\\
&\dot{V}_A C_A + \dot{V}_B C_B + \dot{V}_{AB} C_{AB} = 0,
\end{aligned} \tag{8.6}$$

which coincides with the expressions obtained in [1] written without elastic deformations for a two-component medium.

8.2.2 Helmholtz Free Energy

The free energy is assumed by the following function

$$\psi = \psi\left(C_i, H, \boldsymbol{\varepsilon}^e\right),$$
$$\psi = \sum_{k\in\Psi} F_k^{\text{mix}}(\xi_A, \xi_B)C_k + \psi_H(C_i, H) + \psi_e(C_i, H, \varepsilon^e), \tag{8.7}$$

where $\Psi = \{A, B, AB\}$, F_k^{mix} characterize the mixing energy, ψ_H is the free microstructural energy, and ψ_e is the free elastic energy. Differentiating (8.7) in time gives

$$\dot{\psi} = \sum_{k\in\Psi}\left(\dot{F}_k^{mix}C_k + F_k^{\text{mix}}\dot{C}_k + \frac{\partial\psi_H}{\partial C_k}\dot{C}_k + \frac{\partial\psi_e}{\partial C_k}\dot{C}_k\right)$$
$$+\frac{\partial\psi_H}{\partial H}\dot{H} + \frac{\partial\psi_e}{\partial H}\dot{H} + \frac{\partial\psi_e}{\partial\boldsymbol{\varepsilon}^e} : \dot{\boldsymbol{\varepsilon}}^e,$$
$$\dot{\psi} = \sum_{k\in\Psi}\frac{\partial\psi}{\partial C_k}\dot{C}_k + \frac{\partial\psi}{\partial H}\dot{H} + \frac{\partial\psi}{\partial\boldsymbol{\varepsilon}^e} : \dot{\boldsymbol{\varepsilon}}^e \tag{8.8}$$

The compatibility condition of Eq. (8.8) has the following form

$$F_k = \frac{\partial\psi}{\partial C_k} = F_k^{\text{mix}} + \frac{\partial\psi_H}{\partial C_k} + \frac{\partial\psi_e}{\partial C_k},\ k\in\Psi = \{A, B, AB\},$$
$$F_H = \frac{\partial\psi}{\partial H} = \frac{\partial\psi_H}{\partial H} + \frac{\partial\psi_e}{\partial H},\ \mathbf{F}_e = \frac{\partial\psi_e}{\partial\boldsymbol{\varepsilon}^e} = \frac{\partial\psi}{\partial\boldsymbol{\varepsilon}^e}, \tag{8.9}$$

where $F_k,\ k\in\Psi$ are the mixing partial energies; F_H is the microstructural partial energy; and $\mathbf{F}_e$ is the elastic partial energy.

8.2.3 Thermodynamic Inequality

According to the second law of thermodynamics, the free Helmholtz energy of an isolated system does not increase

$$\int_V \dot{\psi}\mathrm{d}V + \int_S (\mu_A\mathbf{J}_A + \mu_B\mathbf{J}_B + \mu_{AB}\mathbf{J}_{AB})\cdot\mathbf{N}\mathrm{d}S - \int_V \boldsymbol{\sigma} : \dot{\boldsymbol{\varepsilon}}\Omega dV \le 0, \tag{8.10}$$

where $\mu_k,\ k\in\Psi$ are the chemical potentials; $\mathbf{J}_k,\ k\in\Psi$ are the diffusion fluxes; $\boldsymbol{\sigma}$ is the Cauchy stress tensor; S and V are surface and volume of a body in the reference configuration.

The conservation equations of atoms are written as

$$\frac{\mathrm{d}C_k}{\mathrm{d}t} = -\nabla \cdot \mathbf{J}_k + P_k,\ k \in \Psi, \tag{8.11}$$

where $P_k,\ k \in \Psi$ is the production of component k during evolution of a chemical reaction.

Applying the divergence theorem to (8.10) with (8.11) gives

$$\int\limits_V \left(\Omega \boldsymbol{\sigma} : \dot{\boldsymbol{\varepsilon}} - \dot{\psi} + \sum_{k \in \Psi} \left(\mu_k \dot{C}_k - \mathbf{J}_k \cdot \nabla \mu_k - P_k \mu_k \right) \right) \mathrm{d}V \geq 0. \tag{8.12}$$

The microstructural variable satisfies a balance equation without conductive term

$$\frac{\mathrm{d}H}{\mathrm{d}t} = P_H, \tag{8.13}$$

where P_H is the production of the microstructural variable.

Taking into account (8.8), (8.9) and (8.13), we localize (8.12) to any arbitrary volume of material that gives

$$\Omega \boldsymbol{\sigma} : \dot{\boldsymbol{\varepsilon}} + \sum_{k \in \Psi} \left((\mu_k - F_k)\dot{C}_k - \mathbf{J}_k \cdot \nabla \mu_k - P_k \mu_k \right) - F_H P_H - \mathbf{F}_e : \dot{\boldsymbol{\varepsilon}}^e \geq 0. \tag{8.14}$$

Given $\boldsymbol{\sigma} : \dot{\boldsymbol{\varepsilon}} = \sigma_m \dot{\varepsilon}_m + \mathbf{s} : \dot{\mathbf{e}}$ and (8.1) where the Cauchy stress tensor is represented by the expansion into the mean and deviatoric parts $\boldsymbol{\sigma} = \sigma_m \mathbf{I} + \mathbf{s}$ the thermodynamic inequality (8.14) takes the form

$$\begin{aligned} &\sum_{k \in \Psi} \left(\left(\frac{\mu_k - F_k}{V_k} + \sigma_m \right) \frac{V_k \dot{C}_k}{\Omega} - \frac{\mathbf{J}_k \cdot \nabla \mu_k}{\Omega} - \frac{P_k}{\Omega} \mu_k \right) - F_H \frac{P_H}{\Omega} + \mathbf{s} : \dot{\mathbf{e}}^v \\ &+ \mathbf{s} : \dot{\mathbf{e}}^e + \sigma_m \dot{\varepsilon}^e_m - \frac{\mathbf{F_e} : \dot{\boldsymbol{\varepsilon}}^e}{\Omega} \geq 0. \end{aligned} \tag{8.15}$$

Further, we write down the thermodynamic inequality in the terms of the current configuration following [1]. We enter the following definitions in the current configuration: concentrations c_k, $C_k = c_k \Omega$, $k \in \Psi$, diffusion flows $\mathbf{J}_k \cdot \nabla \mu_k = \Omega \mathbf{j}_k \cdot \hat{\nabla} \mu_k$, $k \in \Psi$, production of microstructure $p_H = P_H / \Omega$, production of particles during a chemical reaction $p_k = P_k / \Omega = \nu_k \dot{\xi}$, $k \in \Psi$, where $\dot{\xi}$ is the rate of chemical reaction, the rate of volumetric insertion $i_k = V_k \dot{C}_k / \Omega = V_k \dot{c}_k + c_k V_k \hat{\nabla} \cdot \mathbf{v}$, $k \in \Psi$, the chemical affinity $A = \sum \nu_k \mu_k$. Expression (8.15) is rewritten as

$$\sum_{k\in\Psi}\left(\left(\frac{\mu_k - F_k}{V_k} + \sigma_m\right) i_k - \mathbf{j}_k \cdot \hat{\nabla}\mu_k\right) - F_H p_H - A\dot{\xi} + \mathbf{s} : \dot{\mathbf{e}}^v$$
$$+ \mathbf{s} : \dot{\mathbf{e}}^e + \sigma_m \dot{\varepsilon}^e_m - \frac{\mathbf{F_e} : \dot{\boldsymbol{\varepsilon}}^e}{\Omega} \geq 0. \tag{8.16}$$

Physical relations describing dissipative and conservative processes must satisfy (8.16).

8.2.4 Partial Energies and Elastic Equations

In the absence of irreversible processes, (8.16) degenerates into equality

$$\mathbf{s} : \dot{\mathbf{e}}^e + \sigma_m \dot{\varepsilon}^e_m - \frac{1}{\Omega}\mathbf{F}_e : \dot{\boldsymbol{\varepsilon}}^e = 0. \tag{8.17}$$

Taking into account the small deformations, the expression for the elastic strain energy ψ_e can be written as:

$$\psi_e = G\mathbf{e}^e : \mathbf{e}^e + \frac{1}{2}K\varepsilon^{e2}_m, \tag{8.18}$$

where G is the shear modulus, K is the bulk modulus. Further we assumed that $G =$ const and $K =$ const.

Substituting (8.18) with (8.9) into equality (8.17) leads to the relation

$$\left(\mathbf{s} - 2G\mathbf{e}^e\right) : \dot{\mathbf{e}}^e + \left(\sigma_m - K\varepsilon^e_m\right)\dot{\varepsilon}^e_m = 0. \tag{8.19}$$

The independence of a volume velocity and shear elastic strains means the validity of the generalized Hooke law

$$\mathbf{s} = 2G\mathbf{e}^e, \ \sigma_m = K\varepsilon^e_m. \tag{8.20}$$

Following [2] for the case of a three-component medium under the assumption of an ideal mixing of particles, we can write expressions for the mixing energies as

$$F^{\text{mix}}_k = kT\left(\frac{\xi_n\left(V_n - V_k\right) + \xi_j\left(V_j - V_k\right)}{V_m} + \ln\frac{\xi_k V^0_k}{V_m}\right), \tag{8.21}$$

where the indexes can take values $k = A,\ j = AB,\ n = B$ or $k = B,\ j = AB,\ n = A$ or $k = AB,\ j = B,\ n = A$; $V_k = V_k(\xi_A, \xi_B),\ k \in \Psi$ are the partial volumes; $V_m = \xi_A V_A + \xi_B V_B + \xi_{AB} V_{AB}$ is the mean partial volume; $\xi_{AB} = 1 - \xi_A - \xi_B$ is the number fraction of species AB; $V^0_A = V_A(1,0),\ V^0_B = V_B(0,1),\ V^0_{AB} = V_{AB}(0,0)$; k is the

Boltzmann constant; and T is the absolute temperature. The assumption of an ideal mixing of particles is strong and is used here solely for the reason of its simplicity.

In the vicinity of the equilibrium state

$$\psi_H \approx \frac{f_H (H - H_0)^2}{2} - \sum_{k \in \Psi} f_k \left(C_k - C_k^0\right)(H - H_0), \tag{8.22}$$

where $f_A, f_B, f_{AB}, f_H > 0$. Using the condition of small deformations and $c_A V_A + c_B V_B + c_{AB} V_{AB} = 1$ following from (8.4), we write with precision to constants

$$\frac{\partial \psi_H}{\partial H} = f_H h - f_A c_A - f_B c_B - f_{AB} c_{AB}, \quad \frac{\partial \psi_H}{\partial C_k} = -f_k h, \ k \in \Psi, \tag{8.23}$$

where $h = H/\Omega$ is the microstructural variable in the current configuration. Using (8.9) the thermodynamic inequality is finally wrote down

$$F_k = F_k^{mix} - f_k h, \ F_H = f_H h - f_A c_A - f_B c_B - f_{AB} c_{AB}, \ k \in \Psi. \tag{8.24}$$

The constants f_A, f_B, f_{AB} provide an energy connectivity of the model.

8.2.5 Dissipative Equations

We denote scalar thermodynamic flows $J_1^s = i_A, J_2^s = i_B, J_3^s = i_{AB}, J_4^s = p_H, J_5^s = \dot{\xi}$ and corresponding scalar thermodynamic forces $X_1^s = (\mu_A - F_A)/V_A + \sigma_m$, $X_2^s = (\mu_B - F_B)/V_B + \sigma_m$, $X_3^s = (\mu_{AB} - F_{AB})/V_{AB} + \sigma_m$, $X_4^s = -F_H$, $X_5^s = -A$. Similarly we denote vector and tensor flows $\mathbf{J}_1^v = \mathbf{j}_A$, $\mathbf{J}_2^v = \mathbf{j}_B$, $\mathbf{J}_3^v = \mathbf{j}_{AB}$, $\mathbf{J}^t = \dot{\mathbf{e}}^v$ and corresponding forces $\mathbf{X}_1^v = -\hat{\nabla}\mu_A$, $\mathbf{X}_2^v = -\hat{\nabla}\mu_B$, $\mathbf{X}_3^v = -\hat{\nabla}\mu_{AB}$, $\mathbf{X}^t = \mathbf{s}$. The thermodynamic inequality (8.16) with (8.17) and the accepted designations has the form $\sum_{k=1}^5 J_k^s X_k^s + \sum_{k=1}^3 \mathbf{X}_k^v \cdot \mathbf{J}_k^v + \mathbf{X}^t : \mathbf{J}^t \geq 0$. Taking into account the principles of Curie and Onsager, a particular solution of this thermodynamic inequality is found in a quasi-linear form.

For shear flow kinetics, the rheology of a linearly viscous Newtonian fluid is accepted

$$\mathbf{s} = 2\eta \dot{\mathbf{e}}^v, \tag{8.25}$$

where η is the coefficient of shear viscosity. According to (8.20) and (8.25), the Maxwell rheological model is accepted for the deviatoric part of the stress tensor $\mathbf{s}$ in the absence of other processes.

To satisfy the inequality $\sum_{k=1}^3 \mathbf{X}_k^v \cdot \mathbf{J}_k^v \geq 0$, we write the diffusion kinetics without regard to the cross connections

$$\mathbf{j}_k = -c_k M_k \hat{\nabla}\mu_k, \ k \in \Psi, \tag{8.26}$$

where $M_k \geq 0,\ k \in \Psi$ are the mobility coefficients. For scalar thermodynamic forces, quasi-linear solving of inequality $\sum_{k=1}^{5} J_k^s X_k^s \geq 0$ has the form $X_i^s = \sum_{k=1}^{5} \beta_{ik} J_k^s$. Therefore, the kinetics of a chemical reaction is described by the equation

$$A = -\beta_{\xi}\dot{\xi} - \beta_{H\xi} p_H - \sum_{k \in \Psi} \beta_{\xi k} i_k. \tag{8.27}$$

Similarly, the kinetics of microstructure is written as

$$F_H = -\beta_H p_H - \beta_{H\xi}\dot{\xi} - \sum_{k \in \Psi} \beta_{Hk} i_k. \tag{8.28}$$

The last scalar constituative equation provides the connection of diffusion and rheological kinetics as well as the evolution of microstructure and kinetics of a chemical reaction

$$\frac{\mu_k - F_k}{V_k} + \sigma_m = \beta_k i_k + \beta_{\xi k}\dot{\xi} + \beta_{Hk} p_H,\ k \in \Psi. \tag{8.29}$$

To satisfy inequality (8.16), the coefficient matrix β must be positively defined. Coefficients $\beta_k,\ \beta_{\xi k},\ \beta_{H\xi},\ \beta_{Hk},\ k \in \Psi$ are called the volumetric viscosities.

8.2.6 Balance Equations

To close the constitutive equations (8.20), (8.21), (8.24)–(8.29), three balance equations are written. In the current configuration, mass balance equations are

$$\frac{\mathrm{d}c_k}{\mathrm{d}t} + c_k \hat{\nabla} \cdot \mathbf{v} = -\hat{\nabla} \cdot \mathbf{j}_k + \nu_k \dot{\xi},\ k \in \Psi. \tag{8.30}$$

Further we write the balance equation for microstructural variable without conductive term in the current configuration

$$\frac{\mathrm{d}h}{\mathrm{d}t} + h\hat{\nabla} \cdot \mathbf{v} = p_H. \tag{8.31}$$

The last balance equation is the equilibrium equation which is written without taking into account the small inertial term

$$\hat{\nabla} \cdot \boldsymbol{\sigma} = 0. \tag{8.32}$$

System of equations (8.20), (8.21), (8.24)–(8.32) with the relationship $\sum c_k V_k = 1$ is a nonlinear statement of the coupled diffusion equations with accompanying rheological and chemical processes and microstructure evolution.

8.3 Analysis of the Relaxation of Spatial Perturbations

8.3.1 Model Problem

In order to study the relaxation of small spatial perturbations described by the coupled equations, the model problem [1, 11] is considered. The following assumptions are made:

- Particles can only diffuse along a spatial coordinate x

$$c_k = c_k(x, t), \;\; k \in \Psi. \tag{8.33}$$

- All components of the total strain tensor are zero except of

$$\varepsilon_{xx} = \varepsilon(x, t) \neq 0. \tag{8.34}$$

- The components of the stress tensor are zero except of

$$\sigma_{yy} = \sigma_{zz} = \sigma(x, t) \neq 0. \tag{8.35}$$

- Microstructural changes can only pass along x:

$$h = h(x, t). \tag{8.36}$$

With the using hypotheses (8.33)–(8.36) to equations (8.20), (8.21), (8.24)–(8.32), we give a simpler nonlinear one-dimensional statement of the problem which allows to effectively apply the perturbation method. The equilibrium equation (8.32) is satisfied identically by virtue of condition (8.35).

8.3.2 Field Equations

The nonzero components of the elastic and viscous strain deviator are written as $e^e_{xx} = \frac{2}{3}(\varepsilon^e - \varepsilon_\perp)$, $e^e_{yy} = e^e_{zz} = -\frac{1}{3}(\varepsilon^e - \varepsilon_\perp)$ and $e^v_{xx} = \frac{2}{3}(\varepsilon^v + \varepsilon_\perp)$, $e^v_{yy} = e^v_{zz} = -\frac{1}{3}(\varepsilon^v + \varepsilon_\perp)$, that is given with $\varepsilon = \varepsilon^e + \varepsilon^v$, (8.34) and the definition $\varepsilon_\perp = \varepsilon^e_{yy} = \varepsilon^e_{zz} = -\varepsilon^v_{yy} = -\varepsilon^v_{zz}$. Similarly, the nonzero components of the deviatoric part of stress tensor are $s_{xx} = -\sigma_m$, $s_{yy} = s_{zz} = \frac{1}{2}\sigma_m$. Further with using (8.20) and (8.25) expressions are defined $\varepsilon^e = (1/(3K) - 1/(3G))\,\sigma_m$, $\varepsilon_\perp = (1/(3K) + 1/(3G))\,\sigma_m$, $\dot{\varepsilon}^v = -\,(1/(3K) - 1/(3G))\,\dot{\sigma}_m - 3/(4\eta)\sigma_m$, $\dot{\varepsilon}^v_m = -\,(1/K + 3/(4G))\,\dot{\sigma}_m - 3/(4\eta)\sigma_m$, $\dot{\varepsilon} = -3\dot{\sigma}_m/(4G) - 3\sigma_m/(4\eta)$. The last equation is written in terms of the velocity $v(x, t)$ of the medium material points along the spatial coordinate x

$$\frac{\partial v}{\partial x} = -\frac{3}{4G}\frac{\partial \sigma_m}{\partial t} - \frac{3\sigma_m}{4\eta}. \tag{8.37}$$

Because the velocity v is a small value we accept $d/dt \approx \partial/\partial t$.

From the solution analysis of the coupled diffusion problem with chemical reactions and viscoelastic deformations of a three-component medium, it follows that chemical reactions do not affect on the fast interdiffusion coefficient. Therefore, the diffusion problem for a two-component viscoelastic medium with microstructure evolution is considered further. Field equations (8.21), (8.24), (8.26), (8.30) and (8.31) take the form of

$$\frac{\partial c_k}{\partial t} + c_k\frac{\partial v}{\partial x} = -\frac{\partial j_k}{\partial x},\ i_k = V_k\frac{\partial c_k}{\partial t} + c_k V_k\frac{\partial v}{\partial x},\ k \in \Psi^1 = \{A, B\}, \tag{8.38}$$

$$\frac{\partial h}{\partial t} + h\frac{\partial v}{\partial x} = p_H,\ \frac{\partial v}{\partial x} = -\frac{3}{4G}\frac{\partial \sigma_m}{\partial t} - \frac{3\sigma_m}{4\eta},\ j_k = -c_k M_k\frac{\partial \mu_k}{\partial x},\ k \in \Psi^1, \tag{8.39}$$

$$F_k^{\text{mix}} = kT\left(\frac{\xi_j(V_j - V_k)}{V_m} + \ln\frac{\xi_k V_k^0}{V_m}\right),\ \{k, j\} \in \Psi^2 = \{A, B;\ B, A\}, \tag{8.40}$$

$$F_k = F_k^{\text{mix}} - f_k h,\ F_H = f_H h - f_A c_A - f_B c_B,\ k \in \Psi^1. \tag{8.41}$$

They are closed by the material equations (8.28), (8.29) with $\beta_{H\xi},\ \beta_{\xi k} = 0$ and $\sum c_k V_k = 1$.

8.3.3 Perturbation Method

The above system of coupled nonlinear equations has a homogeneous stationary solution

$$\sigma_m(x, t) \equiv 0,\ h(x, t) \equiv h_0,\ c_A(x, t) \equiv c_A^0,\ c_B(x, t) \equiv c_B^0, \tag{8.42}$$

which corresponds to some equilibrium state.

The spectrum of relaxation times is determined using the perturbation method according to which the system of coupled equations is linearized in the neighborhood of the state (8.42). From the solution of a nonlinear homogeneous stationary problem, it follows $h_0 = (f_A c_A^0 + f_B c_B^0)/f_H$. Further we assume $\beta_k = \beta,\ \beta_{Hk} = \beta_{Hi},\ k \in \Psi^1$. As a result, we get a system of three linear differential equations with constant coefficients which at $\beta,\ \beta_{iH} \equiv 0$ has the form

$$\frac{1}{\phi_A^0}\frac{\partial \phi_A}{\partial t} - \frac{3}{4G}\left(\frac{\partial \sigma_m}{\partial t} + \frac{G\sigma_m}{\eta}\right)$$
$$-kT\varXi\frac{M_A}{\phi_A^0}\frac{V_m}{V_B}\frac{\partial^2 \phi_A}{\partial x^2} + M_A V_A\frac{\partial^2 \sigma_m}{\partial x^2} + f_A M_A\frac{\partial^2 h}{\partial x^2} = 0,$$
$$\frac{1}{\phi_B^0}\frac{\partial \phi_A}{\partial t} + \frac{3}{4G}\left(\frac{\partial \sigma_m}{\partial t} + \frac{G\sigma_m}{\eta}\right) + kT\varXi\frac{M_B}{\phi_B^0}\frac{V_m}{V_A}\frac{\partial^2 \phi_A}{\partial x^2}$$
$$-M_B V_B\frac{\partial^2 \sigma_m}{\partial x^2} - f_B M_B\frac{\partial^2 h}{\partial x^2} = 0,$$
$$\frac{\partial h}{\partial t} - \frac{3h_0}{4G}\left(\frac{\partial \sigma_m}{\partial t} + \frac{G\sigma_m}{\eta}\right)$$
$$+\left(\frac{f_B}{V_B} - \frac{f_A}{V_A}\right)\frac{\phi_A}{\beta_H} + \frac{f_H}{\beta_H}h = 0,$$

where $\phi_A = c_A V_A$ is the volume fraction of species A, $\varXi = (V_A\phi_A^0 + V_B\phi_B^0)/V_m$.

In the vicinity of a homogeneous stationary solution (8.42), perturbations are superimposed on the unknown variables

$$\begin{aligned}\phi_A(x,t) &= \phi_A^0 + \hat{\phi}\exp\left(-\tfrac{t}{\tau}\right)\sin\left(\tfrac{2\pi x}{\lambda}\right),\\ \sigma_m(x,t) &= \hat{\sigma}\exp\left(-\tfrac{t}{\tau}\right)\sin\left(\tfrac{2\pi x}{\lambda}\right),\\ h(x,t) &= h_0 + \hat{h}\exp\left(-\tfrac{t}{\tau}\right)\sin\left(\tfrac{2\pi x}{\lambda}\right),\end{aligned} \tag{8.43}$$

where $\hat{\phi}$, $\hat{\sigma}$, $\hat{h} \in R$, $\left|\hat{\phi}\right|$, $\left|\hat{\sigma}\right|$, $\left|\hat{h}\right| \ll 1$, τ is the relaxation time, λ is the wavelength of the perturbation. The imposed thermodynamic constraints provide the real value of eigenvalues and eigenvectors.

Substituting relations (8.43) into a linearized system of equations leads to an eigenvalue problem whose solution is characterized by three branches $\tau = \tau_k(\lambda)$ and the corresponding eigenvectors $\mathbf{u}_k$. Nonzero components of the eigenvector for a given λ characterize the presence of the influence of the corresponding process on the system relaxation behavior. If $u_{k1} \neq 0$, the relaxation process is influenced by the diffusion kinetics. If $u_{k2} \neq 0$, it is influenced by the stress state and if $u_{k3} \neq 0$, the kinetics of microstructure influences on the relaxation process. The perturbation method technique allows to study the physics of relaxation processes in coupled systems depending on various factors without resorting to numerical analysis.

8.3.4 Relaxation Times of Perturbations and Their Asymptotes

Dependencies $\tau = \tau_k(\lambda)$, $k = 1, 2, 3$ are obtained but they are not given here because of their bulkiness. Further reasoning is carried out under the assumption

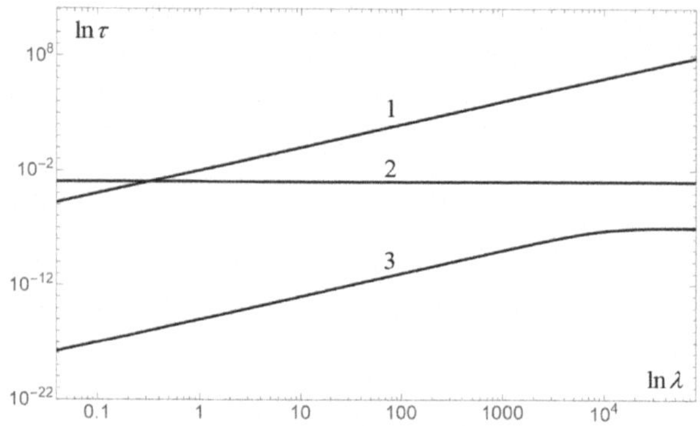

Fig. 8.1 Relaxation times spectrum of the coupled system depending on perturbation wavelength

that the Lyapunov stability condition fulfills. The plot $\ln\tau = \ln\tau_k(\ln\lambda),\ k = 1, 2, 3$ at $\beta,\ \beta_{Hi} \equiv 0$ is shown in Fig. 8.1. Diffusive and viscous behavior in the coupled system is observed in the asymptotic cases at $\lambda \to 0$ and $\lambda \to \infty$.

The diffusion asymptote corresponds to the sloping line with a tangent equal to 2 in Fig. 8.1 and the model

$$\frac{\partial \mathbf{u}_p}{\partial t} = D_p \frac{\partial^2 \mathbf{u}_p}{\partial x^2},$$

where $\mathbf{u}_p$ is the corresponding eigenvector and D_p is the coefficient defined by dependence $\tau_p(\lambda)$. This coefficient in coupled systems is called interdiffusion coefficient [1, 3, 8, 9]. The viscous asymptote corresponds to the horizontal line in Fig. 8.1 and the model

$$\frac{\partial \mathbf{u}_p}{\partial t} = -\frac{\mathbf{u}_p}{\tau_p},$$

where $\mathbf{u}_p$ is the corresponding eigenvector and τ_p is the coefficient defined by dependence $\tau_p(\lambda)$.

The volumetric viscosities $\beta,\ \beta_{Hi}$ control the relaxation process at $\lambda \to 0$ that leads to the appearance of viscous horizontal asymptotes in the small neighborhood of $\lambda = 0$. For $\lambda \to \infty$, these viscosities do not influence on perturbation relaxation in any way. Further we assume that $\beta,\ \beta_{Hi} \equiv 0$.

For $\tau = \tau_1(\lambda)$ at $\lambda \to \infty$, the stress gradient does not affect on the diffusion flows of the components in any way but the interdiffusion coefficient equals to

$$D_1^{\infty} = kT\,\Xi\,(M_A\xi_B^0 + M_B\xi_A^0) + \left(\frac{f_B}{V_B} - \frac{f_A}{V_A}\right)\frac{\phi_A^0\phi_B^0}{f_H}(M_A f_A - M_B f_B), \qquad (8.44)$$

where $\Xi = (V_A\phi_A^0 + V_B\phi_B^0)/V_m$. The eigenvector is $\mathbf{u}_1^{\infty} = (\hat{\phi}_1^{\infty},\ 0,\ \hat{h}_1^{\infty})$. The coefficient (8.44) corresponds to the sequential connection of diffusion structural elements. If $M_A \gg M_B$, the rate of interdiffusion is determined by the fastest component A. According to (8.44), a change in the microstructure can both accelerate and slow down the process of interdiffusion at long-wave perturbations. In the case of low energy coupling $h_0 f_k \ll kT,\ k \in \Psi^1$, the expression (8.44) is reduced to the Darken's interdiffusion coefficient [3].

If the characteristic thermal kT and elastic GV_m energies are comparable, the interdiffusion coefficients D_2^0, D_3^0 have a complex dependence on the microstructure, elastic, and diffusion properties of the system and therefore, they are not given here. The corresponding eigenvectors have the form $\mathbf{u}_{2,3}^0 = \left(\hat{\phi}_{2,3}^0, \hat{\sigma}_{2,3}^0, \hat{h}_{2,3}^0\right)^{\mathrm{T}}$. If $GV_m \gg kT$, the interdiffusion coefficients

$$D_2^0 = \frac{kT\,\Xi M_A M_B V_m\left(\frac{4G}{3} + f_H h_0^2\right)}{\frac{4G}{3}(M_A V_A \phi_A^0 + M_B V_B \phi_B^0) + h_0(f_A M_A \phi_A^0 + f_B M_B \phi_B^0)}, \tag{8.45}$$

$$D_3^0 = \frac{4G}{3}(M_A V_A \phi_A^0 + M_B V_B \phi_B^0) + h_0(f_A M_A \phi_A^0 + f_B M_B \phi_B^0), \tag{8.46}$$

at that $D_3^0 \gg D_2^0$. Here, diffusion flows are moderated by stress and microstructure density gradients. So (8.46) predicts much faster homogenization of components then conventional diffusion mechanism. The coefficient (8.46) is determined by the sequential connection of the diffusion structural elements and (8.45) is defined by their parallel connection. If $M_A \gg M_B$, then the rate of diffusion is limited by the slow component B. For $h_0 f_k \ll GV_m,\ k \in \Psi^1$, the expression (8.45) is reduced to the Nazarov-Gurov interdiffusion coefficient [8, 9]. In nanostructured metal alloys, there is a large energy coupling $h_0 f_k \gg GV_m,\ k \in \Psi^1$ that allows to overwrite (8.45) and (8.46) in the following form

$$D_2^0 = \frac{kT\,\Xi M_A M_B V_m f_H h_0}{f_A M_A \phi_A^0 + f_B M_B \phi_B^0},\quad D_3^0 = h_0\left(f_A M_A \phi_A^0 + f_B M_B \phi_B^0\right). \tag{8.47}$$

The interdiffusion coefficients (8.47) are determined only by the microstructure energy.

For $\lambda \to 0$, $kT \gg GV_m$, the interdiffusion coefficients are

$$D_{2,3}^0 = \frac{kT\,\Xi}{2}(M_A \xi_B^0 + M_B \xi_A^0) + \frac{h_0}{2}(f_A M_A \phi_A^0 + f_B M_B \phi_B^0)$$
$$\mp\frac{1}{2}\sqrt{(kT\,\Xi(M_A \xi_B^0 + M_B \xi_A^0) + h_0(f_A M_A \phi_A^0 + f_B M_B \phi_B^0))^2 - 4kTM_A M_B \Xi V_m f_H h_0^2}.$$

Although the coefficients G and η are lacked in $D_{2,3}^0$, the stress gradient affects on mass transfer. Additionally assuming a small energy coupling $h_0 f_k \ll kT,\ k \in \Psi^1$, we have

$$D_2^0 = \left(\frac{4G}{3} + f_H h_0^2\right)\frac{M_A M_B V_m}{M_A \xi_B^0 + M_B \xi_A^0}, \tag{8.48}$$

$$D_3^0 = kT\,\Xi(M_A \xi_B^0 + M_B \xi_A^0), \tag{8.49}$$

where $D_2^0 \ll D_3^0$ that characterizes slow interdiffusion. The coefficient (8.48) corresponds to the parallel connection of the Fick bodies.

Asymptotes of the second and the third branches at $\lambda \to \infty$ are characterized by viscous relaxation times

$$\tau_2^\infty = \frac{\beta_H}{f_H}, \quad \tau_3^\infty = \frac{\eta}{G}$$

with the corresponding eigenvectors $\mathbf{u}_2^\infty = (0, 0, \hat{h}_2^\infty)^{\mathrm{T}}, \mathbf{u}_3^\infty = (0, \hat{\sigma}_3^\infty, 0)^T$, i.e., there is not coupling processes here.

For asymptote of the first branch at $\lambda \to 0$, the viscous relaxation time

$$\tau_1^0 = \frac{3kT}{4G} \frac{\beta_H (V_A \phi_A^0 + V_B \phi_B^0) \left(\frac{4G}{3} + f_H h_0^2\right)}{kT f_H (V_A \phi_A^0 + V_B \phi_B^0) \left(1 + \frac{3\beta_H}{4\eta} h_0^2\right) - \left(\frac{f_B}{V_B} - \frac{f_A}{V_A}\right)^2 \phi_A^0 \phi_B^0} \tag{8.50}$$

exists with the eigenvector $\mathbf{u}_1^0 = (\hat{\phi}_1^0, \hat{\sigma}_1^0, \hat{h}_1^0)^T$. Although the rate of relaxation in (8.50) does not depend on the diffusion characteristics of the system, the concentrations of substances relax to an equilibrium state through diffusion flows.

8.4 Conclusion

The perturbation method is applied to a one-dimensional nonlinear problem based on a model of coupled quasi-static rheological and diffusion processes with elastic properties and changes in the microstructure in a two-component medium. It gives a spectrum of relaxation times of small spatial perturbations and corresponding eigenvectors that allow to analyze possible physical processes in asymptotic cases. The sloping asymptotes of relaxation times can be compared with the interdiffusion coefficients. The simplest of them conforms to a sequential or parallel connection of diffusion structural elements. Horizontal asymptotes are controlled by the viscous properties of the medium. Mathematical analysis of the problem is effectively carried out with using computer algebra systems. This technique was used earlier by Stephenson [11] and Brassart et al. [1] for models of coupled diffusion and rheological processes.

The presence of changes in the microstructure as well as elastic deformations significantly affected the rate of diffusion. When the characteristic microstructure energy and the characteristic elastic energy significantly exceed the characteristic thermal energy, the fast interdiffusion is realized. The fast interdiffusion occurs together with the conventional thermal interdiffusion and the viscous relaxation. In the opposite case, when the characteristic thermal energy significantly exceeds the characteristic elastic energy and the microstructure energy, a slow interdiffusion occurs. Fast and slow diffusion flows are determined by the gradient of the microstructural parameter and the mean stress value.

Acknowledgements This work was performed in the framework of the state assignment (reg. No. AAAA-A16-116121410009-8) and under partial financial support by the grant of the Russian Foundation for Basic Research (project No. 17-08-01085).

References

1. Brassart, L., Liu, Q., Suo, Z.: Mixing by shear, dilation, swap and diffusion. J. Mech. Phys. Solids **112**, 253–272 (2018)
2. Brassart, L., Liu, Q., Suo, Z.: Shear, dilation and swap: mixing in the limit of fast diffusion. J. Mech. Phys. Solids **96**, 48–64 (2016)
3. Darken, L.S.: Diffusion, mobility and their interrelation through free energy in binary metallic systems. Trans. Am. Inst. Min. Metall. Eng. **175**, 184–201 (1948)
4. Fomin, L.V., Basalov, Y.G., Lokoshchenko, A.M.: On accouning the influence of corrosion processes on delayed fracture of a rod under creep condition. Mech. Compos. Mater. Struct. **25**(3), 327–335 (2019)
5. Knyazeva, A.G.: Cross effects in solid media with diffusion. J. Appl. Mech. Techn. Phys. **44**(3), 373–384 (2003)
6. Knyazeva, A.G.: Model of medium with diffusion and internal surfaces and some applied problems. Mater. Phys. Mech. **7**(1), 29–36 (2004)
7. Knyazeva, A.G.: Diffusion by the vacancy mechanism in the materials with the large number of internal surfaces. Chem. Sustain. Dev. **2**, 233–242 (2005)
8. Mehrer, H.: Diffusion in Solids. Solid State Sciences, Heidelberg (2007)
9. Nazarov, A.V., Gurov, K.P.: The kinetic theory of interdiffusion in binary system. Concentration of vacancies during mutual diffusion. Phys. Metals Metall. **37**, 496–503 (1974)
10. Straumal, B.B., Baretzky, B., Mazilkin, A.A., et al.: Formation of nanograined structure and decomposition of supersaturated solid solution during high pressure torsion of Al-Zn and Al-Mg alloys. Acta Mater. **52**(15), 4469–4478 (2004)
11. Stephenson, G.B.: Deformation during interdiffusion. Acta Metall. **36**, 2663–2683 (1988)
12. Wilmanski, K.: Continuum Thermodynamics. Part I: Foundations. Series on advances in mathematics for applied sciences, Singapore (2008)

Chapter 9
Linear and Nonlinear Problems of Wave Resistance to the Movement of Objects Along Elastic Guides

Vladimir I. Erofeev, Sergey I. Gerasimov, Elena E. Lisenkova, Alexey O. Malkhanov, and Vladimir M. Sandalov

Abstract The work is devoted to the study of wave resistance to the movement of the load along a flexible guide lying on a linearly elastic foundation or on a nonlinearly elastic foundation. In the case of a rigid type of nonlinearity, the frequency of transmission and critical speeds of movement of the load are determined, when passing through which the picture of wave formation changes qualitatively. At the source frequency lying in the reject band, the constant component of the wave resistance is found and its dependence on the speed of the load is investigated. The problem of an elastic guide experiencing a moving object, as a dynamic controlled system, is posed and solved. The general patterns inherent in the waves propagating in one-dimensional elastic systems are revealed. The local laws of energy transfer and wave momentum are given in the case when the Lagrangian of the elastic system depends on the generalized coordinates and their derivatives up to the second order inclusive. It is shown that in a reference frame moving with a phase velocity, the ratio of the energy flux density to the wave pulse flux density is equal to the phase velocity. It is established that for systems whose dynamic behavior is described by linear equations or nonlinear with respect to an unknown function, the ratio of the average values of

V. I. Erofeev (✉) · E. E. Lisenkova · A. O. Malkhanov
Mechanical Engineering Research Institute of RAS, 85 Belinskogo Str, Nizhny Novgorod, Russian Federation 603024
e-mail: erof.vi@yandex.ru

E. E. Lisenkova
e-mail: EELissen@yandex.ru

A. O. Malkhanov
e-mail: alexey.malkhanov@gmail.com

S. I. Gerasimov
Russian Federal Nuclear Center–All-Russian Research Institute of Experimental Physics, 37, Mira Ave, Sarov, Nizhny Novgorod Region, Russian Federation 607188
e-mail: s.i.gerasimov@mail.ru

V. M. Sandalov
National Research Lobachevsky State University of Nizhny Novgorod, 23 Gagarin Avenue, Nizhny Novgorod, Russian Federation 603950
e-mail: sandalovvm@gmail.com

H. Altenbach et al. (eds.), *Multiscale Solid Mechanics*,
Advanced Structured Materials 141,
https://doi.org/10.1007/978-3-030-54928-2_9

the energy flux density to the wave pulse density is equal to the product of the phase and group wave velocities.

Keywords Flexible guide · Nonlinear elastic foundation · Wave · Moving load · Wave resistance

9.1 Introduction

The movement of loads along elastic guides has traditionally been given much attention by researchers [1–15]. This interest is associated with a wide range of practical applications. The problem itself arose more than 150 years ago when applied to the problems of bridge dynamics [16]. In the nineteenth century, the corresponding problems were solved in a quasistatic approximation, without taking into account the inertial properties of the elastic system. From the perception of the dynamics of elastic systems, these problems in the first third of the twentieth century began to be solved by Krylov [17] and Timoshenko [18]. They proposed to replace the action of a moving system with a moving force that reflects in the appearance in the right side of the equation of dynamics of the elastic system the Dirac delta function. All subsequent successes of the theory and its technical applications were largely due to this approach (see, for example, [1–3, 5]).

The drawback of the Krylov–Timoshenko approach is that, replacing the moving system with a moving force, we are obliged to make a number of, generally speaking, unsubstantiated assumptions, which are usually the following:

- The magnitude of the force is assumed to be known (e.g., equal to the weight of the moving system).
- Its direction is perpendicular to the unperturbed surface of the elastic guide.
- Law of motion along the elastic guide is assumed to be known.
- Width of the zone of the moving contact is also assumed to be known.

In the early 1980s, A.I. Vesnitsky and his school proposed an approach [13, 19] that made it possible for the first time to physically and mathematically correctly formulate the problems of self-consistent motion of distributed elastic and concentrated systems, which moves along the latter. This problem statement allows to reject the above assumptions and do not replace them with others.

The new approach led to a new look at the problem, revealing two fundamentally important factors that were not previously taken into account: the presence of convective (inertial) forces and pressure forces of elastic waves in a moving contact. Rayleigh [20] and Nikolai [21] pointed out the existence of the latter. Considering these factors not only made it possible to clarify the previously existing theory of forced oscillations and resonance in elastic structures carrying moving loads, but also opened up solutions to a number of other urgent problems. Among them are sliding and rolling resistance due to wave formation; collision of bodies taking into account slippage; stability of systems moving along elastic guides; motion of bodies under the influence of wave pressure.

9.2 Resistance to the Movement Along an Elastic Load Guide with Its Own Degrees of Freedom

The appearance of wave resistance to the movement of loads is directly related to the emission of elastic waves. This issue was studied in [19, 22], where constant or variable sources of forces were used as loads. Below, we consider the resistance to the movement of the load (crew), which has its own degrees of freedom, along the elastic guide. A string lying on an elastic basement acts as a guide. Let us suppose that crew represents two elastically connected masses M and m, located one above the other. The movement of the crew happens by the law $x = \ell(t)$. Let us also suppose that vertical displacement of the upper mass M is defined by the function $y(t)$, and similar displacements of the lower mass are always equal to the deflections of the string $u(x, t)$ in the points of its contact $x = \ell(t)$ with the crew ${}^0u(t)$, i.e.,

$$
{}^1u(\ell(t), t) = {}^2u(\ell(t), t) = {}^0u(t), \tag{9.1}
$$

The equations of motion of such a system can be written in the following form:

$$
\begin{aligned}
&u_{tt} - c_0^2 u_{xx} + \omega_*^2 u = 0,\\
&\ddot{y} + \Omega_0^2\left(y - {}^0 u\right) = P_0 e^{i(\Omega t + \phi)},\\
&{}^0\ddot{u} + \omega_0^2\left({}^0u - y\right) = \mu\left[c_0^2 u_x + \dot{\ell} u_t\right] + Q_0 e^{i\Omega t},
\end{aligned} \tag{9.2}
$$

$$
m\ddot{\ell} = -\left[\frac{1}{2}\rho u_t^2 + \frac{1}{2} N u_x^2 - \frac{1}{2} k u^2 + \rho\dot{\ell} u_t u_x\right] + R \tag{9.3}
$$

supplemented with conditions given at infinity:

$$
\left|{}^\nu u(x, t)\right| < W < \infty, \quad x \to \pm\infty, \quad {}^1S/{}^1h < \dot{\ell}, \quad {}^2S/{}^2h > \dot{\ell}
$$

which guarantees the absence of any other sources, including at infinity. Here, $c_0 = \sqrt{N/\rho}$—the speed of wave propagation in a string without taking into account the basement, ω_*—lowest string vibration frequency due to elastic base, $\Omega_0 = \sqrt{k_0/M}$, $\omega_0 = \sqrt{k_0/m}$—partial vibration frequencies of the object, determined without taking into account the influence of the string, $\mu = \rho/m$ ρ—linear string density, Ω—the frequency of forcing impact, $Q_0 = q_{01}/m$, $P_0 = q_{02}/M$, q_{01}, q_{02}—driving force amplitudes, applied to the lower and upper masses, respectively, ϕ—phase shift, S/h—energy transfer velocity [21], h—energy density, S—energy flux density. Square brackets mean the difference between the limit values to the right and left of the contact point.

Equation (9.3) describes the movement of the crew along the string under the action of an external force R and the longitudinal component of the reaction force from the side of the guide (wave pressure forces [19])

$$F_{\text{pr}} = -\left[\frac{1}{2}\rho u_t^2 + \frac{1}{2}N u_x^2 - \frac{1}{2}k u^2 + \rho \dot{\ell} u_t u_x\right].$$

Consequently, moving at a constant speed ($\dot{\ell} = V = \text{const}$), the crew experiences resistance to movement from the side of the elastic system with force

$$F = -F_{\text{pr}} = \left[\frac{1}{2}\rho u_t^2 + \frac{1}{2}N u_x^2 - \frac{1}{2}k u^2 + \rho V u_t u_x\right]. \tag{9.4}$$

We will search for functions that describe the steady-state forced oscillations of the system in the form

$$^{\nu}u(x,t) = {}^{\nu}A \exp\{i({}^{\nu}\omega t - {}^{\nu} kx)\},\ y(t) = B\exp(i\Omega t).$$

We obtain the following exact solution to problem (9.1)–(9.3) depending on the speed of the crew.

When $V < V_* = c_0\sqrt{1 - (\Omega/\omega_*)^2}$, $0 \le \Omega < \omega_*$, the wave numbers and frequencies of the waves to the left and right of the moving source are complex. Consequently, the field of transversal displacements is localized near the crew decreases with distance growth:

$$u(x,t) = \begin{cases} A\exp\left(-\omega_*\sqrt{V_*^2 - V^2}\left(c_0^2 - V^2\right)^{-1}(Vt - x)\right)\cos\left(\Omega\left(c_0^2 t - Vx\right)\left(c_0^2 - V^2\right)^{-1} + \psi_1\right), x < Vt \\ A\exp\left(-\omega_*\sqrt{V_*^2 - V^2}\left(c_0^2 - V^2\right)^{-1}(x - Vt)\right)\cos\left(\Omega\left(c_0^2 t - Vx\right)\left(c_0^2 - V^2\right)^{-1} + \psi_1\right), x > Vt \end{cases},$$

$$A = \Delta_A \Delta^{-1},$$

$$\Delta_A = \sqrt{P_0^2\omega_0^4 + 2Q_0P_0\omega_0^2(\Omega_0^2 - \Omega^2)\cos\phi + Q_0^2(\Omega_0^2 - \Omega^2)^2},$$

$$\Delta = -\omega_0^2\Omega^2 + (\Omega_0^2 - \Omega^2)\left(-\Omega^2 + 2\mu\omega_*\sqrt{V_*^2 - V^2}\right),$$

$$\psi_1 = 2\operatorname{arctg}\left(\frac{\Delta_A - Q_0(\Omega_0^2 - \Omega^2) - P_0\omega_0^2\cos\phi}{P_0\omega_0^2\sin\phi}\right).$$

Oscillation of the upper mass of the crew is described with the help of the function:

$$y(t) = \Delta_B \Delta^{-1}\cos(\Omega t + \psi),$$

$$\Delta_B = \left\{P_0^2\left(\omega_0^2 - \Omega^2 + 2\mu\omega_*\sqrt{V_*^2 - V^2}\right)^2\right.$$

$$+2P_0Q_0\Omega_0^2\left(\omega_0^2-\Omega^2+2\mu\omega_*\sqrt{V_*^2-V^2}\right)\cos\phi+Q_0^2\Omega_0^4\Big\}^{1/2},$$

$$\psi=2\mathrm{arctg}\left(\frac{\Delta_B-Q_0\Omega_0^2-P_0\left(\omega_0^2-\Omega^2+2\mu\omega_*\sqrt{V_*^2-V^2}\right)\cos\phi}{P_0\left(\omega_0^2-\Omega^2+2\mu\omega_*\sqrt{V_*^2-V^2}\right)\sin\phi}\right).$$

Note that the amplitudes of the oscillations of masses M and m can take both unlimited and zero values, i.e., for them in the range of parameters $V < V_* = c_0\sqrt{1-(\Omega/\omega_*)^2}$ resonance and antiresonance phenomena are possible. From the above formulas, it follows that unlimited displacements are achieved at speeds

$$V=\sqrt{V_*^2-\frac{\Omega^4\left(\omega_0^2+\Omega_0^2-\Omega^2\right)^2}{4\mu^2\omega_*^2\left(\Omega_0^2-\Omega^2\right)^2}}<V_*.$$

The absence of fluctuations in the lower mass of the crew is possible in the following cases:

- Transverse force does not act on the upper mass of the crew $P_0=0$, and the frequency of the force on the lower one coincides with the natural frequency of the upper mass $\Omega=\Omega_0$.
- Reduced amplitudes of forces acting on both masses are related by the relation $\frac{P_0}{Q_0}=\pm\frac{\Omega^2-\Omega_0^2}{\omega_0^2}$.

The upper sign corresponds to the situation when there is no phase shift between the forces ($\phi=0$) and $\Omega>\Omega_0$, and the bottom—$\phi=\pi$ and $\Omega<\Omega_0$.

Dynamic damping of fluctuations in the upper mass of the crew occurs, for example, in the case when the transverse force does not act on the lower mass $Q_0=0$, and the crew moves at a speed

$$V=\sqrt{V_*^2-\frac{\Omega^4}{4\mu^2\omega_*^2}}<V_*.$$

The deflection profile of the string is symmetrical with respect to moving crew with velocity $V<V_*$, as a result, the constant component of the wave resistance force is zero.

In all other cases, the constant component of the resistance to movement is nonzero. This is due to the asymmetry of the profile of the deflection of the guide due to the excitation of waves, the frequencies and wave numbers of which are determined by the formulas

$$\omega_{1,2}=\frac{c_0^2\Omega\pm V\sqrt{V^2\omega_*^2-c_0^2\omega_*^2+\Omega^2c_0^2}}{c_0^2-V^2},$$

$$k_{1,2} = \frac{V\Omega \pm \sqrt{V^2\omega_*^2 - c_0^2\omega_*^2 + \Omega^2 c_0^2}}{c_0^2 - V^2}.$$

When $V_* < V < c_0$ $(0 \le \Omega < \omega_*)$, the so-called deceleration emission occurs when a stationary source does not excite waves, since $\Omega < \omega_*$, but moving at a speed $V > V_*$, begins to excite them which leads to the appearance of the breaking effect. One of the excited waves (high-frequency) runs in front of the source, and the other (low-frequency) follows him. As a result, after finding the amplitudes of these waves, the deflection function will have the form:

$$u(x,t) = \begin{cases} A\cos(\omega_1 t - k_1 x + \psi_1) \text{when } x < Vt \\ A\cos(\omega_2 t - k_2 x + \psi_1) \text{when } x > Vt \end{cases}. \quad (9.5)$$

Here,

$$A = \Delta_A \Delta^{-1},$$

$$\Delta_A = \sqrt{P_0^2\omega_0^4 + 2Q_0 P_0 \omega_0^2(\Omega_0^2 - \Omega^2)\cos\phi + Q_0^2(\Omega_0^2 - \Omega^2)^2},$$

$$\Delta = \left\{(\omega_0^2 + \Omega_0^2 - \Omega^2)^2\Omega^4 + 4\mu^2\omega_*^2(\Omega_0^2 - \Omega^2)^2(V^2 - V_*^2)\right\}^{1/2},$$

$$\psi_1 = 2\text{arctg}\frac{\Delta_A\Delta - (\alpha_1\gamma + \beta_1\delta)}{\beta_1\gamma - \alpha_1\delta}.$$

The following law guides the oscillations of the upper mass:

$$y(t) = \sqrt{\alpha^2 + \beta^2}\Delta^{-1}\cos(\Omega t + \psi),$$

where

$$\begin{aligned} \alpha_1 &= Q_0(\Omega_0^2 - \Omega^2) + P_0\omega_0^2\cos\phi, \\ \beta_1 &= P_0\omega_0^2\sin\phi, \\ \gamma &= -(\omega_0^2 + \Omega_0^2 - \Omega^2)\Omega^2, \\ \delta &= 2\mu\omega_*(\Omega_0^2 - \Omega^2)\sqrt{V^2 - V_*^2}, \end{aligned}$$

$$\psi = 2\text{arctg}\frac{\sqrt{\alpha^2 + \beta^2}\Delta - (\alpha\gamma + \beta\delta)}{\beta\gamma - \alpha\delta},$$

$$\alpha = Q_0\Omega_0^2 + (\omega_0^2 - \Omega^2)P_0\cos\phi - 2\mu\omega_*\sqrt{V^2 - V_*^2}P_0\sin\phi,$$

$$\beta = \left(\omega_0^2 - \Omega^2\right) P_0 \sin\phi + 2\mu\omega_* \sqrt{V^2 - V_*^2} P_0 \cos\phi.$$

Since when $V_* < V$ in the system, there is the radiation of energy and, consequently, its loss, this leads to a limitation of resonance peaks.

Substituting (5) into (4), and averaging over the object oscillation frequency, we obtain the expression for the constant component of the wave resistance force

$$\langle F \rangle = \frac{\rho \Omega V \omega_* \sqrt{V^2 - V_*^2} \Delta_A^2}{\left(c_0^2 - V^2\right) \Delta^2}. \tag{9.6}$$

When $0 \le \Omega < \omega_*$, $V > c_0$, unlike the previous case, both waves run after the source, followed by it, and the solutions will look like:

$$u(x,t) = \begin{cases} A\{\cos(\omega_1 t - k_1 x + \psi_1) - \cos(\omega_2 t - k_2 x + \psi_1)\}, & x < Vt \\ 0, & x > Vt \end{cases},$$

$$A = \Delta_A \Delta^{-1},$$

$$\Delta_A = \left\{ Q_0^2 \left(\Omega_0^2 - \Omega^2\right)^2 + 2 P_0 Q_0 \omega_0^2 \left(\Omega_0^2 - \Omega^2\right) \cos\phi + P_0^2 \omega_0^4 \right\}^{1/2},$$

$$\Delta = 2\left(\Omega_0^2 - \Omega^2\right) \mu \omega_* \sqrt{V^2 - V_*^2},$$

$$\psi_1 = 2\mathrm{arctg}\left(\frac{P_0 \omega_0^2 \sin\phi - \Delta_A}{\left(\Omega_0^2 - \Omega^2\right) Q_0 + P_0 \omega_0^2 \cos\phi} \right),$$

$$y(t) = P_0 \left(\Omega_0^2 - \Omega^2\right)^{-1} \cos(\Omega t + \phi).$$

Note that this also includes the case corresponding to the Vavilov–Cherenkov effect [23], when the waves are excited by a source of zero frequency ($\Omega = 0$), moving at supercritical speed ($V > c_0$).

The average value of the wave resistance to motion is:

$$\langle F \rangle = \frac{\rho \left\{ Q_0^2 \left(\Omega_0^2 - \Omega^2\right)^2 + 2 P_0 Q_0 \omega_0^2 \left(\Omega_0^2 - \Omega^2\right) \cos\phi + P_0^2 \omega_0^4 \right\}}{4\mu^2 \left(V^2 - c_0^2\right) \left(\Omega_0^2 - \Omega^2\right)^2}. \tag{9.7}$$

Let us introduce dimensionless variables:

$$\bar{V} = \frac{V}{c_0}, \bar{\Omega} = \frac{\Omega}{\mu c_0}, \bar{\Omega}_0 = \frac{\Omega_0}{\mu c_0}, \bar{\omega}_0 = \frac{\omega_0}{\mu c_0}, \bar{\omega}_* = \frac{\omega_*}{\mu c_0}, \bar{F} = \frac{\mu^2 c_0^2 \langle F \rangle}{Q_0^2 \rho}$$

Figure 9.1a, b depicts the dependences of the constant component of the force wave resistance versus crew speed at various ratios between parameter values $\bar{\omega}_*$, $\bar{\Omega}_0$, $\bar{\omega}_0$

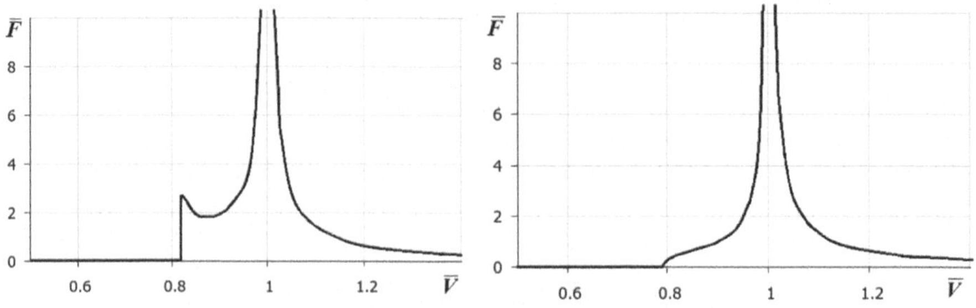

Fig. 9.1 Dependences of the constant component of the force wave resistance

characterizing the elastic-inertial properties of the string and the crew. Here, for certainty, $\bar{\omega}_0 < \bar{\Omega}_0$, and the value of $\bar{\omega}_*$ varies in the plane $(\bar{V}, \bar{\Omega})$.

For the soft basement $\omega_* < \sqrt{\omega_0^2 + \Omega_0^2}$, there are possible two situations: $\omega_* < \Omega_0$, $\omega_* > \Omega_0$. If $\omega_* < \Omega_0 < \sqrt{\omega_0^2 + \Omega_0^2}$ at velocities $V_* < V < c_0$, there is a single frequency for which it is characteristic that if the frequency of external influence does not exceed it, then there is a falling section (Fig. 9.1a) on the characteristic of the dependence of the wave resistance on the speed of the crew, indicating the possible existence of instability [23]. When this frequency is exceeded, the incident section in the region $V_* < V < c_0$ is absent (Fig. 9.1b). If $\Omega_0 < \omega_* < \sqrt{\omega_0^2 + \Omega_0^2}$, then there are two frequencies (there is no incident section on the characteristic if the frequency of the external influence falls in the interval between these frequencies). With a sufficiently rigid basement $\omega_* > \sqrt{\omega_0^2 + \Omega_0^2}$, there are three such.

In case $\Omega > \omega_*$ when $0 \le V < c_0$, the usual Doppler effect takes place, namely a moving source emits one wave along its course with a frequency $\omega_1 > \Omega$ and one in the opposite direction with a frequency $\omega_2 < \Omega$.

The increase of the speed of the crew $V > c_0$ leads to the fact that both waves propagate behind the object, in the region $x < Vt$, which corresponds to the so-called complex Doppler effect.

Determining the constant component of the wave resistance force of the movement of the crew, we obtain

$$\langle F \rangle = \begin{cases} \dfrac{\rho \Omega V \sqrt{V^2\omega_*^2 + \Omega^2 c_0^2 - c_0^2\omega_*^2}\left\{Q_0^2\left(\Omega_0^2-\Omega^2\right)^2 + 2P_0 Q_0 \omega_0^2\left(\Omega_0^2-\Omega^2\right)\cos\phi + P_0^2\omega_0^4\right\}}{\left(c_0^2-V^2\right)\left\{\left(\Omega^4-\Omega^2\Omega_0^2-\Omega^2\omega_0^2\right)^2 + 4\mu^2\left(\Omega_0^2-\Omega^2\right)^2\left(V^2\omega_*^2+\Omega^2c_0^2-c_0^2\omega_*^2\right)\right\}} & V < c_0 \\ \dfrac{\rho\left\{Q_0^2\left(\Omega_0^2-\Omega^2\right)^2 + 2P_0\Omega_0\omega_0^2\left(\Omega_0^2-\Omega^2\right)\cos\phi + P_0^2\omega_0^2\right\}}{4\mu^2\left(V^2-c_0^2\right)\left(\Omega_0^2-\Omega^2\right)^2} & V > c_0 \end{cases}. \tag{9.8}$$

The dependence corresponding to Formula (9.8) is shown in Fig. 9.2.

The frequency $\Omega = \Omega_0$ and velocity $V = c_0$ are resonant values, accompanied by unlimited growth of F. Neglecting the transverse force acting on the upper mass at the speed of the object $V > c_0$, the average value of the resistance to movement does

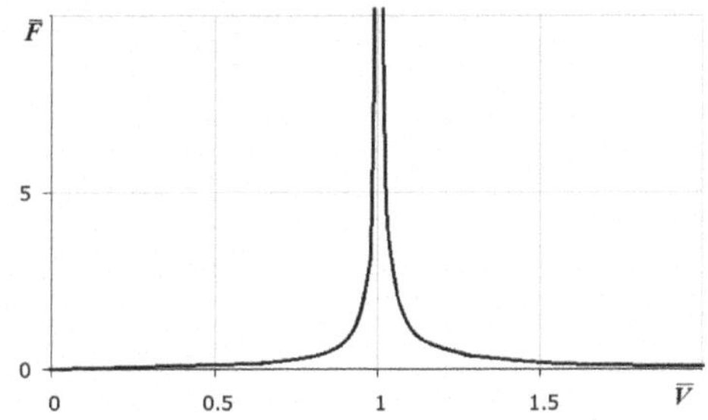

Fig. 9.2 Dependences of the constant component of the force wave resistance

not depend on the frequency of the source Ω and natural frequency of oscillations of the upper mass of the crew Ω_0, and similar to resistance to movement of the load, the action of which is characterized by a transverse force.

9.3 Some General Relations for Waves Propagating in One-Dimensional Elastic Systems

Let us consider one-dimensional elastic system with the density of Lagrange function $\lambda(x, t, \boldsymbol{u}, \boldsymbol{u}_x, \boldsymbol{u}_t, \boldsymbol{u}_{xx}, \boldsymbol{u}_{xt})$. The Lagrangian function depends on generalized coordinates $\boldsymbol{u}(x, t) = \{U_1(x, t), \ldots, U_n(x, t)\}$, and its derivatives up to the second order inclusively. This is the case, for example, to describe the beams of Bernoulli–Euler and Rayleigh models [2]. Let us suppose that the function λ twice continuously differentiable by the totality of its arguments. We calculate the partial derivative with respect to time t:

$$\frac{\partial \lambda}{\partial t} = \lambda_t + \sum_{j=1}^{n} \left(\lambda_{U_j} U_{jt} + \lambda_{U_{jx}} U_{jxt} + \lambda_{U_{jt}} U_{jtt} + \lambda_{U_{jxx}} U_{jxxt} + \lambda_{U_{jxt}} U_{jxtt} \right). \quad (9.9)$$

Let us denote λ_t the derivative on time which explicitly included.

With the help of the equations of vibration of the elastic system [23]

$$\lambda_{U_j} - \frac{\partial}{\partial x} \lambda_{U_{jx}} - \frac{\partial}{\partial t} \lambda_{U_{jt}} + \frac{\partial^2}{\partial x^2} \lambda_{U_{jxx}} + \frac{\partial^2}{\partial x \partial t} \lambda_{U_{jxt}} = 0, \quad j = 1, \ldots, n \quad (9.10)$$

which are the followed from the stationary condition of the action integral in the absence of non-potential forces acting on the system, and identities

$$U_{jt} \frac{\partial}{\partial t} \lambda_{U_{jt}} \equiv \frac{\partial}{\partial t} \left(U_{jt} \lambda_{U_{jt}} \right) - U_{jtt} \lambda_{U_{jt}}$$

$$U_{jt} \frac{\partial}{\partial x} \lambda_{U_{jx}} \equiv \frac{\partial}{\partial x} \left(U_{jt} \lambda_{U_{jx}} \right) - U_{jtx} \lambda_{U_{jx}}$$

$$U_{jt}\frac{\partial^2}{\partial x\partial t}\lambda_{U_{jxt}} \equiv \frac{\partial}{\partial x}\left(U_{jt}\frac{\partial}{\partial t}\lambda_{U_{jxt}}\right) - \frac{\partial}{\partial t}\left(U_{jxt}\lambda_{U_{jxt}}\right) + U_{jxtt}\lambda_{U_{jxt}}$$

$$U_{jt}\frac{\partial^2}{\partial x^2}\lambda_{U_{jxx}} \equiv \frac{\partial}{\partial x}\left(U_{jt}\frac{\partial}{\partial x}\lambda_{U_{jxx}} - U_{jxt}\lambda_{U_{jxx}}\right) + U_{jxxt}\lambda_{U_{jxx}}$$

we transform equality (9.9) to the energy transfer equation

$$\frac{\partial W}{\partial t} + \frac{\partial S}{\partial x} = -\lambda_t \tag{9.11}$$

where

$$W(x,t) = \sum_{j=1}^{n}\left(U_{jt}\lambda_{U_{jt}} + U_{jxt}\lambda_{U_{jxt}}\right) - \lambda \tag{9.12}$$

$$S(x,t) = \sum_{j=1}^{n}\left\{U_{jt}\left(\lambda_{U_{jx}} - \frac{\partial}{\partial x}\lambda_{U_{jxx}} - \frac{\partial}{\partial t}\lambda_{U_{jxt}}\right) + U_{jxt}\lambda_{U_{jxx}}\right\} \tag{9.13}$$

$W(x,t)$—Hamilton function density (energy density) [13].

For the stationary system ($\lambda_t = 0$) in the absence of non-potential forces, relation (9.11) expresses the law of energy change in an element of a distributed system due to its flow through the boundaries of the element. Thus, the value $S(x,t)$ should be considered as the wave energy flux density.

In the general case, the energy transfer equation has the form

$$\frac{\partial W}{\partial t} + \operatorname{div}\boldsymbol{S} = -\lambda_t \tag{9.14}$$

The $\boldsymbol{S}$—The energy flux density, often called the Umov–Poynting vector, is widely used in continuum mechanics and electrodynamics [13, 24, 25] to describe wave processes.

For non-stationary systems, when the right-hand side of Eqs. (9.13) and (9.14) is nonzero, it is possible to change (increase or decrease) the wave field energy due to the work of external forces that change the parameters of the system.

In order to obtain the equations of transfer of the wave pulse, we differentiate the Lagrangian with respect to the spatial coordinate x:

$$\frac{\partial\lambda}{\partial x} = \lambda_x + \sum_{j=1}^{n}\left(\lambda_{U_j}U_{jx} + \lambda_{U_{jx}}U_{jxx} + \lambda_{U_{jt}}U_{jxt} + \lambda_{U_{jxx}}U_{jxxx} + \lambda_{U_{jxt}}U_{jxxt}\right),$$

where λ_x—the derivative on x variable which explicitly included. Using Eqs. (9.10) and grouping the terms into an equation of divergence type, we obtain

$$\frac{\partial p}{\partial t} + \frac{\partial T}{\partial x} = \lambda_x \tag{9.15}$$

Here,

$$p(x,t) = -\sum_{j=1}^{n} \left(U_{jx} \lambda_{U_{jt}} + U_{jxx} \lambda_{U_{jxt}} \right) \tag{9.16}$$

$$T(x,t) = \lambda - \sum_{j=1}^{n} \left\{ U_{jx} \left(\lambda_{U_{jx}} - \frac{\partial}{\partial x} \lambda_{U_{jxx}} - \frac{\partial}{\partial t} \lambda_{U_{jxt}} \right) + U_{jxx} \lambda_{U_{jxx}} \right\} \tag{9.17}$$

where $p(x,t)$—wave pulse density, $T(x,t)$—wave pulse flux density [5].

If the parameters of the system clearly depend on the coordinate, as, for example, in a bar of variable cross section, then the right-hand side of Eq. (9.15) is nonzero. This indicates the possibility of a change (amplification or attenuation) of the wave pulse in an inhomogeneous system. Equations (9.11) and (9.15) express the local laws of change in energy and wave momentum in a one-dimensional distributed system. To obtain global laws, it is necessary to integrate these equations over the size of the system.

Note that in the literature, there is more often a different way of obtaining the equations of energy transfer and wave momentum, namely by multiplying the equations of motion (9.10) by U_{jt} and U_{jx}, respectively [3]. The result will not change.

In the future, we will consider $U_j = U(x,t)$. Let us multiply the energy flux density by U_x, energy density by U_t and summarize them. Further, multiplying the ratio thus obtained by U_x and using the expressions for the density of the wave pulse and its flux, we will have

$$U_x(U_x S + U_t W) + U_t(U_x T + U_t p) = (U_x U_{xt} - U_t U_{xx})\left(U_x \lambda_{U_{xx}} + U_t \lambda_{U_{xt}}\right) \tag{9.18}$$

If the Lagrangian of an elastic system contains derivatives of generalized coordinates of only the first order, then relation (9.18) is simplified and its right-hand side vanishes.

For a harmonic wave with parameters slowly varying in time and space (quasiharmonic wave)

$$U(x,t) = A(\varepsilon x, \varepsilon t) \exp[i(\omega t - kx)] + c.c.$$

where $A(\varepsilon x, \varepsilon t)$—complex amplitude, ω—frequency, k—wave number, $c.c.$—complex conjugate expression, $(i\omega A)^{-1} \partial A/\partial t \sim (-ikA)^{-1} \partial A/\partial x \sim \varepsilon << 1$, and right side of Eq. (9.18) vanishes, due to the fact that for the given approximation $U_x U_{xt} - U_t U_{xx} = 0$. In this case, relation (9.10) can be rewritten in the form

$$\frac{S - v_{\text{ph}} W}{\omega} = \frac{T - v_{\text{ph}} p}{k}, \text{ or } S - v_{ph} W = v_{\text{ph}}\left(T - v_{\text{ph}} p\right); \; v_{\text{ph}} = \frac{\omega}{k} \tag{9.19}$$

The second equality (9.19) can be interpreted in such a way that in a reference frame moving with a phase velocity, the energy flux carried by the wave is related to the density of the momentum flux through the phase velocity. Indeed, setting in Eqs. (9.11) and (9.15) $x' = x - v_{\text{ph}}t$, $t' = t$, thus if we turn into a moving system coordinate with the speed v_{ph}, we can see that the density of energy and momentum fluxes will be, respectively, equal $S - v_{\text{ph}}W$ and $T - v_{\text{ph}}p$.

As examples, consider the equations

$$U_{tt} - c_0^2 U_{xx} + \omega_*^2 U = 0, \rho F U_{tt} + IEU_{xxxx} + hU = 0 \tag{9.20}$$

The first of these describes oscillations with an additional restoring force proportional to the displacement $U(x, t)$, in particular, transverse vibrations of a string lying on a linear elastic foundation; in quantum field theory, this equation is known as the Klein–Gordon equation [11]. In the framework of the Bernoulli–Euler model, the second Eq. (9.20) describes the bending vibrations of a beam lying on a linear elastic foundation, where ρF—linear density, IE—bending stiffness of the beam, h—elastic coefficient of the basement.

Lagrangians λ which correspond to these examples are defined with the relations:

$$\lambda = \left(U_t^2 - c_0^2 U_x^2 - \omega_*^2 U^2\right)/2, \tag{9.21}$$

$$\lambda = \left(\rho F U_t^2 - IEU_{xx}^2 - hU^2\right)/2. \tag{9.22}$$

Substituting expression (9.22) into Eqs. (9.12), (9.13), (9.16), (9.17), we obtain

$$\begin{aligned} W(x,t) =& \left(U_t^2 + c_0^2 U_x^2 + \omega_*^2 U^2\right)/2, \quad S(x,t) = -c_0^2 U_x U_t, \\ p(x,t) =& -U_x U_t, \quad T(x,t) = \left(U_t^2 + c_0^2 U_x^2 - \omega_*^2 U^2\right)/2. \end{aligned} \tag{9.23}$$

For a wave packet slowly varying in space and time, we have:

$$U \sim a\cos(\psi + \phi); a = |A|, \phi = \arg A. \tag{9.24}$$

Values of energy, momentum densities and their fluxes neglecting the derivatives of a and ϕ have the following form:

$$\begin{aligned} &W \sim \left\{c_0^2 k^2 + \omega_*^2 - c_0^2 k^2 \cos 2(\psi + \phi)\right\} a^2/2, \\ &S \sim \{1 - \cos 2(\psi + \phi)\} c_0^2 \omega k a^2/2, \\ &p \sim \{1 - \cos 2(\psi + \phi)\} \omega k a^2/2, \\ &T \sim \left\{c_0^2 k^2 - \left(\omega_*^2 + c_0^2 k^2\right) \cos 2(\psi + \phi)\right\} a^2/2 \end{aligned}$$

Here, we account that $\partial\psi/\partial t = \omega$, $\partial\psi/\partial x = -k$, and frequency ω and the wave number k are connected through dispersion equation $\omega^2 = c_0^2 k^2 + \omega_*^2$, which follows from the first Eq. (9.20).

Calculating the expressions corresponding to the left and right sides of the first equality (9.19), we verify the validity of the equalities:

$$\frac{S - v_{ph}W}{\omega} = \frac{T - v_{ph}p}{k} = -\frac{\omega_*^2 a^2}{2k}.$$

In the case of a beam, the values obtained as a result of: (1) substituting expression (9.23) into relations (9.12), (9.13), (9.16), (9.17), and then into relation (9.24), (2) neglect the derivatives of slowly varying a, ϕ, ω and k, (3) with the help of the dispersion equation following from the second Eq. (9.20)

$$\rho F\omega^2 = IEk^4 + h, \tag{9.25}$$

thus:

$$\begin{aligned} &W \sim \left(IEk^4 + h\right)a^2/2, S \sim IE\omega k^3 a^2, \\ &p \sim \rho F\{1 - \cos 2(\psi + \phi)\}\omega k a^2/2, \\ &T \sim \left\{2IEk^4 - \rho F\omega^2 \cos 2(\psi + \phi)\right\}a^2/2. \end{aligned} \tag{9.26}$$

From these relations, it follows that:

$$\left(S - v_{ph}W\right)/\omega = \left(T - v_{ph}p\right)/k = \left(IEk^3 - h/k\right)a^2/2.$$

Within the framework of the assumptions made, we calculate the period averages of the energy flux density and wave pulse density

$$\langle S\rangle = IE\omega k^3 a^2 \quad \langle p\rangle = \rho F\omega k a^2/2. \tag{9.27}$$

By the definition of the group velocity, we have $v_{gr} = \frac{d\omega}{dk}$, and then from the (9.25), it follows $v_{gr} = 2IEk^3/\rho F\omega$. From relations (9.27), we find that

$$\frac{\langle S\rangle}{\langle p\rangle} = v_{gr}v_{ph}. \tag{9.28}$$

This expression is also valid for transverse vibrations of the string described by the first Eq. (9.20).

We consider a one-dimensional distributed system with stationary and homogeneous parameters, whose Lagrangian has the form

$$\lambda = \alpha_1 U_t^2 + \alpha_2 U_{xt}^2 - \beta_1 U^2 - \beta_2 U_x^2 - \beta_3 U_{xx}^2 - \gamma_1 U^4 - \gamma_2 U_x^4 - \gamma_3 U_{xx}^4. \tag{9.29}$$

Substituting expression (9.29) into Eqs. (9.10), we find that in the absence of distributed losses and sources, the oscillations of the distributed system will be described by the equation

$$\alpha_1 U_{tt} - \alpha_2 U_{xxtt} - \beta_2 U_{xx} + \beta_3 U_{xxxx} + \beta_1 U + 2\gamma_1 U^3 - 6\gamma_2 U_x^2 U_{xx}$$
$$+ 12\gamma_3 U_{xxx}^2 U_{xx} + 6\gamma_3 U_{xx}^2 U_{xxxx} = 0. \tag{9.30}$$

The constant coefficients included here are determined through the parameters of the distributed system, $U(x, t)$—generalized coordinate, which, for example, has the meaning of transverse or longitudinal displacements, or a twist angle, depending on which vibrations are considered.

For example, if $\alpha_1 = \rho F/2$, $\alpha_2 = \rho J/2$, $\beta_3 = EJ/2$, $\gamma_2 = 5EF/8$, $\beta_1 = \beta_2 = \gamma_1 = \gamma_3 = 0$, and $U(x, t)$—transverse displacement of the midline of the beam, then Eq. (9.30) describes the bending vibrations of the beam in the Rayleigh model [23] (Bernoulli–Euler beam, when $\alpha_2 = 0$) with geometric nonlinearity. Here F—the square of the cross section, J—axial moment of inertia of the section, E—elasticity module, ρ—the density of the material.

If $\alpha_1 = \rho I_0/2$, $\alpha_2 = \rho I_\phi/2$, $\beta_2 = \mu I_k/2$, $\beta_3 = EI_\phi/2$, $\gamma_2 = \mu I_\phi/4$, $\beta_1 = \gamma_1 = \gamma_3 = 0$, where I_0—polar moment of inertia, I_k—moment of inertia during torsion, I_ϕ—moment of twist, and $U(x, t)$—angle of rotation of the cross section, then Eq. (9.30) is an equation of torsional vibrations of a rod with cubic nonlinearity.

Similarly, by setting the corresponding values of the coefficients of Eq. (9.30), one can obtain from it a large set of technical and refined theories of transverse, longitudinal and torsional vibrations of homogeneous rectilinear rods, both in the linear approximation and taking into account cubic nonlinearity.

When deriving Eq. (9.30), it was assumed that a one-dimensional distributed system is infinite. Such idealization is permissible if optimal damping devices are located at its boundaries, i.e., the parameters of the boundary fixing are such that disturbances incident on it will not be reflected. Based on exact solutions of model problems for elastic systems, it was substantiated [19] the existence of matched end dampers of various types of vibrations that do not produce reflected disturbances in the system. This allows us to consider model (9.30) without taking into account the boundary conditions and consider vibrations propagating in the system as traveling elastic waves.

Note that the consideration of joint, for example, bending-torsional or longitudinal-bending vibrations of the rods in the framework of Eq. (9.30) is not possible and requires specifying the Lagrangian of an even more general form than (9.29).

In accordance with Eq. (9.30), the frequency ω and the wave number k are connected through the relation:

$$\left(\alpha_1 + \alpha_2 k^2\right)\omega^2 = \beta_3 k^4 + \beta_2 k^2 + \beta_1 + 3\left(\gamma_1 + \gamma_2 k^4 + \gamma_3 k^8\right)|U_0|^2/2, \tag{9.31}$$

which is the so-called nonlinear dispersion equation, U_0—the wave amplitude. It can also be obtained by varying the amplitude of the averaged Lagrangian [25].

In the case of a linear system ($\gamma_1 = \gamma_2 = \gamma_3 = 0$), the account of coefficient β_1 leads to the presence of an area of non-transmission, when waves in the system can propagate only starting from frequencies $> \omega_*$, and besides

$$\omega_* = \begin{cases} \sqrt{\beta_1/\alpha_1} & \text{when } \beta_1/\alpha_1 < \beta_2/\alpha_2 \\ \left\{2\left[\sqrt{\alpha_1^2\beta_3^2 + \alpha_2\beta_3(\alpha_2\beta_1 - \alpha_1\beta_2)} - \alpha_1\beta_3\right] + \alpha_2\beta_2\right\}/\alpha_1^2 & \text{when } \beta_1/\alpha_1 > \beta_2/\alpha_2 \end{cases}$$

In a system with "rigid" nonlinearity ($\gamma_1 > 0$), the area of non-transmission increases, and with "soft" ($\gamma_1 < 0$)—decreases.

From the relation (9.31), it follows that the phase $v_{ph} = \frac{\omega}{k}$ and group $v_{gr} = \frac{d\omega}{dk}$ velocities depend on the wave amplitude U_0. Of course, strictly speaking, the concept of group velocity is valid until the packet is distorted, i.e., for relatively small time intervals and for signals with a narrow spectral range [25].

We express the energy density, energy flux density, wave pulse density and wave pulse flux density through the generalized coordinate and the first and second derivatives of it. The average values of these quantities over the wave period will have the form:

$$\begin{aligned} \langle W\rangle &= \left(\beta_1 + \beta_2k^2 + \beta_3k^4 + 9\left(\gamma_1 + \gamma_2k^4 + \gamma_3k^8\right)|U_0|^2/8\right)|U_0|^2, \\ \langle S\rangle &= \omega k\left(-\alpha_2\omega^2 + 2\beta_3k^2 + \beta_2 + 3k^2\left(\gamma_2 + 2\gamma_3k^4\right)|U_0|^2/2\right)|U_0|^2, \\ \langle p\rangle &= \omega k\left(\alpha_1 + \alpha_2k^2\right)|U_0|^2, \\ \langle T\rangle &= \left(-\alpha_2\omega^2k^2 + 2\beta_3k^4 + \beta_2k^2 + 3\left(\gamma_1 + 5\gamma_2k^4 + 9\gamma_3k^8\right)|U_0|^2/8\right)|U_0|^2. \end{aligned} \tag{9.32}$$

These expressions, neglecting second-order derivatives ($\alpha_2 = \beta_3 = \gamma_3 = 0$), consistent with the expressions obtained on the basis of the averaged variational principle [25].

The value of the ratio of the wave energy flux density to the energy density determines the magnitude of the energy transfer rate [3], i.e.,

$$v_{\text{en}} = \frac{\langle S\rangle}{\langle W\rangle}, \tag{9.33}$$

and the transfer velocity of the wave pulse

$$v_{\text{imp}} = \frac{\langle T\rangle}{\langle p\rangle}. \tag{9.34}$$

If $\gamma_j \to 0$, $(j = 1, 2, 3)$, then from expressions (9.33)–(9.34) with the dispersion equation taken into account, we obtain the known result $\langle S\rangle = v_{\text{gr}}\langle W\rangle$, $\langle T\rangle = v_{\text{gr}}\langle p\rangle$, and consequently, $v_{\text{en}} = v_{\text{imp}} = v_{\text{gr}}$. In this regard, the group velocity is interpreted as the rate of energy transfer by the wave field. It can be seen that, in the presence of nonlinearity in this approximation, the energy and wave momentum transfer rates differ from the group velocity.

As was established above, in a reference frame moving with a phase velocity, the ratio of the average values of the density of energy fluxes and the wave pulse is equal to the phase velocity of the wave:

$$\frac{\langle S\rangle - v_{\mathrm{ph}}\langle W\rangle}{\langle T\rangle - v_{\mathrm{ph}}\langle p\rangle} = v_{\mathrm{ph}}.$$

The relation can be rewritten in the following form:

$$\frac{\langle S\rangle - v_{ph}\langle W\rangle}{\omega} = \frac{\langle T\rangle - v_{ph}\langle p\rangle}{k}$$
$$= \left(-\alpha_2\omega^2 k + \beta_3 k^3 - \frac{\beta_1}{k} - \frac{3}{8}\left(\frac{3\gamma_1}{k} - \gamma_2 k^3 - 5\gamma_3 k^7\right)|U_0|^2\right)|U_0|^2$$

In conclusion, we note that for average values$\langle S\rangle$ and $\langle p\rangle$, relation (9.28) remains the same for all systems whose dynamic behavior is described by linear equations and nonlinear with respect to the unknown function. To verify this, just put $\gamma_2 = \gamma_3 = 0$.

9.4 Resistance to the Movement of the Load Along a Guide Lying on a Nonlinear Elastic Basement

When studying the dynamic behavior of one-dimensional elastic systems under the action of moving loads, they most often use a string or a beam lying on a linear Winkler base as guides [10, 17]. An exception is the work [26], where the uniform motion of a constant load on a string lying on a nonlinear elastic foundation was considered. The analysis of the problem in [26] was carried out using the phase plane. It was found that in the case of rigid nonlinearity of the elastic base, there are no qualitative differences with the linear case. Based on this, we solve the problem of resistance to the uniform movement of the load along a guide lying on a nonlinear elastic foundation, using the procedure that is usually used for linear systems.

Let us suppose that the influence of the load is characterized with the help of vertical variable force $P(t) = \frac{1}{2}P_0\exp(i\Omega t) + \text{c.c.}$ with frequency Ω, amplitude P_0 (c.c.–complex conjugate quantity) moving at a constant velocity V. The motion of a constant load is obtained by simple degeneration at $\Omega \to 0$. As a guide model, we choose a string, transverse vibrations $U(x,t)$ which are described by the density of the Lagrange functions $\lambda = \frac{1}{2}\left(\rho U_t^2 - NU_x^2 - h_0U^2 - \frac{1}{2}h_1U^4\right)$, where ρ and N—linear density and string tension, nonlinearity parameter h_1 is a small addition to the rigidity of the basement $h_0 \geq 0$. Depending on the sign h_1, we have a system with either "hard" ($h_1 > 0$), either with a "soft" ($h_1 < 0$) type of nonlinearity.

According to [11], the formulation of the boundary value problem in this case will have the form

$$\rho U_{tt} - NU_{xx} + h_0U + h_1U^3 = 0, \tag{9.35}$$

$$U(Vt-0,t) = U(Vt+0,t) = U(Vt,t) = U_0(t), \tag{9.36}$$

$$[NU_x + \rho V U_t]_{Vt+0}^{Vt-0} = P(t), \tag{9.37}$$

$$|U(x,t)| < W < \infty, x \to \infty,$$
$$S/h\langle V \text{ when } x\langle Vt,\ S/h\rangle V \text{ when } x\rangle Vt \tag{9.38}$$

Relations (9.36) determine the conditions of continuity, (9.37)–the balance of forces on a moving boundary $x = Vt$, (9.38)—conditions specified at infinity (which guarantees the absence of any other sources), under $R(t)$ we assume the force necessary to maintain a uniform law of motion, S/h—energy transfer rate [13, 19], h—energy density, S—energy flux density. Square brackets mean the difference between the limit values, the values in them.

Considering the process of excitation of vibrations in the string to be steady, the solution of problem (9.35)–(9.36) on the left ($j = 1$) and on the right ($j = 2$) from the load we will look in the form

$$U(x,t) = \frac{1}{2}A^j \exp\{i(\omega^j t - k^j x)\} + c.c., \tag{9.39}$$

where A^j—complex amplitude, $\frac{\partial A^j}{\partial t}/(i\omega^j A^j) \sim \frac{\partial A^j}{\partial x}/(-ik^j A^j) \sim \varepsilon << 1$, and frequencies$\omega^j$ and wave numbers k^j are determined from the system of equations (index j is omitted)

$$\begin{cases} -\omega^2 + c_0^2 k^2 + \omega_*^2 + \mu_*^2 |A|^2 = 0 \\ \omega - kV = \Omega \end{cases}. \tag{9.40}$$

Here $c_0 = \sqrt{N/\rho}$—the speed of wave propagation in a string without taking into account the basement, $\omega_* = \sqrt{h_0/\rho}$—lowest vibration frequency of the string due to the presence of a linear elastic base,$\mu_* = \sqrt{|h_1|/2\rho}$. The first of the equations of system (9.40) is a nonlinear dispersion equation, which for $h_1 \to 0$ degenerates into the dispersion equation for transverse waves in a string on a linear Winkler base, and the second is a kinematic invariant [13, 19].

Taking into account the linear elastic base, as is known [13, 19], leads to the presence of a transmission region [27], when waves in the system can propagate, starting from frequencies $\omega > \omega_*$. In a system with hard nonlinearity, the area of non-transmission increases and, with soft, it decreases. In the future, we will consider only a system with a hard type of nonlinearity and a situation where the frequency of the source lies in the bandwidth.

Note that if the frequency of the source lies in the transmission band, then the stationary wave source does not emit, but moving at a certain speed $V > V_*$, it begins to radiate them. Such radiation is usually called deceleration emission, and the speed V_*, upon transition through which the picture of wave formation changes critically [13, 19].

Fig. 9.3 Dependence of speed on frequency

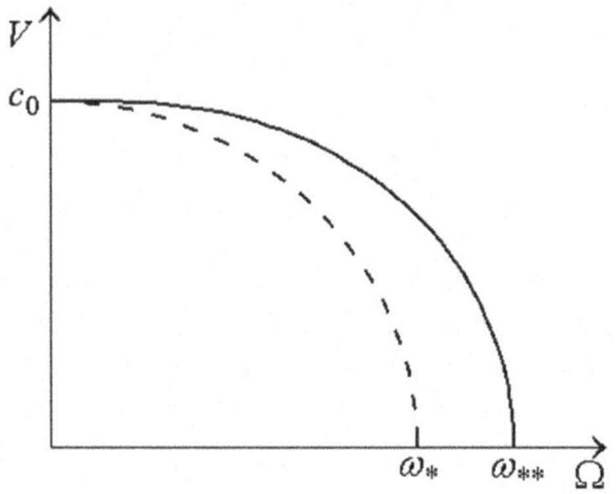

For a string on a linearly elastic foundation, the frequency source $\Omega < \omega_*$ will emit waves at speed $V > V_* = c_0\sqrt{1 - (\Omega/\omega_*)^2}$ [13, 19].

For a string lying on a nonlinear base, studies have shown that the critical velocity depends not only on the frequency, but also on the amplitude of the forcing effect and the nonlinearity parameter

$$V_{**} = \left\{ c_0^2\left(1 - \frac{\Omega^2}{\omega_*^2}\right) - \frac{\mu_*^2 P_0^2}{2\omega_*^4\rho^2} + \frac{\mu_* P_0}{\omega_*^2\rho}\sqrt{c_0^2\frac{\Omega^2}{\omega_*^2} + \frac{\mu_*^2 P_0^2}{4\omega_*^4\rho^2}} \right\}^{1/2}$$

Figure 9.3 is shown the qualitative dependence of speed on frequency, taking into account (solid line) and in the absence (dashed line) the nonlinearity parameter. It can be seen that in a system with a rigid type of nonlinearity of the elastic base, the critical velocity V_{**} and skip rate $\omega_{**} = \sqrt{\omega_*^2 + \frac{\mu_1 P_0}{c_0\rho}}$ higher than neglecting nonlinearity.

From the solution of problem (9.35)–(9.38) for $\Omega < \omega_*$, it follows that the average value of the wave resistance force has the form

$$\langle R \rangle = \begin{cases} 0, & V < V_{**} \\ \dfrac{\Omega V P_0\sqrt{c_0^2\Omega^2 - \left(c_0^2 - V^2\right)\omega_*^2 - \sqrt{\left(c_0^2\Omega^2 - \left(c_0^2 - V^2\right)\omega_*^2\right)^2 - \left(c_0^2 - V^2\right)\frac{\mu_*^2 P_0^2}{\rho^2}}}}{2\sqrt{2}\left(c_0^2 - V^2\right)^{3/2}\mu_*}, & V_{**} < V < c_0 \\ \sqrt{\begin{aligned}&\frac{\rho}{16\left(V^2 - c_0^2\right)^2\mu_*^2}\left\{\left(c_0^2\Omega^2 + \left(V^2 - c_0^2\right)\omega_*^2\right)\sqrt{\left(c_0^2\Omega^2 + \left(V^2 - c_0^2\right)\omega_*^2\right)^2 + \left(V^2 - c_0^2\right)\frac{\mu_*^2 P_0^2}{\rho^2}} - \right.\\ &\left. - \left(c_0^2\Omega^2 + \left(V^2 - c_0^2\right)\omega_*^2\right)^2 + \left(V^2 - c_0^2\right)\frac{\mu_*^2 P_0^2}{\rho^2}\right.,\end{aligned}} & V > c_0 \end{cases}$$

Corresponding to this formula, dependence in dimensionless variables $\bar{P}_0 = \frac{P_0}{\rho c_0^2}$, $\bar{V} = \frac{V}{c_0}$, $\bar{\Omega} = \frac{\Omega}{\omega_*}$, $\bar{\mu}_* = \frac{\mu_* c_0}{\omega_*^2}$, $\bar{R} = \frac{\langle R \rangle}{\rho c_0^2}$ is depicted in Fig. 9.4. The graph is plotted with the following parameter values $\bar{P}_0 = 0.2$, $\bar{\Omega} = 0.6$, $\bar{\mu}_* = 0.1$.

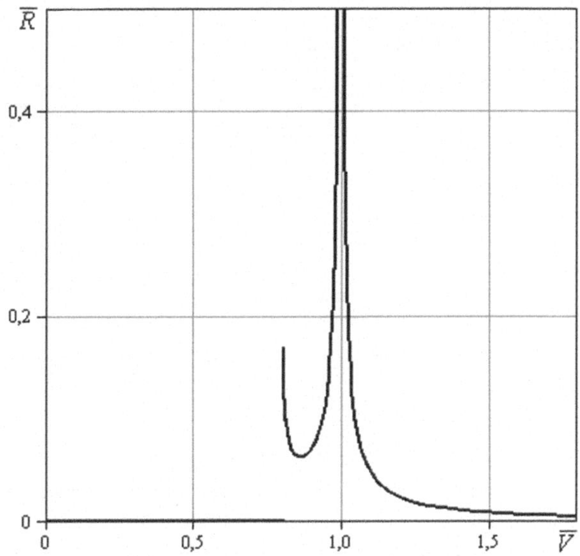

Fig. 9.4 Dependence of resistance force on velocity

Velocities close to $c_0 = \sqrt{N/\rho}$ impossible since for $V \to c_0$ resistance to movement $\langle R \rangle \to \infty$. According to calculations, the resistance to movement increases sharply not only when $\to c_0$ $\left(V/c_0 \to 1, \bar{R} \to \infty\right)$, but also when the speed of the load moves V to critical. Moreover, if in the absence of nonlinearity $\langle R \rangle \to \infty$ when $V \to V_*$, then taking into account the nonlinearity of the elastic base leads to the fact that for $V = V_{**}$, the resistance force is finite, although there is a gap of the first kind (Fig. 9.4).

It can be assumed that the stability of uniform motion is possible, i.e., excitation of longitudinal load oscillations. The fact is that the pressure of the waves creates a braking force, which, as calculations show, has a falling section of the characteristic (Fig. 9.4). In such a section, according to the canons of the theory of oscillations [27], dynamic friction is negative and a self-oscillating regime can occur.

9.5 Stabilization of Transverse Vibrations of a Mass Moving Along a String

Below, we consider the introduction of control to stabilize the lateral vibrations of an object moving along an elastic string. That is, the statement is carried out and the problem of an elastic guide (string) carrying a moving object (mass) is carried out as a dynamic controlled system.

Consider the mass m, moving with a speed V along an endless string (Fig. 9.5):

The system of equations describing the transverse vibrations of the string, mass, as well as the corresponding boundary conditions is :

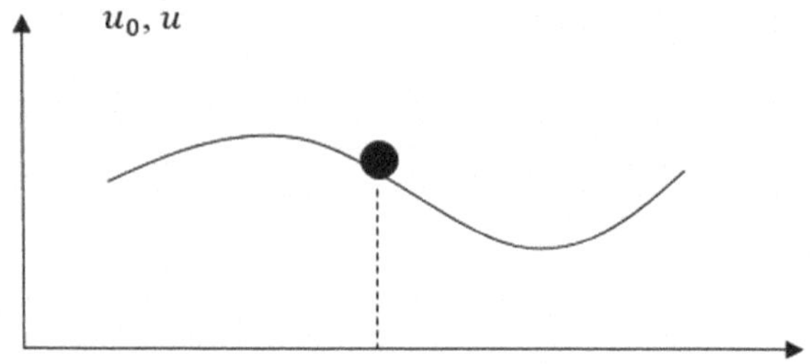

Fig. 9.5 A mass moving along an endless string

$$\frac{\partial^2 u}{\partial t^2} - c_0^2 \frac{\partial^2 u}{\partial x^2} = 0, \tag{9.41}$$

$$m\frac{d^2 u_0}{dt^2} = \left[N\frac{\partial u}{\partial x} + \rho\frac{dl}{dt}\frac{\partial u}{\partial t}\right]_{x=l(t)}, \tag{9.42}$$

$$[u(x,t)]_{x=l(t)} = u(\ell(t)+0,t) - u(\ell(t)-0,t) = 0, \tag{9.43}$$

$$u_0(t) = u(\ell(t)-0,t) = u(\ell(t)+0,t), \tag{9.44}$$

$$\ell(t) = Vt, \tag{9.45}$$

$$\lim_{|x|\to\infty} |u(x,t)| < \infty. \tag{9.46}$$

$$c_0 = \sqrt{\frac{N}{\rho}}, \tag{9.47}$$

$$F(u_0) = \alpha u_0 + \beta\frac{du_0}{dt} \tag{9.48}$$

Here, $u(x,t)$, $u_0(t)$—functions describing lateral deviations of the string and mass, $x = \ell(t)$—the law of mass motion in the longitudinal direction (in our case $\ell(t) = V(t)$, N—string tension, ρ—string density.

It is known that the dynamics of the string-mass system will have a fundamentally different character depending on the velocity V. In case when $V < c_0$, the constant deflection of the string at the point of contact with the mass propagates at a constant velocity V together with the mass. This case is not of particular interest.

The case when the mass velocity exceeds the wave velocity in a string is interesting by wave radiation due to the Vavilov–Cherenkov effect. We will assume that the waves arising as a result of the supersonic motion of the mass in the longitudinal direction excite its transverse vibrations, which can grow unlimitedly. To damp the transverse vibrations of the mass, we introduce an external force $F(u_0)$, under the assumption that it linearly depends on the transverse movement of the mass and on the speed of this movement. Let us find the parameters α and β of the external force $F(u_0)$ to suppress the instability effect. Then, Eq. (9.42') takes the form

$$m\frac{d^2u_0}{dt^2} = \left[N\frac{\partial u}{\partial x} + \rho\frac{dl}{dt}\frac{\partial u}{\partial t}\right]_{x=l(t)} + F(u_0), \tag{9.42'}$$

The transverse vibrations of the string and mass will be sought in the form

$$u(x,t) = Ae^{i(\omega t - kx)},\ u_0(t) = Be^{i\Omega t}. \tag{9.49}$$

Substituting solutions in (9.41) in this form, we obtain the dispersion equation:

$$\omega^2 - c_0^2k^2 = 0, \tag{9.50}$$

We supplement this equation with a kinematic invariant expressing the condition for the equality of phases at the contact point:

$$\omega - kV = \Omega, \tag{9.51}$$

The solution of the kinematic problem (9.50)–(9.51) allows for a visual graphic interpretation. On the plane (k, ω), the dispersion Eq. (9.50) determines some curves (in this case, straight lines), and the kinematic invariant (9.51) corresponds to a straight line, the slope of which is determined by the speed of movement (Fig. 9.6):

As can be seen from Fig. 9.6, in this case, two traveling waves are excited, propagating behind the mass and traveling in different directions.

Solving Eqs. (9.50) and (9.51) together, we obtain:

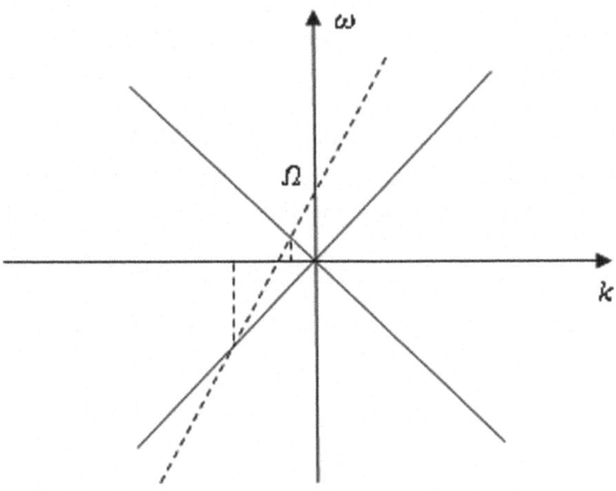

Fig. 9.6 Dispersion law and the kinematic invariant of problem (9.41)–(9.45)

$$k_1 = \frac{\Omega}{c_0 - V},$$
$$k_2 = -\frac{\Omega}{c_0 + V} \quad (9.52)$$

Next, we consider the case when the transverse oscillations of the mass increase. This is possible if we assume that in solution (9.52), $\mathrm{Im}(\Omega) < 0$. Then, for the supercritical velocity $V > c_0$, we get:

$$\mathrm{Im}(k_1) > 0,$$
$$\mathrm{Im}(k_2) > 0 \quad (9.53)$$

Using the condition of finiteness of the magnitude of the transverse movements of the string at infinity, we take the following: A wave with a wave number k_1 propagates behind the mass to the positive direction of the axis, and with k_2—behind $(x < Vt)$ to the negative direction of the axis. Thus, the transverse movement of the string behind the mass will be composed of $u^1(x, t)$ and $u^2(x, t)$, which have the following form:

$$u^1(x, t) = A_1 e^{i(\omega_1 t - k_1 x)},$$
$$u^2(x, t) = A_2 e^{i(\omega_2 t - k_2 x)}, \quad u^-(x, t) = u^1 + u^2, \quad (9.54)$$

where ω_j can be found from (9.47) after the substitution of the corresponding k.

In the front of the mass, there is no disturbance:

$$u^+(x, t) = 0.$$

Note that a wave propagating in the positive direction of the axis does not decay at infinity when conditions (9.53) are satisfied, i.e., condition (9.46) will not be satisfied. We turn here to the statement of the problem of finding the parameters of an external force $F(u_0)$ for suppression of the oscillations of the mass. It turns out that if we find proper values for parameters α and β, thereby translating the imaginary part of the mass oscillation frequency $\mathrm{Im}(\Omega)$ from the negative half-plane to the positive, we will automatically achieve the fulfillment of condition (9.46).

Let us differentiate condition (9.43) with respect to time, we obtain

$$\left[\frac{\partial u}{\partial x} \cdot \frac{dl}{dt} + \frac{\partial u}{\partial t}\right]_{x=l(t)} = 0, \quad (9.55)$$

Then, it follows

$$\left[\frac{\partial u}{\partial t}\right]_{x=l(t)} = -\frac{dl}{dt}\left[\frac{\partial u}{\partial x}\right]_{x=l(t)}. \quad (9.56)$$

Taking into account (9.45), relation (9.42) will take the form:

$$m\frac{d^2u_0}{dt^2} = \left(N - V^2\rho\right)\left[\frac{\partial u}{\partial x}\right]_{x=l(t)} + \alpha u_0 + \beta\frac{du_0}{dt}, \tag{9.57}$$

Substituting the found displacements of the string and mass, we obtain

$$-mB\Omega^2 = \mathrm{i}\left(N - V^2\rho\right)(A_1k_1 + A_2k_2) + \alpha B + \mathrm{i}\beta B\Omega, \tag{9.58}$$

or, taking into account condition (9.44), which can be rewritten in the form

$$A_1 = A_2 = B, \tag{9.59}$$

where A_1—the amplitude of the wave which propagates in front of the mass, A_2—the amplitude of the wave propagating behind the mass, and the expression of the wave numbers in terms of the mass oscillation frequency, relation (9.48) can be rewritten in the form

$$-m\Omega^2 = i\left(N - V^2\rho\right)\left(\frac{\Omega}{c_0 - V} - \frac{\Omega}{c_0 + V}\right) + \alpha + i\beta\Omega, \tag{9.60}$$

or

$$m\Omega^2 + \mathrm{i}(\beta + 2V\rho)\Omega + \alpha = 0. \tag{9.61}$$

Thus, the problem of damping the growing oscillations is reduced to the problem of finding the parameters α and β, to make the condition $\mathrm{Im}(\Omega) \geq 0$ true. If we substitute $\Omega = -\Theta$, the problem can be reduced to searching for the above parameters such that the complex polynomial

$$m\Theta^2 - \mathrm{i}(\beta + 2V\rho)\Theta + \alpha = 0 \tag{9.62}$$

would be stable, i.e., $\mathrm{Im}(\Theta) < 0$.

To solve this problem, it is convenient to use the Hermite–Hurwitz criterion. For polynomial (9.62), the Hermite–Hurwitz matrix has the form:

$$M = \begin{bmatrix} m & 0 & \alpha & 0 \\ 0 & -(\beta + 2V\rho) & 0 & 0 \\ 0 & m & 0 & \alpha \\ 0 & 0 & -(\beta + 2V\rho) & 0 \end{bmatrix} \tag{9.63}$$

According to the Hermite–Hurwitz criterion, for the stability of polynomial (9.62), it is necessary and sufficient that a series composed of determinants:

$$\Delta_0 = m,$$

$$\Delta_2 = \begin{vmatrix} m & 0 \\ 0 & -(\beta + 2V_\rho) \end{vmatrix},$$

$$\Delta_4 = \begin{vmatrix} m & 0 & \alpha & 0 \\ 0 & -(\beta + 2V_\rho) & 0 & 0 \\ 0 & m & 0 & \alpha \\ 0 & 0 & -(\beta + 2V_\rho) & 0 \end{vmatrix}$$

should be alternating. Where do we get:

$$-m(\beta + 2V\rho) < 0, -m(\beta + 2V\rho)^2\alpha > 0, \quad (9.64)$$

or

$$\beta > -2V\rho, \alpha < 0 \quad (9.65)$$

Thus, the introduction of an external force $F(u_0)$, whose parameters are in the region determined by relations (9.65) it is possible to suppress mass oscillations, and at the same time, to satisfy conditions (9.46) for the string. Physically, the appearance of a stabilizing force similar to $F(u_0)$ is equivalent to introducing a visco-elastic element for the mass.

9.6 Conclusions

Thus, it has been studied the phenomenon of wave resistance for a load (crew), which has its own degrees of freedom, and moves along an elastic guide. The problem has been considered in linear and nonlinear formulations. In order to avoid undesirable effects, the results, which have been obtained in this research, should be taking into account when constructing systems consisting of a moving load along the elastic guides like high-speed trains, pantographs, etc.

Acknowledgements The work was supported of the Russian Science Foundation (project No. 20-19-00613).

References

1. Filippov, A.P.: Vibrations of Distorted Systems. Mashinostroenie, Moscow (1970). (in Russian)
2. Filippov, A.P., Kokhmanyuk, S.S., Vorobyov, Y.S.: The Impact of Dynamic Loads on Structural Elements. Naukova Dumka, Kiev (1974). (in Russian)
3. Fryba, L.: Vibrations of Solids and Structures Under Moving Loads. Noordhoff International Publishing, Groningen (1972)

4. Ivanchenko, I.I.: Dynamics of Transport Facilities: High-Speed Mobile and Shock Loads. Nauka, Moscow (2011). (in Russian)
5. Kokhmanyuk, S.S., Yanyutin, V.G., Romanenko, L.G.: Oscillations of Deformable Systems Under Pulsed and Moving Loads. Naukova Dumka, Kiev (1980). (in Russian)
6. Krylov, V.V. (ed.): Ground Vibrations from High-Speed Railways: Prediction and Mitigation. ICE Publishing, London (2019)
7. Krylov, V.V. (ed.): Noise and Vibration from High Speed Trains. Thomas Telford Publishing, London (2001)
8. Krysov, S.V.: Forced Vibrations and Resonance in Elastic Systems with Moving Loads. GSU, Educational Guidance, Gorky (1985). (in Russian)
9. Metrikine, A.V., Verichev, S.N., Vostrukhov, A.V.: Fundamental Trasks of High-Speed Land Transport. Lambert Academic Publishing, Saarbrücken (2014). (in Russian)
10. Suiker, A.S.J.: The Mechanical Behaviour of Ballasted Railway Tracks. Delft Univ. Press, Delft (2002)
11. Verigo, M.F., Kogan, A.Y.: The Interaction of the Track and Rolling Stock, Transport, Moscow (1987) (in Russian)
12. Veritcev, S.N.: Instability of a Vehicle Moving on an Elastic Structure. Delft Univ. Press, Delft (2002)
13. Vesnitsky, A.I.: Waves in Systems with Moving Boundaries and Loads. Fizmatlit, Moscow (2001). (in Russian)
14. Vostroukhov, A.V.: Three-Dimensional Dynamic Models of a Railway Track for High-Speed Trains. Delft Univ. Press, Delft (2002)
15. Wulfert, A.R.M.: Wave Effects Interacting With Moving Objects. Delft Univ. Press, Delft (1999)
16. Panovko, Y.G., Gubanova, N.I.: Stability and Vibrations of Elastic Systems. Nauka, Moscow (1967). (in Russian)
17. Krylov, A.N.: Vibration of Ships. ONTI, Moscow (1936). (in Russian)
18. Weaver Jr., W., Timoshenko, S.P., Young, D.H.: Vibration Problems in Engineering, 5th edn. Wiley, New York (1990)
19. Vesnitsky, A.I.: Selected Works on Mechanics. Nash Dom, Nizhny Novgorod (2010). (in Russian)
20. Raileigh, D. Strutt, D.W.: On the pressure of vibrations. Phil. Mag. Ser. **6**, 3 (15), 338–350 (1902)
21. Nicolai, E.L.: To the question of vibration pressure, Bulletin of the St. Petersburg Polytechnic Institute. **18**(1), 49–60 (1912) (in Russian)
22. Bychenkov, V.A., Krysov, S.V.: Wave resistance to rolling of a rigid wheel over 1D visco-elastic support. Mashinovedenie **3**, 60–66 (1988). (in Russian)
23. Bolotovskii, B.M., Ginzburg, V.L.: The Vavilov-Cerenkov effect and Doppler effect in the motion of sources with superluminal velocity in vacuum. Sov. Phys. Usp. **15**, 184–192 (1972)
24. Landau, L.D., Lifshitz, E.M.: Electrodynamics of Continuous Media. Pergamon Press, Oxford (1984)
25. Witham, G.B.: Linear and Nonlinear Waves. Wiley, New York (1974)
26. Metrikin, A.V.: Stationary waves in a nonlinear elastic system interacting with a moving load. Acoust. Phys. **40**(4), 647–650 (1994)
27. Rabinovich, V.I., Trubetskov, D.I.: Oscillations and Waves in Linear and Nonlinear Systems. Springer, Berlin (2012)

Chapter 10
Dispersion, Attenuation and Spatial Localization of Thermoelastic Waves in a Medium with Point Defects

Vladimir I. Erofeev, Anna V. Leonteva, and Ashot V. Shekoyan

Abstract This paper deals with the study of the propagation of plane longitudinal waves in an unbounded point-defected medium located in a nonstationary inhomogeneous temperature field. The problem is considered in the self-consistent formulation making allowance for both the influence of an acoustic wave on the formation and movement of defects and the effect of defects on the propagation specificity of an acoustic wave. It is shown that with no heat diffusion, the system of equations is reduced to a nonlinear evolutionary equation, which is the formal generalization of the Korteweg–de Vries–Burgers equation (KdVB). Using the truncated expansion method, a specific solution of an evolutionary equation has been found in the form of a stationary shock wave with a monotonic decrease. It is noted that dissipative effects induced by available defects prevail over the dispersion associated with the migration of defects in the medium. The influence of the initial temperature and type of defects have been herein studied on the main parameters of a stationary wave such as velocity, amplitude and width. Three limiting cases of the evolutionary equation have been considered, and some generalizations of the known equations have been obtained, namely the equations of Korteweg–de Vries (KdV), Burgers and Riemann. The obtained equations as well as the generalized KdVB equation have solutions in the form of stationary shock waves. The propagation of a harmonic wave in a thermoelastic defective medium is herein analyzed. It is shown that the availability of defects in the medium promotes the occurrence of the frequency-dependent dissipation and dispersion. The influence of diffusion parameters and the type of defects on the harmonic wave propagation are studied.

V. I. Erofeev (✉) · A. V. Leonteva
Mechanical Engineering Research Institute of Russian Academy of Sciences, 85 Belinskogo str., Nizhny Novgorod 603024, Russia
e-mail: erof.vi@yandex.ru

A. V. Leonteva
e-mail: aleonav@mail.ru

A. V. Shekoyan
Institute of Mechanics of the National Academy of Sciences of the Republic of Armenia, 24B Baghramyan Ave., Yerevan 0019, Armenia
e-mail: ashotshek@mechins.sci.am

H. Altenbach et al. (eds.), *Multiscale Solid Mechanics*,
Advanced Structured Materials 141,
https://doi.org/10.1007/978-3-030-54928-2_10

Keywords Thermoelastic medium · Point defects · Longitudinal wave · Stationary shock wave · Harmonic wave · Dispersion

10.1 Introduction

When a material is exposed to the laser radiation or a stream of particles (e.g., in ion implantation), point defects (vacancies, interstices) are formed therein [15]. The passage of an intense longitudinal acoustic wave promotes a change in areas of the tension and compression of energy activating the generation of point defects, thus causing their spatial redistribution. Defects migrating through the material are recombined at various centers. Dislocations, interstitial impurities, etc., may perform the role of such centers.

Wave effects in ensembles of dislocations were studied in [1–3, 5, 9, 10, 18, 19]. In [1], the problems of the elastic wave propagation in a viscothermoelastic composite with spherical inhomogeneities were examined.

It was demonstrated in [16] that the problem on the acoustic wave propagation in a material with point defects should be regarded as self-consistent one comprising in addition to the dynamical equation of the theory of elasticity the kinetic equation for the density of defects.

In [4], the interaction of a nonlinear deformation wave with the field of point defects (vacancies, interstices) concentration was studied, which caused both the wave scattering and changing the energy activating the generation of defects and their spatial redistribution.

It was therewith assumed that the main processes, which determine the defect behavior, were the generation, recombination and diffusion processes. The volume mutual recombination of heteronymous defects was not taken into consideration.

The propagation of nonlinear longitudinal waves in a plate with an allowance for the interaction of the longitudinal component of the medium displacements with temperature fields and concentration of nonequilibrium atomic point defects was studied in [14].

In [8], the effect of dislocations and point defects is studied on the spatial localization of nonlinear acoustic waves propagating in materials. The effect of defects on the spatial localization of nonlinear waves with the mutual recombination of heteronymic defects is studied in [6]. It is shown in the said papers that vacancies and interstitials promote the formation of spatially localized nonlinear waves.

In this paper, we consider the propagation of a plane longitudinal wave in a boundless medium with point defects with the formation thereof the medium temperature changes. We suppose that defects of only one type, either vacancy or interstices, emerge in the medium.

10.2 Problem Statement

To obtain a self-consistent dynamic problem for a thermoelastic point-defected medium, to equations of the theory of elasticity

$$\rho \frac{\partial^2 u_i}{\partial t^2} = \frac{\partial \sigma_{ij}}{\partial x_j} \tag{1}$$

and thermal conductivity

$$T \frac{\partial S}{\partial t} = \kappa \Delta \Theta \tag{2}$$

where u_i means the displacement vector components, σ_{ij} are stress tensor components, $\Theta = T - T_0$, T_0 is the initial temperature of the medium, T is the current temperature, κ is the thermal conduction coefficient, ρ is the material density, S is the entropy of the medium volume unit, kinetic equations should be added which describe the change in the number of point defects per a volume unit [15]:

$$\frac{\partial n_1}{\partial t} = q_{01} + q_1 \left(\frac{\partial u_1}{\partial x_1} + \frac{\partial u_2}{\partial x_2} + \frac{\partial u_3}{\partial x_3} \right) + D_1 \Delta n_1 - \beta_1 n_1 - \beta_{12} n_2, \tag{3}$$

$$\frac{\partial n_2}{\partial t} = q_{02} + q_2 \left(\frac{\partial u_1}{\partial x_1} + \frac{\partial u_2}{\partial x_2} + \frac{\partial u_3}{\partial x_3} \right) + D_2 \Delta n_2 - \beta_2 n_2 - \beta_{21} n_1. \tag{4}$$

n_1, n_2 are herein the volume concentrations of vacancies and interstitial atoms per a medium volume unit, respectively, q_{01}, q_{02} are the rates of point defect generation under no deformation, D_1, D_2 are the diffusion coefficients of vacancies and interstitial atoms, respectively, β_{12}, β_{21} are rates of the mutual defect recombination of the "interstitial atoms—vacancy" and "vacancy—interstitial atoms" type, β_1, β_2 are rates of recombination of defects on sinks, q_1, q_2 are coefficients of the deformation and defect interaction.

The free energy per a medium volume unit appears as follows:

$$\begin{aligned} F = {} & F_0 + \frac{\lambda}{2} u_{ll}^2 + \mu \left(u_{ik} - \frac{1}{2} \delta_{ik} u_{ll} \right)^2 - d_1 u_{ll} n_1 - d_2 u_{ll} n_2 - \gamma \Theta u_{ll} \\ & + m_1 n_1 \Theta + m_2 n_2 \Theta - \frac{c_p}{T_0^2} \Theta^2 + \frac{A}{2} u_{ik} u_{il} u_{kl} + B u_{ik}^2 u_{ll} \\ & + \frac{C}{3} u_{ll}^3 + \frac{q_3}{2} u_{ll}^2 n_1 + q_4 u_{ll} n_1^2 + \frac{q_5}{2} u_{ll}^2 n_2 + q_6 u_{ll} n_2^2 \end{aligned} \tag{5}$$

where F_0 is the free energy per a medium volume unit before the perturbation, λ and μ are Lamé constants, d_1, d_2 are coefficients characterizing the elastic wave and defect interaction, γ is the thermal coefficient, m_1, m_2 are coefficients characterizing the defect and temperature change interaction, c_p is the specific heat of the material, A, B, C are Landau third-order constants, q_3, q_4, q_5 are (as the previously introduced q_1, q_2) coefficients characterizing the deformation and defect interaction.

The components of the strain u_{ik} and displacement u_i tensor by correlations allowing for the geometric nonlinearity [17]:

$$u_{ik} = \frac{1}{2}\left(\frac{\partial u_i}{\partial x_k} + \frac{\partial u_k}{\partial x_i} + \frac{\partial u_l}{\partial x_i}\frac{\partial u_l}{\partial x_k}\right). \tag{6}$$

The stresses and entropy included in (1), (2) are determined from expression (5):

$$\sigma_{ik} = \frac{\partial F}{\partial\left(\frac{\partial u_i}{\partial x_k}\right)}, \tag{7}$$

$$S = \frac{\partial F}{\partial T}. \tag{8}$$

When describing the plane wave propagation along the coordinate x_3, the system (1)–(8) is transformed to:

$$\rho\frac{\partial^2 u_3}{\partial t^2} - a\frac{\partial^2 u_3}{\partial x_3^2} - \beta_N\frac{\partial u_3}{\partial x_3}\frac{\partial^2 u_3}{\partial x_3^2} = -\gamma\frac{\partial\theta}{\partial x_3} - d_j\frac{\partial n_j}{\partial x_3}, \tag{9}$$

$$c_\varepsilon\frac{\partial\theta}{\partial t} - \chi\frac{\partial^2\theta}{\partial x_3^2} = -\gamma T_0\frac{\partial^2 u_3}{\partial x_3\partial t} - T_0 p_j\frac{\partial n_j}{\partial t}, \tag{10}$$

$$\frac{\partial n_j}{\partial t} = q_\varepsilon\frac{\partial u_3}{\partial x_3} + D_j\frac{\partial^2 n_j}{\partial x_3^2} - \beta_j n_j \tag{11}$$

where $u_3(x_3, t)$ is the medium particle displacement along the coordinate x_3, $\theta(x_3, t)$ is the temperature difference ($\theta = T - T_0$), T_0 is the natural state temperature, $n_j(x_3, t)$ is the volume concentration of point defects of the jth type ($j = V$ is for vacancies, $j = i$ is for interstices), ρ is density, $a = \lambda + 2\mu$, λ, μ are Lame constants, $c_l = \sqrt{a/\rho}$ is the longitudinal wave velocity in case of no defects, β_N is the nonlinearity factor ($\beta_N = 3\lambda + 6\mu + 2A + 6B + 2C$), A, B, C are Landau third-order modules, γ is the thermal factor, $d_j = \left(\lambda + \frac{2}{3}\mu\right)\Omega_j$, Ω_j is the dilatation parameter characterizing the change in the medium volume with the formation of one point defect therein ($\Omega_V < 0$, $\Omega_i > 0$), c_ε is the thermal conductivity under the constant deformation, χ is the thermal conductivity factor, p_j is the thermal capacity of the defect of type j, q_ε is the rate of the point defect generation with available deformation, D_j is the defect type j diffusion factor and β_j is the recombination rate stocks of the defect of type j.

The system of Eqs. (9)–(11) describes the propagation of a plane longitudinal wave in the medium, both with defects of the "vacancy" type and with defects of the "interstice" type. Systems differ from each other by signs before the last component of the first equation in the system. In the further analysis, the both cases are considered.

Let the thermal conductivity factor be a small value ($\chi = 0$), then we obtain from the second equation of the system the expression for the temperature

$$\theta(x_3, t) = -\frac{T_0}{c_\varepsilon}\left(\gamma\frac{\partial u_3}{\partial x_3} + p_j n_j\right). \tag{12}$$

Taking into account (12) the system (9), (11) may be reduced to a single equation with respect to displacements of the medium particles u_3

$$\begin{aligned} &\frac{\partial^2 u_3}{\partial t^2} - \left(\left(\frac{a}{\rho} + \frac{\gamma^2 T_0}{\rho c_\varepsilon}\right) + \frac{q_\varepsilon}{\rho\beta_j}\left(\frac{\gamma T_0 p_j}{c_\varepsilon} - d_j\right)\right)\frac{\partial^2 u_3}{\partial x_3^2} \\ &+ \frac{1}{\beta_j}\left(\frac{\partial}{\partial t} - D_j\frac{\partial^2}{\partial x_3^2}\right)\left(\frac{\partial^2 u_3}{\partial t^2} - \left(\frac{a}{\rho} + \frac{\gamma^2 T_0}{\rho c_\varepsilon}\right)\frac{\partial^2 u_3}{\partial x_3^2}\right) \\ &- \frac{\beta_N}{\rho}\left(\frac{\partial u_3}{\partial x_3}\frac{\partial^2 u_3}{\partial x_3^2} + \frac{1}{\beta_j}\frac{\partial}{\partial x_3}\left(\frac{\partial u_3}{\partial x_3}\frac{\partial^2 u_3}{\partial x_3 \partial t}\right)\right) + \frac{\beta_N D_j}{\rho\beta_j}\frac{\partial^2}{\partial x_3^2}\left(\frac{\partial u_3}{\partial x_3}\frac{\partial^2 u_3}{\partial x_3^2}\right) = 0. \end{aligned} \tag{13}$$

In Eq. (13), we introduce dimensionless values for the longitudinal displacement, coordinate and time, respectively $U = u_3/u_0$, $z = x_3/X$, $\tau = t/T$. The equation takes the form:

$$\begin{aligned} &\frac{\partial^2 U}{\partial \tau^2} - \frac{\partial^2 U}{\partial z^2} + \frac{\partial}{\partial \tau}\left(\frac{\partial^2 U}{\partial \tau^2} - a_1\frac{\partial^2 U}{\partial z^2}\right) - a_2\frac{\partial^2}{\partial z^2}\left(\frac{\partial^2 U}{\partial \tau^2} - a_1\frac{\partial^2 U}{\partial z^2}\right) \\ &- a_3\left(\frac{\partial U}{\partial z}\frac{\partial^2 U}{\partial z^2} + \frac{\partial}{\partial z}\left(\frac{\partial U}{\partial z}\frac{\partial^2 U}{\partial z \partial \tau}\right)\right) + a_2 a_3\frac{\partial^2}{\partial z^2}\left(\frac{\partial U}{\partial z}\frac{\partial^2 U}{\partial z^2}\right) = 0 \end{aligned} \tag{14}$$

where a_i—are dimensionless parameter complexes, which looks as follows

$$a_1 = \left(\frac{a}{\rho} + \frac{\gamma^2 T_0}{\rho c_\varepsilon}\right)\left(\frac{a}{\rho} + \frac{\gamma^2 T_0}{\rho c_\varepsilon} + \frac{q_\varepsilon}{\rho\beta_j}\left(\frac{\gamma T_0 p_j}{c_\varepsilon} - d_j\right)\right)^{-1},$$

$$a_2 = D_j\beta_j\left(\frac{a}{\rho} + \frac{\gamma^2 T_0}{\rho c_\varepsilon} + \frac{q_\varepsilon}{\rho\beta_j}\left(\frac{\gamma T_0 p_j}{c_\varepsilon} - d_j\right)\right)^{-1},$$

$$a_3 = \frac{\beta_j\beta_N u_0}{\rho}\left(\frac{a}{\rho} + \frac{\gamma^2 T_0}{\rho c_\varepsilon} + \frac{q_\varepsilon}{\rho\beta_j}\left(\frac{\gamma T_0 p_j}{c_\varepsilon} - d_j\right)\right)^{-3/2}.$$

As characteristic values of the length and time, there are taken respectively

$$X = \frac{1}{\beta_j}\sqrt{\frac{a}{\rho} + \frac{\gamma^2 T_0}{\rho c_\varepsilon} + \frac{q_\varepsilon}{\rho \beta_j}\left(\frac{\gamma T_0 p_j}{c_\varepsilon} - d_j\right)},\ T = \frac{1}{\beta_j}$$

In Eq. (14), dissipative components (third-order derivatives) are conditioned by available defects in the medium and dispersion ones (fourth-order derivatives) by diffusion processes occurring in the medium. We assume that such diffusion processes proceed slower than processes of the defects and an acoustic wave interaction.

10.3 Linearized Equation

Discarding the nonlinear part of Eq. (14), we obtain the equation

$$\frac{\partial^2 U}{\partial \tau^2} - \frac{\partial^2 U}{\partial z^2} + \frac{\partial}{\partial \tau}\left(\frac{\partial^2 U}{\partial \tau^2} - a_1\frac{\partial^2 U}{\partial z^2}\right) - a_2\frac{\partial^2}{\partial z^2}\left(\frac{\partial^2 U}{\partial \tau^2} - a_1\frac{\partial^2 U}{\partial z^2}\right) = 0, \quad (15)$$

which enables us to study the dispersion properties of thermoelastic waves.

Searching for the solution to Eq. (15) in the form of a traveling harmonic wave $U(z,\tau) = U_0 e^{i(\omega\tau - kz)} + \kappa.c.$, where ω is frequency, k is the wave number, $\kappa.c.$ is the complex conjugate value, we obtain the complex dispersion equation

$$i\omega^3 + \left(1 + a_2 k^2\right)\omega^2 - i a_1 k^2 \omega - a_1 a_2 k^4 - k^2 = 0$$

or

$$\left(\omega^2 - a_1 k^2\right)\left(i\omega + a_2 k^2\right) = \left(k^2 - \omega^2\right), \quad (16)$$

where $k = k_1 + ik_2$ is the complex wave number. The real part of the wave number corresponds to the wave propagation and the imaginary component corresponds to the wave attenuation during the propagation. Selecting the real and imaginary parts in Eq. (16), we shall come to the following system of algebraic equations

$$\begin{cases} \left(1 + a_2\left(k_1^2 - k_2^2\right)\right)\omega^2 + 2a_1 k_1 k_2 \omega \\ \quad + a_1 a_2\left(4k_1^2 k_2^2 - \left(k_1^2 - k_2^2\right)^2\right) - \left(k_1^2 - k_2^2\right) = 0 \\ \omega^3 + 2a_2 k_1 k_2 \omega^2 - a_1\left(k_1^2 - k_2^2\right)\omega \\ \quad - 2k_1 k_2\left(1 + 2a_1 a_2\left(k_1^2 - k_2^2\right)\right) = 0 \end{cases}. \quad (17)$$

The graphs of dependences of the real and imaginary parts of the wave number are shown in Fig. 10.1 ($a_1 > 0$, $a_2 > 0$). This figure shows two pairs of curves

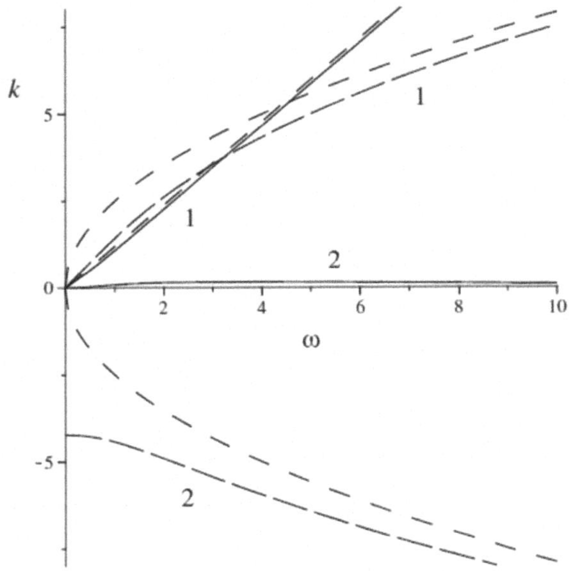

Fig. 10.1 Dependences $k_1(\omega)$ (1) and $k_2(\omega)$ (2)

$(k_1(\omega); k_2(\omega))$. We consider herein only those pairs the real part of the wave number thereof has positive values. The branch of the imaginary part $k_2(\omega)$ with a horizontal zero asymptote $k_2 = 0$ corresponds to the branch of the real part curve $k_1(\omega)$ having an inclined asymptote $\omega - \sqrt{a_1}k_1 = 0$. Two remaining branches of the real and imaginary parts of the wave number have different signs and a common nonlinear asymptote in the form of quadratic parabola $\omega - 2a_2k_{1,2}^2 = 0$. The imaginary part branch has a cutoff on the wave number $k_2 = -\sqrt{\frac{1}{a_1 a_2}}$. A dashed line marks asymptotes hereinafter in the figures.

The attenuation factor dependence on the frequency $\gamma(\omega)$ is shown in Fig. 10.2. The value of the attenuation factor $\gamma = \frac{k_2}{k_1}$ shows the availability (at $|\gamma| < 1$) or the lack (at $|\gamma| > 1$) of propagating waves. The chart has two branches: a limited and an unlimited one. The limited branch of dependence $\gamma(\omega)$ corresponds to a propagating thermoelastic wave and a pair of branches $k_1(\omega)$, $k_2(\omega)$ with inclined and horizontal asymptotes correspond in turn to the said limited branch. The $k_1(\omega)$ dependence is not linear.

The dependences of the phase $v_{\text{ph}} = \frac{\omega}{k_1}$ and group $v_{\text{gr}} = \frac{d\omega}{dk_1}$ velocities on the frequency with fixed values of parameters presented in Fig. 10.3 show the propagating wave dispersion. At $\omega = 0$, the values of the phase and group velocities are congruent and equal to unity $v_{\text{ph}} = v_{\text{gr}} = 1$. At infinity, the velocity values tend to the common value $v_{\text{ph}} = v_{\text{gr}} = \sqrt{a_1}$ (it is shown by the dashed line in Fig. 10.3).

Equation (15) contains two positive parameters a_1, a_2, which characterize available defects, their type and degree of diffusion in the medium. If such defects are vacancies, the parameter takes values $0 < a_1 < 1$. If defects are interstitials, the

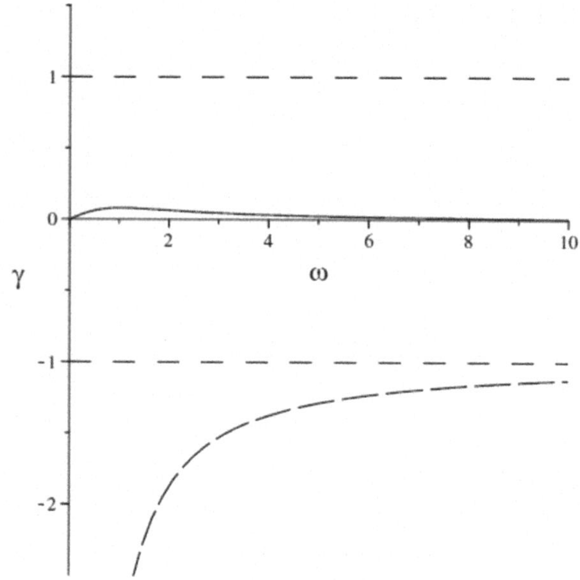

Fig. 10.2 Dependence $\gamma(\omega)$

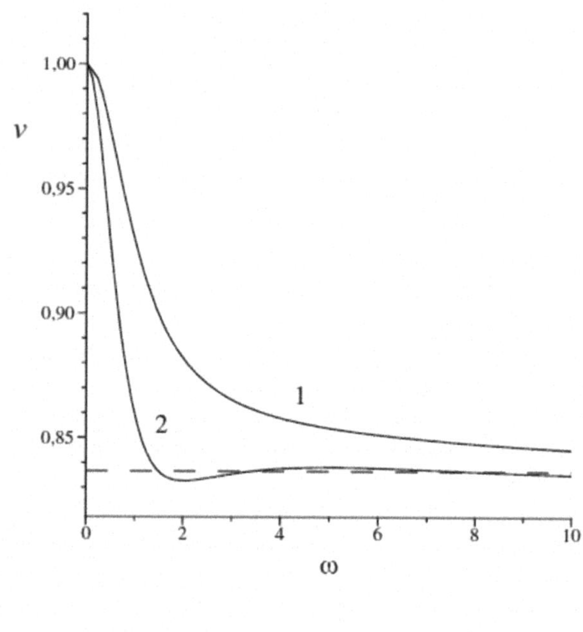

Fig. 10.3 Dependences $v_{\mathrm{ph}}(\omega)$ (1), $v_{\mathrm{gr}}(\omega)$ (2)

parameter may have any value within the half-interval $a_1 > 0$. If there is no attenuation to be caused by available defects in the medium, the parameter values are equal to $a_1 = 1, a_2 = 0$.

Let us study the effect of the said parameters on the dispersion curves of a thermoelastic wave. The location of the dispersion curve branches depends on the parameter

a_1 value as follows: For the parameter values within the $0 < a_1 < 1$ interval, the both branches (the real and imaginary ones) of the propagating wave are in one quarter and for $a_1 > 1$ they are in different ones (Figs. 10.4 and 10.5). The straight line $\omega = k_1$ is marked by a dash-dotted line in Fig. 10.4.

With the increasing parameter a_1 value, the inclination angle of the real branch asymptote decreases (Fig. 10.4), the amplitude of the attenuation factor curve first

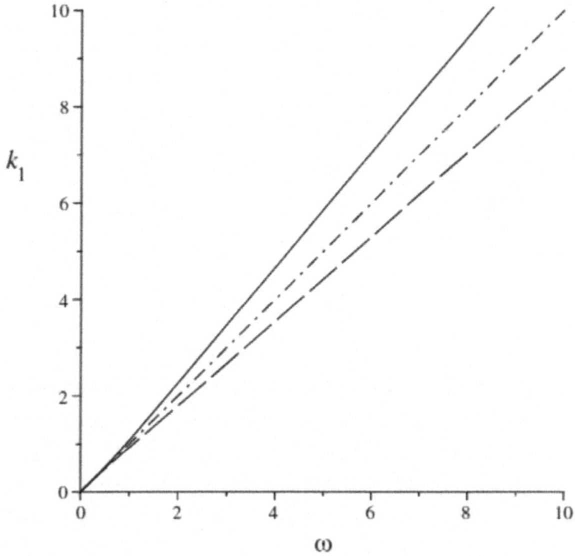

Fig. 10.4 Dependences $k_1(\omega)$ at various values of parameter a_1 ($a_2 \geq 0$): $0 < a_1 < 1$ (solid line), $a_1 = 1$ (dash-dot line), $a_1 > 1$ (long dash line)

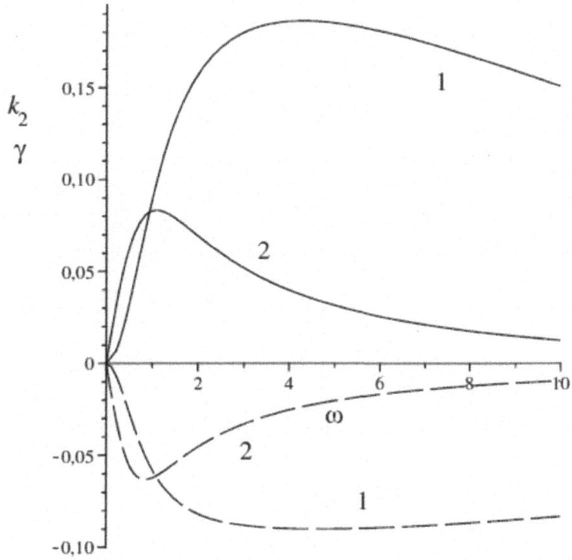

Fig. 10.5 Dependences $k_2(\omega)$ (1), $\gamma(\omega)$ (2) at various values of parameter a_1 ($a_2 > 0$): $0 < a_1 < 1$ (solid line), $a_1 > 1$ (long dash line)

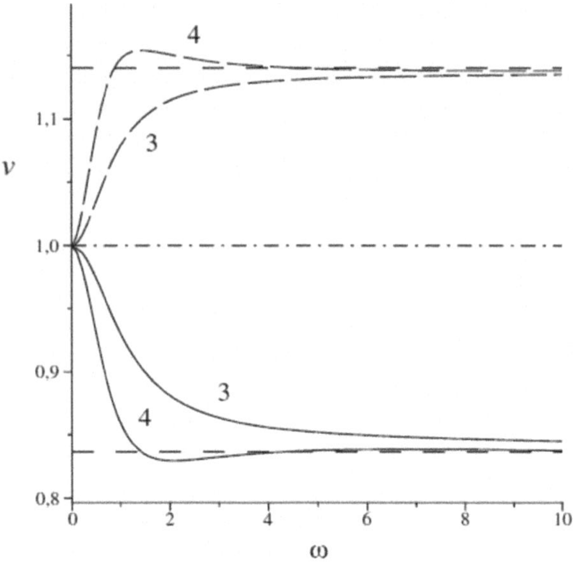

Fig. 10.6 Dependences $v_{ph}(\omega)$ (3), $v_{gr}(\omega)$ (4) at various values of parameter a_1 ($a_2 \geq 0$): $0 < a_1 < 1$ (solid line), $a_1 = 1$ (dash-dot line), $a_1 > 1$ (long dash line)

drops to zero (at $a_1 = 1$) and then increases by changing its sign (Fig. 10.5). The ranges of the phase and group velocities are reduced to zero at an increasing parameter value to unity. The phase velocity curve is located above the group velocity curve; the velocity values are limited from above by unity $v_{gr} < v_{ph} < 1$. At $a_1 = 1$, the velocity values are congruent $v_{ph} = v_{gr} = 1$. Any further increase in the parameter value promotes an increase in the velocity ranges, the group velocity curve is thereat located above the phase velocity curve (Fig. 10.6), and the velocity values are limited from below $1 < v_{ph} < v_{gr}$.

A change in the parameter a_2 value does not affect the asymptotic behavior of curve $k_1(\omega)$ but affects the behavior of curve $k_2(\omega)$, which with $a_2 = 0$ has the nonzero asymptote $k_2 = \frac{1-a_1}{2a_1\sqrt{a_1}}$. With an increasing parameter a_2 ($0 < a_1 < 1$) value, the attenuation factor value decreases across the entire frequency range (Fig. 10.7); the values of the phase and group velocities decrease at low frequencies and increase at high frequencies (Fig. 10.8). And at $a_1 > 1$, on the contrary at low frequencies the values of velocities increase and at high frequencies decrease.

In particular case, at $a_2 \to 0$ the system (17) assumes the following form

$$\begin{cases} \omega^2 + 2a_1k_1k_2\omega - \left(k_1^2 - k_2^2\right) = 0 \\ \omega^3 - a_1\left(k_1^2 - k_2^2\right)\omega - 2k_1k_2 = 0 \end{cases},$$

and enables you to find the frequency dependences

$$k_1 = \frac{\sqrt{2}\omega}{2}\sqrt{\frac{1 + a_1\omega^2 + \sqrt{\left(1+\omega^2\right)\left(1+a_1^2\omega^2\right)}}{1 + a_1^2\omega^2}},$$

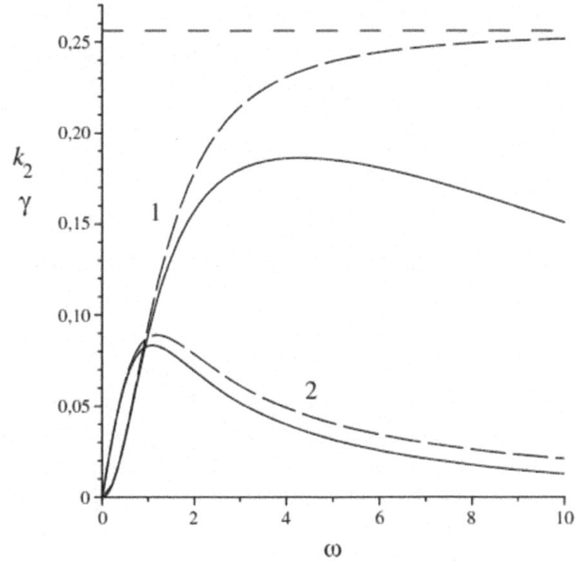

Fig. 10.7 Dependences $k_2(\omega)$ (1), $\gamma(\omega)$ (2) at various values of parameter a_2 $(0 < a_1 < 1)$: $a_2 > 0$ (solid line), $a_2 = 0$ (long dash line)

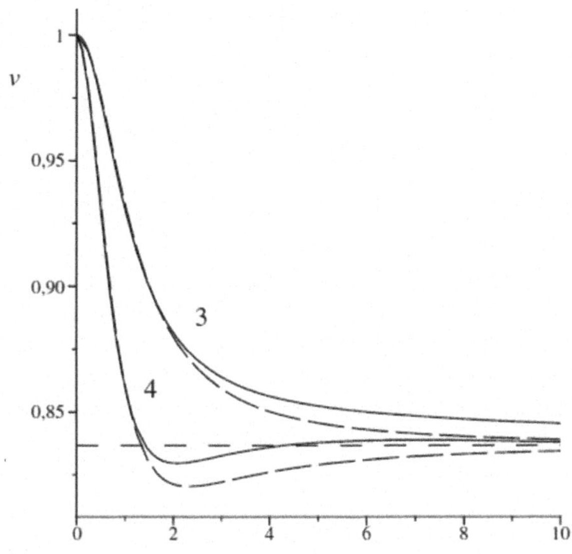

Fig. 10.8 Dependences $v_{\text{ph}}(\omega)$ (3), $v_{\text{gr}}(\omega)$ (4) at various values of parameter a_2 $(0 < a_1 < 1)$: $a_2 > 0$ (solid line), $a_2 = 0$ (long dash line)

$$k_2 = -\frac{\sqrt{2}}{2}\frac{(a_1 - 1)\omega^2}{\sqrt{\left(1 + a_1^2\omega^2\right)\left(1 + a_1\omega^2 + \sqrt{\left(1 + \omega^2\right)\left(1 + a_1^2\omega^2\right)}\right)}}.$$

Whence it follows that high-frequency perturbations do not have dispersion and propagate with attenuation, which does not depend on the frequency $\left(k_1 = \frac{\omega}{a_1},\right.$ $\left.k_2 = -\frac{(a_1-1)}{2a_1\sqrt{a_1}}\right)$, low-frequency perturbations do not have dispersion either but have the frequency-dependent attenuation ($k_1 = \omega$, $k_2 = \frac{1-a_1}{2}\omega^2$).

10.4 Evolutionary Equation and Stationary Wave

Let us proceed to the moving coordinate system $\xi = z - c\tau$, $\eta = \varepsilon\tau$ in Eq. (14), where c is the wave velocity earlier unknown, ε is a small parameter ($\varepsilon \ll 1$). The choice of variables is explained by the fact that the perturbation when propagating with the velocity c along the axis z slowly evolves in the course of time because of the nonlinearity, dispersion and dissipation. We assume that in Eq. (14) all the nonlinear and dissipative components are small values of the ε order, we obtain in the first approximation on ε the evolutionary equation with respect to the longitudinal deformation function $W = \frac{\partial U}{\partial \xi}$:

$$\frac{\partial W}{\partial \eta} + \frac{(1-a_1)}{2\varepsilon}\frac{\partial^2 W}{\partial \xi^2} + \frac{a_2(1-a_1)}{2\varepsilon}\frac{\partial^3 W}{\partial \xi^3} + \frac{a_3}{2\varepsilon}\left(W\frac{\partial W}{\partial \xi} - \frac{\partial}{\partial \xi}\left(W\frac{\partial W}{\partial \xi}\right)\right) = 0. \tag{18}$$

From the zeroth-order approximation, the wave velocity $c = 1$ is determined (measured in relative units), which is recorded in (18). The equation obtained may be classified as the Korteweg–de Vries–Burgers equation with an additional nonlinear component. In Eq. (18), the last component is additional as compared to the classical KdV–Burgers equation. Equation (18) comprises dissipative and nonlinear components; therefore, the equation solution may be in the form of shock waves.

Among all possible solutions of Eq. (18), the greatest interest is solutions from the class of stationary waves. Such waves propagate at a constant velocity and do not change their shape in the course of propagation. The equation of stationary waves is written as:

$$\frac{dW}{d\chi} - \frac{(1-a_1)}{2\varepsilon v}\frac{d^2 W}{d\chi^2} - \frac{a_2(1-a_1)}{2\varepsilon v}\frac{d^3 W}{d\chi^3} - \frac{a_3}{2\varepsilon v}\left(W\frac{dW}{d\chi} - \frac{d}{d\chi}\left(W\frac{dW}{d\chi}\right)\right) = 0, \tag{19}$$

or, after integration,

$$W - \frac{(1-a_1)}{2\varepsilon v}\frac{dW}{d\chi} - \frac{a_2(1-a_1)}{2\varepsilon v}\frac{d^2 W}{d\chi^2}$$

$$-\frac{a_3}{2\varepsilon v}\left(\frac{W^2}{2}-W\frac{\mathrm{d}W}{\mathrm{d}\chi}\right)=0 \tag{20}$$

where $W = W(\chi)$, χ is the running coordinate, $\chi = \xi - v\eta$, v is the velocity of a stationary wave moving in the positive direction of the coordinate axis. The integration constant is taken as zero.

We seek the solution of Eq. (20) by the truncated expansion method described in detail in [12, 13]. This method is used to search for the exact solutions of nonlinear differential equations.

Since the order of the pole for the general solution is equal to unity, the solution of Eq. (20) may be sought in the form:

$$W(\chi) = b_0 Y(\chi) + b_1 \tag{21}$$

where $Y(\chi) = \sqrt{B_0}\mathrm{th}\left(\sqrt{B_0}\chi\right)$ is the solution of Riccati's equation $Y'(\chi) = -Y^2(\chi) + B_0$. By substituting the solution of (21) into Eq. (20) and taking into account Riccati's equation, we obtain a third-degree polynomial relative to $Y(\chi)$. Putting the polynomial factors to zero, we find unknown factors:

$$b_0 = -\frac{2a_2}{a_3}(1-a_1),\ b_1 = \frac{1}{a_3}(1-a_1)(1+a_2),\ B_0 = \frac{1}{4a_2^2}(1+a_2)^2.$$

Simultaneously with the said factors b_0, b_1, B_0, we find an expression for the stationary wave velocity

$$v = \frac{1}{2\varepsilon}(1-a_1)(1+a_2). \tag{22}$$

Only at this value of the velocity, the solution of Eq. (20) may be found by the analytical method. The solution of (21) takes the form:

$$W(\chi) = \frac{1}{a_3}(1-a_1)(1+a_2)\left(1-\mathrm{th}\left(\frac{1+a_2}{2a_2}\chi\right)\right). \tag{23}$$

The dependency graph (23) is shown in Fig. 10.9. The analysis of functions of the first and second derivatives ($W'(\chi)$ and $W''(\chi)$) shows their parity and oddness, respectively. It indicates the symmetry of the function graphs of relative to the ordinate and the origin of coordinates. The dependency graph $W(\chi)$ has the form of a monotonic (steady) symmetric, with respect to the inflection point, kink. The point of the kink symmetry is on the ordinate axis. The amplitude A and the kink width Δ are equal to

$$A = \frac{2}{a_3}(1-a_1)(1+a_2),\ \Delta = \frac{2a_2}{1+a_2}. \tag{24}$$

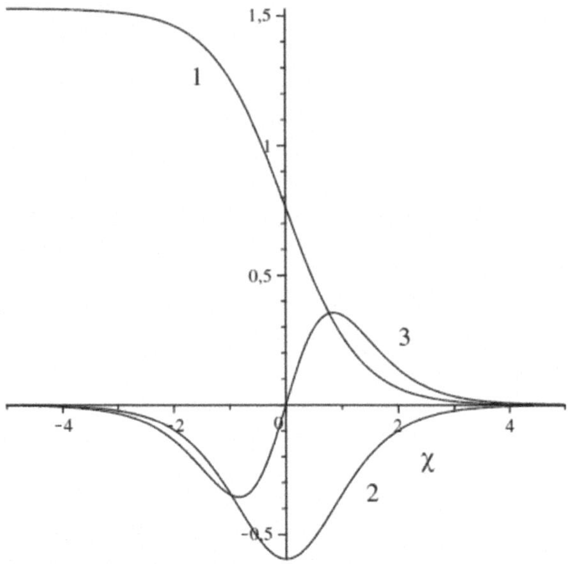

Fig. 10.9 Dependences $W(\chi)$ (1), $W'(\chi)$ (2), $W''(\chi)$ (3) with fixed parameter values

The phase portrait of Eq. (20) is given in Fig. 10.10. The solution of Eq. (23) is marked in the figure with a dashed line. It can be seen in the figure that there are two equilibrium states on the phase portrait. The examination of these equilibrium states for stability shows that one of the equilibrium states is a "node" and the other is a "saddle." The motion occurs from the unstable state of equilibrium (node) (at

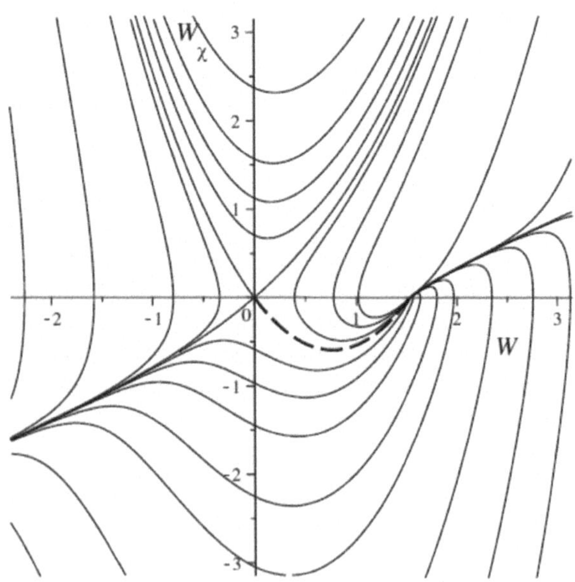

Fig. 10.10 Phase portrait in plane (W, W')

$\chi \to -\infty$) to the saddle point (at $\chi \to +\infty$). The solution is a shock wave similar to a stationary shock wave of the KdV–Burgers equation [11].

The availability of an additional nonlinear component in Eq. (19) gives no rise to the emergence of any fundamentally new solutions. Dissipative effects and dispersion in the medium under consideration balance nonlinear effects. The existence of additional nonlinearity does not give any qualitative changes.

Based on physical considerations, the evolutionary Eq. (18) has a number of particular cases that are dealt with by the authors in [7]. It is shown that only the availability and interaction of two types of nonlinearities, one of which introduces the dissipation into the system, makes possible the existence of a stationary shock wave.

Let us combine the initial parameters of the problem into the following dimensionless complexes

$$m_1 = \frac{q_\varepsilon d_j}{a\beta_j},\quad m_2 = \frac{q_\varepsilon p_j}{\gamma\beta_j},\quad m_3 = \frac{D_j\beta_j\rho}{a},\quad m_4 = \frac{\beta_j\beta_N u_0}{a}\sqrt{\frac{\rho}{a}}$$

which characterize the type of defects, their diffusion in the medium, the thermal properties of the medium volume unit and the system nonlinearity. Dimensionless parameters a_i are rewritten as follows

$$a_1 = \frac{1+T_0}{(1-m_1)+(1+m_2)T_0},\quad a_2 = \frac{m_3}{(1-m_1)+(1+m_2)T_0}$$
$$a_3 = \frac{m_4}{((1-m_1)+(1+m_2)T_0)^{3/2}}$$

where T_0 is the dimensionless value of temperature and its characteristic value is equal to $T_* = \frac{ac_\varepsilon}{\gamma^2}$.

The dependences of the velocity (22), amplitude and front width (24) of the stationary wave on the initial temperature are shown in Figs. 10.11, 10.12 and 10.13 for media with defects of the "vacancy" (1) and "interstitial" (2) types. Asymptotes are marked in the figures with a short dash. A long dash shows the limiting case ($m_1 \to 0$), which corresponds to infinitesimal changes in the medium volume when one point defect is formed therein.

With an increasing value of parameter m_1 in the vacancy medium, the wave velocity and amplitude increase and the front width decreases; in the interstitial medium, on the contrary, the wave velocity and amplitude decrease and the front width increases.

In the vacancy medium, a stationary wave exists at any positive temperature values, and in the interstitial medium, it is only at $T_0 > T_0^*$, where T_0^* is the temperature value there under the wave velocity and amplitude are equal to zero. At high temperatures, T_0 waves in the both media propagate at almost identical velocities

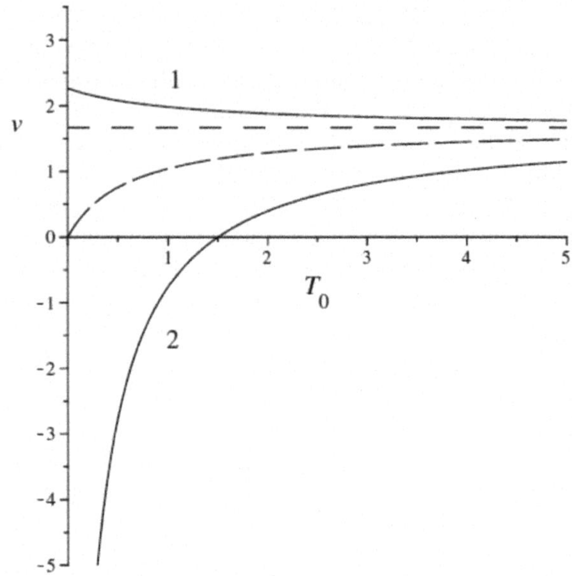

Fig. 10.11 Dependences $v(T_0)$ at $m_1 < 0$ (1) and $m_1 > 0$ (2)

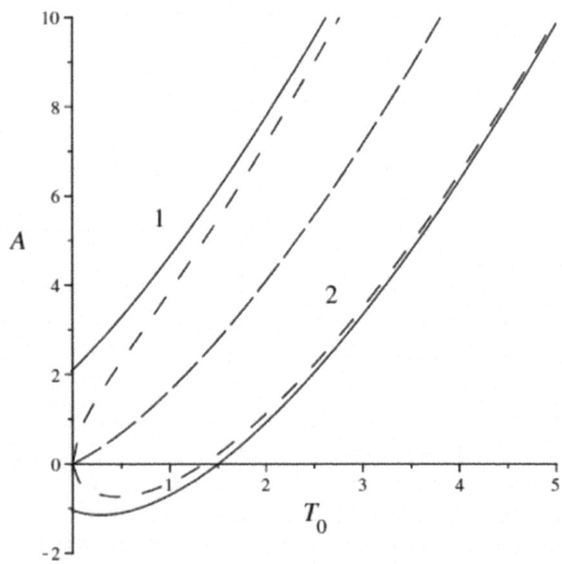

Fig. 10.12 Dependences $A(T_0)$ at $m_1 < 0$ (1) and $m_1 > 0$ (2)

$$v = \frac{m_2}{2\varepsilon(1 + m_2)}.$$

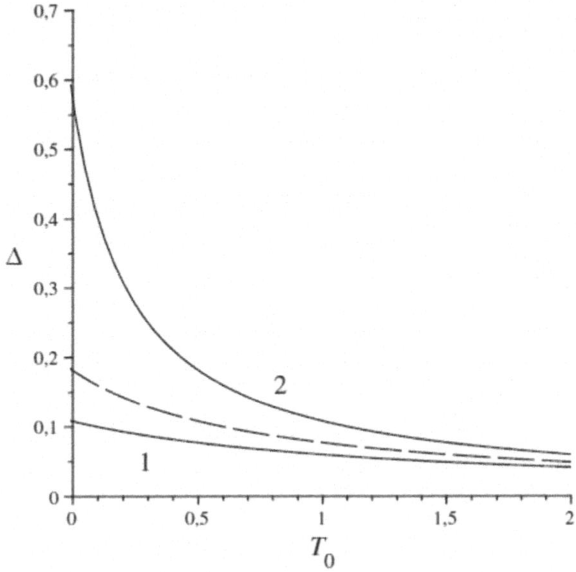

Fig. 10.13 Dependences $\Delta(T_0)$ at $m_1 < 0$ (1) and $m_1 > 0$ (2)

10.5 Particular Cases of the Evolutionary Equation

If in Eq. (18) the dissipative term is small as compared to all the other terms of the equation (case 1), it may be neglected and the Korteweg–de Vries equation will be obtained with an additional nonlinear term

$$\frac{\partial W}{\partial \eta} + \frac{a_2(1-a_1)}{2\varepsilon}\frac{\partial^3 W}{\partial \xi^3} + \frac{a_3}{2\varepsilon}\left(W\frac{\partial W}{\partial \xi} - \frac{\partial}{\partial \xi}\left(W\frac{\partial W}{\partial \xi}\right)\right) = 0. \tag{25}$$

Equation (25) has an analytical solution in the form of a stationary wave

$$W(\chi) = \frac{a_2}{a_3}(1-a_1)\left(1 - \operatorname{th}\left(\frac{\chi}{2}\right)\right) \tag{26}$$

where χ, as before, is the running coordinate, $\chi = \xi - v\eta$, v is the stationary wave velocity equal to $v = a_2(1-a_1)/2\varepsilon$. The solution (26) has been found by the same method as the previously obtained solution (23). The wave profile has the shape similar to that shown in Fig. 10.9 as a smooth jump (kink) between two values of the longitudinal strain function $W(\chi)$. The kink has a monotonic decrease and symmetry with respect to the inflection point.

When examining the equilibrium states $W_0 = 0$, $W_0 = \frac{2a_2}{a_3}(1-a_1)$ for stability, we find that in one case, the characteristic equation has real roots of different signs and it corresponds to the equilibrium state of "the saddle," in the other case the solution of the characteristic equation are real equal positive roots and it corresponds to the equilibrium state of "unstable degenerate node." The solution for Eq. (25) is a stationary shock wave.

It is known that the solution of the Korteweg–de Vries equation is a stationary soliton-type wave. The movement on the phase plane occurs from the "saddle" to the "saddle." In this case, a stationary wave results from the interaction of the following two mechanisms: nonlinearity and dispersion. The nonlinearity causes the steepening of the wave profile with the further rollover (Riemann wave) and the dispersion gives the blurring of the profile since different harmonics of the wave (frequency) propagate at different speeds.

The solution for Eq. (25) is a stationary shock wave, which results from the interaction of the nonlinearity and dissipation effects notwithstanding that the equation contains only the dispersion and nonlinear terms, and there is no dissipative term in the equation.

It is shown from the linearized equation in respect to small perturbations $\left(W(\chi) = W_0 + \tilde{W}(\chi)\right)$:

$$\frac{\partial \tilde{W}}{\partial \eta} + \frac{a_2(1-a_1)}{2\varepsilon}\frac{\partial^3 \tilde{W}}{\partial \xi^3} + \frac{a_3 W_0}{2\varepsilon}\left(\frac{\partial \tilde{W}}{\partial \xi} - \frac{\partial^2 \tilde{W}}{\partial \xi^2}\right) = 0 \tag{27}$$

that the dissipation in Eq. (25) is implicit. The last term in Eq. (27) is dissipative.

Thus, the additional term in (25) is dissipative and nonlinear. The availability of this term in the equation makes possible the existence of solutions in the form of stationary shock waves.

If in Eq. (18) the dispersion term is small as compared to all other terms of the equation (case 2), we obtain the Burgers equation with an additional nonlinear term

$$\frac{\partial W}{\partial \eta} + \frac{(1-a_1)}{2\varepsilon}\frac{\partial^2 W}{\partial \xi^2} + \frac{a_3}{2\varepsilon}\left(W\frac{\partial W}{\partial \xi} - \frac{\partial}{\partial \xi}\left(W\frac{\partial W}{\partial \xi}\right)\right) = 0. \tag{28}$$

It is possible, when $a_2 \to 0$, i.e., the diffusion is very small ($D_j \to 0$).

Equation (28) may be attributed to the generalized unperturbed Burgers equations. As distinct from the Burgers equation [11], there is one more type of the quadratic nonlinearity, and the both nonlinearities manifest equally themselves. Equation (28) has a stationary solution in the form of a shock wave. Taking into account the boundary conditions (the function at infinity has different meanings)

$$W(\chi) = \begin{cases} W_1, \ \chi \to +\infty \\ W_2, \ \chi \to -\infty \end{cases} \quad (W_2 > W_1),$$

we find the solution for the equation of stationary waves

$$\chi = \frac{2(1-a_1-a_3W_1)}{a_3(W_2-W_1)}\ln(W-W_1) - \frac{2(1-a_1-a_3W_2)}{a_3(W_2-W_1)}\ln(W_2-W)$$

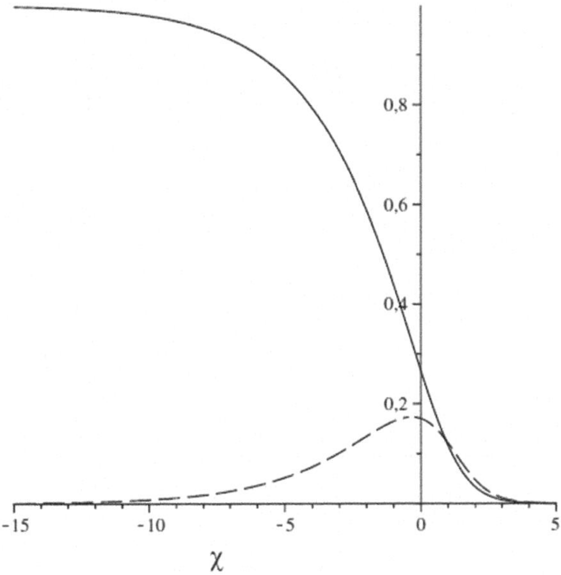

Fig. 10.14 Dependences $W(\chi)$ (solid line), $W'(\chi)$ (dash line)

where $W = W(\chi)$, χ is the running coordinate, $\chi = \xi - v\eta$, the nonlinear wave velocity is determined by the expression $v = \frac{a_3}{4\varepsilon}(W_1 + W_2)$. The profile of the stationary shock wave $W(\chi)$ and the derivative graph is shown in Fig. 10.14. The derivative graph is mapped into the upper half-plane and shifted along the ordinate by W_1.

Parameters of a shock wave resulting from the mutual compensation of the nonlinearity and dissipation effects are bound by the correlation

$$\frac{a_3 A}{1 - a_1}\left(\frac{\Delta}{2} - 1\right) = \text{const}$$

where $A = W_2 - W_1$ is the shock wave amplitude, Δ is the characteristic width of the shock wave front. The nonlinear wave velocity depends on the wave amplitude and width.

Let us consider the case when in Eq. (18) parameter $a_1 = 1$ (case 3). It is possible, if defects are interstitials ($d_i > 0$). For vacancies $0 < a_1 < 1$. For interstitials $a_1 > 0$, moreover, if the temperature effect is small, $a_1 >> 1$.

If $a_1 = 1$, we obtain an equation wherein there are no dissipative and dispersion terms but nonlinear terms are preserved

$$\frac{\partial W}{\partial \eta} + \frac{a_3}{2\varepsilon}\left(W\frac{\partial W}{\partial \xi} - \frac{\partial}{\partial \xi}\left(W\frac{\partial W}{\partial \xi}\right)\right) = 0. \tag{29}$$

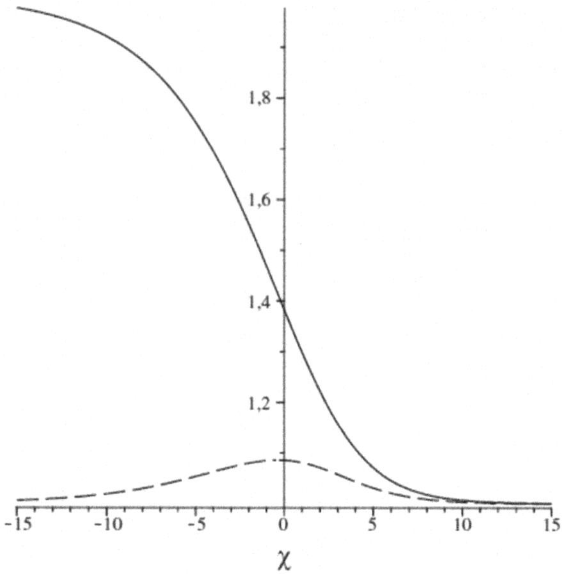

Fig. 10.15 Dependences $W(\chi)$ (solid line), $W'(\chi)$ (dash line)

Equation (29) has an additional nonlinear term with respect to the Riemann equation. Nonlinearities are included in the equation with the same coefficients and different signs.

Equation (29) has the solution

$$\chi = \frac{2}{(W_2 - W_1)}[W_2 \ln(W_2 - W) - W_1 \ln(W - W_1)]$$

in the form of a stationary shock wave propagating with velocity $v = \frac{a_3}{4\varepsilon}(W_1 + W_2)$ along the coordinate $\chi = \xi - v\eta$. The wave profile is shown in Fig. 10.15.

The shock wave width is a constant value, i.e., it is independent of the wave amplitude and initial parameters of the system. In this case, the interaction of heterogeneous nonlinearities causes the emergence of a shock wave. The linearization of Eq. (29) with respect to small perturbations

$$\frac{\partial \tilde{W}}{\partial \theta} + b_3\left(\frac{\partial \tilde{W}}{\partial \xi} - \frac{\partial^2 \tilde{W}}{\partial \xi^2}\right)W_0 = 0,$$

shows that the second nonlinear term introduces the attenuation. This attenuation promotes the emergence of a stationary shock wave in the system. Equation (29) has the property of the Burgers equation notwithstanding that are no explicit dissipative terms therein.

10.6 Conclusion

In the thermoelastic defect-free medium in the formulation under consideration (the lack of heat diffusion in the medium), a harmonic wave has no dispersion, its velocity is equal to unity. In a medium with defects, this wave propagates with dispersion and attenuation depending on the frequency.

At low frequencies close to zero, there is nearly no wave attenuation and waves propagate at a constant speed close to unity irrespective of the type of defects or their availability. At high frequencies, waves also propagate at a constant speed, but it depends on the type of defects.

Low-frequency perturbations have a pronounced dispersion, and high-frequency perturbations propagate without dispersion and losses, if there is the defect diffusion in the system. If there is no defect diffusion, disturbances at high frequencies have a constant attenuation value.

An increasing value of the defect diffusion coefficient enhances the decreasing of the wave attenuation and (if defects are vacancies) the decrease in the wave propagation velocity at low frequencies and an increase thereof at high frequencies. In a medium with vacancy-type defects, the dispersion is normal, and the abnormal dispersion may occur in a medium with interstitial-type defects. In interstitial media, harmonic waves have a longer length and velocity than in a media with vacancies.

Based on the analysis of the exact analytical solution of the evolutionary equation with respect to the displacements of the medium particles, it is shown that vacancies and interstices promote the generation of stationary shock waves in the medium.

Nonlinear waves propagate faster in a vacancy media and have the larger amplitude and the smaller width than in an interstitial media. An increase in the initial temperature results in an increase in the stationary wave velocity, if defects are interstitial, to a decrease if defects are vacancies; the amplitudes increase in the both media, and the front widths decrease forming there at a thin shock front.

An additional quadratic-nonlinear term in the generalized equations (KdVB, KdV, Burgers, and Riemann) displays itself as a dissipative term as it introduces the attenuation into the system. In case of the generalized Riemann equation, this attenuation promotes the emergence of stationary shock waves and the lack of solitons in the case of the generalized KdV equation.

Acknowledgements The work was carried out within the Russian state task for fundamental scientific research for 2019-2022 (the topic No. 0035-2019-0027, the state registration No. 01201458047).

References

1. Bagdoev, A., Erofeyev, V., Shekoyan, A.: Wave Dynamics of Generalized Continua, p. 274. Springer, Berlin (2016)
2. Burlak, G.N., Ostrovskii, I.V.: Acoustic hysteresis phenomena due to the dislocation nonlinearity in crystals. Tech. Phys. Lett. **23**(9), 725–726 (1997)
3. Erofeev, V.I.: Self-modulation of an acoustic wave in a dislocated solid. Tech. Phys. Lett. **34**(2), 150–152 (2008)
4. Erofeev, V.I., Artamonova, O.A.: Vliyaniye tochechnykh defektov v materiale na rasprostraneniye nelineynoy akusticheskoy volny [The effect of point defects in a material on the propagation of a nonlinear acoustic wave]. Trudy XXII Sessii Rossiyskogo akusticheskogo obshchestva i Nauchnogo soveta po akustike RAN, Moskva, GEOS, vol. 1, p. 159 (2010)
5. Erofeev, V.I., Kazhaev, V.V.: Nonlinear stationary acoustic wave in a solid with dislocations. Tech. Phys. **55**(4), 580–583 (2010)
6. Erofeev, V.I., Leonteva, A.V., Malhanov, A.O.: Effect of defects on the spatial localization of nonlinear acoustic waves. Bull. Russ. Acad. Sci. Phys. **82**(5), 520–525 (2018)
7. Erofeev, V.I., Leonteva, A.V., Shekoyan, A.V.: Shock waves in a thermoelastic medium with point defects. Tech. Phys. Russ. J. Appl. Phys. **90**(1) 2020
8. Erofeev, V.I., Leontieva, A.V., Malkhanov, A.O.: Spatial localization of nonlinear waves spreading in materials in the presence of dislocations and point defects. In: IOP Conference Series: Materials Science and Engineering, vol. 208, No. 012017, pp. 1–8 (2017)
9. Erofeev, V.I., Malkhanov, A.O.: Dispersion and self-modulation of waves propagating in a solid with dislocations. Phys. Mesomech. **22**(3), 173–180 (2019)
10. Erofeev, V.I., Romashov, V.P.: Effect of dislocations on the dispersion and damping of ultrasound in solids. Tech. Phys. Lett. **28**(3), 218–220 (2002)
11. Kadomtsev, B.B., Karpman, V.I.: Nonlinear waves. Sov. Phys. Usp. **14**(1), 40–60 (1971)
12. Kudryashov, N.A.: Analiticheskaia teoriia nelineinykh differentsial'nykh uravnenii [Analytical Theory of Nonlinear Differential Equations], p. 360. Institute for Computer Research, Moscow-Izhevsk (2004)
13. Kudryashov, N.A.: Metody nelineynoy matematicheskoy fiziki: Uchebnoye posobiye [Methods of nonlinear mathematical physics: Textbook], p. 368. Intellekt, Dolgoprudnyy (2010)
14. Mirzade, F.K., Shelepin, L.A.: Nonlinear longitudinal waves in the presence of the interaction between the strain and temperature fields and the field of nonequilibrium atomic defects. Acoust. Phys. **53**(6), 676–682 (2007)
15. Mirzoev, F.K., Panchenko, V.Y., Shelepin, L.A.: Laser control of processes in solids. Phys. Usp. **39**(1), 1–29 (1996)
16. Mirzoev, F.K., Shelepin, L.A.: Nonlinear strain waves and densities of defects induced in metal plates by external energy fluxes. Tech. Phys. **46**(8), 952–955 (2001)
17. Porubov, A.V.: Lokalizatsiia nelineinykh voln deformatsii [Localization of Nonlinear Deformation Waves], p. 208. Fizmatlit, Moscow (2009)
18. Sarafanov, G.F.: Kollektivnye i volnovye effekty v ansamble dislokatsii pri plasticheskoi deformatsii metallov [Collective and Wave Effects in a Dislocation Ensemble in the Process of Plastic Deformation of Metals], p. 359. Litera, Nizhny Novgorod (2010)
19. Shekoyan, A.V.: Wave beams in crystals with dislocations and quadratic nonlinearity. Tech. Phys. Lett. **35**(4), 337–339 (2009)

Chapter 11
Problem Statement Options for the Problem of the Corrosion Processes Influence on the Delayed Fracture of the Rod Under Creep Conditions

Leonid V. Fomin

Abstract The relevance of choosing an adequate version of the problem statement on the influence of corrosion processes on ensuring the safe operation of materials and structural elements is beyond doubt. Problem statement options for the delayed fracture of a rod stretched under creep conditions are considered. Two problem statement options are suggested: taking into account the multistage propagation of the diffusion–corrosion process throughout the entire thickness of the rod and taking into account the conjugation of the solutions at the boundary of the corrosion layer. To solve the problem, a mechanical-and-mathematical model has been developed, including the modified diffusion equation, the kinetic equation for damage accumulation, and the relation for the chemical interaction parameter. The parameters of this model are determined based on the experimental dependence of the corrosion film thickness on time. The multistage process of the corrosion layers fracture under the influence of increasing effective stress is taken into account.

Keywords Problem statement · Delayed fracture · Corrosive medium · Modified diffusion equation · Kinetic parameters · Chemical interaction parameter · Damage · Concentration · Rod

11.1 Introduction

Ensuring the safe operation of materials and structural elements subject to the influence of a corrosive medium is an important and urgent task to ensure the safety and reliability of critical structures throughout the entire life cycle. This influence is due to both diffusion penetration and the chemical interaction of the active medium with the material. Of particular importance is the study of such processes during high-temperature delayed fracture of metals under creep conditions [2–9]. In this

L. V. Fomin (✉)
Research Institute of Mechanics, Lomonosov Moscow State University, Moscow, Russia
e-mail: fleonid1975@mail.ru

H. Altenbach et al. (eds.), *Multiscale Solid Mechanics*,
Advanced Structured Materials 141,
https://doi.org/10.1007/978-3-030-54928-2_11

regard, high-tech studies of the high-temperature strength of materials and structures, including those exposed to active media, are quite relevant. The research proposed in this article is a continuation of the creep and long-term strength research [4] on creep and long-term strength taking into account the corrosive interaction of an aggressive medium with materials and structural elements begun at the laboratory of creep and long-term strength of the Institute of Mechanics of Moscow State Lomonosov University.

11.2 Aspects of Existing Models of Interaction of the Active Medium with the Material

Today, there is a huge number and variety of models for the interaction of the active medium with materials in the literature [7]. Due to the complexity of the phenomena under consideration, an interdisciplinary scientific approach is used, because of which the existing models can be divided into two large groups: physicochemical models and phenomenological models. Research of the authors who use physicochemical models is based on a deep atomic–molecular study of physicochemical processes in a material when exposed to an active medium, in particular, taking into account ion fluxes (anionic–cationic processes) and the motion of electrons involved in corrosion processes. Because of a detailed study, the authors construct physicochemical models that relate the basic thermodynamic and physicochemical parameters of the processes under study at a higher level than the atomic–molecular level. Thus, as a result, these models establish a relationship between medium parameters (temperature, humidity, chemical composition of the medium, duration of exposure, etc.) and the parameters of the corrosion process. Physicochemical models feature strict distinctness. When switching to other grades of material, with different shapes and sizes of cross sections of structural elements, other parameters of the corrosive medium, etc., the same model can give essentially different results. The author of this article considers the use of models under a phenomenological approach more promising.

The paramount importance of the fundamental approach is associated with the construction of mathematical phenomenological models based on the constitutive and kinetic equations and describing the processes of deformation and fracture of structures and the processes of interaction of materials and structures with these media. The identifications of these models, i.e., the determination of the values of the coefficients (material parameters) in these equations and the selection of suitable material functions are based on the processing of experimental data, because of solving the obtained set of equations, the behavior of structures under various conditions is analyzed.

In order to proceed to the construction of mathematical (phenomenological) models, it is necessary to choose phenomenological parameters, the change of which could be observed during the experiment, or structural parameters—the so-called kinetic parameters of the process of interaction of the material with the medium penetrating into it. The introduction of kinetic parameters into the model allows describing real experimental data using the mathematical model under consideration for various temperature–force loading programs. Phenomenological parameters include the layer thickness of a material that has already been subjected to the destructive influence of the medium (e.g., corrosion wear), and kinetic parameters also, according to some authors, include the layer thickness, the parameter of corrosion damage, the concentration of elements of an active medium in a metal, etc. Here it is necessary to note the universality of such models and their applicability to engineering calculations.

11.3 The Model of Corrosion Interaction of the Active Medium with the Material

When diffusion is accompanied by chemical reactions, the ordinary one-dimensional diffusion equation can be modified to take this into account and becomes [1]

$$\frac{\partial C}{\partial t} = D\frac{\partial^2 C}{\partial z^2} - \frac{\partial S}{\partial t},\ D = \text{const}, \tag{11.1}$$

where $C = C(z,\ t)$—the concentration of the substance penetrated into the material through physical diffusion, $S = S(z,t)$ — chemical concentration, t — time, z — coordinate along which the active medium penetrates, and D—diffusion coefficient.

Concentration S is related to concentration C by the parameter of the chemical interaction R, and $S = RC$. In the general case, these characteristics are functions of spatial coordinates and time, in the case of the one-dimensional process under consideration $C = C(z,t)$, $R = R(z,t)$, $S = S(z,t)$. The substitution of S into (11.1) results in:

$$\frac{\partial C}{\partial t} = \frac{D}{(1+R)}\frac{\partial^2 C}{\partial z^2},\ C = C(z,t). \tag{11.2}$$

If R is small, then physical diffusion is the predominant process. And vice versa, if R is big, then chemical interaction is the predominant process. To determine the chemical interaction parameter R, the time dependence of the corrosion film thickness δ is analyzed.

We accept the hypothesis that the rate of change of the corrosion film thickness is proportional to the rate of change of the concentration of a substance bound by a chemical reaction:

$$\frac{\partial \delta}{\partial t} = \gamma\frac{\partial S}{\partial t},\quad S = RC, \tag{11.3}$$

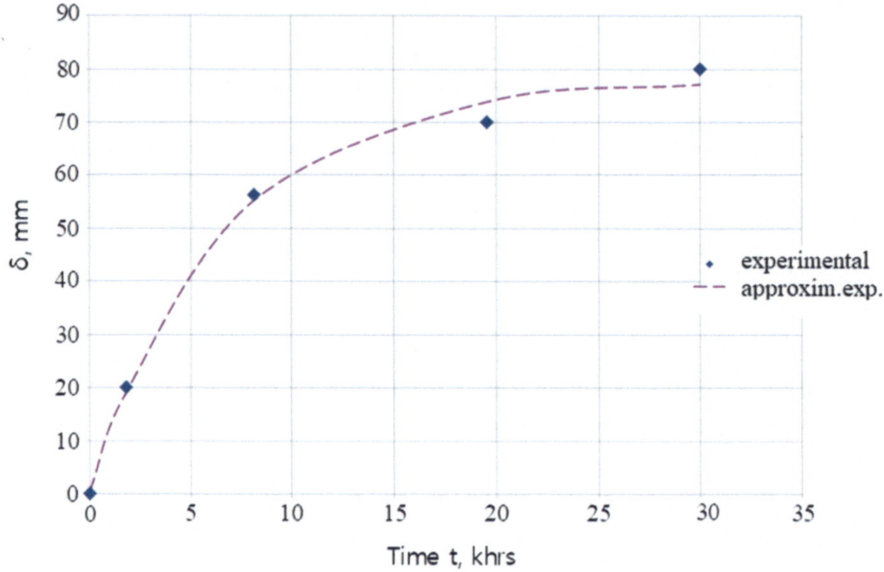

Fig. 11.1 Dependence of the oxide film thickness δ versus the test duration t of steel 10Cr15Ni9Si3Nb

where $\gamma = \text{const}$ — characteristic process constant. Thus, the ratio for R can be determined from the dependence $\delta(t)$.

As an example, there is the dependence of the corrosion film thickness on time considered, obtained in an experiment by the Central Research Institute of Structural Engineering "Prometey" (St. Petersburg) and published in [5]. Figure 11.1 shows the dependence of the oxide film on the duration of testing steel 10Cr15Ni9Si3Nb in the medium of a lead-based liquid metal coolant. The dots show the experimental data, and the dashed lines show the approximation of the experimental data using the dependence $\delta = A - B\exp(-\lambda t)$, where the constants have values $\lambda = 1.5 \cdot 10^{-4}\ (\text{h})^{-1}$, $A = 78.00$ mm, $B = 77.27$ mm.

In general, a corrosion layer may occur at the initial time ($\delta(t_0 = 0) \neq 0$).

Taking into account the accepted approximation for δ and hypothesis (11.3), the expression for the chemical interaction parameter has the following form:

$$R = \lambda B \exp(-\lambda t)\left[\gamma \frac{\partial C}{\partial t}\right]^{-1}. \tag{11.4}$$

Note that this approach can be applied to any nature and type of continuous dependence of the development of the corrosion film thickness over time. Only the type of approximation will change, which will lead to a different (in contrast to (11.4)) form of the dependence of the chemical interaction parameter on time and material constants or functions.

11.4 The System of Delayed Fracture Equations

To solve the problem stated, a system of ratios has been developed for modeling the influence of an active corrosive medium on the long-term strength of materials and structural elements [2–9]. This system of equations includes a modified diffusion equation [1], a kinetic equation for damage accumulation during creep, taking into account the influence of an active medium, and a relation for the chemical interaction parameter. It should be noted that the selected damage accumulation law (the second equation of system (11.5)) takes into account the influence of both the diffusion process and the corrosion process on damage accumulation. However, it is possible that in the further analysis the form of damage accumulation law can be clarified or modified.

The system of equations has the following form:

$$\begin{cases} \frac{\partial C}{\partial t} = \frac{D}{(1+R)}\frac{\partial^2 C}{\partial z^2}, \\ \frac{\partial \omega}{\partial t} = \frac{B_1(\sigma)^n}{(1-\omega)^n}(1 + aC(1+R)), \\ R = \lambda B \exp(-\lambda t)\left[\gamma \frac{\partial C}{\partial t}\right]^{-1}, \end{cases} \tag{11.5}$$

where σ — effective stress, $C = C(z, t)$ — the concentration of the substance penetrated into the material due to physical diffusion, $\omega = \omega(z, t)$ — damage, $R = R(z, t)$ — chemical interaction parameter, D — diffusion coefficient, λ, B — constants determined from the experimental dependence [5] of the corrosion film thickness on time $\delta(t)$, γ — representative constant of the corrosion process, B_1, n — steady creep constants, and a — material constant determined from the long-term strength experiments [2]. Further in the study, this system is considered in a dimensionless form.

11.5 The Multistage Process of Delayed Fracture of the Rod

Further, the article discusses the formulation of the delayed fracture problem for a stretchable rod under creep conditions taking into account diffusion and chemical interaction with an active medium. A long rod of rectangular cross section with thickness H_0 and width b (whereas $H_0 \ll b$) resides in a corrosive medium, the influence of which is characterized by diffusion penetration of its elements and corrosion interaction. Assume that relation (11.1) describes these processes. The diffusion of the medium into the material of the considered rod under consideration is characterized by a diffusion coefficient D, and the formation of a corrosion layer occurs in accordance with the process which is described by relations (11.3) and (11.4). Since $H_0 \ll b$ the characteristics of the medium are the same on both sides of the rod, we consider a symmetric one-dimensional diffusion–corrosion process along the thickness of the rod. The rod is subjected to tensile stress under creep conditions, during which damage accumulates.

A multistage fracture process is assumed, namely sequential destruction of the corrosion layers on the surface, and damage accumulation occurs under the influence of an increased effective stress at the stage of the fracture process under consideration. Here, a hypothesis is introduced about a certain incubation period of time that determines the duration of each stage of the corrosion process during which damage accumulates. The criterion for the destruction of the corrosion layer at each stage is the achievement by the parameter ω of a value equal to unity under the influence of the effective stress σ_{eff}.

Thus, at the first stage:

$$0 < t \le t_1^* \quad \rightarrow \quad \sigma = \sigma_{\text{eff\_1}} = \sigma_0 \quad \rightarrow \quad \omega_1^* = \omega\left(t_1^*, \delta_{k1}\right) = 1.$$

At the second stage:

$$t_1^* < t \le t_2^* \quad \rightarrow \quad \sigma = \sigma_{\text{eff\_2}} = \frac{\sigma_{\text{eff\_1}}}{\left(1 - 2\delta_{k1} / H_0\right)} \quad \rightarrow \quad \omega_2^* = \omega\left(t_2^*, \delta_{k2}\right) = 1$$

Here t_1^* and t_2^* — end times of the first and second stages of the corrosion process, respectively, and δ_{k1} and δ_{k2} — respective corrosion layer values. And so forth, until the rod is completely destroyed.

The total time to fracture of the rod t^{**} is determined based on either of the following two criteria.

(1) The effective stress will increase so much that the entire rod will completely collapse because of a short-term increased load. In fact, this means that the effective stress reaches the short-term strength limit $\sigma = \sigma_b|_{T=T_{\text{test}}}$ at the considered high test temperature $T = T_{\text{test}}$. In this case, the corrosion processes do not have time to influence the destruction at the final stage. Thus, the total time to fracture in this case t^{**} is determined by the criteria approach.

(2) The accumulation of damage inside the entire rod has a significant effect, and the sample will collapse when the damage level reaches the limit value $\omega^{**} = \omega\left(t^{**}\right) = 1$. Thus, the total time to fracture t^{**} in this case is determined by the kinetic approach.

11.6 Problem Statement Taking into Account the Multistage Distribution of the Diffusion–Corrosion Process Throughout the Entire Thickness of the Rod

Dimensionless variables are introduced:

$$\bar{z} = \frac{2z}{H_0}, \bar{C} = \frac{C}{C_0}, \bar{\sigma} = \frac{\sigma}{\sigma_b}, \bar{t} = t \cdot \frac{4D_0}{H_0^2}, \bar{D} = \frac{D}{D_0},$$
$$\bar{B}_1 = \frac{H_0^2 B_1 \sigma_b^n}{4D_0}, \bar{a} = aC_0, \bar{\lambda} = \lambda \frac{H_0^2}{4D_0}, \gamma_1 = \gamma \frac{4D_0 C_0}{H_0^2} \quad (11.6)$$

where C_0 — steady-state concentration on the surface, σ_b — short-term strength limit, and D_0— diffusion coefficient at room temperature. Taking into account the variables (11.6), the system of Eqs. (11.5) takes the following form:

$$\begin{cases} \frac{\partial \bar{C}}{\partial \bar{t}} = \frac{\bar{D}}{(1+R)} \frac{\partial^2 \bar{C}}{\partial \bar{z}^2}, \\ \frac{\partial \omega}{\partial \bar{t}} = \bar{B}_1 \frac{(\bar{\sigma})^n}{(1-\omega)^n} \left(1 + \bar{a}\bar{C}(1+R)\right), \\ R = \lambda B \exp\left(-\bar{\lambda}\bar{t}\right) \left[\gamma_1 \frac{\partial \bar{C}}{\partial \bar{t}}\right]^{-1}. \end{cases} \quad (11.7)$$

Substitution of the third equation of system (11.7) (equation for R) into the first equation of system (11.7) will result in the following equation:

$$\frac{\partial \bar{C}}{\partial \bar{t}} = \bar{D} \frac{\partial^2 \bar{C}}{\partial \bar{z}^2} - \frac{\lambda B}{\gamma_1} \exp\left(-\bar{\lambda}\bar{t}\right) \quad (11.8)$$

A coordinate system is chosen as follows: $\bar{z} = 0$ — center of symmetry of the rod cross section, and $\bar{z} = 1$ — outer border of the rod cross section. The initial and boundary conditions have the form depending on the stage of the corrosion fracture process.

(1) At the first stage of the process of corrosion fracture at $0 < \bar{t} \le \bar{t}_1^*$:

$\bar{C}(\bar{z}, 0) = 0$ — initial condition,
$\bar{C}\left(1, \bar{t}\right) = 1$ — boundary condition on the outer lateral surface of the rod.
From the symmetry conditions of the diffusion process relative to the middle of the rod cross section:

$$\frac{\partial \bar{C}}{\partial \bar{z}}\left(0, \bar{t}\right) = 0.$$

The zero distribution of damage at the initial time is assumed.

$$\omega(\bar{z}, 0) = 0.$$

The condition for the destruction of the corrosion layer $\bar{\delta}_{k1}$ at the end of the first stage $\bar{t}_1^*$:

$$\omega_1^* = \omega(\bar{z} = 1 - \bar{\delta}_{k1}, \bar{t} = \bar{t}_1^*) = 1.$$

The boundary of the corrosion layer $\bar{\delta}_{k1}$ is determined by the equation $\bar{S} = R\bar{C}$ characterizing the proportion of the substance that has entered into the chemical interaction. On the part of the metal, not subject to corrosion, the concentration of a substance associated with a chemical reaction is zero:$\bar{S}(1 - \bar{\delta}_{k1}, \bar{t}) = 0$.

(2) At the second stage of the process of corrosion fracture at $\bar{t}_1^* < \bar{t} \leq \bar{t}_2^*$:

$\bar{C}(\bar{z}, \bar{t}_1^*) = \bar{C}_1$ — concentration distribution along coordinate $\bar{z}$ obtained at the end of the first stage,

$\bar{C}(1 - \bar{\delta}_{k1}, \bar{t}) = 1, \quad \bar{t} > \bar{t}_1^*$ — the concentration value at the new outer boundary of the rod formed as a result of the corrosion layer fracture at the first stage,

$\frac{\partial \bar{C}}{\partial \bar{z}}(0, \bar{t}) = 0$ — boundary condition in the middle of the rod cross section (from the symmetry condition of the diffusion–corrosion process),

$\omega(\bar{z}, \bar{t}_1^*) = \omega_1$ — damage distribution along coordinate $\bar{z}$, obtained at the time $\bar{t}_1^*$ of the end of the first stage of corrosion fracture,

$\omega_2^* = \omega(\bar{z} = 1 - \bar{\delta}_{k1} - \bar{\delta}_{k2}, \bar{t} = \bar{t}_2^*) = 1$ — corrosion layer fracture $\bar{\delta}_{k2}$ at the time $\bar{t}_2^*$ of the end of the second stage,

$\bar{S}(1 - \bar{\delta}_{k1} - \bar{\delta}_{k2}, \bar{t}) = R(1 - \bar{\delta}_{k1} - \bar{\delta}_{k2}, \bar{t})\bar{C}(1 - \bar{\delta}_{k1} - \bar{\delta}_{k2}, \bar{t}) = 0$ — condition for determining the boundary of the corrosion layer on the side of the metal, not subject to corrosion.

11.7 Problem Statement Taking into Account the Conjugation of Solutions at the Boundary of the Corrosion Layer

Equation (11.8) with an exponential term is obtained taking into account the corrosion layer that extends from the surface into the interior of the material and considering the approximation of the dependence of the corrosion layer thickness on time. Next, the following problem statement is taken into account: Eq. (11.8) acts only in the area of the corrosion layer. Therefore, this Eq. (11.8) should not be considered over the entire thickness from 0 to 1, but only over the thickness of the corrosion layer. This corrosion layer is called area I.

In area II, which is adjacent to area I inside the material, there are no corrosion processes, but there is diffusive penetration of an active medium (classical parabolic diffusion equation). At the interface between these areas, with a sufficient degree of certainty, there can be the condition set for the equality of the concentrations of the diffusion component of concentration C. But the equality of the derivatives of concentrations with respect to the coordinate at the conjugation point is a condition whose approval requires additional analysis. The chemical component S in area I at this boundary is obviously 0 (since there is no longer a corrosion interaction deeper than this boundary in area II). The issue of the spread of damage across the border of areas I and II also requires additional reflection and analysis.

Given the features of the adopted statement, there are the corresponding equations written with initial and boundary conditions that determine the diffusion–corrosion process under consideration.

I—area of the corrosion layer from the surface deep into the material. In this area, there occur diffusion and chemical interaction (corrosion).

$$\frac{\partial \bar{C}_{\mathrm{I}}}{\partial \bar{t}} = \bar{D}\frac{\partial^2 \bar{C}_{\mathrm{I}}}{\partial \bar{z}^2} - \frac{\lambda B}{\gamma_1}\exp(-\bar{\lambda}\bar{t})$$
$$\bar{C}_{\mathrm{I}} = \bar{C}_{\mathrm{I}}(\bar{z}, \bar{t})$$

The initial and boundary conditions are as follows:

$$\bar{C}_{\mathrm{I}}(\bar{z}, 0) = 0,$$

$\frac{\partial \bar{C}_{\mathrm{I}}}{\partial \bar{z}}(1, \bar{t}) = \bar{\gamma}_m[\bar{C}_{\mathrm{I}}(1, \bar{t}) - 1]$ — accepted condition of mass transfer with the surrounding medium at the outer ($\bar{z} = 1$) boundary of the rod.

$$\bar{C}_{\mathrm{I}}(\bar{\xi}, \bar{t}) = \bar{C}_{\mathrm{II}}(\bar{\xi}, \bar{t}), \quad \bar{\xi} = 1 - \bar{\delta},$$

where $\bar{\xi} = \bar{\xi}(\bar{t})$ — coordinate of the border of the conjugation of areas I and II, $\bar{\delta}$ — surface corrosion layer at the thickness $\bar{z} = 1$.

$\frac{\partial \bar{C}_{\mathrm{I}}}{\partial \bar{z}}(\bar{\xi}, \bar{t}) = \frac{\partial \bar{C}_{\mathrm{II}}}{\partial \bar{z}}(\bar{\xi}, \bar{t})$ — as noted earlier, the adoption of this condition at the interface $\bar{\xi}$ between areas I and II requires additional analysis.

II—area of the material with no corrosion. In this area, only diffusion processes occur. Areas I and II conjugate on the border with the coordinate $\bar{\xi}$.

$$\frac{\partial \bar{C}_{\mathrm{II}}}{\partial \bar{t}} = \bar{D}\frac{\partial^2 \bar{C}_{\mathrm{II}}}{\partial \bar{z}^2}$$
$$\bar{C}_{\mathrm{II}} = \bar{C}_{\mathrm{II}}(\bar{z}, \bar{t})$$

The initial and boundary conditions are as follows:

$$\bar{C}_{\mathrm{II}}(\bar{z}, 0) = 0$$
$$\bar{C}_{\mathrm{I}}(\bar{\xi}, \bar{t}) = \bar{C}_{\mathrm{II}}(\bar{\xi}, \bar{t}), \quad \bar{\xi} = 1 - \bar{\delta},$$
$$\frac{\partial \bar{C}_{\mathrm{II}}}{\partial \bar{z}}(0, \bar{t}) = 0$$

Note that the interface between areas I and II is mobile, and its coordinate $\bar{\xi}$ depends on time $\bar{\xi} = \bar{\xi}(\bar{t})$.

11.8 Conclusion

This study has as a fundamental value in the development of the kinetic theory of creep and long-term strength of Rabotnov [9] taking into account the influence of a corrosive medium, as the applied nature. It can be used in the design of composite structural elements operated in corrosive media, for example, in power and petrochemical engineering.

Acknowledgements The author of this article is grateful to Professor Alexander Mikhailovich Lokoshchenko for his attention to this work and useful advice. This work was supported and partially funded by a grant from the Russian Foundation for Basic Research (RFBR) No. 20-08-00387.

References

1. Crank, J.: The Mathematics of Diffusion, 2nd edn. Clarendon Press, Oxford (1975)
2. Fomin, L.V.: Description of the long-term strength of stretchable rods of rectangular and round cross-sections in high-temperature air. J. Samara State Tech. Univ. Ser. Phys. Math. Sci. **3**(32), 87–97 (2013)
3. Fomin, L.V.: Steady-state creep of a composite rod in tension in the presence of an aggressive environment. Mech. Compos. Mater. **52**(6), 741–750 (2017)
4. Fomin, L.V., Basalov, YuG, Lokoshchenko, A.M.: On taking into account the influence of corrosion processes on the long-term fracture of a rod under creep conditions. Mech. Compos. Mater. Struct. **25**(3), 327–335 (2019)
5. Kudryavtsev, A.S., Markov, V.G., Lavrukhin, V.S.: Long-term strength of steel in a lead-based liquid metal coolant. Voprosy materialovedeniya **4**(48), 89–94 (2006)
6. Lokoshchenko, A.M.: Creep and Long-Term Strength of Metals. CISP. CRC Press. Taylor & Francis Group. Boca. Raton, London. New York (2018)
7. Lokoshchenko, A.M., Fomin, L.V.: Modeling the behavior of materials and structural elements exposed to aggressive media (review). Strength Plast. Prob. **80**(2), 145–179 (2018)
8. Lokoshchenko, A.M., Fomin, L.V.: Delayed fracture of plates under creep condition in unsteady complex stress state in the presence of aggressive medium. Appl. Math. Model. **60**, 478–489 (2018)
9. Rabotnov, Y.N.: Creep Problems in Structural Members. North-Holland, Amsterdam (1969)

Chapter 12
Quasistatic and Dynamic Deformation of an Asymptotically Thin Perfectly Rigid-Plastic Spherical Layer

Dimitri Georgievskii and Ravil Shabaykin

Abstract The axisymmetric meridional flow of an incompressible perfectly rigid-plastic medium between two concentric rough spheres, such that the outer sphere is fixed and the surface of the inner one is uniformly expanded, is studied in the case of a sink. Two problem statements are considered: quasistatic and dynamic. An asymptotic integration of a boundary value problem with a natural small geometric parameter equal to the ratio of the distance between concentric spheres to the inner radius is performed. The dynamic statement of the problem includes one more dimensionless parameter which does not depend on time (in contrast to geometric one) and equals to the inverse Euler number. This value is also taken much less than one. Depending on the ratio of these parameters, i.e., at different time intervals, using the asymptotic integration procedure, the coefficients of several principal terms in the expansions for both velocities and stresses are obtained in analytical form.

Keywords Quasistatic and dynamic deformation · Rigid-plastic medium · Asymptotic integration procedure · Concentric rough spheres

12.1 Introduction

The classical Prandtl problem [17] has obtained numerous generalizations due to its applicability in pressure metal treatment theory. For example, Prandtl solution is generalized in Ref. [14] for the case of maximum tangential stress linear dependence on average pressure and for the case of layer compression with tilted rough plates as well as plates bent in the form of concentric circles. Solution of the problem concerning shaft extrusion from compressing rough sleeve is shown in Ref. [7], and process of plastic flow along the elastically deformable surfaces is analyzed in Ref.

D. Georgievskii (✉) · R. Shabaykin
Mechanical and Mathematical Department, Moscow State University, Moscow 119991, Russia
e-mail: georgiev@mech.math.msu.su

R. Shabaykin
e-mail: rr.shabaykin@outlook.com

H. Altenbach et al. (eds.), *Multiscale Solid Mechanics*,
Advanced Structured Materials 141,
https://doi.org/10.1007/978-3-030-54928-2_12

[10]. Paper [16] shows solution taking multiple layers and thermal diffusion into account, and compressible material case is examined in Ref. [11].

Generalizations with dynamic setting should be highlighted apart from the others. Thus, Ref. [8] states importance of inertia forces consideration when simulating high-speed plastic flows. Prandtl dynamic problem analytic solution is derived in Refs. [1, 5, 12]. Paper [6] shows problem of ideally plastic disk compression with rough plates taking inertia forces into account. In work [18], solution of plastic theory plane dynamic problem upon condition of power-law hardening was derived.

Detailed description can be found in studies and manuals [7, 9, 14].

12.2 The Formulation of Problem

In the spherical coordinate system (r, θ, φ), where θ is the polar angle, we consider the axisymmetric meridional flow of an incompressible perfectly rigid-plastic medium with a yield stress σ_s (the Saint-Venant flow) in the thin spherical layer

$$\Omega_t = \{R(t) < r < R(t) + h(t); 0 \le \theta < \pi; 0 \le \varphi < 2\pi\}, \ h \ll R. \tag{12.1}$$

The process of compression between the rigid rough spheres $r = R$ and $r = R + h$ is studied. The layer Ω_t is displaced through the sink $\theta = \pi$. It is assumed that the outer sphere is fixed, and the radial velocity V of the inner sphere surface is not time-dependent. The kinematic impermeability boundary conditions are of the form

$$v_r|_{r=R} = V, \ v_r|_{r=R+h} = 0. \tag{12.2}$$

As is well-known [15], the velocity tangential component v_θ should not be specified at the boundaries given in (12.2). In this case, the sink flow rate is equal to $-4\pi R^2 V$.

Now, we consider the following system of equations expressed in the spherical coordinates and used in the theory of axisymmetric plasticity with the Mises–Hencky criterion:

$$-p_{,r} + s_{rr,r} + \left(s_{r\theta,\theta} + 3s_{rr} + s_{r\theta}\cot\theta\right)/r = \rho_m \left(v_{r;t} + v_r v_{r,r} + \left(v_\theta v_{r,\theta} - v_\theta^2\right)/r\right) \tag{12.3}$$

$$-p_{,\theta}/r + s_{r\theta,r} + \left(s_{\theta\theta,\theta} + 3s_{r\theta} + (s_{rr} + 2s_{\theta\theta})\cot\theta\right)/r$$
$$= \rho_m \left(v_{\theta;t} + v_r v_{\theta,r} + \left(v_\theta v_{\theta,\theta} - v_r v_\theta\right)/r\right), \tag{12.4}$$

$$s_{rr}^2 + s_{\theta\theta}^2 + s_{rr}s_{\theta\theta} + s_{r\theta}^2 = \sigma_s^2 \equiv \tau_s^2, \tag{12.5}$$

$$s_{rr}\left(v_{\theta,\theta} + v_r\right)/r = s_{\theta\theta} v_{r,r}, \ s_{rr}\left(v_{\theta,r} + \left(v_{r,\theta} - v_\theta\right)/r\right) = 2s_{r\theta} v_{r,r} \tag{12.6}$$

$$v_{r,r} + \left(2v_r + v_{\theta,\theta} + v_\theta \cot\theta\right)/r = 0. \tag{12.7}$$

Here, ρ_m is medium density; p is pressure; s_{rr}, $s_{\theta\theta}$, and $s_{r\theta}$ are the stress deviator components; and v_r and v_θ are the velocity components. In addition to the fulfillment of conditions (12.2), we require that the shear stress modulus $s_{r\theta}$ be maximal in r on the rigid spherical surfaces:

$$|s_{r\theta}|_{r=R} = |s_{r\theta}|_{r=R+h} = m(\theta)\,\tau_s \tag{12.8}$$

Here, m is the roughness of the press in use. The absolute roughness corresponds to $m = 1$; in other words, in this case, the press is in the total coupling with the flowing material.

12.3 Quasistatic Formulation

The quasistatic process is characterized by the absence of inertial terms in the motion equations. Therefore, the right-hand parts of (12.3) and (12.4) will be zero.

Let us consider the asymptotically small geometric parameter $\alpha = h/R$. Then, we expand the six unknown functions of (12.3)–(12.7) in power series in this small parameter. The structure of these expansions is chosen to be similar to that of the solution to the classical Prandtl problem [9]:

$$v_r(r,\theta) = V\sum_{k=0}^{\infty}\alpha^k\tilde{v}_r^{\{k\}},\quad v_\theta(r,\theta) = V\sum_{k=-1}^{\infty}\alpha^k\tilde{v}_\theta^{\{k\}}, \tag{12.9}$$

$$p(r,\theta) = \tau_s\sum_{k=-1}^{\infty}\alpha^k\tilde{p}^{\{k\}},\quad s_{\beta\gamma}(r,\theta) = \tau_s\sum_{k=0}^{\infty}\alpha^k\tilde{s}_{\beta\gamma}^{\{k\}} \tag{12.10}$$

Here, $(\beta;\gamma) = \{(r;r), (\theta;\theta), (r;\theta)\}$. The dimensionless overlined coefficients of these series are functions of the dimensionless coordinates θ and $\rho = (r-R)/h$, $0 < \rho < 1$. In (12.9) and (12.10), the terms $\alpha^{-1}\tilde{v}_\theta^{\{-1\}}$ and $\alpha^{-1}\tilde{p}^{\{-1\}}$ indicate that v_θ and p tend to infinity as $\alpha \to 0$; the other unknowns remain finite outside the interval $\pi - \varepsilon < \theta \le \pi$.

Substituting (12.9) and (12.10) into (12.3)–(12.7) and equating the coefficients at the first two higher powers of α, we come to the following system of equations:

$$\tilde{p}_{,\rho}^{\{-1\}} = 0,\ \tilde{v}_{\theta,\rho}^{\{-1\}} = 0,\ -\tilde{p}_{,\rho}^{\{0\}} + \tilde{s}_{rr,\rho}^{\{0\}} = 0,\ -\tilde{p}_{,\theta}^{\{-1\}} + \tilde{s}_{r\theta,\rho}^{\{0\}} = 0, \tag{12.11}$$

$$\left(\tilde{s}_{rr}^{\{0\}}\right)^2 + \left(\tilde{s}_{\theta\theta}^{\{0\}}\right)^2 + \tilde{s}_{rr}^{\{0\}}\tilde{s}_{\theta\theta}^{\{0\}} + \left(\tilde{s}_{r\theta}^{\{0\}}\right)^2 = 1,\ \tilde{s}_{rr}^{\{0\}}\tilde{v}_{\theta,\theta}^{\{-1\}} = \tilde{s}_{\theta\theta}^{\{0\}}\tilde{v}_{r,\rho}^{\{0\}}, \tag{12.12}$$

$$\tilde{s}_{rr}^{\{0\}}\left(\tilde{v}_{\theta,\rho}^{\{0\}} - \tilde{v}_\theta^{\{-1\}}\right) = 2\tilde{s}_{r\theta}^{\{0\}}\tilde{v}_{r,\rho}^{\{0\}},\ \tilde{v}_{r,\rho}^{\{0\}} + \tilde{v}_{\theta,\theta}^{\{-1\}} + \tilde{v}_\theta^{\{-1\}}\cot\theta = 0. \tag{12.13}$$

Here, the unknown quantities are the coefficients of (12.9) and (12.10) indicated by the superscripts $\{-1\}$ and $\{0\}$. From the boundary conditions (12.2) and (12.8), it follows that

$$\tilde{v}_r^{\{0\}}|_{\rho=0} = 1,\ \tilde{v}_r^{\{0\}}|_{\rho=1} = 0,\ |\tilde{s}_{r\theta}^{\{0\}}|_{\rho=0} = |\tilde{s}_{r\theta}^{\{0\}}|_{\rho=1} = m(\theta) \tag{12.14}$$

It is not known in advance on which sphere the value of $\tilde{s}_{r\theta}^{\{0\}}$ is equal to m or $-m$. In [2], it is shown that, for the stress $\tilde{s}_{r\theta}^{\{0\}}$, the choice of its sign is associated with the convexity of the profile of the velocity $\tilde{v}_r^{\{0\}}$ (in other words, with the sign of $\tilde{v}_r^{\{0\}}$ dependent on whether the above spheres are approaching or moving apart).

System (12.11)–(12.13) with the boundary conditions (12.14) can be asymptotically integrated according to the procedure discussed in [3–6]. Applying this procedure, we can represent the analytic solution to problem (12.11)–(12.13), (12.14) in the following form for the case of compression and the sink $\theta = \pi$:

$$\tilde{v}_\theta^{\{-1\}} = \tan\frac{\theta}{2},\ \tilde{v}_r^{\{0\}} = 1 - \rho,\ \tilde{p}^{\{-1\}} = \tilde{p}_0^{\{-1\}} - 2\int_0^\theta m(\xi)\,\mathrm{d}\xi, \tag{12.15}$$

$$\tilde{s}_{rr}^{\{0\}} = -\frac{\sin^2\theta}{\sqrt{(1-\cos\theta)\left(1-\cos^3\theta\right)}}\sqrt{1 - m^2(2\rho-1)^2}, \tag{12.16}$$

$$\tilde{s}_{\theta\theta}^{\{0\}} = \sqrt{\frac{1-\cos\theta}{1-\cos^3\theta}}\sqrt{1-m^2(2\rho-1)^2},\ \tilde{s}_{r\theta}^{\{0\}} = -m(2\rho-1),\ \tilde{p}^{\{0\}} = \tilde{s}_{rr}^{\{0\}} + f(\theta), \tag{12.17}$$

$$\tilde{v}_\theta^{\{0\}} = \rho\tan\frac{\theta}{2} + \frac{\sqrt{(1-\cos\theta)\left(1-\cos^3\theta\right)}}{m\sin^2\theta}\sqrt{1-m^2(2\rho-1)^2} + g(\theta) \tag{12.18}$$

The integration functions $f(\theta)$ and $g(\theta)$ can be found by considering a system under the next degree of α. Equating the coefficients at α^1, we come to the following system of equations:

$$-\rho\tilde{p}_{,\theta}^{\{-1\}} - \tilde{p}_{,\theta}^{\{0\}} + \tilde{s}_{r\theta,\rho}^{\{1\}} + \tilde{s}_{\theta\theta,\theta}^{\{0\}} + 3\tilde{s}_{r\theta}^{\{0\}} + \left(\tilde{s}_{rr}^{\{0\}} + 2\tilde{s}_{\theta\theta}^{\{0\}}\right)\cot\theta = 0, \tag{12.19}$$

$$\tilde{v}_{r,\rho}^{\{1\}} + 2\tilde{v}_r^{\{0\}} + \tilde{v}_{\theta,\theta}^{\{0\}} - \rho\tilde{v}_{\theta,\theta}^{\{-1\}} + \left(\tilde{v}_\theta^{\{0\}} - \rho\tilde{v}_\theta^{\{-1\}}\right)\cot\theta = 0 \tag{12.20}$$

Using boundary condition (12.8), we can write $\int_0^1 \tilde{s}_{r\theta,\rho}^{\{1\}} d\rho = 0$, and substituting $\tilde{s}_{r\theta,\rho}^{\{1\}}$ from (12.19), we come to the following:

$$\int_0^1 \left(\rho\tilde{p}_{,\theta}^{\{-1\}} + \tilde{p}_{,\theta}^{\{0\}} - \tilde{s}_{\theta\theta,\theta}^{\{0\}} - 3\tilde{s}_{r\theta}^{\{0\}} - \left(\tilde{s}_{rr}^{\{0\}} + 2\tilde{s}_{\theta\theta}^{\{0\}}\right)\cot\theta\right)\mathrm{d}\rho$$

$$= \frac{\partial}{\partial\theta}\int_0^1 \left(\rho\tilde{p}^{\{-1\}} + \tilde{p}^{\{0\}} - \tilde{s}_{\theta\theta}^{\{0\}} - \int\left(3\tilde{s}_{r\theta}^{\{0\}} + \left(\tilde{s}_{rr}^{\{0\}} + 2\tilde{s}_{\theta\theta}^{\{0\}}\right)\cot\theta\right)d\theta\right)\mathrm{d}\rho = 0 \tag{12.21}$$

This equation allows us to express unknown function $f(\theta)$:

$$f(\theta) = \tilde{p}_0^{\{0\}} + \int_0^1 \left(\int\left(3\tilde{s}_{r\theta}^{\{0\}} + \left(\tilde{s}_{rr}^{\{0\}} + 2\tilde{s}_{\theta\theta}^{\{0\}}\right)\cot\theta\right)d\theta - \rho\tilde{p}^{\{-1\}} - \tilde{s}_{rr}^{\{0\}} + \tilde{s}_{\theta\theta}^{\{0\}}\right)\mathrm{d}\rho \tag{12.22}$$

Similarly, using the boundary condition (12.2) and Eq. (12.20), the function $g(\theta)$ can be found.

The solution expressed by (12.15) is valid outside the interval $\pi - \varepsilon < \theta \le \pi$; on this interval, the values of $\tilde{v}_\theta^{\{-1\}}$ and $\tilde{v}_\theta^{\{0\}}$ are no more finite, and series (12.9) are no more asymptotic. This solution is also valid outside the interval $0 < \theta \le \varepsilon$. These validity conditions are similar to those of the classical Prandtl problem whose solution is valid at a large distance from the boundaries and from the midsection of the layer.

12.4 Dynamic Formulation

Now, unknown functions of (12.3)–(12.7) are time-dependent, and dimensionless overlined coefficients of series (12.9, 12.10) are functions of the dimensionless coordinates θ, $\rho = (r - R)/h$, $0 < \rho < 1$ and $\tau = Vt/h$.

Substituting these expansions into (12.3) and (12.4) and taking $v_{x;t} = v_{x,\alpha}\dot{\alpha} + v_{x,\rho}\dot{\rho} + v_{x,\tau}\dot{\tau}$ into account with being mindful of the fact that derivatives with respect to time are expressed as evolutionary functions in the following manner:

$$\dot{\alpha} = -\frac{V}{h}\alpha(\alpha + 1), \quad \dot{\rho} = -\frac{V}{h}(1 - \rho), \quad \dot{\tau} = \frac{V}{h}(1 + \tau), \tag{12.23}$$

we come to multiplier in right-hand side equals to $\rho_m V^2/\tau_s$. This value equals reciprocal of Euler number. It is supposed to be small—just as geometric parameter—but in contrast to the latest, it does not change during compression. Thus, the process can be divided into stages taking proposed parameters comparability into account:

$$\mathrm{Eu}^{-1} = O\left(\alpha^\beta\right) = C_\beta \alpha^\beta, \; C_\beta = O(1). \tag{12.24}$$

For exponent β, following interval is of interest: $0 < \beta \le 2$. It includes two whole-number values $\beta = 2$ and $\beta = 1$.

12.4.1 *Case $\beta = 2$*

Equating the coefficients at the first, two higher powers of α give us the system (12.11)–(12.13), and therefore, the expressions for unknowns will coincide with (12.15)–(12.18) up to the integration constant and functions, which will now depend on time

$$\tilde{p}_0^{\{-1\}} \to \tilde{p}_0^{\{-1\}}(\tau), \quad f(\theta) \to f(\theta, \tau), \quad g(\theta) \to g(\theta, \tau). \tag{12.25}$$

Equating the coefficients at α^1, we come to the following system of equations:

$$-\rho\tilde{p}^{\{-1\}}_{,\theta} - \tilde{p}^{\{0\}}_{,\theta} + \tilde{s}^{\{1\}}_{r\theta,\rho} + \tilde{s}^{\{0\}}_{\theta\theta,\theta} + 3\tilde{s}^{\{0\}}_{r\theta} + \left(\tilde{s}^{\{0\}}_{rr} + 2\tilde{s}^{\{0\}}_{\theta\theta}\right)\cot\theta$$
$$= C_2\left(\tilde{v}^{\{-1\}}_{\theta} + \tilde{v}^{\{-1\}}_{\theta}\tilde{v}^{\{-1\}}_{\theta,\theta}\right), \tag{12.26}$$

$$\tilde{v}^{\{1\}}_{r,\rho} + 2\tilde{v}^{\{0\}}_{r} + \tilde{v}^{\{0\}}_{\theta,\theta} - \rho\tilde{v}^{\{-1\}}_{\theta,\theta} + \left(\tilde{v}^{\{0\}}_{\theta} - \rho\tilde{v}^{\{-1\}}_{\theta}\right)\cot\theta = 0 \tag{12.27}$$

Using the procedure described in the previous section and denoting quasistatic solution with upper index 'qs,' we find the expression for $f(\theta,\tau)$:

$$f(\theta,\tau) = f^{\mathrm{qs}}|_{\tilde{p}^{\{0\}}_0 \to \tilde{p}^{\{0\}}_0(\tau)} + C_2\left(2\ln\cos\frac{\theta}{2} - \frac{1}{2}\sec^2\frac{\theta}{2}\right). \tag{12.28}$$

Since Eqs. (12.20) and (12.27) coincide, the expression for function g will remain unchanged.

12.4.2 Case $\beta = 1$

If $\beta = 1$, dynamic effects will manifest themselves in more significant terms of the expansions. Equating the coefficients in the motion Eqs. (12.3) and (12.4) at the α^0, we come to the following system of equations:

$$-\tilde{p}^{\{0\}}_{,\rho} + \tilde{s}^{\{0\}}_{rr,\rho} = C_1\left(\tilde{v}^{\{-1\}}_{\theta}\right)^2, \quad -\tilde{p}^{\{-1\}}_{,\theta} + \tilde{s}^{\{0\}}_{r\theta,\rho} = C_1\left(\tilde{v}^{\{-1\}}_{\theta} + \tilde{v}^{\{-1\}}_{\theta}\tilde{v}^{\{-1\}}_{\theta,\theta} + (1+\tau)\,\tilde{v}^{\{-1\}}_{\theta,\tau}\right). \tag{12.29}$$

The remaining equations obtained by equating the coefficients at the first two higher powers of α will coincide with the system (12.11)–(12.13).

Similarly to the previous section, the solution will coincide with the quasistatic one up to the integration functions, with the exception of pressure members, which will take the following form:

$$\tilde{p}^{\{-1\}} = \tilde{p}^{\{-1\}}_0(\tau) + C_1\left(2\ln\cos\frac{\theta}{2} - \frac{1}{2}\sec^2\frac{\theta}{2}\right) - 2\int_0^{\theta} m(\xi)\,d\xi, \tag{12.30}$$

$$\tilde{p}^{\{0\}} = \tilde{s}^{\{0\}}_{rr} - C_1\rho\tan^2\frac{\theta}{2} + f(\theta,\tau) \tag{12.31}$$

Due to its bulkiness, the expressions for the function $f(\theta,\tau)$ are omitted here.

12.4.3 Analysis of Results

The obtained solutions do not differ from quasi-static case (12.15)–(12.18) except for pressure function in which summand $C_\beta\left(2\ln\cos\frac{\theta}{2}-\frac{1}{2}\sec^2\frac{\theta}{2}\right)$ appears in terms with α^{-1} and α^0 in modes $\beta=1$ and $\beta=2$ correspondingly. It rises total force acting on the sphere from the layer side.

It is natural to require pressure on layer boundaries to coincide with atmosphere pressure p_{atm} (it is assumed hereunder that p_{atm} is value of α^0 order); however, this summand with $\theta\to\pi$ tends to infinity. Thus, if the layer fills the complete area, it is impossible to fulfill this boundary condition not breaking the series asymptotic nature in terms of Poincare [13].

References

1. Bikovtsev, G.I.: On plastic layer compression with rigid rough plates taking inertia forces into account. Izv. AN SSSR. OTN. Mekhanika i Mashinostroenie **6**, 140–2 (1960). [in Russian]
2. Georgievskii, D.V.: Asymptotic expansions and the possibilities to drop the hypotheses in the Prandtl problem. Mech. Solids **44**(1), 70–8 (2009)
3. Georgievskii, D.V.: Asymptotic analysis of plastic flow along the generatrix in a thin cylindrical layer. J. Appl. Mech. Tech. Phys. **51**(5), 713–20 (2010)
4. Georgievskii, D.V.: Compression and outflow of an asymptotically thin perfect rigid-plastic spherical layer. Moscow Univ. Mech. Bull. **66**(6), 148–150 (2011)
5. Georgievskii, D.V.: Asymptotic integration of the Prandtl problem in dynamic statement. Mech. Solids **48**(1), 79–85 (2013)
6. Georgievskii D.V., Shabaykin, R.R.: Quasistatic and dynamic compressing of a plane circular perfect plastic layer by rigid plates. Math. Model. Exp. Mech. Deform. Solid **1**, 56–63 (2017) [in Russian]
7. Hill, R.: The Mathematical Theory of Plasticity. Clarendon Press, Oxford (1950)
8. Ilyushin, A.A.: Works. Vol. IV. Simulation of Dynamic Processes in Solids and Engineering Applications. Fizmatlit, Moscow (2009) [in Russian]
9. Ishlinskii, A.Yu, Ivlev, D.D.: Mathematical Theory of Plasticity, 2nd edn. Fizmatlit, Moscow (2001) [in Russian]
10. Kiiko, I.A.: On compressed plastic thin layer effect on elastic surfaces. Izv. AN SSSR. OTN. Mekhanika i Mashinostroenie **6**, 1082–5 (1961). [in Russian]
11. Kiiko, I.A.: A generalization of Prandtl's problem on contraction of a strip to the case of compressible materials. Moscow Univ. Mech. Bull. **57**(4), 16–21 (2002)
12. Kiiko, I.A., Kadymov, V.A.: Generalization of the Prandtl problem on the compression of a band. Moscow Univ. Mech. Bull. **58**(4), 31–6 (2003)
13. Kravchenko, V.F., Nesenenko, G.A., Pustovoit, V.I.: Poincare Asymptotics of Solution to Problems of Irregular Heat and Mass Transfer. Fizmatlit, Moscow (2006) [in Russian]
14. Nadai, A.: Theory of Flow and Fracture of Solids. Wiley, New York (1950)
15. Pobedrya, B.E.: Numerical Methods in the Theory of Elasticity and Plasticity. Moscow State Univ., Moscow (1995)
16. Pobedrya, B.E., Guzei, I.L.: Mathematical modeling of deformation of composites with thermodiffusion taken into account. Math. Simul. Syst. Process. **6**, 82–91 (1998) [in Russian]
17. Prandtl, L.: Anwendungsbeispiele zu einem Henckyschen Satz ber das plastiche Gleichgewicht. ZAMM **3**(6), 401–406 (1923)
18. Zadoyan, M.A.: Spatial Problems of the Theory of Plasticity. Nauka, Moscow (1992). [in Russian]

Chapter 13
Modeling of Fiber Composite Wear

Irina G. Goryacheva and Yulia Yu. Makhovskaya

Abstract A model is proposed to describe the steady-state regime of wear for a fibrous composite material in contact with a rigid counter body. The composite material is modeled by an elastic half-space with embedded elastic fibers arranged parallel to each other and distributed uniformly. The fibers hardness is assumed to be different from that of the matrix, whereas their elastic moduli are equal. Analytic relations are obtained, and analysis is performed for the characteristics of wear (worn surface shape, effective wear rate, contact pressure distribution) depending on the composite structure parameters such as fiber size and density, relative hardness of the matrix and fibers. In particular, it is shown that the effective wear rate as a function of the fiber density increases if the fibers are harder than the matrix and decreases in the opposite case. The results obtained can be used to control the wear resistance of fibrous composites by choosing appropriate microstructure parameters.

Keywords Fiber composites · Worn surface shape · Effective wear rate

13.1 Introduction

Composite materials are widely used in aerospace tribological units such as disks of aircraft brake systems made from carbon-carbon fiber composites. Such disks are manufactured from carbon fibers distributed in a carbon matrix. To provide effective braking process, these materials must have a high wear resistance and low wear of the counter body.

Most of the existing theoretical studies of composite materials are aimed on modeling their mechanical and strength characteristics based on properties of their components and on prediction of their behavior under small strains [1, 11].

I. G. Goryacheva (✉) · Y. Yu. Makhovskaya
Ishlinsky Institute for Problems in Mechanics RAS, Prospekt Vernadskogo 101-1, Moscow 119526, Russian Federation
e-mail: goryache@ipmnet.ru

Y. Yu. Makhovskaya
e-mail: makhovskaya@mail.ru

H. Altenbach et al. (eds.), *Multiscale Solid Mechanics*,
Advanced Structured Materials 141,
https://doi.org/10.1007/978-3-030-54928-2_13

Also, numerical methods of modeling the stress-deformation behavior of composite materials based on mechanical characteristics of their components are developed [6, 10, 13].

Experimental results on the wear resistance of fiber composites under various interface conditions are presented in [2, 9, 12], where the wear rate is analyzed depending on the load and sliding velocity.

Available theoretical models of wear of composites are focused on particular wear mechanisms under given friction conditions [4, 14] and on analysis of the structure effect on the wear resistance.

In the present study, a model of wear process of fiber composites is developed for steady-state conditions. The model is used to analyze the influence of the composite structure (fiber size and density, relative hardness of fiber and matrix materials) on the effective wear rate of the composite material, its worn surface shape, and contact pressure distribution in the wear process.

13.2 Model of Steady-State Wear of Fiber Composites

Consider a composite material consisting of a matrix with embedded identical fibers. Let (x, y, z) be the coordinate system, such that the xy-plane coincides with the contact surface of the composite material, and the z-axis is directed opposite to the normal to this surface (Fig. 13.1). The composite material is modeled by an elastic half-space, fibers being distributed uniformly and directed along the y-axis. The

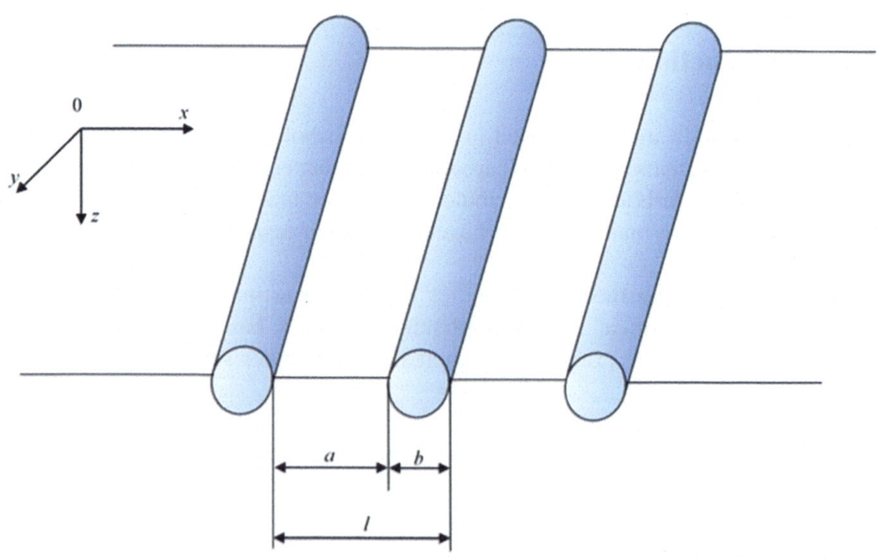

Fig. 13.1 Surface structure of the composite material

length of each fiber is L, the distance between fibers in the plane $z = 0$ in the direction of the x-axis is a, and the fiber diameter is b.

A plane surface of a rigid counter body slides over the composite material surface with the constant velocity V in the direction of the y-axis and wears the composite material.

The composite material is modeled by a homogeneous elastic half-space with a variable wear coefficient. These assumptions are based on the experimental measurements of the elastic moduli and hardness of fibers and matrix of carbon-carbon fiber composites, the results of which indicate the Young's moduli of fiber and matrix being close to each other, whereas their hardness generally different from each other [8]. Since the value of hardness usually correlates with the value of the wear coefficient [7], we consider the composite half-space to have a variable wear coefficient.

Taking into account the assumptions made above, we consider the wear contact problem in 2D formulation in the plane $y = 0$. The wear coefficient is the periodic function $K_\omega(x)$ with period l: $K_\omega(x + l) = K_\omega(x)$. The wear equation is adopted in the form [3]:

$$\frac{\partial \omega_*(x,t)}{\partial t} = K_\omega(x)\left[\frac{p(x,t)}{p^*}\right]^\alpha \tag{1}$$

where $K_\omega(x)$ is a stepwise function determined by the equation:

$$K_\omega(x) = \begin{cases} K_{\omega 1}, \ x \in [nl, a + nl] \\ K_{\omega 2}, \ x \notin [nl, a + nl] \end{cases} \tag{2}$$

Here $K_{\omega 1}$ and $K_{\omega 2}$ are the wear coefficients of matrix and fiber, respectively. The exponent α in Eq. (1) is assumed to be the same for fibers and matrix.

The contact problem under consideration is periodic with the period l. At the initial instant of time $t = 0$, contact between the composite material and the rigid counter body is continuous and the pressure $p(x,t)$ is distributed uniformly along the contact surface, so $p(x, 0) = P(0)/l. x \in (-\infty; +\infty)$. Here $P(0)$ is the specific load (applied on a segment of periodicity) at the initial instant of time. During the wear process, the contact pressure is redistributed and the surface shape of the composite material changes. Since the counter body slides in the direction perpendicular to the xz-plane, we neglect the influence of the shear stress on the contact pressure. The elastic $u_z(x,t)$ and wear $\omega_*(x,t)$ surface displacements as well as the contact pressure $p(x,t)$ are periodic functions of the coordinate x. They are determined from the solution of the wear contact problem for the elastic half-plane with the following contact condition:

$$u_z(x,t) + \omega_*(x,t) = D(t) \tag{3}$$

where $D(t)$ is the approach of the contacting bodies due to wear. The elastic displacements $u_z(x,t)$ are related with contact pressure $p(x,t)$ by the equation [5]:

$$u_z(x,t) = -\frac{2(1-\nu^2)}{\pi E}\int_0^l p(x',t)\ln 2\left|\sin\frac{\pi(x'-x)}{l}\right|dx' \quad (4)$$

The analysis of the complete system of Eqs. (1), (3), and (4) of the wear contact problem makes it possible to conclude (see [3]) that there exists the steady-state solution of the problem for $t \to +\infty$ which is presented by the following relationships for the steady-state pressure $p_\infty(x)$, wear rate $D_\infty = d\omega_*/dt$, and shape $f_\infty(x)$ of the worn surface:

$$p_\infty(x) = \begin{cases} p^*\left(\frac{D_\infty}{K_{\omega 1}}\right)^{1/\alpha}, & x \in [nl, a+nl] \\ p^*\left(\frac{D_\infty}{K_{\omega 2}}\right)^{1/\alpha}, & x \notin [nl, a+nl] \end{cases} \quad (5)$$

$$D_\infty = \left\{\frac{P_\infty}{p^*\left[aK_{\omega 1}^{-1/\alpha} + (l-a)K_{\omega 2}^{-1/\alpha}\right]}\right\}^\alpha \quad (6)$$

$$f_\infty(x) = -\frac{2(1-\nu^2)p^*}{\pi E} \times \left[\left(\frac{D_\infty}{K_{\omega 1}}\right)^{1/\alpha}\int_0^a \ln\left(2\left|\sin\frac{\pi(x'-x)}{l}\right|\right)dx' + \left(\frac{D_\infty}{K_{\omega 2}}\right)^{1/\alpha}\int_a^l \ln\left(2\left|\sin\frac{\pi(x'-x)}{l}\right|\right)dx'\right] \quad (7)$$

where P_∞ is the constant specific load in the steady-state wear process.

Note that the steady-state solution exists if the linear wear during the running-in process is much smaller than the characteristic size of the fiber in its cross-section.

Introduce the following dimensionless parameters:

$$\tilde{a} = \frac{a}{l}, \quad m = \frac{K_{\omega 2}}{K_{\omega 1}} \quad (8)$$

By using the Lobachevski function

$$L(y) = -\int_0^y \ln\cos t\,dt, \quad |y| \le \frac{\pi}{2}$$

Equation (7) is rewritten in the following form:

$$f_\infty(x) = \frac{2(1-\nu^2)D_\infty^{1/\alpha}p^*l}{\pi E K_{\omega 2}^{1/\alpha}}\left(m^{1/\alpha}-1\right) \times \left\{(1-\tilde{a})\ln 2 - \frac{1}{\pi}\left[L\left(\frac{\pi}{2} - \frac{\pi x}{l}\right) + L\left(\frac{\pi}{2} - \pi\tilde{a} + \frac{\pi x}{l}\right)\right]\right\} \quad (9)$$

The function $L(y)$ in Eq. (9) can be represented by the series:

$$L(y) = y \ln 2 - \frac{1}{2} \sum_{n=1}^{\infty} (-1)^{n-1} \frac{\sin 2ny}{n^2} \tag{10}$$

From Eqs. (6), (9), and (10) it follows that:

$$f_\infty(x) = -\frac{2P_\infty(1-v^2)(1-m^{1/\alpha})}{\pi^2 E[1-\tilde{a}(1-m^{1/\alpha})]} \sum_{n=1}^{\infty} \frac{\sin \pi n \tilde{a}}{n^2} \cos \pi n \left(\tilde{a} - \frac{2x}{l}\right) \tag{11}$$

We can calculate the values of the function (11) at the points $x = 0$, $x = a$, and $x = l$:

$$f_\infty(0) = f_\infty(a) = f_\infty(l) = -\frac{P_\infty(1-v^2)(1-m^{1/\alpha})}{\pi^2 E[1-\tilde{a}(1-m^{1/\alpha})]} \sum_{n=1}^{\infty} \frac{\sin 2n\pi\tilde{a}}{n^2} \tag{12}$$

To find the amplitude of the function $f_\infty(x)$ defined by Eq. (11), the values of this function are determined at the extrema points $x = a/2 + kl$ and $x = (a+l)/2 + kl$, $(k = 0, \pm 1, \pm 2, \ldots)$, where $f'_\infty(x) = 0$:

$$f_\infty\left(\frac{a}{2}\right) = -\frac{2P_\infty(1-v^2)(1-m^{1/\alpha})}{\pi^2 E[1-\tilde{a}(1-m^{1/\alpha})]} \sum_{n=1}^{\infty} \frac{\sin n\pi\tilde{a}}{n^2}$$

$$f_\infty\left(\frac{a+l}{2}\right) = -\frac{2P_\infty(1-v^2)(1-m^{1/\alpha})}{\pi^2 E[1-\tilde{a}(1-m^{1/\alpha})]} \sum_{n=1}^{\infty} (-1)^n \frac{\sin n\pi\tilde{a}}{n^2}$$

So the amplitude of the function $f_\infty(x)$ can be calculated from the following expression:

$$\Phi = \left| f_\infty\left(\frac{a}{2}\right) - f_\infty\left(\frac{a+l}{2}\right) \right| = \left| \frac{4P_\infty(1-v^2)(1-m^{1/\alpha})}{\pi^2 E[1-\tilde{a}(1-m^{1/\alpha})]} \sum_{n=1}^{\infty} \frac{\sin(2n-1)\pi\tilde{a}}{(2n-1)^2} \right| \tag{13}$$

The relationship for the wear rate in the steady-state stage follows from Eq. (6) and has the form:

$$D_\infty = mK_{\omega 1} \left\{ \frac{P_\infty}{p^* l[1-\tilde{a}(1-m^{1/\alpha})]} \right\}^\alpha \tag{14}$$

The effective wear coefficient in the steady-state stage of the wear process is determined from Eqs. (1) and (14) as follows:

$$\tilde{K}_\omega = \frac{D_\infty}{K_{\omega 1}} \left(\frac{p^* l}{P_\infty} \right)^\alpha = \frac{m}{[1-\tilde{a}(1-m^{1/\alpha})]^\alpha} \tag{15}$$

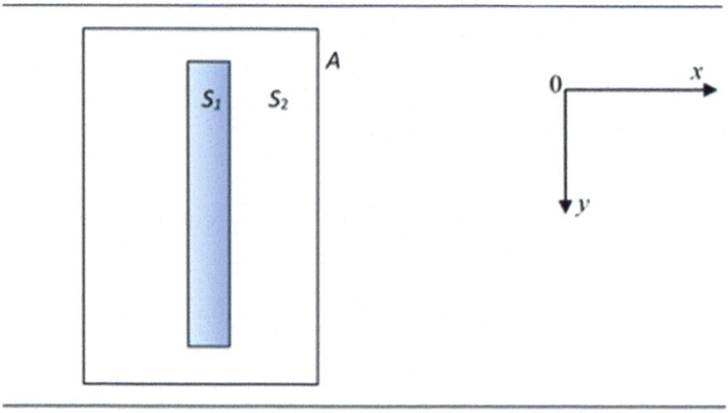

Fig. 13.2 Scheme of a fiber lying at the surface $z = 0$ (a cell of periodicity)

The pressures acting upon the fibers and the matrix in the steady-state regime can be calculated by using the equilibrium equation and the geometrical characteristics of fibers in the matrix (Fig. 13.2). It follows from the equilibrium conditions that

$$P_\infty S = p_1 S_1 + p_2 S_2 \tag{16}$$

Here p_1 and p_2 are the pressures acting upon the fibers and the matrix, respectively, $S_1 = bL$ and S_2 are the areas occupied by the fibers and matrix at the surface (in a cell of periodicity), S is the area of a cell of periodicity, and $p_\infty = P_\infty/S$ is the nominal pressure.

The following relationship between the nominal pressures and wear coefficients of the fibers and the matrix in the steady-state regime of wear process is obtained from Eqs. (1) and (5):

$$\left(\frac{p_1}{p_2}\right)^\alpha = \frac{K_{\omega 1}}{K_{\omega 2}} = \frac{1}{m} \tag{17}$$

From Eqs. (16) and (17), the pressures acting upon the fibers and the matrix are calculated as

$$\begin{cases} p_1 = \frac{P_\infty \cdot S}{S_1 + S_2 \cdot m^{1/\alpha}} \\ p_2 = p_1 \cdot m^{1/\alpha} \end{cases} \tag{18}$$

Equations (11), (14), and (18) allow us to analyze the shape of the worn surface, wear rate, and pressure distribution in the steady-state regime of wear process depending on the mechanical, geometrical, and strength characteristics of the structural elements of the composite material.

Note that the solution of the wear contact problem developed above is also valid for any directions of relative sliding of contacting bodies under the assumption that the shear stress does not influence the contact pressure distributions.

13.3 Calculation Results and Their Analysis

For calculations, we use the following numerical values of the geometrical and mechanical characteristics of structural elements, which are characteristic for a carbon-carbon fiber material:

- fiber length $L = 3$–10 mm,
- fiber diameter $b = 5$ μm (for model simplification we consider the fiber having a square cross-section),
- Young's modulus of the fiber and matrix $E = 15$ GPa, Poisson's ratio $\nu = 0.35$,
- power α in the wear Eq. (1) is $\alpha = 1.5$.

13.3.1 Effect of the Ratio of the Wear Coefficients of Fibers and Matrix

We use Eq. (11) to analyze the influence of the parameter m which is the ratio of wear coefficients of fiber to matrix, on the shape of the worn surface of composite material in the steady-state stage of the wear process. The results are presented in Fig. 13.3

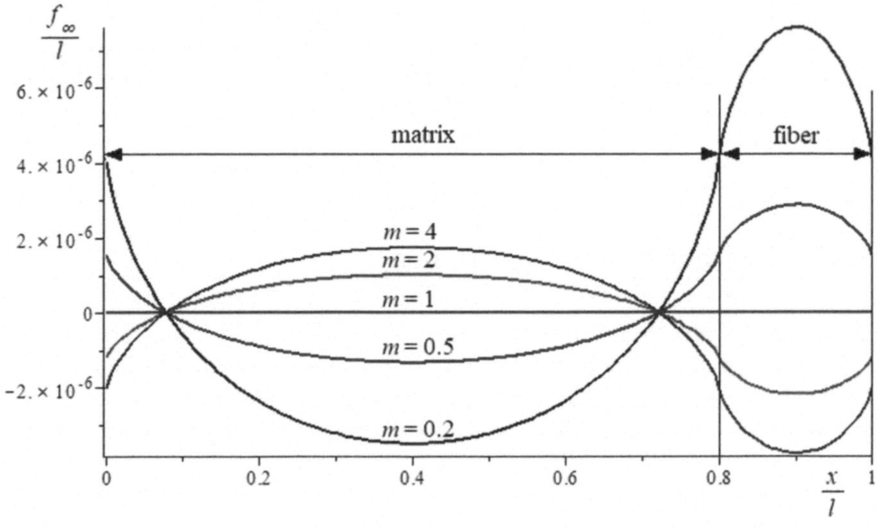

Fig. 13.3 The shape of the worn surface for various values of parameter m

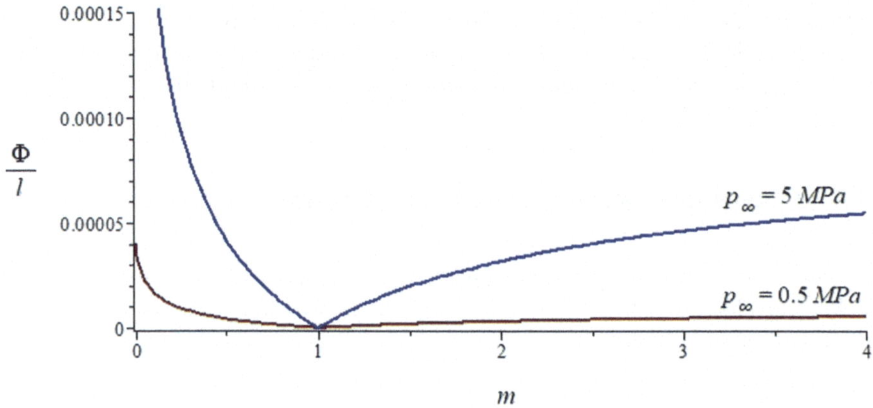

Fig. 13.4 The amplitude of the worn surface shape of the composite material versus the parameter m for two values of the nominal pressure

for a segment of periodicity $x \in [0, l]$ in which the segment $[0, a]$ is occupied by the matrix and $[a, l]$ by the fiber. A composite material with the fiber density $b/l = 0.2$ under the nominal pressure $p_\infty = 0.5$ MPa is considered.

The results show that the worn surface shape has the lowest amplitude, when the wear coefficients of the fiber and the matrix are close to each other ($m \approx 1$). If $m < 1$ then the level of the matrix relative to the fiber is negative (i.e., the fiber is above the matrix). If $m > 1$ then the fibers lie below the level of the matrix on the worn surface.

As follows from Eq. (13), increasing the nominal pressure p_∞ leads to an increase in the amplitude of the worn surface shape (see Fig. 13.4), the influence of the nominal pressure increase being most significant for very small values of m.

Note that as applied to carbon-carbon fibrous composite materials, the case $m < 1$ corresponds to carbonized fibers whose hardness is higher than the hardness of the carbon matrix, and the case $m > 1$ corresponds to graphitized fibers whose hardness is lower than the hardness of the matrix.

13.3.2 *Effect of Fiber Concentration*

Equations (11) and (15) are used to analyze the influence of the fiber concentration on the shape of the worn surface and the effective wear coefficient. Figure 13.5 illustrates the amplitude of the worn surface of the composite material depending on the fiber density b/l. For the results presented in Fig. 13.5 and below, the nominal pressure is taken $p_\infty = 0.5$ MPa.

The results indicate that reducing the surface concentration of fibers from 1 (full surface treatment with fibers) to 0 (untreated surface of the matrix), the volume of valleys on the worn surface first increases and then decreases to almost zero.

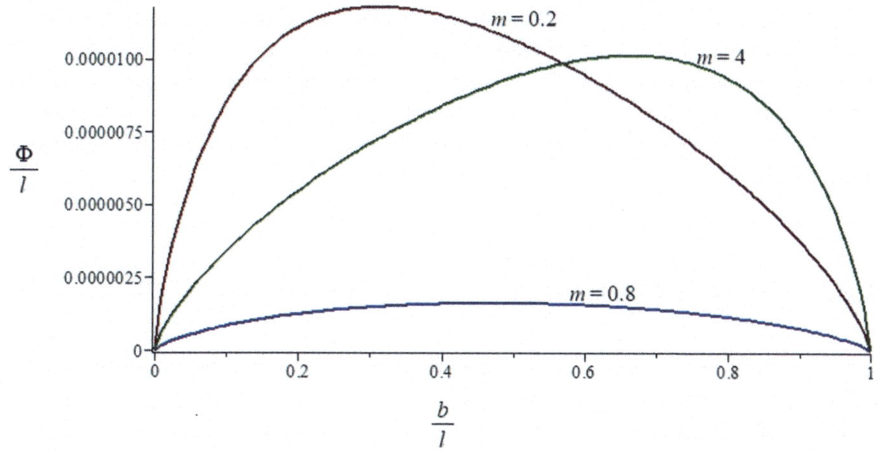

Fig. 13.5 Amplitude value of the worn surface versus fiber density for various values of parameter m

The effective wear coefficient of the composite material as a function of the fiber density b/l is presented in Fig. 13.6. The results indicate that in the case of fibers

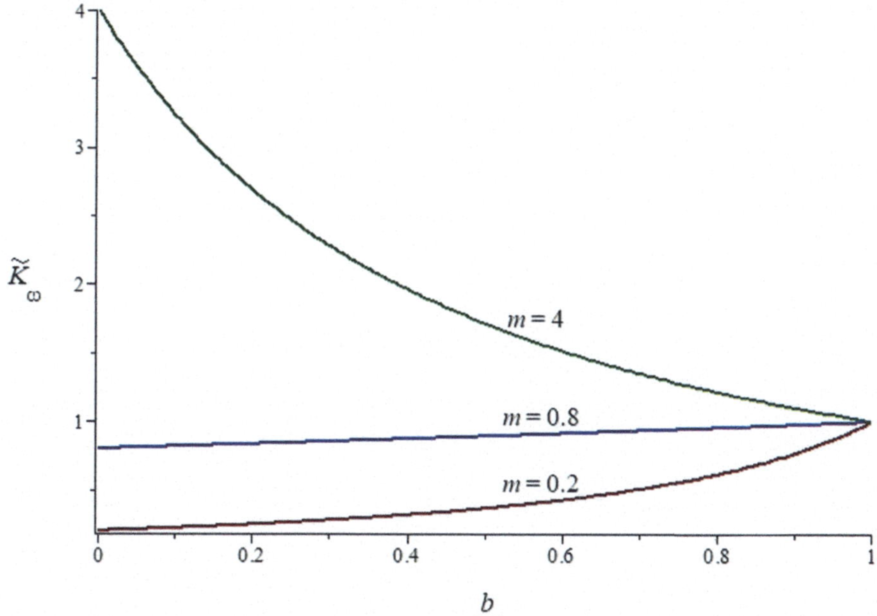

Fig. 13.6 Effective wear coefficient versus fiber density b/l for various values of parameter m

harder than the matrix ($m < 1$), an increase in their surface concentration leads to an increase in the effective coefficient of wear of the composite. In the case of the harder matrix than fibers ($m > 1$), the opposite result takes place: An increase in the fiber concentration leads to a decrease in the effective wear coefficient. Note that the rate of this decrease is the higher, the greater the ratio m of the wear coefficients of the matrix and fiber.

13.3.3 Effect of Fiber Length

In order to study the effect of the fiber length L on the pressure distribution between the fibers and the matrix, the pressures p_1 and p_2 are also calculated from Eq. (18) for various fiber lengths.

The values of these pressures as a function of the fiber length for the fixed fiber density $b/l = 0.8$ in composite are shown in Fig. 13.7. The results are obtained for $m = 0.1$; i.e., the fiber is 10 times harder than the matrix, so it takes upon itself higher value of the pressure. From the obtained results, it follows that the use of longer fibers in the production of composite materials allows one to reduce the pressure acting both on the fiber and on the matrix itself.

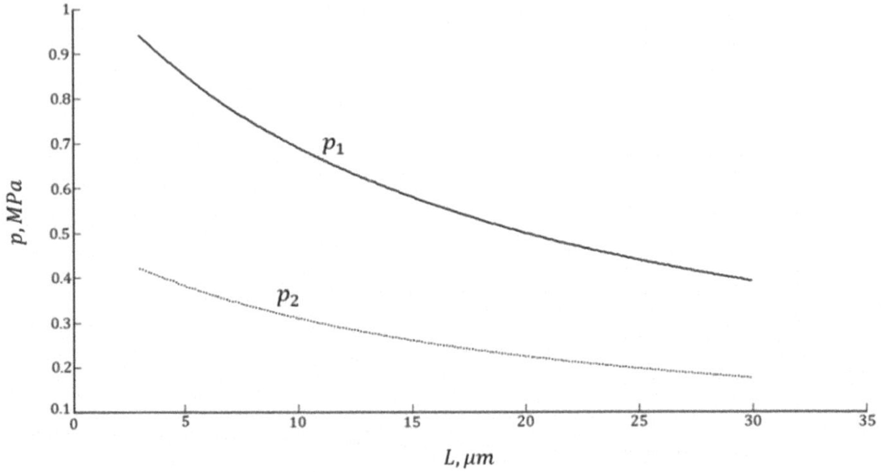

Fig. 13.7 Pressures acting upon fiber p_1 and matrix p_2 in the steady-state wear process versus fiber length

13.4 Conclusion

A model of wear for a fibrous composite material in the steady-state regime is constructed. Relations are obtained for calculating the shape of the worn surface, the effective wear rate of the composite, and the magnitude of contact pressures acting on the fiber and matrix. The effect on these values of the parameters of the composite microstructure is investigated. In particular, it is established that

- the worn surface shape of composite is not flat, and the depth of the valleys strongly depends on the size and concentration of fibers as well as on the ratio of the wear coefficients of fibers to matrix (or inverse ratio of their hardness);
- the effective wear coefficient decreases as the ratio of wear coefficients of the fiber to matrix decreases;
- the effective wear rate as a function of the fiber concentration increases if the fibers are harder than the matrix and decreases in the opposite case.

Analysis of the pressures acting upon the fibers and matrix in the steady-state wear process makes it possible to predict pulling fibers out of the matrix during wear. The results of modeling can be used to improve the wear resistance of fibrous composites by controlling their microstructures.

Acknowledgements The work was carried out under the financial support of the Russian Science Foundation (grant No. 19-19-00548).

References

1. Christensen, R.M.: Mechanics of Composite Materials. Dover Publications (2005)
2. Gomes, J.R., Silva, O.M., Silva, C.M., Pardini, L.C., Silva, R.S.: The effect of sliding speed and temperature on the tribological behaviour of carbon-carbon composites. Wear **249**, 240–245 (2001)
3. Goryacheva, I.G.: Contact Mechanics in Tribology. Kluwer (1997)
4. Lee, G.Y., Dharan, C.K.H., Ritchie, R.O.: A physically-based abrasive wear model for composite materials. Wear **252**, 322–331 (2002)
5. Johnson, K.L.: Contact Mechanics. Cambridge University Press (1985)
6. Kormanikova, E., Kotrasova, K.: Elastic mechanical properties of fiber reinforced composite materials. Chem. Listy **105**(17), 758–776 (2011)
7. Kragelsky, I.V., Dobychin, M.N., Kombalov, V.S.: Friction and Wear: Calculation Methods. Pergamon Press (2013)
8. Marx, D.T., Riester, L.: Mechanical properties of carbon-carbon composite components determined using nanoindentation. Carbon **37**(11), 1679–1684 (1999)
9. Ozcan, S., Filip, P.: Microstructure and wear mechanisms in C/C composites. Wear **259**(1), 642–650 (2005)
10. Piat, R., Böhlke, T., et al.: Modeling of effective elastic properties of carbon/carbon laminates. In: Proceedings of the 17th International Conference on Composite Materials, Edinburgh (2009)
11. Pobedrya, B.E.: Mechanics of Composite Materials. Moscow State University, Moscow (1984). (In Russian)
12. Shin, H.K., Lee, H.B., Kim, K.S.: Tribological properties of pitch-based 2-D carbon-carbon composites. Carbon **39**(7), 959–970 (2001)

13. Younes, R., Hallal, A., Fardoun, F., Hajj Chehade, F.: Comparative review study on elastic properties modeling for unidirectional composite materials. In: Composites and Their Properties. Intech Open Access Publisher, pp. 391–408 (2012)
14. Zhao, S., Hilmas, G.E., Dharani, L.: Numerical simulation of wear in a C/C composite multidisc clutch (preprint). Carbon **47**(9), 2219–2225 (2009)

Chapter 14
Shear Wave Propagation in Soft Biological Tissues: A Comparison of Numerical and Physical Modeling

Sergey Gurbatov, Igor Demin, Artem Lisin, Nikolay Pronchatov-Rubtsov, and Aleksey Spivak

Abstract The paper presents the results on the physical and numerical simulation of the propagation of a shear wave in soft biological tissues. The shear wave velocity and Young's modulus were experimentally determined on a Verasonics acoustic system (Verasonics, Inc., Kirkland, WA, USA) in calibrated polymer CIRS Model 049 Elasticity QA Phantom. The method of shear wave elasticity imaging (SWEI) allows one to determine the viscoelastic characteristics of an object under investigation by measuring the speed of a shear wave launched in a medium. The numerical analysis of the generation of the acoustic radiation force is realized in the toolbox k-Wave. The results of physical and numerical simulations of determining the velocity of shear waves in soft tissues are compared.

Keywords Elastography · Shear wave velocity · Verasonics · k-Wave toolbox · Numerical modeling

---

S. Gurbatov (✉) · I. Demin · A. Lisin · N. Pronchatov-Rubtsov · A. Spivak
National Research Lobachevsky State University of Nizhny Novgorod, 23 Gagarin Avenue, Nizhny Novgorod, Russian Federation 603950
e-mail: gurb@rf.unn.ru

I. Demin
e-mail: demin@rf.unn.ru

A. Lisin
e-mail: lisin.artem.a@gmail.com

N. Pronchatov-Rubtsov
e-mail: nikvas@rf.unn.ru

A. Spivak
e-mail: strigino831@gmail.com

H. Altenbach et al. (eds.), *Multiscale Solid Mechanics*,
Advanced Structured Materials 141,
https://doi.org/10.1007/978-3-030-54928-2_14

14.1 Introduction

Ultrasonography has been widely used for diagnosis since it was first introduced in clinical practice in the 1970s. Based on the propagation of mechanical waves and more particularly on high frequency compressional waves aka ultrasound, it allows the construction of morphological images of organs but lacks a fundamental and quantitative information on tissue elastic properties. Since then, new ultrasound modalities have been developed, such as Doppler imaging, which provides new information for diagnosis. Elastography was developed in the 1990's to map tissue stiffness, and reproduces/replaces the palpation performed by clinicians. Ultrasound elastography techniques can be classified based on the type of the external force that induces the tissue deformation. In general, four types of excitation sources are used to induce the deformation in the tissue: static excitation, natural physiological excitation, transient excitation, and harmonic excitation. For example, the shear wave elasticity imaging (SWEI) technique was introduced by Sarvazyan et al. in 1998 [1]. In SWEI, one or more focused or unfocused ultrasound beams are applied for a short period of time to generate tissue deformation. In turn, this deformation produces shear waves. Next, the shear wave speed is estimated from the measured particle displacements, and quantitative shear elasticity parameters are then obtained from the estimated shear wave speed. In clinical practice, SWEI has been used to quantitatively measure the stiffness of liver tissue and determine the stage of liver fibrosis. Other clinical applications for SWEI include quantitative assessment of tissue mechanical properties in thyroid, breast, kidney, and prostate [2, 3].

14.2 Verasonics Research System—Physical Simulation of Shear Waves

The Verasonics acoustic system (Verasonics, Inc., Kirkland, WA, USA) is a universal ultrasonic device designed for prototyping and debugging various algorithms of medical acoustics. The system allows you to work with almost any medical ultrasonic sensors, for example, L7-4, C5-2, and P4-2, which allows you to simulate the algorithms of various expert commercial ultrasound systems. The main advantage of the Verasonics system is its openness, that is, the ability to widely change the parameters of ultrasonic waves, for example, the number of emitted and receiving channels from 64 to 256, the carrier frequency from 1 to 15 MHz, the ultrasound power up to 1000 W, and program them depending on tasks and objects of research. Received echoes are recorded by the device and are available for post-processing in the form of arrays of numerical data. The whole scenario of sending pulses, receiving and processing data, and imaging is programmed by the user in the MATLAB software environment.

In the Laboratory of Biomedical Technologies, Medical Instrument Engineering, and Acoustic Diagnostics (MedLab), Department of Acoustics, Nizhny Novgorod

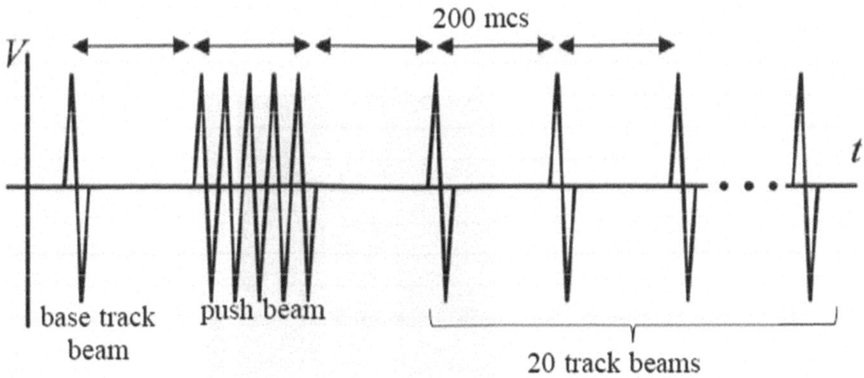

Fig. 14.1 Time diagram of the signals during the experiment

University on the Verasonics ultrasound system, the shear wave elasticity imaging (SWEI) method was implemented [4]. This method allows one to determine the viscoelastic characteristics of an object under investigation by measuring the speed of a shear wave launched in a medium. For this, a focused ultrasonic "pushing" pulse is applied to a given point, which creates a disturbance. Under the influence of a radiation force, a shear wave begins to propagate from this point, resulting in a displacement of particles of the medium along the path of its propagation. To track the displacements, and hence, the propagation of the shear wave, one unfocused "reference" pulse is sent to the pushing pulse to scan the unperturbed medium. After a focused pulse, several unfocused "imaging" pulses, similar to the reference one, with a certain interval, which scan the medium at the time of wave propagation, are fed into the medium (see Fig. 14.1).

Subsequently, when processing the obtained data, the medium displacement ξ *(x, t)* is recorded as a function of time t at various distances x from the focus point of the pushing pulse. Different x values correspond to different curves. Each function has a maximum corresponding to the front of the shear wave, due to which the time of arrival of the front to a certain point is determined. This allows us to calculate the shear wave velocity V, which is converted into Young's modulus E according to formula

$$E = 3\rho V^2 \tag{14.1}$$

where ρ is the density of soft biological tissue.

As mentioned earlier, the received echo signals are recorded by Verasonics and are available for post-processing in the form of arrays of numerical data, which are recorded in a separate file. This allows you to display after processing various data necessary for the study, in contrast to commercial expert ultrasound systems, which can only display information about shear wave velocity and Young's modulus of the medium. To this end, while working with the Verasonics system, several programs were developed in the MATLAB software environment for processing source data.

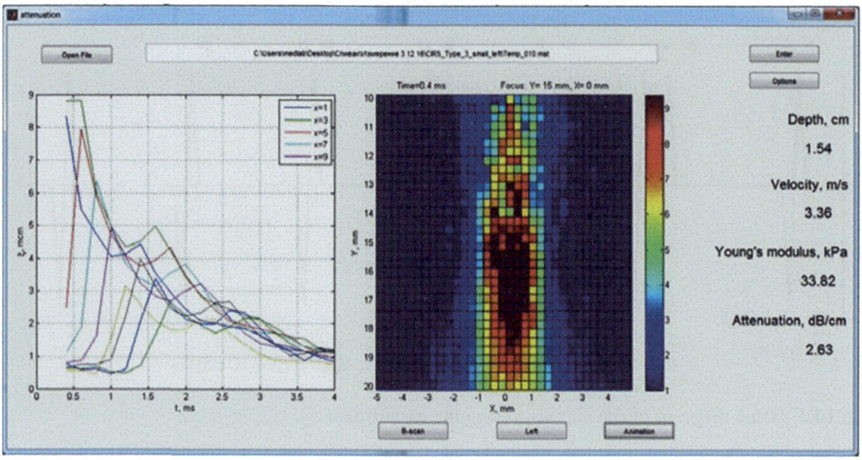

Fig. 14.2 Interface for calculating the shear wave velocity and Young's modulus of the medium with the output of shear wave visualization

One of the programs allows one to obtain the values of the displacement of the medium ξ *(x, t)* at a given point and graphically displays the form of this function. The following program allows you to get the values of the shear wave velocity in the studied object. These programs are combined into an interface that combines the functional of obtaining the medium displacement ξ *(x, t)* at a given point and the shear wave velocity (see Fig. 14.2).

14.3 Numerical Simulation of Shear Waves

To numerically solve the problem of the propagation of shear waves in soft biological tissues, the k-Wave software package for the MATLAB medium was used. It combines the MATLAB optimization for working with matrix operations and a set of tools that allow you to simulate the propagation medium through parameters such as density and speed of sound in a given environment. The numerical model is based on the transition to k-space, where spatial gradients are calculated using the FFT scheme, and time gradients are calculated using the adjusted k-spatial difference scheme [5].

Numerical modeling of a physical experiment allows predicting the measurement result with high accuracy, and therefore, to solve the problem of computer simulation of media with given parameters, a large number of algorithms, software modules, and independent packages, both paid and with a free license, have been developed. Their main difference from each other is the numerical methods used and the approximations allowed for their application. Modern computer technology allows calculation with high accuracy, but makes high demands on the hardware of the machine, so it becomes important to choose an algorithm that effectively uses the power of computing modules with the lowest possible load, while ensuring the necessary and sufficient accuracy.

To solve this problem, it is convenient to use k-Wave—a software package (toolbox) for the MATLAB environment. It combines MATLAB optimization for working with matrix operations and a set of tools that allows you to simulate the propagation medium through such parameters as the density and speed of sound in a given environment. The combination of these factors allows us to simulate 2D and 3D spaces, while maintaining a high computational speed. The numerical model is based on the transition to k-space, where spatial gradients are calculated using the FFT scheme, and time gradients are calculated using the adjusted k-spatial difference scheme. The temporal scheme is accurate in the limit of propagation of a linear wave in a homogeneous medium without losses and significantly reduces numerical scattering in a more general case [6].

The solution to the problem of modeling the SWEI method can be divided into stages, which are an analysis of the initial conditions necessary for the model to work, and the use of these data to simulate a specific stage of the method. The first step is to decide on the initial conditions for the simulation, i.e., the choice of the propagation medium through the density ρ and the speed of sound in medium C. Based on these data, a numerical simulation of the focusing of an ultrasonic beam, which is a source of radiation force, is carried out. According to the pressure data obtained from the simulation results, the radiation force is calculated. From this, shear wave propagation can be modeled. The last step is to visualize the results.

The simulation subtasks involve the following steps: first, the environment in which the simulation is performed (it can be either a linear medium or a nonlinear space) is set, then a sensor is placed (in this case, it is a model of a standard linear L7-4 linear sensor for ultrasound studies), and simulation of wave propagation in the medium. In the case of a linear sensor, the emitter is a phased array of 128 elements. Within the framework of a given numerical model, these are 128 point sources. But to obtain a shear wave, it is necessary to focus the emitters to the point where the wave from each emitter comes in one phase. This is achieved by means of a quadratic phase incursion on each radiator; for zero, we consider the center of the sensor (see Fig. 14.3a). To simulate a shear wave, the source is located in the region of the focusing spot and consists of eight elements located in the shape of a rhombus (see Fig. 14.3b).

The result of this algorithm allows you to calculate the distribution of acoustic pressure, which allows you to evaluate the geometric dimensions of the focusing spot and also serve as input to calculate the amplitude and velocity of the shear wave.

The shear wave velocity and Young's modulus were experimentally determined on a Verasonics system in calibrated polymer CIRS Model 049 Elasticity QA Phantom Spherical with spheres 10 and 20 mm in diameter located at different depths. The spheres in phantoms were of four types with different values of Young's modulus (Type I–IV), indicated in the accompanying documents. These elements were in a polymer medium (matrix), the elastic characteristics of which were also known. The advantage of phantoms is that they are made of Zerdine polymer material, the

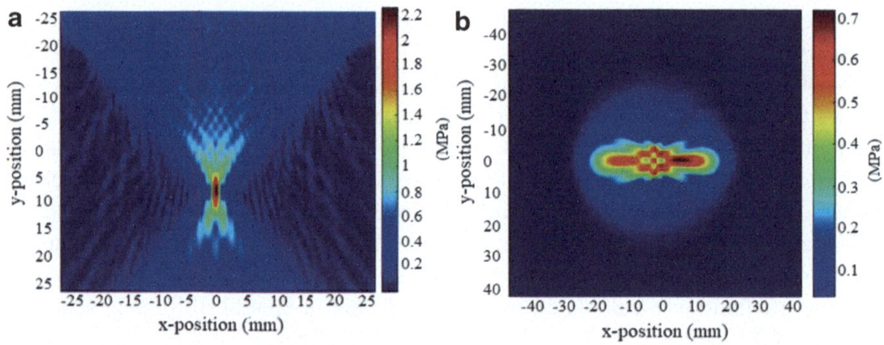

Fig. 14.3 Implementation of steps of numerical simulation of shear waves **a** pressure distribution at the stage of focusing of the ultrasonic beam, **b** pressure distribution during shear wave propagation

characteristics of which are independent of changes in external temperature and applied pressure.

The shear wave velocity is calculated based on the position of the pressure peak characterizing the location of the wavefront. In Fig. 14.4 shows an example of calculating the shear wave velocity using the formula

$$V = S/t,$$

where *S* is the distance between the peaks determined for two times *t1* and *t2*

$$S = (S_2 - S_1)/2, \text{ and } t = t_2 - t_1.$$

Table 14.1 presents a comparison of the results of numerical and physical modeling

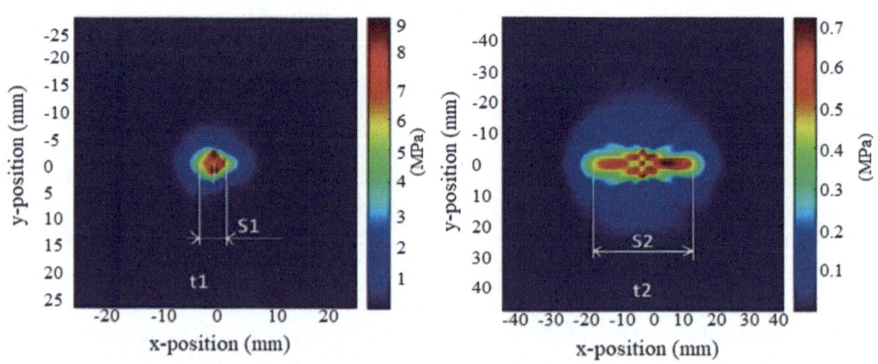

Fig. 14.4 Shear wave velocity determination by wavefront position

Table 14.1 Shear wave velocity (*m/s*)—comparison of physical and numerical simulations

CIRS phantom	Numerical simulations (K-wave)	Physical simulations (verasonics)
Background	4.17	2.68
I	1.84	1.34
II	2.07	1.81
III	4.85	3.46

A comparison of the results shows that the higher the concentration of the polymer substance in the phantom, the closer the results of numerical and physical modeling. In addition, the results of numerical simulations are consistently larger than physical ones. This effect occurs due to the strong heterogeneity of the studied physical environment.

Acknowledgements This work was supported by a grant from the Russian Science Foundation №. 19- 12-00256.

References

1. Sarvazyan, A.P., Rudenko, O.V., Swanson, S.D., Fowlkes, J.B., Emelianov, S.Y.: Shear wave elasticity imaging: a new ultrasonic technology of medical diagnostics. Ultrasound Med. Biol. **24**, 1419–1435 (1998)
2. Safonov, D.V., Rykhtik, P.I., Shatokhina, I.V., Romanov, S.V., Gurbatov, S.N., Demin, IYu.: Shear wave elastography: comparing the accuracy of ultrasound scanners using calibrated phantoms in experiment. Sovremennye Tehnologii v Med. **9**, 51–59 (2017)
3. Ultrasound Elastography for Biomedical Applications and Medicine, ed. by Ivan Z. Nenadic. Hoboken, NJ: Wiley (2019)
4. Khalitov, R.S., Gurbatov, S.N., Demin, IYu.: The use of the verasonics ultrasound system to measure shear wave velocities in CIRS phantoms. Phys. Wave Phenom. **24**, 73–76 (2016)
5. Treeby, B.E.: Modeling nonlinear wave propagation on nonuniform grids using a mapped k-space pseudospectral method. IEEE Trans. Ultrason. Ferroelectr. Freq. Control **35**, 2208–2213 (2013)
6. Prieur, F., Catheline, S.: Simulation of shear wave elastography imaging using the toolbox "k-Wave". Proc. Mtgs. Acoust. (POMA). **29**, 20002 (2016)

Chapter 15
On the Complex Dynamics of Nonlinear Systems with Viscous and Dry Friction with Memory

Leonid A. Igumnov, Vladimir S. Metrikin, and Natalya S. Starodubrovskaya

Abstract A dynamical system consisting of a pair of bodies in frictional contact is considered. One of the bodies moves with a constant velocity, whereas the other is attached to a fixed support. The friction force is assumed as a sum of two friction forces: dry friction and viscous friction. The coefficient of the static friction is represented in the form of a piecewise linear function of the duration of the stick phase, whereas the coefficient of the kinetic friction is assumed to be constant. The qualitative structure of the phase space is studied in detail with qualitative methods, depending on the values of the coefficients of viscous friction and relative rest friction. The study is focused on the effect of the coefficients of viscous and dry friction. It is proven that the dynamical regime that include the stick phase occur only in certain parameter ranges. The mapping function is derived analytically and is shown to be a piecewise continuous function of the stick phase. The main regimes of periodic and stochastic motions are obtained by a semi-analytical method. The bifurcation diagrams are employed in order to study these regimes under various system parameters.

Keywords Friction with memory · Mapping function · Stick phase · Bifurcation diagram · Chaos

---

L. A. Igumnov (✉) · V. S. Metrikin
Research Institute for Mechanics, National Research Lobachevsky State University of Nizhny Novgorod, 23 Gagarin Avenue, Building 6, Nizhny Novgorod 603950, Russian Federation
e-mail: igumnov@mech.unn.ru

V. S. Metrikin
e-mail: v.s.metrikin@mail.ru

N. S. Starodubrovskaya
Nizhny Novgorod State Technical University n.a. R.E. Alekseev, Nizhny Novgorod, Russian Federation
e-mail: nsstarod@mail.ru

H. Altenbach et al. (eds.), *Multiscale Solid Mechanics*,
Advanced Structured Materials 141,
https://doi.org/10.1007/978-3-030-54928-2_15

15.1 Introduction

It is known that in most cases, friction forces play a purely damping role. Due to their effect, free vibrations decay, amplitudes of forced vibrations decrease, stability regions expand, etc. However, for a whole number of dynamical systems, these forces act as a destabilizing factor, i.e., even small friction forces in the system give rise to unstable equilibrium. Friction-induced self-oscillatory systems (FSS) can be referred to this kind of systems. Friction-induced self-oscillations are the most spread type of mechanical self-oscillations connected with instability of the friction process at relatively low sliding velocities. Ishlinsky and Kragelsky [1] was the first to draw attention to this fact, and then followed by numerous studies of both national and foreign researchers such as [2, 7, 11–13] and foreign researchers [3–6, 10]. In their works, [7, 11–13] considered the fundamental fact that mechanical relaxation oscillations of bodies in frictional contact were caused by the elasticity of the system and increasing friction force due to increase in duration of the stick phase. This phenomenon was called "friction with memory." The analysis of FSS of the Coulomb–Amonton model of friction with and without memory showed that in the latter case, there exist only friction-induced self-oscillations [8]. Dry friction forces with memory give rise to both arbitrarily complex periodic motion regimes with a great number of sick intervals and stochastic self-oscillations [7, 11–13].

15.2 Equation of Motion

A physical system consisting of a body of mass m, located on a rough belt moving at a constant velocity V_0, is considered. The body is secured with a rigidity spring k to a fixed support (Fig. 15.1). In the following, it is assumed that the kinetic friction coefficient is constant, whereas the coefficient of static friction is, in accordance with the hypothesis by Ishlinsky and Kragelsky [1], a piecewise linear function of the duration t_k of the stick phase of the process; the latter is equal to.

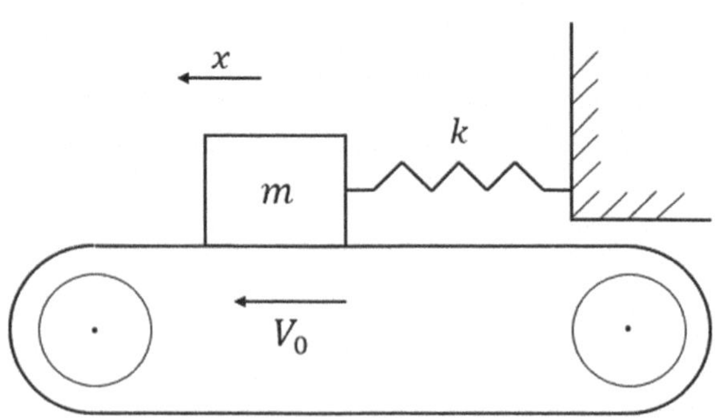

Fig. 15.1 Scheme of the vibration mechanism

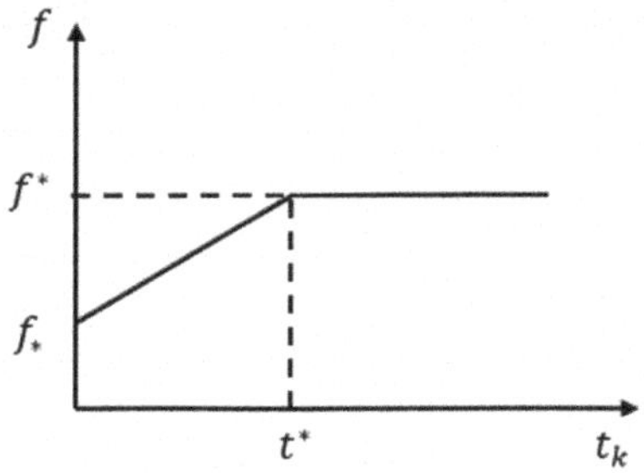

Fig. 15.2 Coefficient of static friction as a function of the duration of stick phase

The equations of motion of the system with stick and slip intervals can be written in the form

$$m\ddot{x} = -kx - f_* P\operatorname{sgn}(\dot{x} - V_0) + \alpha_1(\dot{x} - V_0), \quad \dot{x} \neq V_0, \tag{15.1}$$

$$k|x| \leq f(t_k)P, \quad \dot{x} = V_0, \tag{15.2}$$

where P is body weight, f_* and $f(t_k)$ are coefficients of kinetic friction and static friction, respectively; $\alpha_1 > 0$ is viscous friction coefficient. The coefficient of kinetic friction is assumed to be constant, whereas the coefficient of static friction, in accordance with [1], is a piecewise linear function of the duration t_k of stick phase of the body and the belt and equals to

$$f(t_k) = \begin{cases} f_* + (f^* - f_*)t_k/t^*, & 0 < t_k < t^*, \\ f^*, & t_k \geq t^*. \end{cases} \tag{15.3}$$

Figure 15.2 presents $f(t_k)$ dependence.

We introduce dimensionless variables $\tau = \varepsilon^* t/t^*, \xi = kx/(f_* P)$ and parameters $\varepsilon^* = (f^* - f_*)/f_*, \theta = V_0\sqrt{mk}/(f_* P)$ and $\tilde{\alpha}_1 = \alpha_1/\sqrt{mk}$. Then, Eqs. (15.1, 15.2) will have the form

$$\ddot{\xi} = -\xi - \operatorname{sgn}(\dot{\xi} - \theta) + \tilde{\alpha}_1(\dot{\xi} - \theta), \quad \dot{\xi} \neq \theta, \tag{15.4}$$

$$|\xi| \leq 1 + \varepsilon(\tau_k), \dot{\xi} = \theta, \tag{15.5}$$

where $\varepsilon(\tau_k) = (f(t^*\tau_k/\varepsilon^*) - f_*)/f_*$ is dimensionless characteristic of the coefficient of static friction. It can be easily seen that by virtue (15.3) we have $\varepsilon(\tau_k) = \begin{cases} \tau_k, & 0 < \tau_k < \varepsilon^*, \\ \varepsilon^*, & \tau_k \geq \varepsilon^*. \end{cases}$

15.3 Qualitative Study of the Phase Space Structure

The qualitative form of splitting of the phase plane trajectory $\xi, \dot{\xi}$ of Eq. (15.4) is studied. The phase plane is divided by the straight line $\dot{\xi} = \theta$ into two subspaces $\dot{\xi} > \theta$ and $\dot{\xi} < \theta$, each of them having their own equation of motion, namely

$$\ddot{\xi} = -\xi - 1 + \tilde{\alpha}_1(\dot{\xi} - \theta), \quad \dot{\xi} > \theta \tag{15.6}$$

and

$$\ddot{\xi} = -\xi + 1 + \tilde{\alpha}_1(\dot{\xi} - \theta), \quad \dot{\xi} < \theta. \tag{15.7}$$

Considering the behavior of the trajectories in the vicinity of this straight line, it can be shown that on the straight line $\dot{\xi} = \theta$, there exists a segment of sliding motions $|\xi| \leq 1, \dot{\xi} = \theta$.

The dynamic system (15.6, 15.7) has a unique equilibrium state $A(1 - \tilde{\alpha}_1\theta, 0)$, which is an unstable focus at $0 < \tilde{\alpha}_1 < 2$ and an unstable node at $\tilde{\alpha}_1 \geq 2$. After having fixed θ, we trace the change in phase portrait of this system with increasing $\tilde{\alpha}_1$. For sufficiently small $\tilde{\alpha}_1$, the system (15.6, 15.7) has two stitched limit cycles: unstable L_1 and stable L_2, consisting of an arc CB of the half-path γ_1 and a segment of sliding motions BC (Fig. 15.3).

With increasing $\tilde{\alpha}_1$, half-paths γ_1 and γ_2 are getting closer and for some $\tilde{\alpha}_1 = \tilde{\alpha}_1^{(1)}$, they coincide. For $\tilde{\alpha}_1$, arbitrarily close to $\tilde{\alpha}_1^{(1)}, \tilde{\alpha}_1 > \tilde{\alpha}_1^{(1)}$, half-path γ_1 intersects the

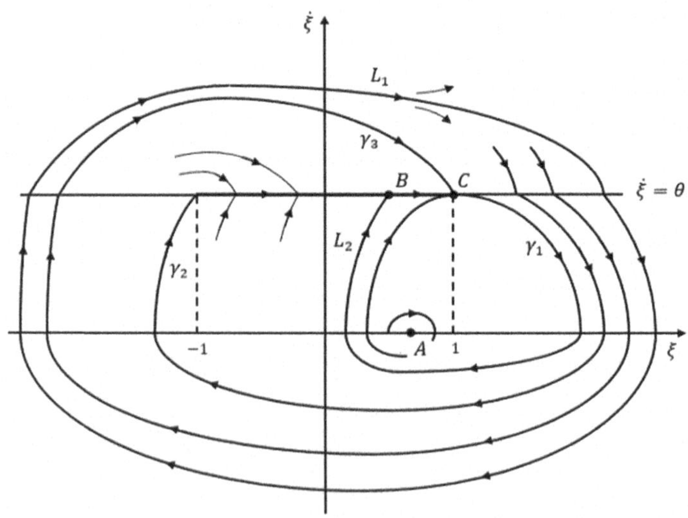

Fig. 15.3 Qualitative form of the phase trajectories of the system (15.6, 15.7) for sufficiently small $\tilde{\alpha}_1$

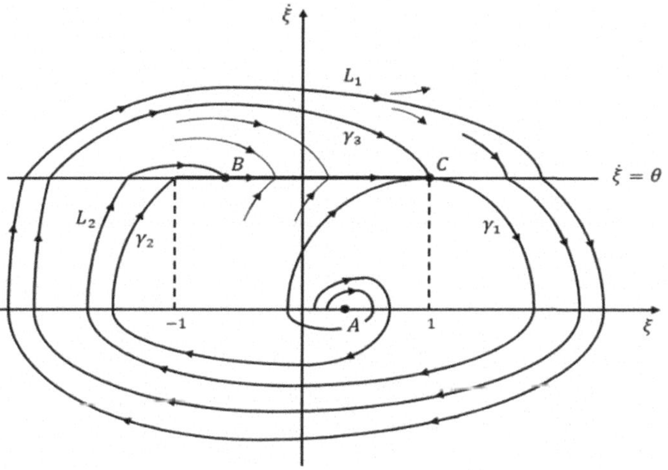

Fig. 15.4 Qualitative form of phase trajectories of the system (15.6, 15.7) for $\tilde{\alpha}_1$, arbitrarily close to $\tilde{\alpha}_1^{(1)}$, $\tilde{\alpha}_1 > \tilde{\alpha}_1^{(1)}$

straight line $\dot{\xi} = \theta$, α-limit set of the half-path γ_2 is equilibrium state A (Fig. 15.4). With a further increase in $\tilde{\alpha}_1$, half-paths γ_1 and γ_3 are getting closer and for some $\tilde{\alpha}_1 = \tilde{\alpha}_1^{(2)}$, they coincide. Cross-linked limit circle L_2 here does not contain the segment of sliding motions.

With increasing $\tilde{\alpha}_1$, unstable limit cycle L_1 contracts, whereas stable limit circle L_2 expands. Let us denote by $\tilde{\alpha}_1^{(3)}$ the bifurcation value of the parameter $\tilde{\alpha}_1$, which corresponds to merging the above limit cycles. Note that $\tilde{\alpha}_1^{(3)} \geq \tilde{\alpha}_1^{(2)}$. Suppose $\tilde{\alpha}_1^{(3)} > \tilde{\alpha}_1^{(2)}$ (the case $\tilde{\alpha}_1^{(3)} = \tilde{\alpha}_1^{(2)}$ is considered similarly). For $\tilde{\alpha}_1$, arbitrarily close to $\tilde{\alpha}_1^{(2)}$, $\tilde{\alpha}_1 > \tilde{\alpha}_1^{(2)}$, half-path γ_3 by α-limit set tends to the equilibrium state A, whereas half-space γ_1 by ω-limit set has stitched limit cycle L_2 (Fig. 15.5). Unstable and stable stitched limit cycles L_1 and L_2 for $\tilde{\alpha}_1 = \tilde{\alpha}_1^{(3)}$ merge into a semi-stable limit cycle. For $\tilde{\alpha}_1 > \tilde{\alpha}_1^{(3)}$ half-path γ_1 tends to infinity at $t \to +\infty$ (Fig. 15.6). The equilibrium state A for $\tilde{\alpha}_1 \geq 2$ is an unstable node (Fig. 15.7).

Table 15.1 presents numerically obtained the above values $\tilde{\alpha}_1^{(1)}$, $\tilde{\alpha}_1^{(2)}$ and $\tilde{\alpha}_1^{(3)}$ as a function of the parameter θ.

15.4 Constructing a Mapping Function

It is assumed that for $-\tau_k < \tau < 0$, the body and the belt are at stick phase, whereas for $\tau = 0$, we have $\xi = \xi_0$, $\xi_0 = 1 + \varepsilon(\tau_k)$, $\dot{\xi} = \theta$. According to the nature of the trajectory splitting of the phase space in Eq. (15.4) for various parameter values $\tilde{\alpha}_1$, the image point moving along the positive half-path from a starting point $M(\xi_0, \theta)$ hits the segment of sliding motions only when $0 < \tilde{\alpha}_1 < \tilde{\alpha}_1^{(2)}$, $0 < \varepsilon^* < \tilde{\varepsilon}^*$,

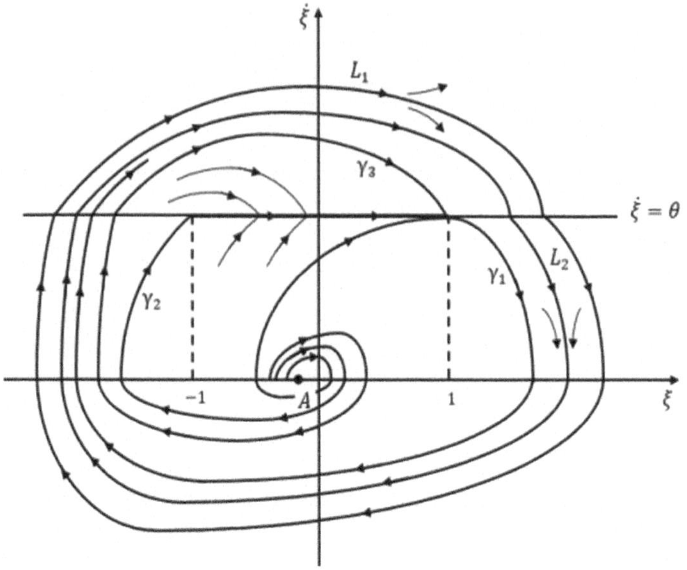

Fig. 15.5 Qualitative form of phase trajectories of the system (15.6, 15.7) for $\tilde{\alpha}_1$, arbitrarily close to $\tilde{\alpha}_1^{(2)}$, $\tilde{\alpha}_1 > \tilde{\alpha}_1^{(2)}$

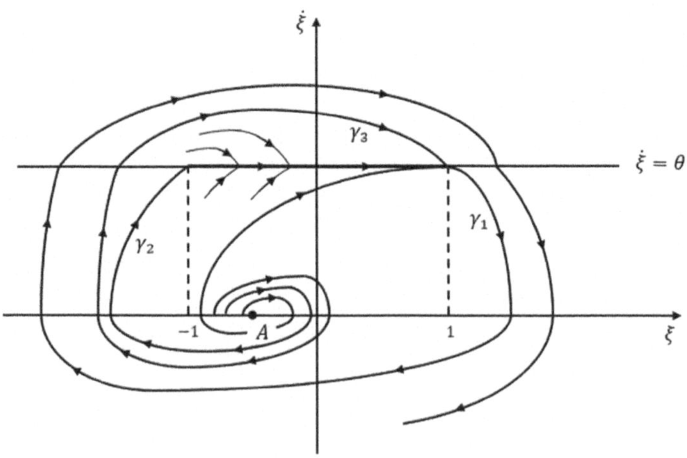

Fig. 15.6 Qualitative form of phase trajectories of the system (15.6, 15.7) at $\tilde{\alpha}_1 > \tilde{\alpha}_1^{(3)}$

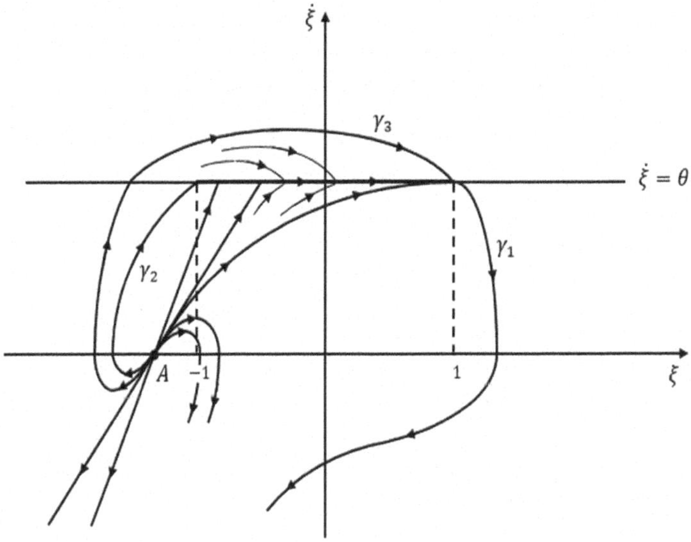

Fig. 15.7 Qualitative form of phase trajectories of the system (15.6, 15.7) at $\tilde{\alpha}_1 > 2$

Table 15.1 Mechanical properties of birch

θ	$\tilde{\alpha}_1^{(1)}$	$\tilde{\alpha}_1^{(2)}$	$\tilde{\alpha}_1^{(3)}$
0.1	1.1666	1.2086	1.2086
0.15	1.0322	1.0964	1.0964
0.2	0.9264	1.0106	1.0106
0.25	0.839	0.9412	0.9412
0.5	0.5526	0.7136	0.71437999
0.75	0.392	0.5772	0.579545
1	0.2908	0.4814	0.4865501
1.25	0.223	0.4092	0.41786898
1.5	0.1754	0.3524	0.36507279
1.75	0.1412	0.3062	0.32334483
2	0.1156	0.2684	0.2896595
2.25	0.0962	0.2368	0.26198
2.5	0.0812	0.21	0.23887697
2.75	0.0692	0.1872	0.21936009
3	0.0598	0.1678	0.20267789

$\tilde{\varepsilon}^* = \xi^* - 1$, where ξ^* is the abscissa of the intersection point of the cross-linked limit cycle L_1 with half-line $\xi > 1$, $\dot{\xi} = \theta$. Further, we assume that $0 < \tilde{\alpha}_1 < \tilde{\alpha}_1^{(2)}$, $0 < \varepsilon^* < \tilde{\varepsilon}^*$.

At $\tau > 0$ the body moves at a lower velocity than that of the belt ($\dot{\xi} < \theta$) and its moving ends at $\tau = \tau_1$, when $\dot{\xi} = \theta$ for the first time. The Eq. (15.7) with initial conditions $\tau = 0, \xi = \xi_0, \dot{\xi} = \theta$ can be written in the form

$$\xi = 1 - \tilde{\alpha}_1\theta + \exp(\tilde{\alpha}_1\tau/2)\left[(\xi_0 - 1 + \tilde{\alpha}_1\theta)\cos\frac{\sqrt{4-\tilde{\alpha}_1^2}}{2}\tau \right.$$
$$\left. + \frac{2\theta - \tilde{\alpha}_1(\xi_0 - 1 + \tilde{\alpha}_1\theta)}{\sqrt{4-\tilde{\alpha}_1^2}}\sin\frac{\sqrt{4-\tilde{\alpha}_1^2}}{2}\tau\right].$$

Then for velocity $\dot{\xi}$, we have

$$\dot{\xi} = \exp(\tilde{\alpha}_1\tau/2)\left[\theta\cos\frac{\sqrt{4-\tilde{\alpha}_1^2}}{2}\tau - \frac{\tilde{\alpha}_1\theta + 2(\xi_0 - 1)}{\sqrt{4-\tilde{\alpha}_1^2}}\sin\frac{\sqrt{4-\tilde{\alpha}_1^2}}{2}\tau\right].$$

By assuming $\tau = \tau_1, \xi(\tau_1) = \xi_1, \dot{\xi} = \theta$ equations considered, we obtain the following relations in order to determine the values τ_1 and ξ_1, respectively.

$$\exp\left(\frac{\tilde{\alpha}_1\tau_1}{2}\right)\left[\theta\cos\frac{\sqrt{4-\tilde{\alpha}_1^2}}{2}\tau_1 - \frac{\tilde{\alpha}_1\theta + 2(\xi_0 - 1)}{\sqrt{4-\tilde{\alpha}_1^2}}\sin\frac{\sqrt{4-\tilde{\alpha}_1^2}}{2}\tau_1\right] = \theta, \quad (15.8)$$

$$\xi_1 = 1 - \tilde{\alpha}_1\theta + \exp\left(\frac{\tilde{\alpha}_1\tau_1}{2}\right)\left[(\xi_0 - 1 + \tilde{\alpha}_1\theta)\cos\frac{\sqrt{4-\tilde{\alpha}_1^2}}{2}\tau_1 \right.$$
$$\left. + \frac{2\theta - \tilde{\alpha}_1(\xi_0 - 1 + \tilde{\alpha}_1\theta)}{\sqrt{4-\tilde{\alpha}_1^2}}\sin\frac{\sqrt{4-\tilde{\alpha}_1^2}}{2}\tau_1\right].$$

It can be easily seen that $\tilde{\tau}_1 < \tau_1 < \tilde{\tau}_2$,

$$\tilde{\tau}_1 = \frac{2}{\sqrt{4-\tilde{\alpha}_1^2}}\left(\pi + \operatorname{arctg}\frac{\theta\sqrt{4-\tilde{\alpha}_1^2}}{\tilde{\alpha}_1\theta + 2(\xi_0 - 1)}\right),$$

$$\tilde{\tau}_2 = \frac{2}{\sqrt{4-\tilde{\alpha}_1^2}}\left(2\pi - \operatorname{arctg}\frac{(\xi_0 - 1)\sqrt{4-\tilde{\alpha}_1^2}}{2\theta + \tilde{\alpha}_1(\xi_0 - 1)}\right).$$

Figure 15.8a presents graphical solution of the Eq. (15.8). (In this figure, the function

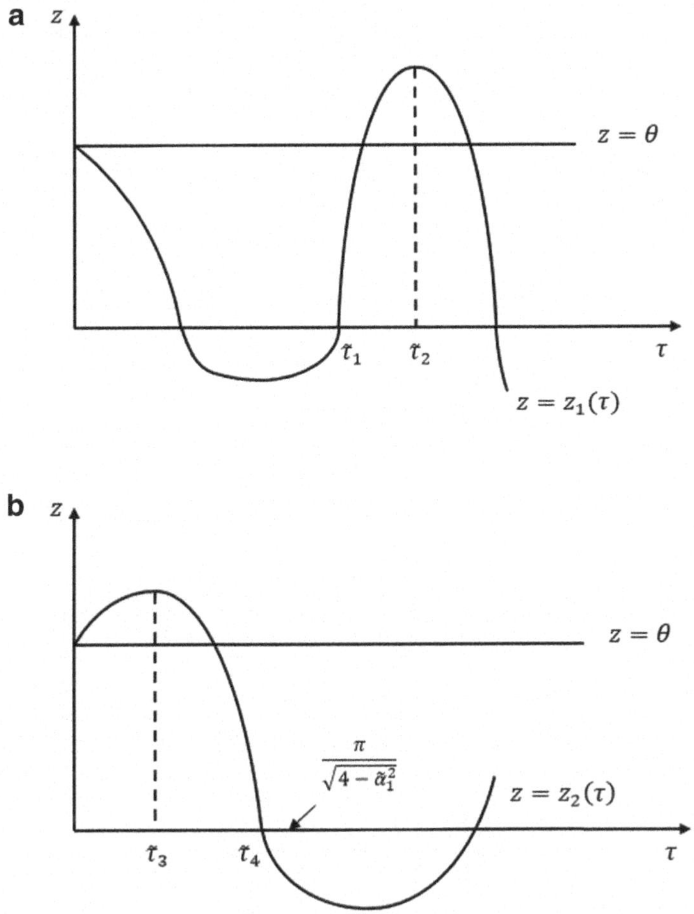

Fig. 15.8 **a** Function graph $z = z_1(\tau)$. **b** Function graph $z = z_2(\tau)$

$$\exp(\tilde{\alpha}_1 \tau/2)\left[\theta \cos \frac{\sqrt{4-\tilde{\alpha}_1^2}}{2}\tau - \frac{\tilde{\alpha}_1\theta + 2(\xi_0 - 1)}{\sqrt{4-\tilde{\alpha}_1^2}} \sin \frac{\sqrt{4-\tilde{\alpha}_1^2}}{2}\tau\right]$$

is denoted by $z_1(\tau)$).

An instantaneous change in the slip direction at the moment $\tau = \tau_1$ occurs if $\xi_1 < -1$. Suppose that this inequality holds. In the subsequent time interval $\tau_1 < \tau < \tau_1 + \tau_2$ (the point $\xi = \xi_2, \dot{\xi} = \theta$ corresponds to the value $\tau = \tau_1 + \tau_2$), the body moves at a higher velocity than that of the belt ($\dot{\xi} > \theta$). According to (15.6) by virtue of the initial conditions $\tau = \tau_1, \xi = \xi_1, \dot{\xi} = \theta$, we obtain

$$\xi = -1 - \tilde{\alpha}_1\theta + \exp(\tilde{\alpha}_1(\tau - \tau_1)/2)\left[(\xi_1 + 1 + \tilde{\alpha}_1\theta) * \cos\frac{\sqrt{4 - \tilde{\alpha}_1^2}}{2}(\tau - \tau_1)\right.$$
$$\left. + \frac{2\theta - \tilde{\alpha}_1(\xi_1 + 1 + \tilde{\alpha}_1\theta)}{\sqrt{4 - \tilde{\alpha}_1^2}}\sin\frac{\sqrt{4 - \tilde{\alpha}_1^2}}{2}(\tau - \tau_1)\right],$$

$$\dot{\xi} = \exp(\tilde{\alpha}_1(\tau - \tau_1)/2)\left[\theta\cos\frac{\sqrt{4 - \tilde{\alpha}_1^2}}{2}(\tau - \tau_1)\right.$$
$$\left. - \frac{\tilde{\alpha}_1\theta + 2(\xi_1 + 1)}{\sqrt{4 - \tilde{\alpha}_1^2}}\sin\frac{\sqrt{4 - \tilde{\alpha}_1^2}}{2}(\tau - \tau_1)\right].$$

Substituting $\tau = \tau_1 + \tau_2, \xi = \xi_2, \dot{\xi} = \theta$ into equations considered, we derive

$$\exp\left(\frac{\tilde{\alpha}_1\tau_2}{2}\right)\left[\theta\cos\frac{\sqrt{4 - \tilde{\alpha}_1^2}}{2}\tau_2 - \frac{\tilde{\alpha}_1\theta + 2(\xi_1 + 1)}{\sqrt{4 - \tilde{\alpha}_1^2}}\sin\frac{\sqrt{4 - \tilde{\alpha}_1^2}}{2}\tau_2\right] = \theta, \quad (15.9)$$

$$\xi_2 = -1 - \tilde{\alpha}_1\theta + \exp(\tilde{\alpha}_1\tau_2/2)\left[(\xi_1 + 1 + \tilde{\alpha}_1\theta)\cos\frac{\sqrt{4 - \tilde{\alpha}_1^2}}{2}\tau_2\right.$$
$$\left. + \frac{2\theta - \tilde{\alpha}_1(\xi_1 + 1 + \tilde{\alpha}_1\theta)}{\sqrt{4 - \tilde{\alpha}_1^2}}\sin\frac{\sqrt{4 - \tilde{\alpha}_1^2}}{2}\tau_2\right].$$

The value τ_2 satisfies the inequality: $\tilde{\tau}_3 < \tau_2 < \tilde{\tau}_4$, where

$$\tilde{\tau}_3 = \begin{cases} \frac{2}{\sqrt{4-\tilde{\alpha}_1^2}}\left(\pi - \operatorname{arctg}\frac{(\xi_1+1)\sqrt{4-\tilde{\alpha}_1^2}}{2\theta+\tilde{\alpha}_1(\xi_1+1)}\right), & \xi_1 \le -1 - \frac{2\theta}{\tilde{\alpha}_1}, \\ -\frac{2}{\sqrt{4-\tilde{\alpha}_1^2}}\operatorname{arctg}\frac{(\xi_1+1)\sqrt{4-\tilde{\alpha}_1^2}}{2\theta+\tilde{\alpha}_1(\xi_1+1)}, & -1 - \frac{2\theta}{\tilde{\alpha}_1} < \xi_1 < -1, \end{cases}$$

$$\tilde{\tau}_4 = \begin{cases} \frac{2}{\sqrt{4-\tilde{\alpha}_1^2}}\left(\pi + \operatorname{arctg}\frac{\theta\sqrt{4-\tilde{\alpha}_1^2}}{\tilde{\alpha}_1\theta+2(\xi_1+1)}\right), & \xi_1 \le -1 - \frac{\tilde{\alpha}_1\theta}{2}, \\ \frac{2}{\sqrt{4-\tilde{\alpha}_1^2}}\operatorname{arctg}\frac{\theta\sqrt{4-\tilde{\alpha}_1^2}}{\tilde{\alpha}_1\theta+2(\xi_1+1)}, & -1 - \frac{\tilde{\alpha}_1\theta}{2} < \xi_1 < -1. \end{cases}$$

Figure 15.8b presents graphical solution of the Eq. (15.9) for $-1-\tilde{\alpha}_1\theta/2 < \xi_1 < -1, 0 < \tilde{\alpha}_1 \le \sqrt{2}$. (In this figure, the function

$$\exp\left(\frac{\tilde{\alpha}_1\tau}{2}\right)\left[\theta\cos\frac{\sqrt{4-\tilde{\alpha}_1^2}}{2}\tau-\frac{\tilde{\alpha}_1\theta+2(\xi_1+1)}{\sqrt{4-\tilde{\alpha}_1^2}}\sin\frac{\sqrt{4-\tilde{\alpha}_1^2}}{2}\tau\right]$$

is denoted by $z_2(\tau)$). Note that instantaneous change in the slip direction at the moment $\tau=\tau_1+\tau_2$ occurs if $\xi_2>1$.

Subsequent slip intervals with the lower or higher velocity than that of the belt are investigated in the same way. This investigation at the ith stage ($i=1,2,3,\cdots$) leads to two relations, one of which determines time τ_i of the stage considered,

$$\exp(\tilde{\alpha}_1\tau_i/2)\left[\theta\cos\frac{\sqrt{4-\tilde{\alpha}_1^2}}{2}\tau_i-\frac{\tilde{\alpha}_1\theta+2(\xi_{i-1}\mp 1)}{\sqrt{4-\tilde{\alpha}_1^2}}\sin\frac{\sqrt{4-\tilde{\alpha}_1^2}}{2}\tau_i\right]=\theta,$$

whereas the other determines the coordinate value ξ_i

$$\xi_i=\pm 1-\tilde{\alpha}_1\theta+\exp(\tilde{\alpha}_1\tau_i/2)\left[(\xi_{i-1}\mp 1+\tilde{\alpha}_1\theta)\cos\frac{\sqrt{4-\tilde{\alpha}_1^2}}{2}\tau_i\right.$$
$$\left.+\frac{2\theta-\tilde{\alpha}_1(\xi_{i-1}\mp 1+\tilde{\alpha}_1\theta)}{\sqrt{4-\tilde{\alpha}_1^2}}\sin\frac{\sqrt{4-\tilde{\alpha}_1^2}}{2}\tau_i\right]$$

(the upper sign corresponds to $i=2n-1$, whereas the lower corresponds to $i=2n$, $n\in N$). Slipping at the ith stage herein exists if $|\xi_{i-1}|>1$, $i=1,2,3,\cdots$.

Suppose that $-1<\xi_j(\tau_k)<1(j=1,2,3,\cdots)$. Then, after j slip phase at the moment $\tau=\bar{\tau},\bar{\tau}=\sum_{i=1}^{j}\tau_i$, the stick phase begins. At this stage by virtue of (15.5), we have

$$\xi=\xi_j(\tau_k)+\theta(\tau-\bar{\tau}). \tag{15.10}$$

The stick phase ends at $\tau=\bar{\tau}+\tau_{k+1}$ when $\xi=1+\varepsilon(\tau_{k+1})$. Substituting the given values of time and coordinates into Eq. (15.10), we obtain an implicit relation connecting times τ_k and τ_{k+1} of successive intervals of relative rest

$$\begin{aligned}&\psi(\tau_{k+1})=\varphi(\tau_k),\quad \psi(\tau)=\theta\tau-\varepsilon(\tau),\\&\varphi(\tau)=1-\xi_j(\tau),\quad -1<\xi_j(\tau)<1,\ j=1,2,3,\cdots.\end{aligned} \tag{15.11}$$

Let $\xi^{(i)}(i=1,2,3,\cdots)$ be the sequence of abscissas of the intersection points of the half-path γ_3 with half-line $\xi>1$, $\dot{\xi}=\theta$. Depending on the parameter ε^*, the considered half-path either does not have common points with the interval $1<$

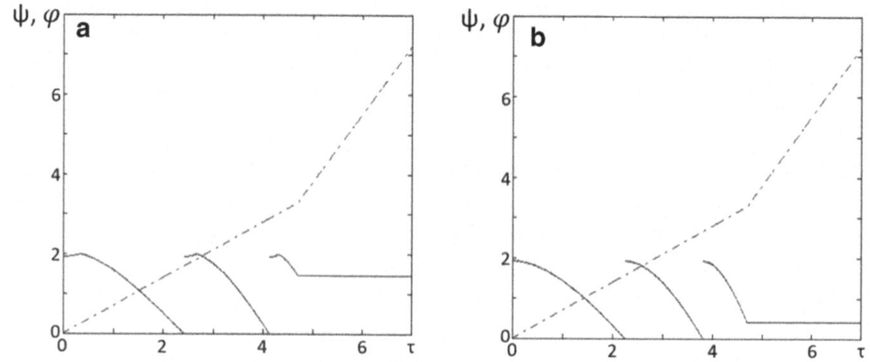

Fig. 15.9 **a** Dependences $\psi(\tau), \varphi(\tau)$ at $\tilde{\alpha}_1 = 0.14, \theta = 1.7, \varepsilon^* = 4.7$. **b** Dependences $\psi(\tau), \varphi(\tau)$ at $\tilde{\alpha}_1 = 0.155, \theta = 1.7, \varepsilon^* = 4.7$

$\xi < 1 + \varepsilon^*, \dot{\xi} = \theta$ of the half-line $\xi > 1, \dot{\xi} = \theta$, or intersects it in a finite number of points with abscissas $\xi^{(1)}, \xi^{(1)}, \ldots, \xi^{(k)}, k \geq 1$. It is easily shown that in the first case, the function $\varphi(\tau)$ is continuous, whereas in the second case it is piecewise continuous. Discontinuity occurs at the points $\tau^{(i)} = \xi^{(i)} - 1$, and $\varphi(\tau^{(i)} - 0) = 0$, $\varphi(\tau^{(i)} + 0) = \varphi(0)$ $(i = \overline{1,k})$. The dependences $\psi(\tau)$ and $\varphi(\tau)$ at $\tilde{\alpha}_1 = 0.14$, $\theta = 1.7$, $\varepsilon^* = 4.7$ ($\tilde{\alpha}_1 = 0.155$, $\theta = 1.7$, $\varepsilon^* = 4.7$) are presented in Fig. 15.9a, b by dash-dotted and solid lines, respectively. Figure 15.9a depicts dependences for $\tilde{\alpha}_1 < \tilde{\alpha}_1^{(1)}$, whereas Fig. 15.9b for $\tilde{\alpha}_1 > \tilde{\alpha}_1^{(1)}$.

The mapping function (15.11) specifies the point mapping of the half-line, with the exception of the discontinuity points $\tau^{(i)}, i = \overline{1,k}$ (if any) into the half-line. For $\tilde{\alpha}_1 = 0$ this function coincides with mapping function [7]. Note that for $\theta < 1$ point mapping (15.11) has the only simple able fixed point τ_j^*, satisfying the inequality $\tau_j^* > \varepsilon^*$ and determined by the relation $\tau_j^* = (1 - \xi_j(\varepsilon^*) + \varepsilon^*)/\theta$ $(j = 1, 2, 3, \cdots)$. The specified point corresponds to $j/2$ reverse periodic motion regimes with duration of the slip phase τ_j^*, $j = 1, 2, 3, \cdots$ [11].

15.5 Numerical Results

As the image point moving along the positive half-path from a starting point $M(\xi, \theta)$, $1 \leq \xi < \xi^*$, always hits (for selected $\tilde{\alpha}_1$ and ε^*) the segment of sliding motions, the investigation of the dynamics of the system considered can be reduced to studying the sequence properties $\tau_k (k = 1, 2, 3, \cdots)$ of the duration of the stick phase. Since for any element of this sequence there must exist a successor, then it is necessary to define point mapping (15.11) in discontinuity points $\tau^{(i)}$, $i = \overline{1,k}$ (if any). We assume that $0 \leq \varphi(\tau^{(i)}) \leq \varphi(0) (i = \overline{1,k})$. The bifurcation diagrams were plotted using so defined mapping function.

Figures 15.10 and 15.11 show bifurcation diagrams plotted against belt velocity

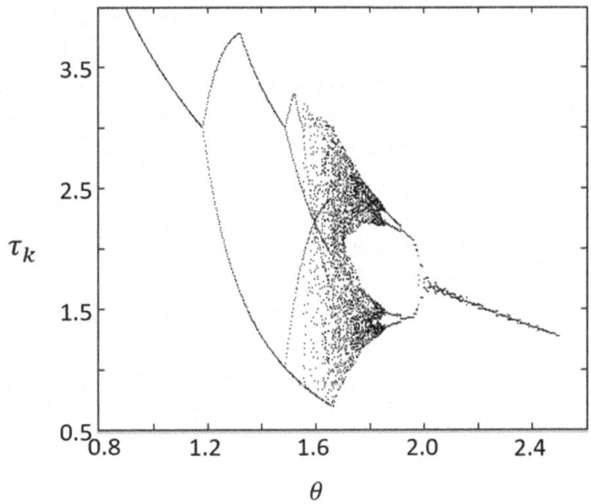

Fig. 15.10 Bifurcation diagram by θ for $\varepsilon^* = 3$, $\tilde{\alpha}_1 = 0.05$

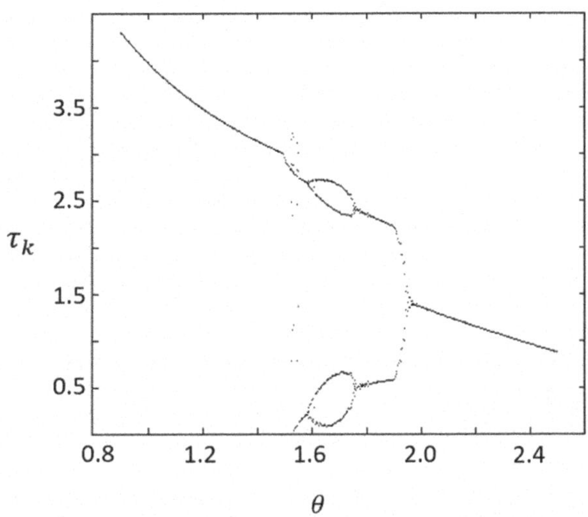

Fig. 15.11 Bifurcation diagram by θ for $\varepsilon^* = 3$, $\tilde{\alpha}_1 = 0.11$

θ for $\varepsilon^* = 3$ and $\tilde{\alpha}_1 = 0.05$, $\tilde{\alpha}_1 = 0.11$ respectively. In these figures, the values of θ parameter ranged from 0.9 to 2.5 are plotted on the abscissa axis, whereas the values of the duration of the stick phase are plotted on the ordinate axis. It is clearly shown that due to an increase in the belt velocity or dependence on the parameter $\tilde{\alpha}_1$, either a period-doubling process occurs leading to a chaotic motion [9] (Fig. 15.10), or there exist only periodic motions of the system with one or several stick phases (Fig. 15.11). Figures 15.10 and 15.11 show that the increase of θ leads to the reduction of the duration of the stick phase in the periodic regimes of motion. The increase of the coefficient of the viscous friction results in the disappearance of the stochastic regimes of motion.

Figures 15.12, 15.13 and 15.14 show bifurcation diagrams plotted against belt velocity θ, $0.9 \leq \theta \leq 2.5$, for $\tilde{\alpha}_1 = 0.15$ and for various parameter values ε^*. Figure 15.12 corresponds to the value $\varepsilon^* = 2$, whereas Figs. 15.13, 15.14 to the values $\varepsilon^* = 4$ and $\varepsilon^* = 5$, respectively. For $\varepsilon^* = 2$ depending on the parameter θ, the system exhibits periodic motion regimes with one or two times of relative rest of the body and the belt. With increasing $\varepsilon^*(\varepsilon^* = 4, \varepsilon^* = 5)$, stochastic motion regimes occur, which turn into periodic motion regimes with one prolonged stop due to further increase in belt velocity. With an increase in $\varepsilon^*(\varepsilon^* = 4, \varepsilon^* = 5)$ stochastic movements arise.

Figures 15.15 and 15.16 show bifurcation diagrams for the viscous friction coef-

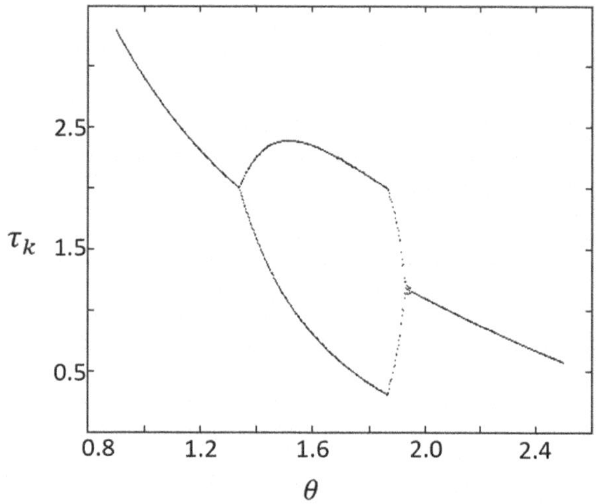

Fig. 15.12 Bifurcation diagram by θ for $\tilde{\alpha}_1 = 0.15$, $\varepsilon^* = 2$

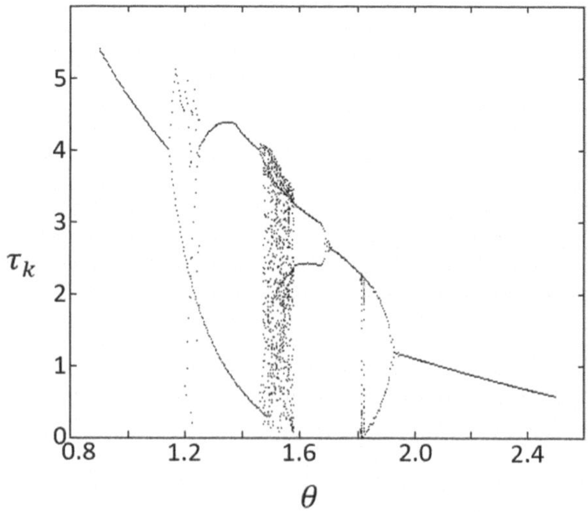

Fig. 15.13 Bifurcation diagram by θ for $\tilde{\alpha}_1 = 0.15$, $\varepsilon^* = 4$

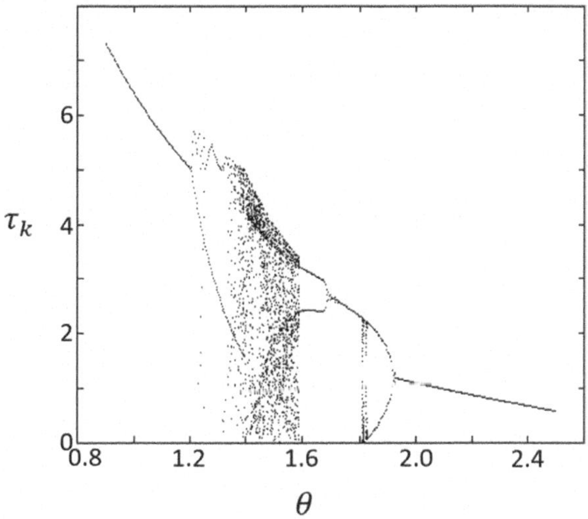

Fig. 15.14 Bifurcation diagram by θ for $\tilde{\alpha}_1 = 0.15$, $\varepsilon^* = 5$

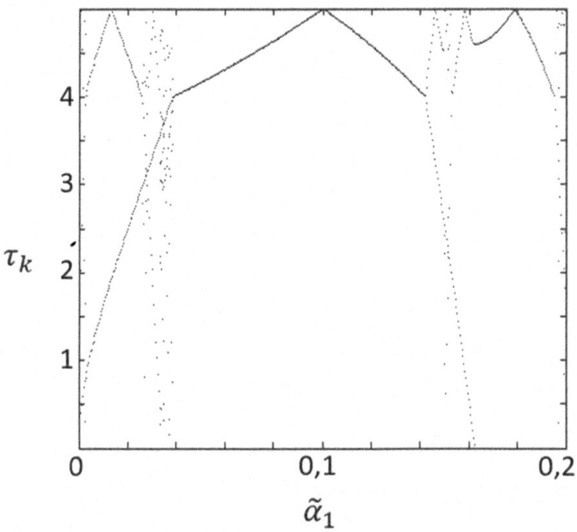

Fig. 15.15 Bifurcation diagram for $\tilde{\alpha}_1$ at $\varepsilon^* = 4$, $\theta = 1.2$

ficient $\tilde{\alpha}_1$ (plotted on the abscissa axis), $0 \leq \tilde{\alpha}_1 \leq 0.2$, for $\varepsilon^* = 4$ and $\theta = 1.2$, $\theta = 1.6$, respectively. These figures show that the parameter $\tilde{\alpha}_1$ affects the dynamics of the system. When $\tilde{\alpha}_1$ is small, the increase of the velocity of the belt leads to the transformation of the periodic regimes with a few stick phases per period (Fig. 15.15) into the stochastic regimes of motion (Fig. 15.16). With increasing $\tilde{\alpha}_1$, the considered motion regimes are interrupted by periodic motions with one ($\theta = 1.2$) or with several ($\theta = 1.6$) stick regimes of the body and the belt.

Figures 15.17, 15.18 and 15.19 show bifurcation diagrams for the parameter ε^* (plotted on the abscissa axis), $0 \leq \varepsilon^* \leq 6$, for $\tilde{\alpha}_1 = 0.07$ and $\theta = 1.2$, $\theta = 1.5$

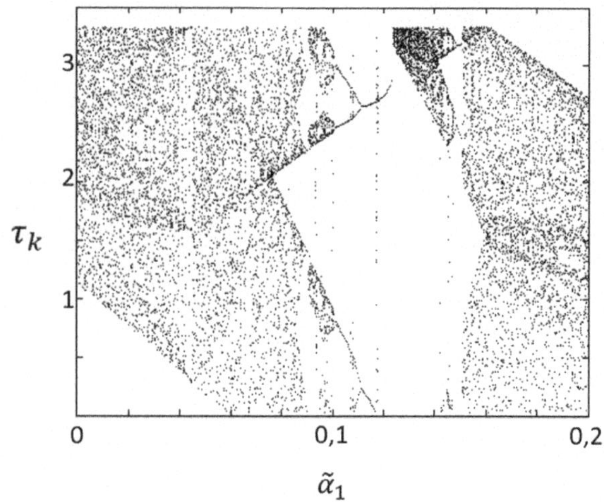

Fig. 15.16 Bifurcation diagram for $\tilde{\alpha}_1$ at $\varepsilon^* = 4$, $\theta = 1.6$

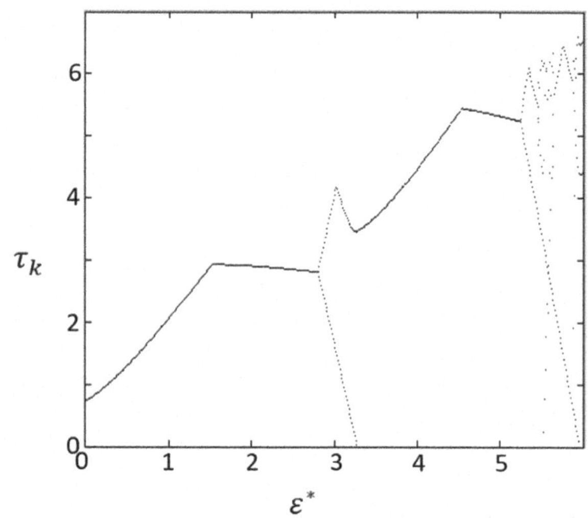

Fig. 15.17 Bifurcation diagram for ε^* at $\tilde{\alpha}_1 = 0.07$, $\theta = 1.2$

and $\theta = 1.8$ respectively. These figures show that for all values of the parameter ε^* within the range determined by the belt velocity θ, the system exhibits periodic motion regimes with one relative rest of the body and the belt. With the increase of ε^*, and depending on θ (see Figs. 15.17, 15.18, 15.19), one observes the period-doubling process that results in chaos. The higher the velocity of the belt, the smaller ε^*, which corresponds to the transition to chaos. Note that with increasing belt velocity, the interval of the values of the parameter ε^* corresponding to the stochastic motion regimes of the system, expands.

Figures 15.20, 15.21 and 15.22 present bifurcation diagrams plotted against the parameter ε^*, $0 \leq \varepsilon^* \leq 6$, for $\theta = 1.7$ and various values of the parameter $\tilde{\alpha}_1$.

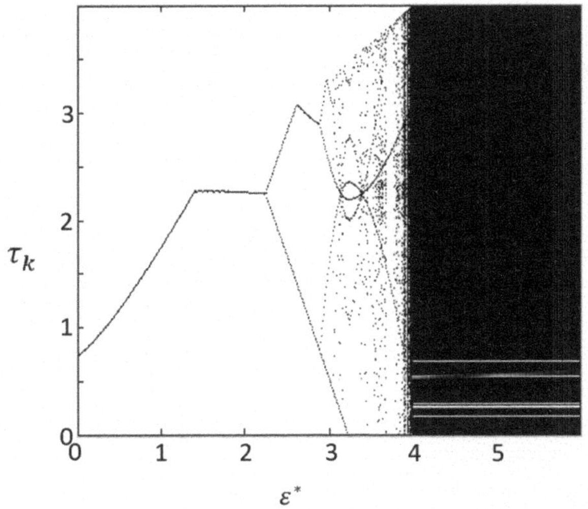

Fig. 15.18 Bifurcation diagram for ε^* at $\tilde{\alpha}_1 = 0.07, \theta = 1.5$

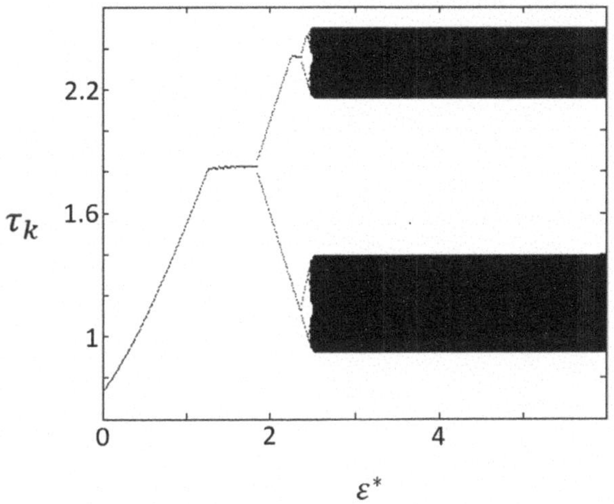

Fig. 15.19 Bifurcation diagram for ε^* at $\tilde{\alpha}_1 = 0.07, \theta = 1.8$

Figure 15.20 corresponds to the value $\tilde{\alpha}_1 = 0.06$, whereas Figs. 15.21 and 15.22 to the values $\tilde{\alpha}_1 = 0.12$ and $\tilde{\alpha}_1 = 0.16$, respectively. For $\tilde{\alpha}_1 = 0.06$ depending on the parameter ε^*, there exist either periodic motion regimes with one or several stick phases of the body and the belt or stochastic motion regimes. With increasing $\tilde{\alpha}_1$. ($\tilde{\alpha}_1 = 0.12, \tilde{\alpha}_1 = 0.16$) stochastic motions disappear and periodic motions with one or two stick phases are observed.

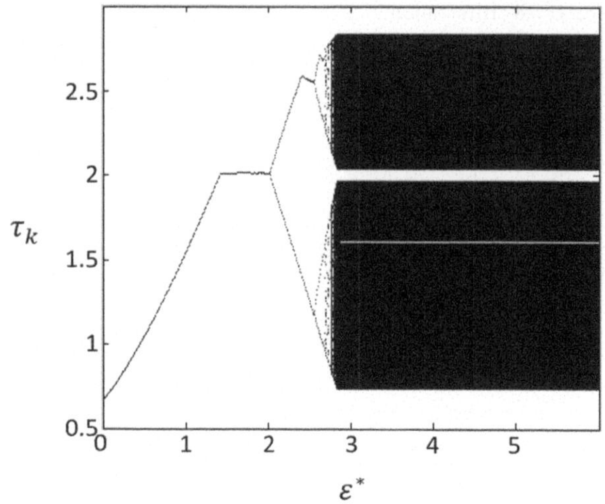

Fig. 15.20 Bifurcation diagram for ε^* at $\theta = 1.7$, $\tilde{\alpha}_1 = 0.06$

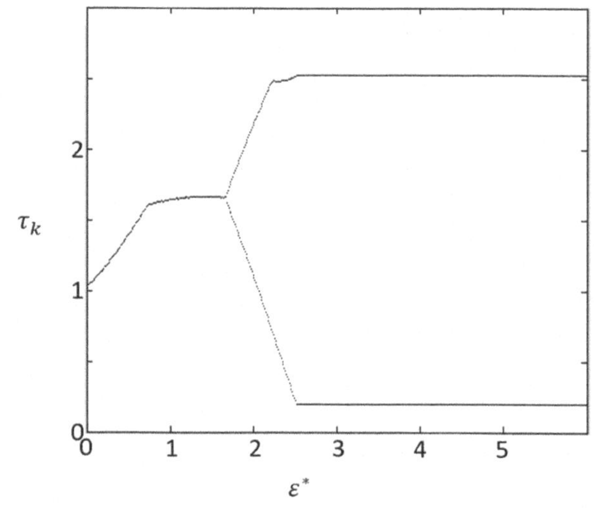

Fig. 15.21 Bifurcation diagram for ε^* at $\theta = 1.7, \tilde{\alpha}_1 = 0.12$

15.6 Conclusion

The main results of the present work are as follows:

- The structure of the phase space of the vibration system consisting of a pair of bodies in frictional contact with account for viscous-dry friction with memory is studied in detail using the methods of qualitative theory of dynamical systems on a plane.
- It is shown for the first time that the regimes that include the stick phase are possible only in certain parameter ranges.

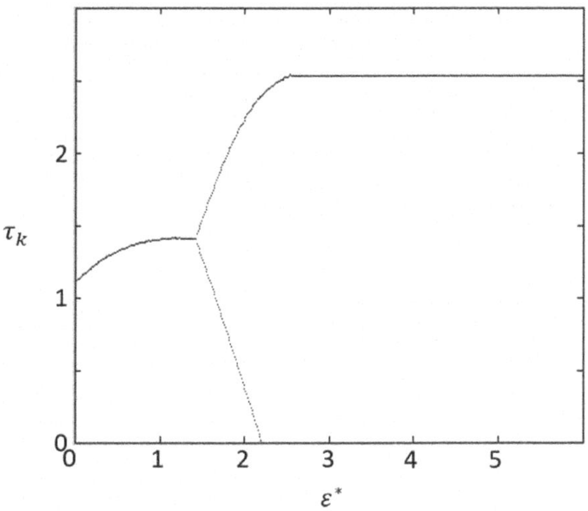

Fig. 15.22 Bifurcation diagram for ε^* at $\theta = 1.7$, $\tilde{\alpha}_1 = 0.16$

- The main transitions between periodic and stochastic motion regimes are revealed depending on the system parameters.

Acknowledgements This work was supported by a grant of the Russian Science Foundation (16-19-10237-P).

References

1. Ishlinsky, AYu., Kragelsky, I.V.: O skachkah pri trenii. Zhurnal Tech. Phys. **14**(4/5), 276–282 (1944)
2. Kashchenevsky, L.Ya.: Stochasticheskie avtokolebania pri suhom trenii. V Eng. Phys. Zhurnale **47**(1), 143–147 (1984)
3. Leine, R.I., van Campen, D.H.: Discontinuous fold bifurcations in mechanical systems. Archive Appl. Mech. **72**, 138–146 (2002)
4. Leine, R.I., van Campen, D.H.: Bifurcation phenomena in non-smooth dynamicalsystems. Eur. J. Mech. A. Solids **25**, 595–616 (2006)
5. Leine, R.I., van Campen, D.H., deKraker, A., van den Steen, L.: Stick-slip vibrations induced by alternate friction models. Int. J. Nonlinear Dyn. **16**, 41–54 (1998)
6. Luo, G.W., Lv, X.H., Ma, L.: Periodic-impact motions and bifurcations in dynamics of a plastic impact oscillator with a frictional slider. Eur. J. Mech. A. Solids **27**, 1088–1107 (2008)
7. Metrikin, V.S., Nagaev, R.F., Stepanova, V.V.: Periodicheskie b stohasticheskie avtokolebania v sisteme s suhim treniem nasledstvennogo tipa. V zhurnale Pr.mat.meh **60**(5), 859–864 (1996)
8. Neymark, Yu.I.: Dinamicheskie sistemy i uplyaemie processi, p. 336. Knizhnii dom Librokom (2019)
9. Schuster, G.: Determinirovannii haos, p. 237. Mir
10. Van de Vrande, B.L., van Campen, D.H., de Kraker, A.: An approximate analysis of dry-friction induced stick-slip vibrations by a smoothing procedure. Int. J. Nonlinear Dyn. **19**, 157–169 (1999)

11. Vetyukov, M.M., Dobroslavsky, S.V., Nagaev, R.F.: Avtokolebania v sisiteme s characteristicoi suhogo treniya nasledstvennogo tipa. V zhurnale Mechanica tverdogo tela (1), 23–28 (1990)
12. Vetyukov, M.M., Platovsky, MYu.: Frikcionnie avtokolebania v sistemah s odnoy I dvumya stepenyami svobody. Sovr.problemy mehaniki I ee prepodavaniya v vuze Tr.Vseros.nauch.-metod.konf **1**, 58–62 (2015)
13. Zaitsev, M.V., Metrikin, V.S.: K teorii neavtonomnoy vibratsionnoy sistemi s treniem nasledstvennogo tipa. V zhurnale Vestnik NNGU Matematicheskoe modelirovanie Optimal'noe upravlenie **3**(1), 141–146 (2012)

Chapter 16
Experimental Studies of Elastic–Plastic Deformation of Structural Materials Under Conditions of Triaxial Loading

Leonid A. Igumnov, Sergey Ye. Vlasov, Dmitry A. Kazakov, Dmitry V. Zhegalov, and Ivan A. Modin

Abstract The behavior of metal tubular specimens under combined tension–torsion, torsion compression and internal torsion pressures is studied experimentally. The tests are necessary for obtaining experimental data in order to study loss of stability and deformation of elastic–plastic shells caused by large deformations under complex stress conditions loaded along multi-link broken polygonal deformation trajectories, as well as for assessing the role of deformation anisotropy. In order to conduct tests under complex stress conditions (internal pressure loading in combination with other types of loading), the testing equipment was modified. Equipment of the test complex with a system for loading of tubular specimens with internal pressure ensures the conditions for achieving destructive stresses, caused by both internal pressure and torsion with tension. The Zwick/Roell Z100 testing machine equipped with the system generating internal pressure in hollow specimens makes it possible to expand significantly the types of tests on determining deformation and strength characteristics of materials. The results can be used to verify mathematical models and program codes for the numerical study of the processes of elastic–plastic deformation and loss of stability of shells of revolution under complex loading.

L. A. Igumnov (✉) · D. A. Kazakov · D. V. Zhegalov · I. A. Modin
Research Institute for Mechanics, National Research Lobachevsky State University of Nizhny Novgorod, 23 Gagarin Avenue, Building 6, Nizhny Novgorod 603950, Russian Federation
e-mail: igumnov@mech.unn.ru

D. A. Kazakov
e-mail: kazakov@mech.unn.ru

D. V. Zhegalov
e-mail: zhegalov@mech.unn.ru

I. A. Modin
e-mail: mianet@mail.ru

S. Ye. Vlasov
Scientific Research Institute for System Analysis of the Russian Academy of Sciences (SRISA), 36 Nakhimovsky Prospekt, Building 1, Moscow 117218, Russian Federation
e-mail: vlasov@niisi.ru

H. Altenbach et al. (eds.), *Multiscale Solid Mechanics*,
Advanced Structured Materials 141,
https://doi.org/10.1007/978-3-030-54928-2_16

Keywords Experiment · Complex stress–strain state · Elastoplastic deformation of a material · Tubular sample · Internal pressure · Anisotropy

16.1 Introduction

The processes of elastic–plastic deformation, loss of stability due to large deformations can be studied in detail in experiments on complex loading (beam, multi-link broken lines and spatial deformation trajectories). In the experiments, the test tubular specimen is simultaneously subjected to tension–compression, alternating torsion and loading by internal pressure [4, 7, 10, 13]. The loading device and measuring systems were finalized based on the universal testing machine Zwick/Roell Z100 of Lobachevsky University. This enables simultaneous loading of the test tubular specimen by tension–compression, alternating torsion and internal pressure. The geometry of the working part and the heads of both solid and tubular specimens have been elaborated for complex loading tests. The experiments on complex combined loading in tension–compression, torsion and internal pressure are presented. The university actively uses both experimental studies of materials and numerical modeling of the studied objects [1–3, 5, 6, 8, 9, 11, 12, 14, 15].

16.2 Test Equipment Modernization

The testing machine Zwick/RoellZ100 allows one to conduct both tension–compression experiments and experiments on alternating torsion. The control unit is equipped with TestXpert II software that provides changes in axial loading rates and alternating torque in a wide range. The system has laser channels for simultaneous measurement of longitudinal displacements and twist angle of the specimen working part and a channel for determining transverse deformation with the help of a digital video camera.

A universal hydraulic unit Zwick/Roell with an internal pressure control system was employed to create internal pressures in the specimens installed in the hydraulic grips of the Z100 testing machine and to conduct complex loading tests according to the (p-q) test program. The unit ensures a maximum pressure of 480 bar and a flow rate of 3.6 l/min. Thus, the hydraulic unit of Z100 testing machine is supplemented by a system that operates in a blind (dead end) mode for internal pressure loading of hollow specimens.

A system for fixing the specimens in the grips of the testing machine has been elaborated. The specimen and grips have square cross section (Fig. 16.1) on which a special sleeve of identical internal cross section is mounted. The end of the specimen has an internal thread necessary for tensile loading of a specimen. For internal pressure test, an intermediate-shaped sleeve with an end groove and a sealing ring is

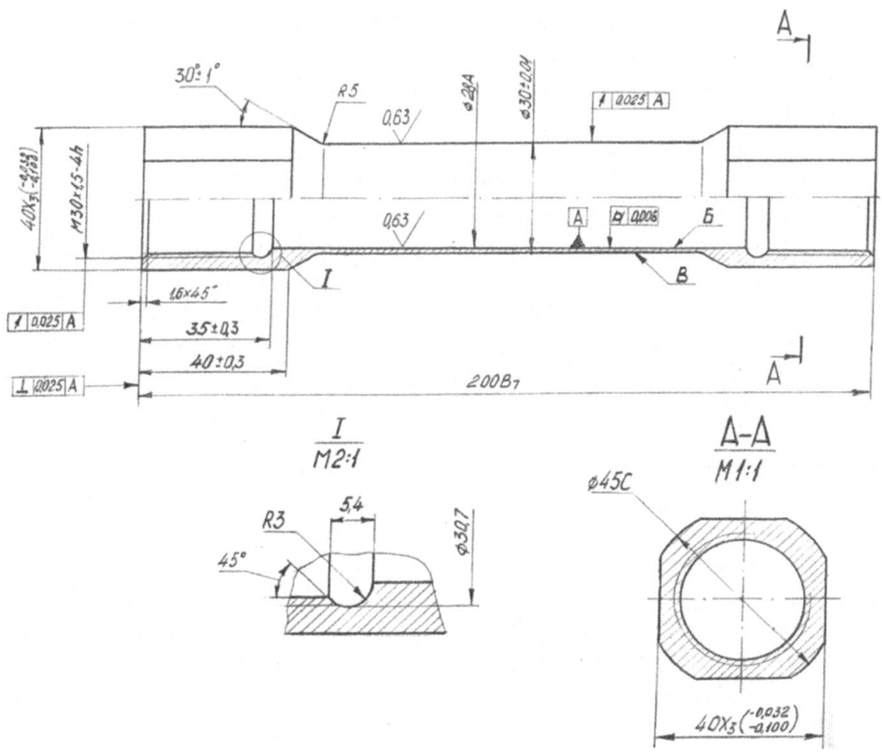

Fig. 16.1 Scheme of tubular specimen

inserted between the specimen end and hydraulic grips of the testing machine. From hydraulic unit, oil flows into internal cavity of the specimen through a sleeve side hole.

In experiments on loading by internal pressure, the internal cavity of a specimen is filled with oil. The maximum pressure generated by the pump within the chosen loading scheme reaches 480 kg/cm^2, which is sufficient enough to cause failure of specimens made of various materials with wall thickness of up to 1.0 mm. A characteristic feature of the specimen design is a pronounced working part and smooth transition to a gripping part, which minimizes the effect of stress concentrators [7].

16.3 Experimental Investigation

An experimental study of deformation processes of metal tubular specimens made of 12X18H10T steel under combined loading in tension–torsion and internal pressure was carried out on the modified test complex Z100. Longitudinal and transverse deformations, twist angle are recorded directly on the specimen working part with a laser extensometer base of 60 mm.

Test 1. The deformation of cylindrical shell under sequential tension–torsion loading.

The shell has the following geometric dimensions: the working part of 100 mm length, the outer diameter of 30 mm, the inner diameter of 28 mm, 1 mm wall thickness and cross-sectional area of 364.24 mm^2. One end of the shell is rigidly fixed, while the second moves with a constant velocity of 3 mm/min until elongation of the shell is reached $\Delta l = 28.4$ mm, $l_{11} = 30.9\%$ (time 570 s, Fig. 16.2a). Then, the shell is unloaded, and its free end twists at a constant angular velocity of 30/min until the loss of stability occurs and an explicit form of wave formation on the specimen appears (Fig. 16.2b). The values of axial force, torque, elongation and twist angle as a function of time are shown in Fig. 16.2.

Test 2. Deformation of a cylindrical shell under sequential loading by internal pressure and torsion.

The shell with the following geometric dimensions was used in the test: the working part of 100 mm length, the outer diameter of 26.9 mm, the inner diameter of 28 mm, 0.8 mm wall thickness and cross-sectional area of 289.38 mm^2.

The shell is subjected to step loading with internal pressure of 100, 150, 200, 250 and 300 bar (point 1 of Fig. 16.3a), followed by unloading to 0 bar (point 2 of Fig. 16.3a). Then, one end of the shell is rigidly fixed, the second is twisted with a constant angular velocity until the specimen exhibits loss of stability under torsion (Fig. 16.3b).

Values of internal pressure, torque, twist angle, changes in the outer diameter and working length of the specimen as a function of time are shown in Fig. 16.3.

Test 3. Loss of stability of a cylindrical shell under proportional loading in compression–torsion.

The shell has the following geometric dimensions: the working part of 100 mm length, the outer diameter of 29.6 mm and the inner diameter of 28 mm, 0.8 mm wall thickness and cross-sectional area of 289.38 mm^2. One end of the shell is rigidly fixed, the other moves and twists at a constant axial velocity of 1 mm/min and angular velocity of 10°/min.

Figure 16.4 shows the values of axial force, elongation, torque and twist angle as a function of time.

16.4 Conclusion

In order to conduct tests under complex stress conditions (internal pressure loading in combination with other types of loading), the testing equipment was modified. Equipment of the test complex with a system for loading of tubular specimens with internal pressure ensures the conditions for achieving destructive stresses, caused by both internal pressure and torsion with tension. The Zwick/Roell Z100 testing

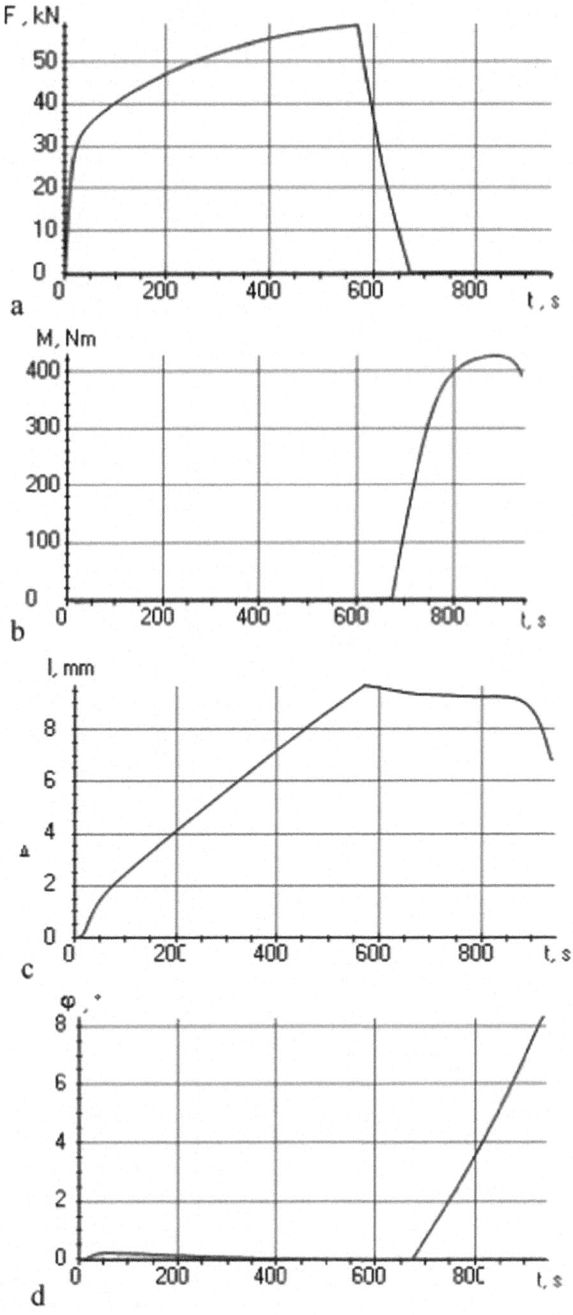

Fig. 16.2 Test 1

Fig. 16.3 Test 2

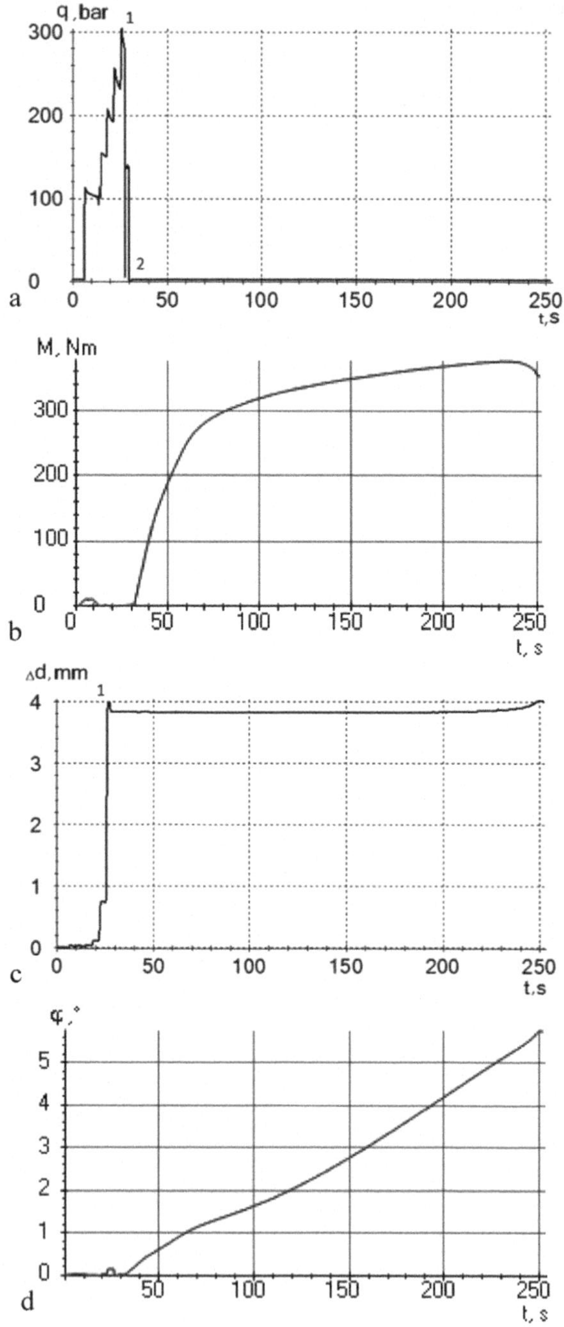

Fig. 16.3 (continued)

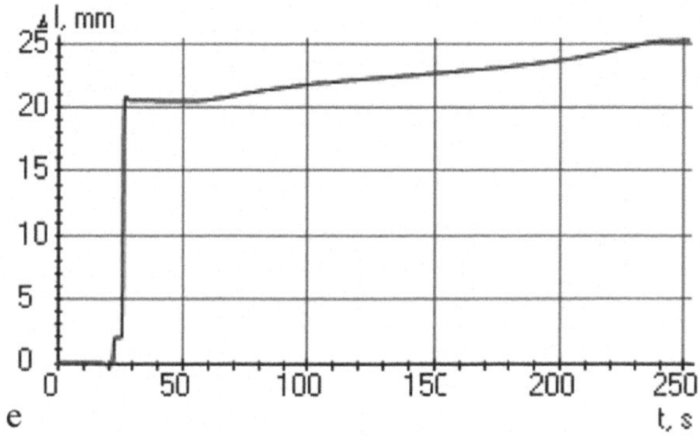

machine equipped with the system generating internal pressure in hollow specimens makes it possible to expand significantly the types of tests on determining deformation and strength characteristics of materials.

The behavior of metal tubular specimens under combined tension–torsion, torsion–compression and internal torsion pressures is studied experimentally. The tests are necessary for obtaining experimental data in order to study loss of stability and deformation of elastic–plastic shells caused by large deformations under complex stress conditions loaded along multi-link broken polygonal deformation trajectories, as well as for assessing the role of deformation anisotropy.

The results can be used to verify mathematical models and program codes for the numerical study of the processes of elastic–plastic deformation and loss of stability of shells of revolution under complex loading.

Fig. 16.4 Test 3

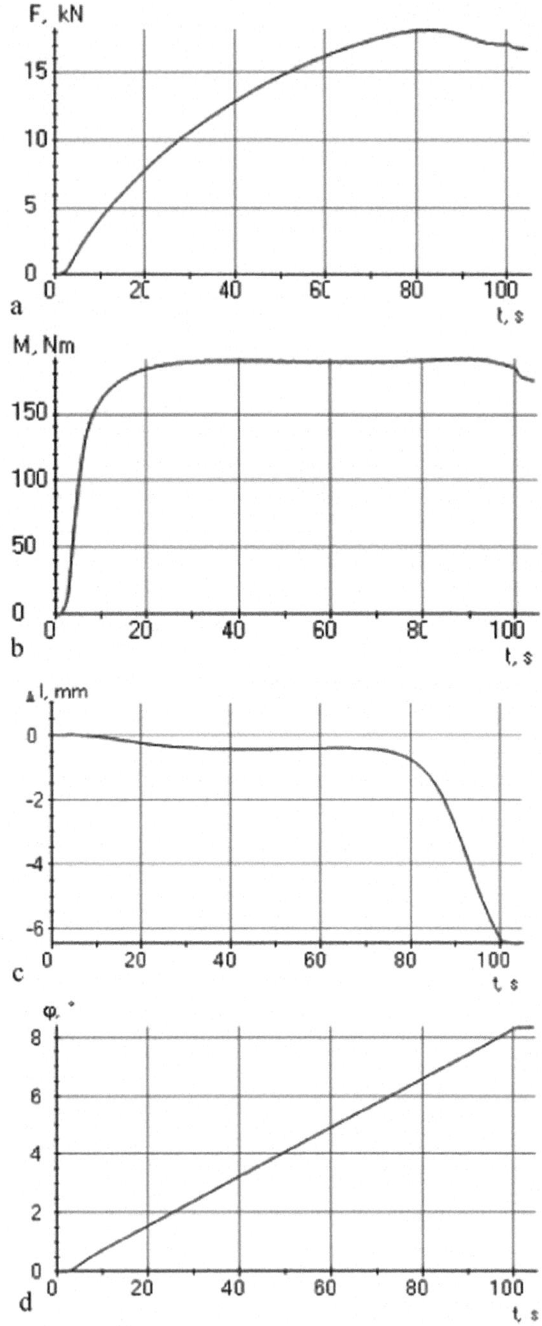

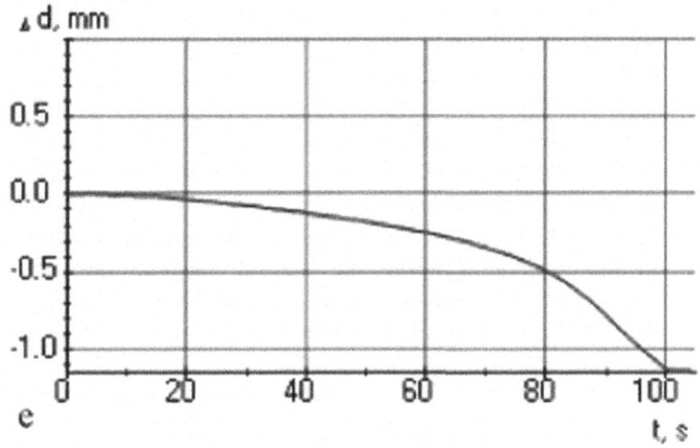

Fig. 16.4 (continued)

Acknowledgements This work was financially supported by the Ministry of Science and Higher Education of the Russian Federation (Agreement No. 075-11-2019-050) regarding the modernization of experimental equipment and grant RSF No. 19-19-00062 regarding experimental studies on complex loading.

References

1. Antipov, A.A., Gorokhov, V.A., Yegunov, V.V., Kazakov, D.A., Kapustin, S.A., Churilov, U.A.: Chislennoye modelirovaniye vysokotemperaturnoy polzuchesti elementov konstruktsiy iz zharoprochnykh splavov s uchetom effektov neytronnogo oblucheniya [Numerical simulation of high-temperature creep of elements of structures from heat-resistant alloys taking into account effects of neutron irradiation]. Problemy prochnosti i plastichnosti [Problems of Strength and Plasticity] T. 81. № 3. C. 345–358 (2019). (In Russian)
2. Balandin, V.V., Kochetkov, A.V., Krylov, S.V., Modin, I.A.: Numerical and experimental study of the penetration of a package of woven metal grid by a steel ball. J. Phys: Conf. Ser. **1214**(1), 012004 (2019)
3. Barchiesi, E., Plasidi, L., Modin, I.A., Zhegalov, D.V., Grigoryev, M.V.: A variational approach to strain gradient damage mechanics with an application to compressed frames. Prob. Strength Plast. **80**(3), 392–401 (2018)
4. Botvina, L.R.: Razrushenie: kinetika, mekhanizmy, obshchie zakonomernosti [Destruction: kinetics, mechanisms, general laws]. M.: Nauka (2008), 334 p. (In Russian)
5. Bragov, A.M., Konstantinov, A.I., Kochetkov A.V., Modin I.A., Savikhin A.O.: Experimental study of deformation properties of a bulk layer from plumbum balls under dynamic and quasistatic loading. PNRPU Mechanics Bulletin **4**, 16–27 (2017)
6. Dell'Isola, F., Igumnov, L.A., Litvinchuk, S.Y., Ipatov, A.A., Petrov, A.N., Modin, I.A.: Surface waves in dissipative poroviscoelastic layered half space: boundary element analyses. Adv. Struct. Mater. **103**, 305–319 (2019)
7. Kazakov, D.A., Kapustin, S.A., Korotkih, Y.G.: Modelirovanie processov deformirovaniya i razrusheniya materialov i konstrukcij [Modeling of the processes of deformation and fracture of materials and structures]. N. Novgorod: Izd-vo Nizhegorodskogo Universiteta (1999), 226 p. (In Russian)
8. Kochetkov, A.V., Leontev N.V., Modin I.A., Savikhin, A.O.: Study of the stress-strain and strength properties of the metal woven grids. Vestnik Tomskogo Gosudarstvennogo Universiteta, Matematika i Mekhanika **52**, 53–62 (2018)

9. Kochetkov, A.V., Modin, I.A., Leontev, N.V., Turigina, I.A., Poverennov, E.Y.: Numerical modeling of nonlinear dynamic and static compression of the metal mesh. PNRPU Mechanics Bulletin **4**, 106–113 (2019)
10. Lokoshchenko, A.M.: Polzuchest' i dlitel'naya prochnost' metallov [Creep and long-term strength of metals]. M.: FIZMATLIT (2016), 504 p. (In Russian)
11. Modin, I.A., Kochetkov, A.V., Leontiev, N.V.: Numerical simulation of quasistatic and dynamic compression of a granular layer. AIP Conf. Proc. **2116**, 270003 (2019)
12. Nagornykh, E.V., Bazhenov, V.G., Kazakov, D.A.: Eksperimentalnoye i teoreticheskoye issledovaniye bolshikh deformatsiy tsilindricheskikh obraztsov iz materialov, chuvstvitelnykh k vidu napryazhennogo [Experimental and theoretical study of large deformations of cylindrical samples from materials sensitive to the type of stress condition, under loading by tension-torsion]. XII Vserossiyskiy sezd po fundamentalnym problemam teoreticheskoy i prikladnoy mekhaniki [XII All-Russian Congress on Fundamental Problems of Theoretical and Applied Mechanics] (2019), pp. 350–352. (In Russian)
13. Volkov, I.A., Igumnov, L.A.: Vvedenie v kontinual'nuyu mekhaniku povrezhdennoj sredy [Introduction to the continuum mechanics of a damaged medium]. M.: FIZMATLIT (2017), 304 p. (In Russian)
14. Volkov, I.A., Igumnov, L.A., Modin, I.A., Boev, E.V.: A model of damaged media for describing a creep-induced damage accumulation process. In: 9th International Conference on Materials Structure and Micromechanics of Fracture. Procedia Struct. Integrity **23**, 281–286 (2019)
15. Volkov, I.A., Tarasov, I.S., Smetanin, I.V., Igumnov, L.A., Kazakov, D.A., Shishulin, D.N.: Constitutive relations of the mechanics of a damaged medium for evaluating the creep-rupture strength of structural alloys. J. Appl. Mech. Tech. Phys. **60**(1), 156–166 (2019)

Chapter 17
Evaluating Fatigue Life of Structures Under Thermocyclic Loading

Leonid A. Igumnov, Ivan A. Volkov, and Sergey Ye. Vlasov

Abstract The issue of evaluating strength and service life is discussed as applied to structures, the performance properties of which are characterized by multi-parametric non-stationary thermal–mechanical effects. The main degradation mechanisms of structural materials (metals and their alloys) are considered. The main requirements to mathematical models describing fatigue damage accumulation are formulated. In the framework of mechanics of damaged media (MDM), a mathematical model is developed that describes thermoplastic deformation and fatigue damage accumulation under combined low-cycle and high-cycle fatigue. The model consists of three interrelated parts: relations that define thermocyclic plastic behavior of the material with account for its dependence on the failure process; evolutionary equations describing the kinetics of damage accumulation; and a strength criterion of the damaged material. A version of the constitutive relations of thermoplasticity is based on the concept of the yield surface and the gradientality principle of the plastic strain rate vector to the yield surface at the loading point. These relations describe the major effects of the cyclic plastic deformation of the material for arbitrary complex loading trajectories. This version of kinetic equations of fatigue damage accumulation is based on a scalar damage parameter and energy principles and considers the main effects of nucleation, growth, and merging of microdefects under arbitrary complex loadings. A generalized form of the evolutionary equation for the fatigue damage accumulation under low-cycle and high-cycle fatigue is proposed. The critical damage value is used as the strength criterion of the damaged material. The effect of the distillate droplet impingement frequency on thermocyclic fatigue life

L. A. Igumnov (✉) · I. A. Volkov
Research Institute for Mechanics, National Research Lobachevsky State University of Nizhny Novgorod, 23 Gagarin Avenue, building 6, Nizhny Novgorod, Russian Federation 603950
e-mail: igumnov@mech.unn.ru

I. A. Volkov
e-mail: pmptmvgavt@yandex.ru

S. Ye. Vlasov
Scientific Research Institute for System Analysis, Russian Academy of Sciences (SRISA), 36 Nakhimovsky Prospekt, building 1, Moscow, Russian Federation 117218
e-mail: vlasov@niisi.ru

H. Altenbach et al. (eds.), *Multiscale Solid Mechanics*,
Advanced Structured Materials 141,
https://doi.org/10.1007/978-3-030-54928-2_17

of the heated pipe is numerically analyzed on the basis of the developed version of the constitutive relations of MDM. The numerical results of the fatigue damage accumulation under thermal pulsation are in good agreement with experimental data. It is shown that the proposed MDM model qualitatively, with the accuracy required for practical calculations, describes the experimental results and can effectively evaluate thermocyclic fatigue damage accumulation in structural alloys under combined multiaxial nonproportional thermal–mechanical loading.

Keywords Modeling · Stress–strain state · Plasticity · Thermal loading · Mechanics of damaged medium · Damage accumulation · Fatigue life · Failure · Service life

17.1 Introduction

The general trend in the development of structures and machines of modern mechanical engineering is characterized by an increase in their performance parameters, by a decrease in metal consumption due to design optimization and application of novel high-strength materials as well as by increase in relative share of non-stationary loading regimes. At the same time, the requirements to reliability and accident-free life of both entire structures and their separate elements are becoming tougher. As a result, the reliable evaluation of service life of newly designed objects and prediction of the individual residual life of existing structures have become crucial tasks in designing new structures and machines.

Sudden strength malfunctions of structural elements result from uncontrolled degradation of initial strength properties of structural materials due to damage accumulation under the effect of various physical fields. Due to locality of degradation processes, service life of structural elements is determined by the service life of their most hazardous zones with the highest rates of degradation processes.

A solution to this problem is possible only if the methodology for monitoring the service life is implemented [1, 2]. This methodology aims at controlling the individual worked out service life of each hazardous zone of the most loaded structural elements, as well as individual material damage of these zones according to the known actual strain history based on mathematical modeling of degradation processes using modern models and methods of mechanics of damaged medium and fracture mechanics (FM) [3–8, and references therein].

The calculated values of material damage degree in accessible zones, estimated by modeling, must be updated using periodic nondestructive testing of the material behavior in accessible zones by modern physical methods during facility shutdowns or maintenance [1, 7]. None of the attempts to quantitatively relate the damage to a change of a measurable physical parameter (magnetic permeability, electrical resistance, hardness, elasticity moduli, etc.) made it possible to obtain the results which could be used in calculations.

Therefore, another approach has been developed on the basis of a macroscopic parameter characterizing the material damage degree at the macro-level. In general, it should be a second or higher rank tensor Ω_{ijkl} as a function of stress–strain state (SSS) history. However, due to the lack of necessary experimental data, a scalar parameter ω is selected, in most cases, as a measure of damage degree. This parameter varies from its initial state ω_0 that corresponds to undamaged material to its limit value ω_f, at which macrocrack nucleation of a certain size in a given volume of the material occurs [7].

By now, a large number of constitutive relations describing the processes of damage in a material have been developed. However, most of them are focused only on certain loading regimes and are not related to specific equations of deformation processes. Therefore, these equations cannot reflect damage accumulation as a function of SSS history, temperature, and strain rate. In fact, the history of viscoplastic deformation (the form of the deformation trajectory, the nature of the temperature change, the type of stress state, the history of its change, etc.) significantly affects the rate of damage accumulation.

The deformation due to thermal–mechanical loading is determined by the physico-mechanical and strength characteristics of structural materials as a function of temperature, the presence of temperature deformation limitations, and the ratio of mechanical and temperature strain rates. As frequencies and cycle phases of temperature variations and mechanical strain tensor components as a rule do not coincide, non-isothermal cyclic loading is generally irregular, nonproportional, and multiaxial and is accompanied by rotation of the main areas of the stress–strain tensors [9, 10].

Fatigue life due to thermal–mechanical loading decreases because of additional degradation mechanisms that are not observed under isothermal loading (a change in the structure of the material, additional deformation aging, and increased environmental influence). Under thermocyclic loading conditions, material durability is significantly affected by the variation rate of load and temperature and by total duration of the loading cycle. Thus, under thermocyclic fatigue loading conditions, the number of cycles to failure is no longer sufficient characteristic of the fatigue life and time to failure is to be accounted for.

In the present work, in the framework of MDM, a mathematical model of MDM is developed that describes fatigue damage accumulation in structural materials (metals and their alloys) under multiaxial stressed states and arbitrarily complex regimes of combined thermal–mechanical loading. This version of constitutive relations describes fatigue life of polycrystalline structural alloys due to material degradation by the mechanism of low- and high-cycle fatigue (taking into account their cross-effect). A developed version of the constitutive relations of MDM is used to assess the fatigue life of structures under thermocyclic loading [11].

17.2 The Model of Damaged Media for Evaluating Fatigue Life of Structural Alloys Under Low- and High-Cycle Fatigue

The model of damaged medium consists of three interrelated parts:

- the relations of thermoplastic material behavior with account for the failure process;
- the evolutionary equations describing the kinetics of fatigue damage accumulation;
- the strength criterion of the damaged material.

(a) The constitutive relations of plasticity

The constitutive relations of plasticity are based on the following basic assumptions [9, 10]:

- The material of the medium is initially isotropic and free of damage (anisotropy due to strain processes is only accounted for, while anisotropy of elastic properties due to damage processes is not accounted for).
- The components of strain tensor e_{ij} and strain rates $\dot{e}_{ij}$ include elastic e^e_{ij}, $\dot{e}^e_{ij}$ and plastic parts e^p_{ij}, $\dot{e}^p_{ij}$, i.e., reversible and irreversible components.
- The initial yield surface for various temperatures is described by a von Mises-type surface. The evolution of the yield surface is described by the variation of its radius C_p and the displacement of its center ρ^p_{ij}.
- The surface of "microplastic" load having at the initial instant of time a common center with the von Mises surface and a constant radius C_m is introduced in the stress space.
- The body changes its volume elastically.
- Scalar parameter ω—damage degree $\left(\omega_o \le \omega \le \omega_f\right)$ is the only structural parameter characterizing material damage degree at the macro-level.
- The effect of the accumulated damage degree on the deformation process of the material is considered by introducing effective stresses.
- The processes characterized by small deformations are considered.

In the elastic region, the relation between spherical and deviatoric components of stress and strain tensors and their rates is established by Hooke's law:

$$\sigma = 3\kappa[e - \alpha(T - T_0)], \sigma_{ij} = 2Ge'^e_{ij},$$
$$\dot{\sigma} = 3\kappa\left(\dot{e} - \dot{\alpha}T - \alpha\dot{T}\right) + \frac{\dot{\kappa}}{\kappa}\sigma, \dot{\sigma}'^e_{ij} = 2G\dot{e}'^e_{ij} + \frac{\dot{G}}{G}\sigma'_{ij}, \quad (17.1)$$

where T is temperature, T_0 is initial temperature, $\kappa(T)$ is the bulk modulus, $G(T)$ is the shear modulus, and $\alpha(T)$ is the coefficient of linear thermal expansion of the material.

The von Mises yield surface is introduced to describe the effects of monotonic and cyclic deformation [9, 10]:

$$F_p = S_{ij}S_{ij} - C_p^2 = 0,\ S_{ij} = \sigma'_{ij} - \rho^p_{ij}, \tag{17.2}$$

To describe the material degradation under high-cycle fatigue, a surface of "microplastic" loading is introduced in the stress space:

$$F_m = S_{ij}S_{ij} - C_m^2 = 0,\ S_{ij} = \sigma'_{ij} - \rho_{ij},\ C_m = \sqrt{2/3\sigma_f} = \text{const}, \tag{17.3}$$

where σ_f is fatigue limit of the material.

To describe complex cyclic deformation regimes in the stress space, a cyclic "memory" surface is introduced. The equation of the "memory" surface is:

$$F_\rho = \rho^p_{ij}\rho^p_{ij} - \rho^2_{\max} = 0, \tag{17.4}$$

where $\rho_{\max}$ is maximal for the loading history value of modulus ρ^p_{ij}.

For the temperature range T where annealing effects are insignificant, isotropic hardening (evolution C_p) is assumed to consist of three types: monotonic, cyclic, and associated with a change in temperature T. For the yield surface radius, the evolutionary equation has the form [9, 10]:

$$\dot{C}_p = \left[q_x H(F_\rho) + a(Q_s - C_p)\Gamma(F_\rho)\right]\dot{\chi}_p + q_3\dot{T}, \tag{17.5}$$

$$C_p = C_p^0 + \int_0^t \dot{C}_p \mathrm{d}t,\ \dot{\chi} = \left(\frac{2}{3}\dot{e}^p_{ij}\dot{e}^p_{ij}\right)^{1/2},\ \chi_p = \int_0^t \dot{\chi}_p \mathrm{d}t,\ \chi_{pm} = \int_0^t \dot{\chi}_p H(F_\rho)\mathrm{d}t, \tag{17.6}$$

$$q_x = \frac{q_2 A\Psi_1 + (1-A)q_1}{A\Psi_1 + (1-A)},\ Q_s = \frac{Q_2 A\Psi_2 + (1-A)Q_1}{A\Psi_2 + (1-A)},\quad 0 \le \Psi \le 1,\ i = 1, \tag{17.7}$$

$$A = 1 - \cos^2\theta,\ \cos\theta = n^e_{ij}n^s_{ij},\ n^e_{ij} = \frac{\dot{e}'_{ij}}{\left(\dot{e}'_{ij}\dot{e}'_{ij}\right)^{1/2}},\ n^s_{ij} = \frac{S_{ij}}{\left(S_{ij}S_{ij}\right)^{1/2}}, \tag{17.8}$$

$$H(F_\rho) = \begin{Bmatrix} 1,\ F_\rho = 0 \wedge \rho_{ij}\dot{\rho}_{ij} > 0 \\ 0,\ F_\rho < 0 \vee \rho_{ij}\dot{\rho}_{ij} \le 0 \end{Bmatrix},\ \Gamma(F_\rho) = 1 - F(F_\rho), \tag{17.9}$$

where q_1, q_2, q_3 are modules of isotropic hardening corresponding to the monotonous radial loading paths (q_1), to a 90° kink of the deformation trajectory (q_2), and to temperature variation of the yield surface radius (q_3); a is a constant defining stabilization process rate of the hysteresis loop shape of cyclic deformation; Q_s is stationary value of the yield surface radius for the given $\rho_{\max}$ and T; χ are lengths of plastic strain trajectories of the material under cyclic and monotonic loading; C_p^0 initial value of the yield surface radius [9, 10].

The first member of Eq. (17.6) describes isotropic hardening as a result of monotonic plastic deformation ($H(F_\rho) = 1$ and $\Gamma(F_\rho) = 0$), the second member is responsible for cyclic hardening of the material ($H(F_\rho) = 0$ and $\Gamma(F_\rho) = 1$), and the third one is responsible for the change of the yield surface radius with temperature. In general, Eq. (17.6) describes the local anisotropy of plastic hardening as a function of the parameter A, characterizing the deviation of the loading vector from the normal to the yield surface at the loading point. Operators $H(F_\rho)$ and $\Gamma(F_\rho)$ make it possible to automatically separate the processes of monotonous and cyclic deformation.

Hereinafter, for any value B, enclosed in angle brackets $\langle\rangle$, the following conditions are held

$$\langle B \rangle = \begin{cases} B \text{ at } B > 0 \\ 0 \text{ at } B \le 0 \end{cases}$$

We postulate that the evolution of an internal variable ρ_{ij} has the form:

$$\dot{\rho}_{ij} = \rho_{ij}^p + \rho_{ij}^m,\ \rho_{ij} = \int_0^t \dot{\rho}_{ij} \mathrm{d}t,, \tag{17.10}$$

$$\rho_{ij}^p = f(\chi_{pm})\left[g_1 \dot{e}_{ij}^p - g_2 \rho_{ij}^p \dot{\chi}_p\right] + g_3 \rho_{ij}^p \langle \dot{T} \rangle + \dot{\rho}_{ij}^*, , \tag{17.11}$$

$$f(\chi_{pm}) = 1 + k_1\left(1 - e^{-k_2 \chi_{pm}}\right),$$
$$\dot{\rho}_{ij}^* = g_4 \dot{e}_{ij}^p H(F_\rho) - g_5 \rho_{ij}^* \dot{\chi}_p \Gamma(F_\rho) \langle \cos\beta \rangle, \tag{17.12}$$

$$\langle \cos\beta \rangle = \frac{\dot{\rho}_{ij}^p \rho_{ij}^*}{\left(\dot{\rho}_{ij}^p \dot{\rho}_{ij}^p\right)^{1/2} \left(\rho_{ij}^* \rho_{ij}^*\right)1/2},\ g_3 = \frac{1}{g_1}\left(\frac{\partial g_1}{\partial T} - \frac{g_1}{g_2}\frac{\partial g_2}{\partial T}\right), \tag{17.13}$$

$$\dot{\rho}_{ij}^m = g_6 \dot{e}_{ij}^m - g_7 \rho_{ij}^m \dot{\chi}_m,\ \dot{\chi}_m = \left(\frac{2}{3} \dot{e}_{ij}^m \dot{e}_{ij}^m\right)^{1/2},\ \chi_m = \int_0^t \dot{\chi}_m \mathrm{d}t, \tag{17.14}$$

where $g_1, g_2, g_3, g_4, g_5, g_6, g_7, k_1$ and k_2 are experimentally determined material parameters.

In Eq. (17.10), the first term describes evolution of ρ_{ij}, associated with the initiation of macroscopic plastic deformations e_{ij}^p, and the second one describes microplastic deformations e_{ij}^m.

In Eq. (17.11), the first and second terms of the equation describe the anisotropic part of strain hardening; the third term is responsible for the variation of ρ_{ij}^p due to change in temperature T, and the fourth one for anisotropic hardening associated with unilateral accumulation of plastic deformation. Weight factor $f(\chi_{pm})$ allows describing evolution of ρ^p_{ij} under block cyclic deformation.

For both non-symmetric hard and soft cyclic loadings, due to terms $\dot{\rho}^*_{ij}$, Eq. (17.11) describes setting and ratcheting of the cyclic plastic hysteresis loop. At $g_3 = g_4 = g_5 = k_1 = 0$ using (17.11), we obtain a special case of equation—the Armstrong–Fredericks–Kadashevich equation.

$$\dot{\rho}^p_{ij} = g_1 \dot{e}^p_{ij} - g_2 \rho^p_{ij} \dot{\chi}_p, , \tag{17.15}$$

To describe the behavior of the "memory" surface, it is necessary to formulate an evolution equation for $\rho_{\max}$:

$$\dot{\rho}^p_{\max} = \frac{\left(\rho^p_{ij}\dot{\rho}^p_{ij}\right)H\left(F_\rho\right)}{\left(\rho^p_{mn}\rho^p_{mn}\right)^{1/2}} - g_2 \rho^p_{\max}\dot{\chi}_p - g_3 \rho^p_{\max}\langle \dot{T}\rangle \tag{17.16}$$

Gradient laws govern the tensor components of macroplastic e^p_{ij} and microplastic e^m_{ij} strain rates:

$$\dot{e}^p_{ij} = \lambda_p S_{ij}, \tag{17.17}$$

$$\dot{e}^m_{ij} = \lambda_m S_{ij}, \tag{17.18}$$

At the stage of the development of defects scattered over the volume, the effect of damage degree on the material physico-mechanical properties is observed. This effect can be taken into account by introducing effective stresses [7]:

$$\tilde{\sigma}'_{ij} = F_1(\omega)\sigma'_{ij} = \frac{G}{\tilde{G}}\sigma'_{ij} = \frac{\sigma'_{ij}}{(1-\omega)\left[1 - \frac{(6K+12G)}{(9K+8G)}\omega\right]}, \tag{17.19}$$

$$\tilde{\sigma} = F_2(\omega)\sigma = \frac{\kappa}{\tilde{\kappa}}\sigma = \frac{\sigma}{4G(1-\omega)/(4G+3\kappa\omega)}, \tag{17.20}$$

where $\tilde{G}$, $\tilde{\kappa}$ are effective moduli of elasticity defined by McKenzie formulas [7]. Effective variable $\tilde{\rho}_{ij}$ is determined in a similar way:

$$\tilde{\rho}_{ij} = F_1(\omega)\rho_{ij}, \tag{17.21}$$

(b) Evolutionary equation of fatigue damage accumulation

In order to formulate damage evolution equations, we use an approach based on relationship between $\omega^{\cdot}$ velocities and some strain-dependent mechanical parameters, the critical value of which defines complete failure of the elementary bulk of a material. An energy approach is considered to be the most effective for solving this problem [1, 7, 8, 12–16].

In [17], V. V. Novozhilov showed that at low-cycle fatigue, the energy, spent on the defect formation under unsteady elastoplastic deformation, correlates fairly well with the work of microstress tensor ρ_{ij}^{p} on plastic strains e_{ij}^{p}. In [16], V. T. Troshchenko suggested a failure criterion for multi-cycle fatigue processes, which establishes a relationship between the total scattered elastic strain energy and the number of cycles to failure and proves itself to be good for a wide range of structural steels.

In the present work, to describe the process of fatigue life of structural materials (metals and their alloys) due to degradation under combined low- and high-cycle fatigue loading, we therefore postulate the following energy relation

$$\dot{W} = \rho_{ij}^{p}\dot{e}_{ij}^{p} + \rho_{ij}^{m}\dot{e}_{ij}^{m},\ W = \int \left(\rho_{ij}^{p}\dot{e}_{ij}^{p} + \rho_{ij}^{m}\dot{e}_{ij}^{m}\right). \tag{17.22}$$

Using (17.22) as the damage energy relation enables us to obtain the generalized equation for fatigue damage accumulation due to material degradation by the mechanisms of low- and high-cycle fatigue (with account for their interaction). In addition, there is no need in simulating the process of damage summation by the mechanisms of low- and high-cycle fatigue. Thus, the construction of a generalized model of fatigue damage accumulation is becoming feasible.

Multiaxial stress state significantly affects the fatigue life of a material, and this effect is manifested in two ways: the effect of the multiaxiality itself under proportional loading (when all the strain tensor components change proportionally with respect to one parameter) and the effect of rotation of the main stress tensor areas (when the stress components change out of phase). Experimental and theoretical data on the effect of loading multiaxiality show a significant effect of "bulk" stress state $\beta = \sigma / \sigma_u$ on fatigue life of the material, characterized by the intensity of the stress tensor σ_u and by its spherical (hydrostatic) component σ [4]. Summarizing the data available in the literature, it can be argued that the rate of damage accumulation depends on the "bulk" stress state, characterized by some function $f_1(\beta)$: $\dot{\omega}$ increases at $\beta > 0$ and decreases at $\beta < 0$. At $\beta = 0$ (pure shear), the normalized function $f_1(\beta)$ should be equal to 1.

The effect of nonproportional loading, under which the components of the stress–strain tensors change out of phase (the main tensor areas rotate), and the stress and plastic strain tensors are not coaxial, is as follows [18–22]:

- The shape of the strain path is a parameter that significantly affects the fatigue life.
- Structural materials exhibit complex cyclic behavior under multiaxial nonproportional loading conditions—additional cyclic hardening or softening.
- Under nonproportional cyclic loading, the criterion of equivalent strains or strain intensity is other than equivalence criterion and may lead to nonconservative evaluations.

Summarizing the above results, the general structure of an evolutionary equation of the damage accumulation in elementary volume of the material with account for combined low- and high-cycle fatigue can have the following form:

$$\dot{\omega} = f_1(\beta) f_2(\omega) f_3(W) f_4\left(\dot{W}\right), \tag{17.23}$$

where functions f_j, $j = 1, \ldots, 4$ account for: bulk stress state ($f_1(\beta)$), level of accumulated damage ($f_2(\omega)$), accumulated relative damage energy, spent on defect nucleation (($f_3(W)$)), and the change rate of damage energy $\left(f_4\left(\dot{W}\right)\right)$.

In (17.23):

$$f_1(\beta) = \exp(\beta), \quad f_2(\omega) = \begin{cases} 0, & W \le W_a \\ \omega^{1/3}(1-\omega)^{2/3}, & W > W_a \wedge \omega \le 1/3 \\ \frac{\sqrt[3]{16}}{9}\omega^{-1/3}(1-\omega)^{-2/3}, & W > W_a \wedge \omega > 1/3 \end{cases}, \tag{17.24}$$

$$f_3(W) = \frac{W - W_a}{W_f - W_a}, \; f_4\left(\dot{W}\right) = \frac{\dot{W}}{W_f - W_a}, \tag{17.25}$$

where W_a is the damage energy at the end of the nucleation stage of scattered defects and W_f is the energy corresponding to a macroscopic crack nucleation. The duration of the microdefect nucleation phase will be related to the value of parameter W_a.

When the size of microdefects becomes comparable with the average distance between them, the process of merging begins (breakage of the remaining continuous spaces between the defects). In the present paper, a detailed model of merging of cavities was not constructed, but to account for this process, a kinetic equation (due to $f_2(\omega)$) was formulated so that, when the damage degree reaches the value of $\omega = 1/3$, the dependence $\dot{\omega} = f_2(\omega)$ accounts for the avalanche-like increase of the damage degree.

(c) Strength criterion of the damaged material

By integrating evolutionary equation of damage accumulation (17.23) along with constitutive relations of thermal viscoplasticity (17.1)–(17.21) for the known loading history, one can obtain conditions for macrocrack nucleation (failure criterion).

For the criterion of the end of the growth phase of scattered microdefects, we assume the condition when the damage degree ω reaches its critical value:

$$\omega = \omega_f \le 1, \tag{17.26}$$

17.3 Investigation Results

Results presented in [21] for modeling thermal fatigue on 12X18H10T steel pipe surface with a temperature of $T = 310$ °C and falling distillate droplets of $T = 20$ °C with a frequency of 1, 2, and 3 Hz are compared with the experimental results obtained in OJSC "NPO CKTI" [11]. In the experiment, the pipe diameter in the zone of droplet impingement was 63 mm and wall thickness of 3 mm (Fig. 17.1). The parameters of fatigue crack nucleation and growth were controlled using ultrasonic and eddy current methods.

After $N = 9.3 \times 10^6$ thermal cycles, thin sections were fabricated and metallographic studies were carried out, which revealed 85–1873 μm cracks coming from the surface of the tubular specimen. According to the ultrasonic signals, at a frequency of 1 Hz, cracks get fixed after $N \approx 1 \times 10^6$ cycles and in accordance with the eddy current method, after $N \approx 2 \times 10^6$ cycles.

The problem of evaluating the fatigue life of a cylindrical shell under thermocyclic loading was numerically solved in two stages:

- solution of non-stationary problem based on the heat equation [23];
- evaluation of the fatigue life of a pipe subjected to block thermocyclic loading under inhomogeneous stress conditions.

At the first stage, the pipe was heated up to $T = 310$ °C with internal pressure raised up to ~10 MPa. Thereafter, the droplet impingement process at a temperature of $T = 20$ °C and frequency of 1, 2, and 3 Hz was simulated on a pipe surface, heated up to $T = 310$ °C. Using the ANSYS software, unsteady thermal state of the cooling

Fig. 17.1 Finite element model of fragment of the pipe around the zone of droplet impingement

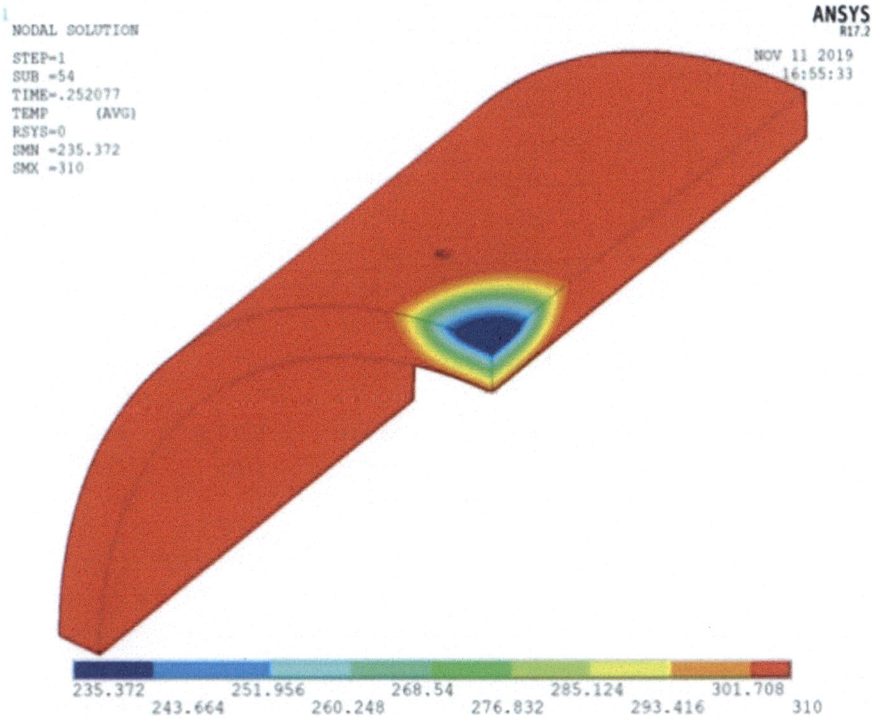

Fig. 17.2 Results of numerical analysis of change in the temperature field at time $t = 0.25$

product was investigated with account for the thermocycle parameters. A numerical analysis of the temperature field of an object on a perforated surface showed that the temperature field is not uniform over the pipe thickness (Fig. 17.2 shows the temperature field at time $t = 0.25$ s).

At the second stage, the problem of evaluating the thermocyclic fatigue life of the pipe was numerically solved using the calculated temperature fields of the object and constitutive relations of MDM. The physico-mechanical characteristics and parameters of MDM model for 12X18H10T steel at temperatures of 20, 150, and 350 °C are shown in Tables 17.1, 17.2, and 17.3.

Figures 17.3 and 17.4 show the intensity distribution of plastic strains, Figs. 17.5 and 17.6 the distribution of stress intensity, and Figs. 17.7 and 17.8 the damage degree of the perforated fragment of the pipe surface for the droplet impingement frequency of 1 and 3 Hz, respectively.

The analysis of the calculation of SSS kinetics and damage degree showed that the outer surface of the pipe central part is the most stressed area. The temperature T and components of the total strain tensor e_{ij} were determined for the central part of the droplet impingement based on the results of three-dimensional calculations for several initial cycles. The history of temperature T and components of total strain tensor e_{ij} for the droplet impingement frequency of 1 and 3 Hz is shown in Figs. 17.9

Table 17.1 Physico-mechanical characteristics and parameters of MDM model

Characteristics	Temperature T		
	20 °C	150 °C	350°C
К, MPa	162,500	154,500	142,500
G, MPa	75,000	71,300	65,800
α, 1/degree	0.0000166	0.0000171	0.0000175
C_p^o, MPa	191	168	152
C_m, MPa	50	40	36,5
g_1, MPa	20,850	15,670	13,900
g_2	296	278	270
g_4, MPa	0	0	0
g_5	0	0	0
g_6, MPa	400,000	300,750	266,150
g_7	8000	7511	7292
k_1	0	0	0
a	5	5	5
W_a, mj/ m^3	0	0	0
W_f, mj/ m^3	6650	6110	5786

Table 17.2 Monotonous isotropic hardening modulus q_1, MPa ($q_2 = 0$)

$T = 20$ °C									
χ_p	0	0.003	0.006	0.009	0.012	0.015	0.018	0.021	0.4
q_1	−4219	−952	1495	1369	1454	1272	1235	824	0
$T = 150$ °C									
χ_p	0	0.003	0.006	0.009	0.012	0.015	0.023	0.3	0.4
q_1	2500	2150	1450	1150	1050	950	300	0	0
$T = 350$ °C									
χ_p	0	0.003	0.004	0.006	0.007	0.008	0.010	0.3	0.4
q_1	4400	4000	3200	2500	2100	1500	1000	0	0

Table 17.3 Cyclic isotropic hardening modulus $Q_1(\rho_{\max})$, MPa ($Q_2 = 0$)

$T = 20$ °C					
$\rho_{\max}$	0	74	105	113	147
Q_S	195	217	230	273	300
$T = 150$ °C					
$\rho_{\max}$	0	50	81	96	150
Q_S	168	195	204	227	230
$T = 350$ °C					
$\rho_{\max}$	0	32	55	66	100
Q_S	152	210	221	248	250

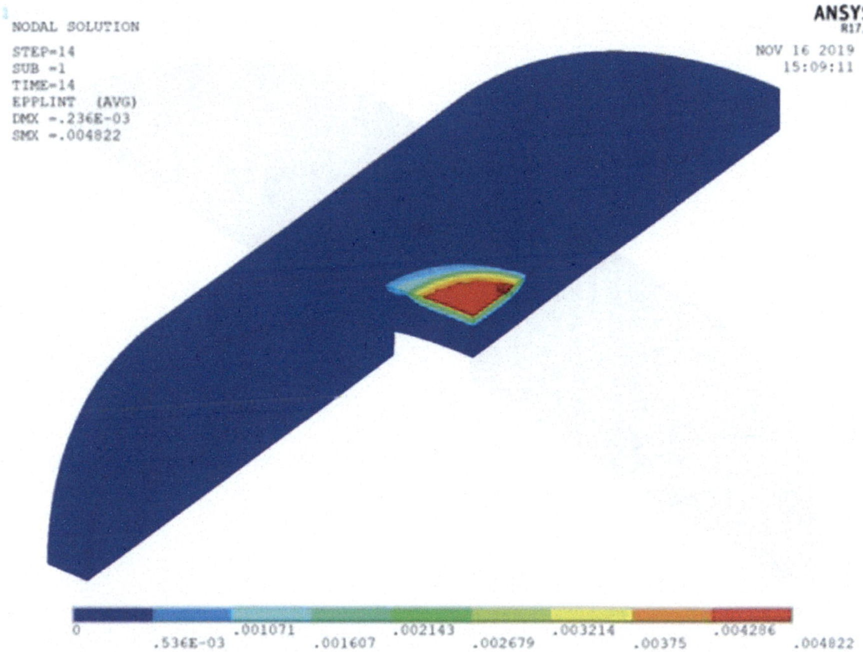

Fig. 17.3 Intensity distribution of plastic strains at the moment of stabilization of the plastic hysteresis loop for the droplet impingement frequency of 1 Hz

and 17.10, respectively.

The obtained histories of temperature T and the components of the total strain tensor e_{ij} are used to calculate the process of thermoplastic deformation and fatigue damage accumulation prior to fatigue macrocrack nucleation in the spot zone of the material (Fig. 17.1).

Figures 17.11, 17.12, 17.13, and 17.14 depict the cyclic hysteresis loops $\sigma_{11} \sim e_{11}$ (Figs. 17.11 and 17.12) and $\sigma_{22} \sim e_{22}$ (Figs. 17.13 and 17.14) for the most loaded zones at the droplet impingement frequency of 1 and 3 Hz, respectively.

Figures 17.15 and 17.16 depict loading trajectories at the droplet impingement frequency of 1 and 3 Hz for the central point, respectively. The type of graphs shows the presence of rotation of the main areas of stress–strain tensors (changes in the components of the stress–strain tensors are disproportional) and misalignment of stress tensors, microstress tensors, and total and plastic strain tensors.

The numerical calculations showed a sharp rise in the stress tensor components σ_{11}, σ_{22}, σ_{33} at a cold droplet of $T = 20$ °C falling on a heated pipe surface ($T = 310$ °C), and the stress state is being close to triaxial tension. Then, stresses σ_{11}, σ_{22}, σ_{33} decrease to negative values as a result of heating of droplet impingement zone due to the heated pipe material around the zone. This cyclic process occupies approximately a hundredth of the total duration of the droplet impingement cycle. The stress–strain state remains practically unchanged until the next droplet falls.

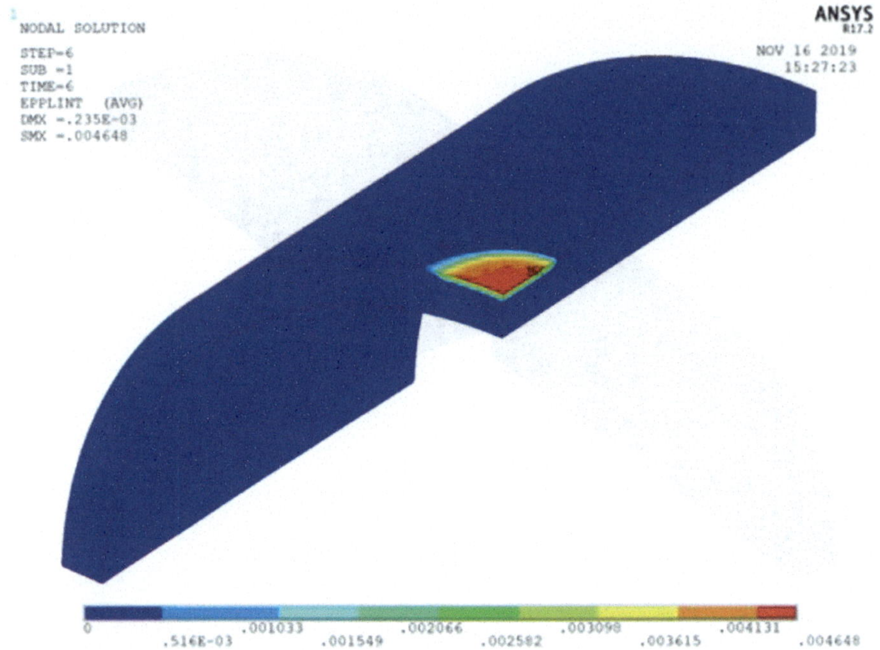

Fig. 17.4 Intensity distribution of plastic strains at the moment of stabilization of the plastic hysteresis loop for the droplet impingement frequency of 3 Hz

The analysis of the deformation process shows its unsteady thermoplastic character at a frequency of 1, 2, and 3 Hz. A sharp "surge" (mainly hydrostatic tension) of the normal components of the stress tensor is observed when a cold droplet falls on a hot surface. With heating, the stress spots tend to zero and thereafter take negative values due to the initial plastic tension of the spot material at droplet impingement.

Integration of evolutionary equations of fatigue damage accumulation (17.23)–(17.25) at the most loaded point allows us to study the effect of the droplet impingement frequency on the thermocyclic fatigue life of the material of the heated pipe. Numerical results were compared with experimental data [11]. The calculation results and their comparison with the available experimental data are shown in Fig. 17.17 and Table 17.4. Figure 17.17 shows the value of damage degree ω as a function of the number of loading cycles for the droplet impingement frequency of 1, 2, and 3 Hz. The calculated number of cycles to crack nucleation in the droplet impingement zone (see Table 17.4) correlates qualitatively and quantitatively with the experimental results.

Ultrasonic and eddy current methods for nondestructive detection of growing cracks in the droplet impingement zone in terms of their reflectivity area record the onset of cracking and their confident fixation.

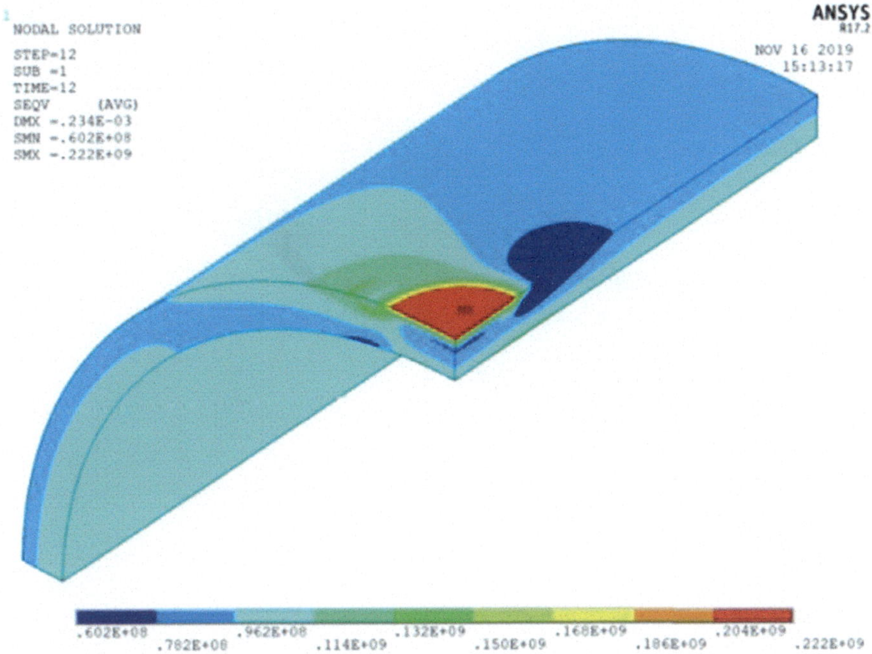

Fig. 17.5 Distribution of stress intensity at the moment of stabilization of the plastic hysteresis loop for the droplet impingement frequency of 1 Hz

17.4 Conclusion

The reliability of the damaged medium model [7, 8] has been assessed by comparing the computational results with the experimental data available in the literature on thermal fatigue life of heated pipe surface in the zones of distillate droplet impingement.

The reliability of the developed constitutive relations of MDM model for thermal fatigue and the possibility of their effective use for assessing the thermocyclic fatigue of materials and structures are proven by using the method of numerical modeling and comparing the obtained results with experimental data.

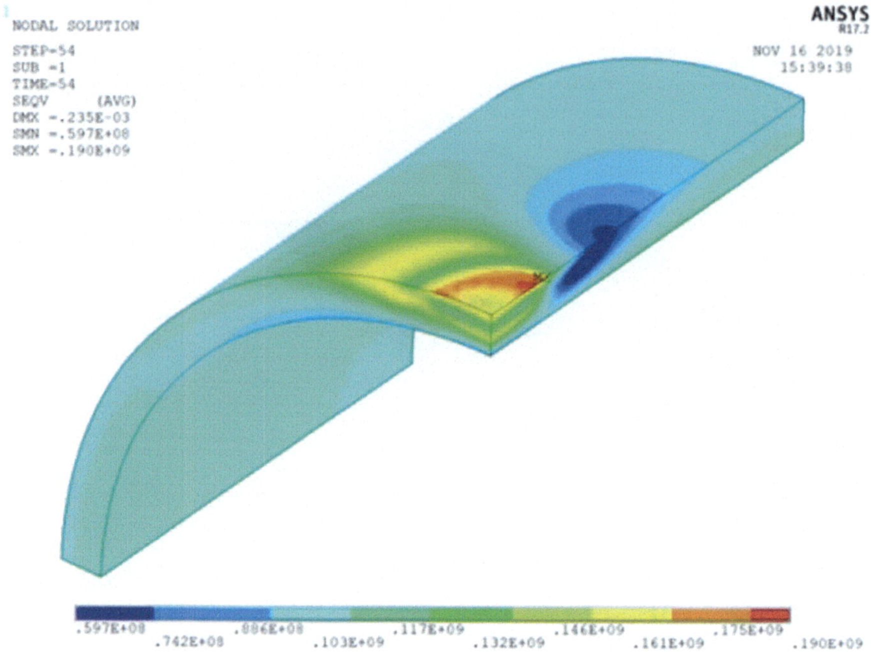

Fig. 17.6 Distribution of stress intensity at the moment of stabilization of the plastic hysteresis loop for the droplet impingement frequency of 3 Hz

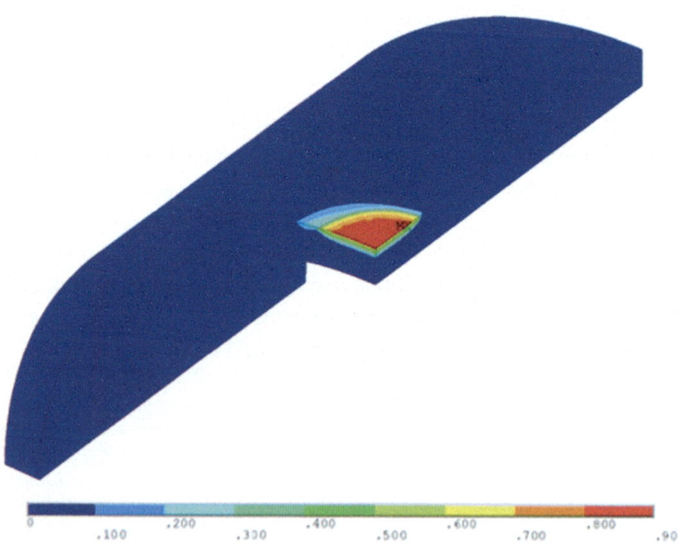

Fig. 17.7 Distribution of the damage degree at the moment of macroscopic crack nucleation for the droplet impingement frequency of 1 Hz

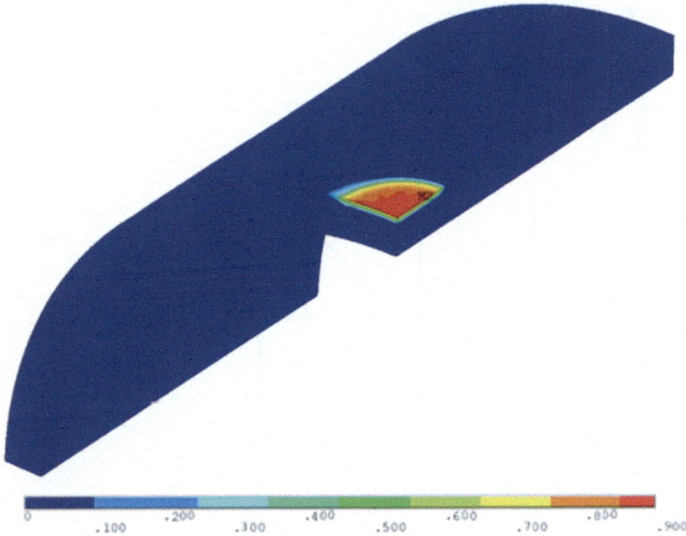

Fig. 17.8 Distribution of the damage degree at the moment of macroscopic crack nucleation for the droplet impingement frequency of 3 Hz

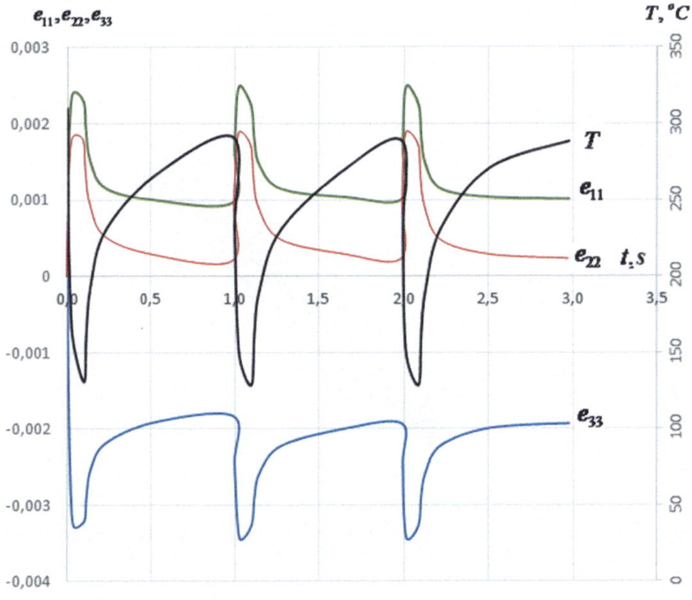

Fig. 17.9 History of temperature and components of the strain tensor at the spot zone of the pipe central part for the droplet impingement frequency of 1 Hz

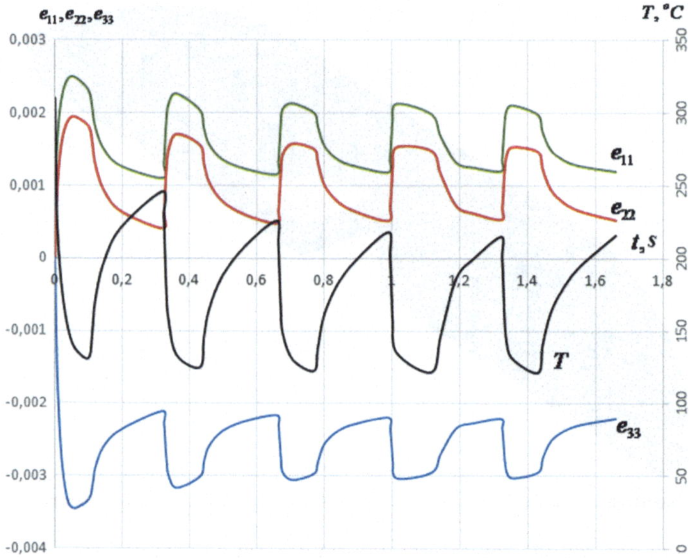

Fig. 17.10 History of temperature and components of the strain tensor at the spot zone of the pipe central part for the droplet impingement frequency of 3 Hz

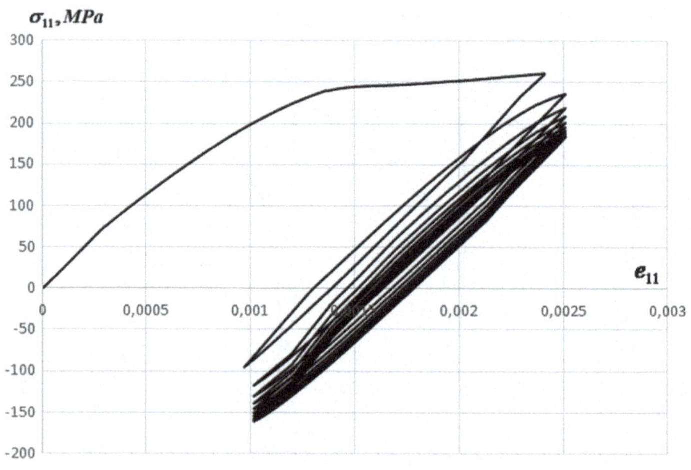

Fig. 17.11 Cyclic hysteresis loops $\sigma_{11} \sim e_{11}$ at the spot zone of the pipe central part for the droplet impingement frequency of 1 Hz

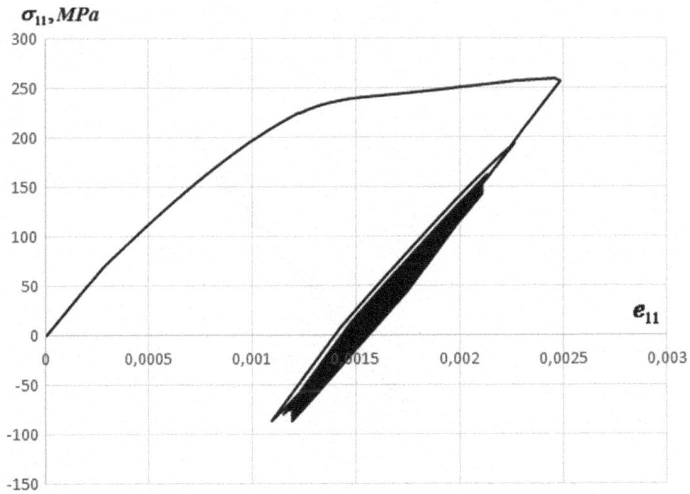

Fig. 17.12 Cyclic hysteresis loops $\sigma_{11} \sim e_{11}$ at the spot zone of the pipe central part for the droplet impingement frequency of 3 Hz

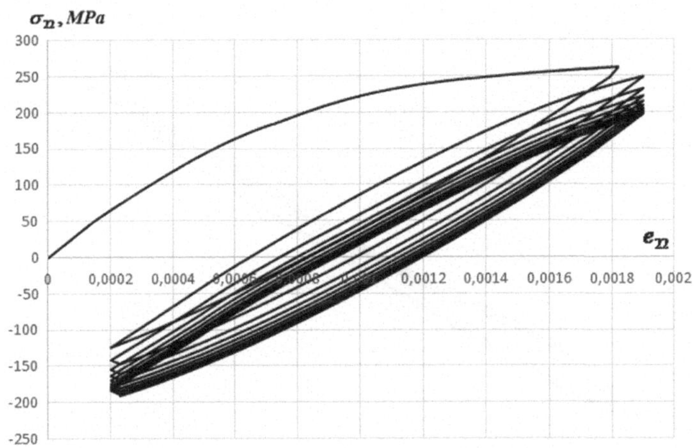

Fig. 17.13 Cyclic hysteresis loops $\sigma_{22} \sim e_{22}$ at the spot zone of the pipe central part for the droplet impingement frequency of 1 Hz

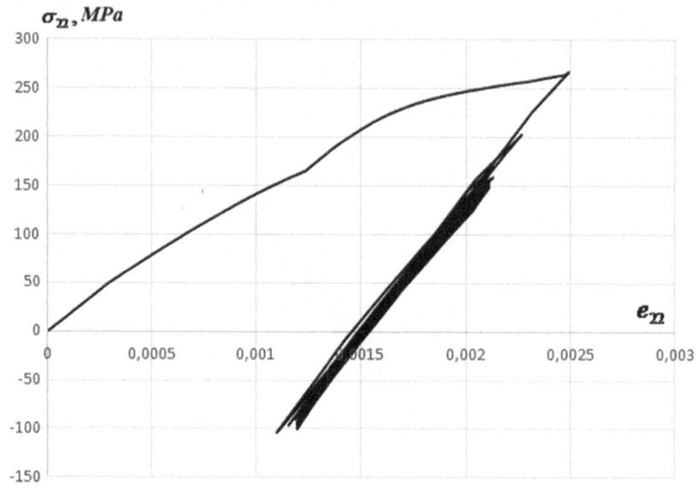

Fig. 17.14 Cyclic hysteresis loops $\sigma_{22} \sim e_{22}$ at the spot zone of the pipe central part for the droplet impingement frequency of 3 Hz

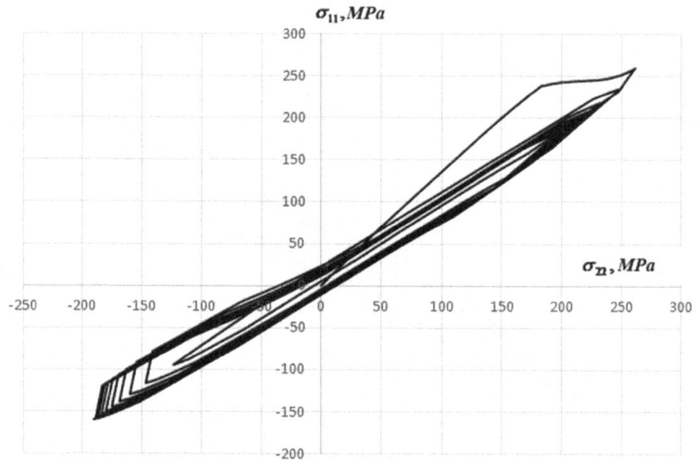

Fig. 17.15 Loading trajectory at the spot zone of the pipe central part for the droplet impingement frequency of 1 Hz

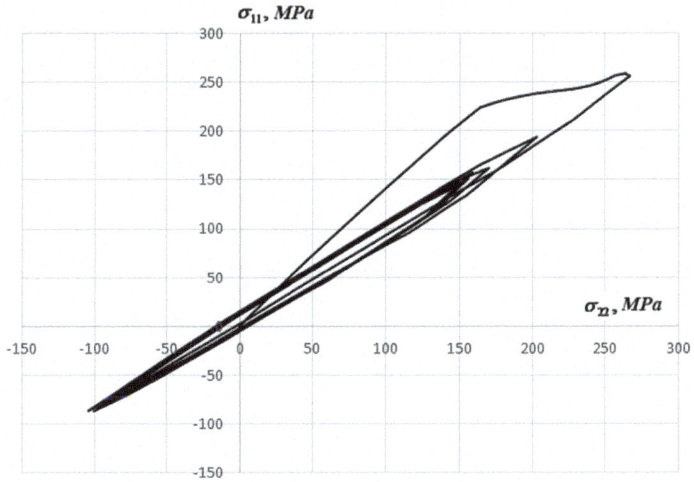

Fig. 17.16 Loading trajectory at the spot zone of the pipe central part for the droplet impingement frequency of 3 Hz

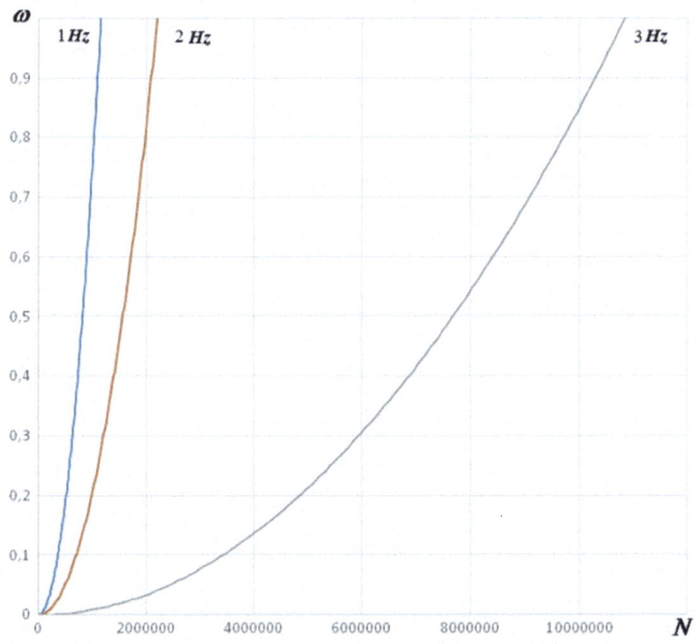

Fig. 17.17 Value of damage degree as a function of the number of loading cycles for the droplet impingement frequency of 1, 2, and 3 Hz

Table 17.4 Number of cycles to macroscopic crack nucleation

Distillate drop frequency, Hz	Number of cycles to macrocrack nucleation	
	Numerical calculations	Experiment
1	1,152,330	1,179,700
2	2,207,560	2,359,400
3	10,848,633	10,267,900

Acknowledgements The work was carried out with the financial support of the Ministry of Education (task 0729-2020-0054).

References

1. Mitenkov, F.M., Kaydalov, V.F., Korotkikh, Y.G.: Metody obosnovaniya resursa yadernyh energeticheskih ustanovok. Mashinostroenie (2007) (in Russian)
2. Mitenkov, F.M., Korotkih, YuG: K Voprosu o sozdanii ekspluatacionnogo monitoringa resursa oborudovaniya i sistem yadernyh energeticheskih ustanovok. Probl. Mashinostroeniya I Nadezhnosti Mashin **4**, 105–116 (2003). (in Russian)
3. Chaboche, J.L.: Continuous damage mechanics a tool to describe phenomena before crack initiation. Eng. Des. **64**, 233–247 (1981)
4. Lemaitre, J.: A continuous damage mechanics model for ductile fracture. ASME J. Eng. Mater. Technol. **107**(1), 83–89 (1985)
5. Murakami, S.: Notion of continuum damage mechanics and its application to anisotropic creep damage theory. ASME J. Eng. Mater. Technol. **105**(2), 99–105 (1983)
6. Perzyna, P.: Constitutive modeling of dissipative solids for postcritical behavior and fracture ASME. J. Eng. Mater. Technol. **106**(4), 410–419 (1984)
7. Volkov, I.A., Igumnov, L.A.: Vvedenie v kontinualnuyu mehaniku povrezhdennoy sredy. Fizmatlit (2017) (in Russian)
8. Volkov, I.A., Korotkikh, Y.G.: Uravneniya sostoyaniya vyazkouprugoplasticheskih sred s povrezhdeniyami. Fizmatlit (2008) (in Russian)
9. Mitenkov, F.M., Volkov, I.A., Igumnov, L.A., Korotkikh, Y.G., Panov, V.A.: Prikladnaya teoriya plastichnosti. Fizmatlit (2015) (in Russian)
10. Volkov, I.A., Igumnov, L.A., Korotkikh, Y.G.: Prikladnaya teoriya vyaskoplastichnosti. NNGU (2015) (in Russian)
11. Bescherov, D.E., Bol'shuhin, M.A., Kozin, A.V., Lebedev, V.V., Panov, V.A., Pahomov, V.A., Sveshnikov, D.N., SHishulin, D.N., Legchanov, M.A., Ryazapov, R.R.: Verifikaciya metodiki rascheta ustalostnoj prochnosti oborudovaniya reaktornyh ustanovok pri nestacionarnyh termociklicheskih nagruzkah. Energeticheskie sistemy i kompleksy, pp. 81–91 (2015) (in Russian)
12. Bodner, S.R., Lindholm, U.S.: An incremental criterion for time-dependent failure of materials. ASME J. Eng. Mater. Technol. **98**(2), 140–145 (1976)
13. Bol'shuhin, M.A., Lebedev, V.V., Kozin, A.V.: Modelirovanie processov nakopleniya povrezhdenij pri termopul'saciyah. Problemy prochnosti i plastichnosti: Mezhvuz. Sbornik NNGU **76**(2), 134–143 (2014) (in Russian)
14. Corum, J.M., Sartory, W.K.: Assessment of current high-temperature design methodology based on structural failure tests. ASME J. Press. Vessel. Technol. **109**(2), 160–168 (1987)
15. Romanov, A.N.: Razrushenie pri malociklovom nagruzhenii. Nauka (1988) (in Russian)
16. Troshchenko, V.T.: Rasseyannye ustalostnye povrezhdeniya metallov i splavov, Soobshch. 3. Deformacionnye energeticheskie kriterii. Problemy prochnosti **1**, 5–31 (2006)

17. Novozhilov, V.V.: O perspektivah fenomenologicheskogo podhoda k probleme razrusheniya. Mekhanika deformiruemyh tel i konstrukcij. M: Mashinostroenie, 349–353 (1975) (in Russian)
18. Lamba, H.S., Sidebottom, O.M.: Cyclic plasticity for nonproportional paths: Part 1—cyclic hardening, erasure of memory, and subsequent strain hardening experiments. ASME J. Eng. Mater. Technol. **100**(1), 96–103 (1978)
19. McDowell, D.L.: An experimental study of the structure of constitutive equations for nonproportional cyclic plasticity. ASME J. Eng. Mater. Technol. **107**(4), 307–315 (1985)
20. Ohashi, Y., Kawai, M., Kaito, T.: Inelastic behavior of type 316 stainless steel under multiaxial nonproportional cyclic stressings at elevated temperature. ASME J. Eng. Mater. Technol. **107**(2), 101–109 (1985)
21. Taleb, L., Cailletaud, G., Sa'i, K.: Experimental and numerical analysis about the cyclic behavior of the 304L and 316L stainless steels at 350 °C. Int. J. Plast. **61**, 32–48 (2014)
22. Tanaka, E., Murakami, S., Ooka, M.: Effects of strain path shapes on nonproportional cyclic plasticity. J. Mech. Phys. Solids **33**(6), 559–575 (1985)
23. Il'yushin, A.A.: Mekhanika sploshnoj sredy. MGU (1990) (in Russian)

Chapter 18
Large Deflections and Stability of Low-Angle Arches and Panels During Creep Flow

Rashit A. Kayumov and Farid R. Shakirzyanov

Abstract The problem of deformation and buckling of a low arch or a panel is considered taking into account creep flow and finiteness of displacements. A technique is proposed that makes it possible to simplify the solution of the problem by using the Lagrangian approach as applied to deformation of creep flow. It is believed that buckling occurs instantly and resiliently. To find the critical load at each time step, an analogy is used between the stress states of a straight rod and an arch. An example of calculating the parameters of the stress-strain state and critical time for a circular arch using the theory of flow is given. Based on numerical experiments, some questions of convergence are investigated both from the time step and from the degree of discretization of an arc.

Keywords Low arch · Creep flow · Stability · Large deflections.

18.1 Introduction

A lot of works, in particular, [1–21] have been devoted to the tasks of bending and stability of thin-walled elements of constructions during creep flow. One of the statements of the stability tasks is to study the process (deformation approach), in which, under a given fixed load, the body changes its shape so that at some point in time t_{cr} a new shape becomes such that the body loses stability with a pop, i.e. resiliently. There is an instant jump from one equilibrium state to another. In this formulation, the tasks are considered in [4, 17]. As noted in [6], the application of this method is associated with significant difficulties. Therefore, a number of other methods have been proposed. For example, a dynamic approach was proposed in

R. A. Kayumov (✉) · F. R. Shakirzyanov
Kazan State University of Architecture and Engeneering, 1 Zelenaya str., Kazan 420043, Russia
e-mail: faritbox@mail.ru

Kazan National Research Technical University named after A.N. Tupolev, 10 K. Marks str., Kazan 420111, Russia
e-mail: kayumov@rambler.ru

H. Altenbach et al. (eds.), *Multiscale Solid Mechanics*,
Advanced Structured Materials 141,
https://doi.org/10.1007/978-3-030-54928-2_18

[19], a quasistatic method—in [12]. In addition to such statements of creep flow stability tasks, other statements are also known, for example, those given in [3]. In particular, in his work [3] Gerard considers such a state to be critical in which the total deformation is equal to the Euler critical deformation. Although there is no rigorous justification for this approach, it gives, as noted in [28], a satisfactory agreement with the experiment. According to the concept, proposed by Shenley [25], the critical stress is calculated by the Euler formula, in which an elastic modulus is replaced by a tangent modulus, which is determined for the given critical time by an isochronous creep flow curve in the axes $\sigma - \varepsilon$.

To simplify the analysis of the constructions' behavior, taking into account creep flow, the elastic part of deformation is often neglected [6]. In particular, in the work [5] this is done when studying the process of bulging of a compressed rod with initial curvature. We note that in the theory of plasticity the solution of the tasks of determining the deformed state of a body, taking into account large deflections and even large deformations in a rigidly plastic formulation, can be obtained quite simply using the Lagrange approach (see, for example, [7]), according to which the problem of the limit equilibrium theory is solved at each step of the process parameter (according to the rigid-plastic scheme). We find the ultimate load, the speeds of displacements of each point, then based on them we determine small additional displacements, and after that we find a new body configuration. The same approach can also be applied in the creep flow tasks [6], since in the both cases the work of inelastic deformations completely dissipates. However, in the creep flow tasks to find the speeds of creep flow deformations we have to know a stress field.

18.2 Calculation Procedure

In the tasks of stability loss in an elastic way it is no longer possible to solve the tasks taking into account an elastic part of the deformations, and both elastic and inelastic deformations must be determined at each moment of the time. If the task is statically definable, then this is not difficult. In this paper, a combined approach is used to study the process of deformation and buckling of the constructions with large deflections in the presence of creep flow. Namely, at each time interval, the previous state is considered to be the initial one due to the fact that all the energy of creep flow deformation dissipates. The stress state is determined for a body of a new configuration according to the linear theory of elastic deformations We will demonstrate this approach by the example of a gentle arch (panel). Let us note that with cylindrical bending of the panel, the task becomes one-dimensional and coincides within the accuracy of notations with the task about the arch. And the difference is that instead of bending stiffness EJ cylindrical stiffness $D = Eh^3/12(1 - \nu^2)$ should be used, instead of the tensile stiffness EA should be used $K = Eh/(1 - \nu^2)$, by concentrated force it is necessary to mean a load q distributed according to the line. Further on we will use the notations EJ and EA.

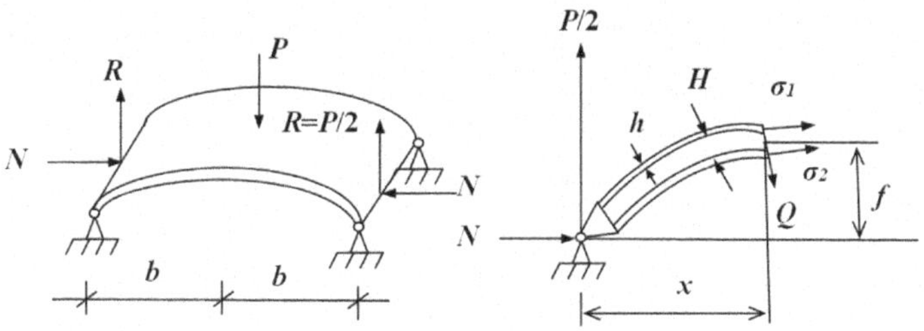

Fig. 18.1 Arch under the action of a centrally applied force

In addition, for simplicity we will consider an arch (panel) formed by two bearing layers and a medium soft layer, using the Kirchhoff-Love hypothesis. It is assumed that the entire load is perceived only by the outer layers. A similar scheme was used in the work [5] to analyze the stability of a compressed rod with initial irregularities under creep flow.

Let us consider a pivotally supported arch (panel) under the action of a centrally applied force P (Fig. 18.1).

The equilibrium equations of a cut off part of the panel have the following form:

$$\frac{\sigma_1 haH}{2} - \frac{\sigma_2 haH}{2} = Nf(x) - \frac{P}{2}x, \tag{18.1}$$

$$\sigma_1 ha + \sigma_2 ha = -N. \tag{18.2}$$

Here H is a distance between the gravitational centers of the cross-sections of the outer layers.

In view of the flatness, we can assume that $N = \text{const}$. Then from (18.1), (18.2) we can find σ_1, σ_2, expressing them in terms of N:

$$\sigma_1 = (2N \cdot f(x) - P \cdot x - NH)\frac{1}{2haH}, \tag{18.3}$$

$$\sigma_2 = (-2N \cdot f(x) + P \cdot x - NH)\frac{1}{2haH}. \tag{18.4}$$

At the first step, we will solve an elastic task in the linear formulation, assuming that the displacements are small (in the case of considerable deflections the solution given in [10] can be used; for the distributed load see the work [11]).

Let us find a bending moment

$$M = -\frac{\sigma_1 ahH}{2} + \frac{\sigma_2 ahH}{2}. \tag{18.5}$$

In the case of an elastic task we have a relation between the deflection w and the bending moment in the following form:

$$M = EJ\frac{\partial^2 w}{\partial x^2}. \tag{18.6}$$

Therefore

$$w = \frac{1}{EJ}\int_0^x \left(\int_0^x M\,\mathrm{d}x\right)\mathrm{d}x + C_1 x + C_2. \tag{18.7}$$

The constants C_1 and C_2 are determined from the conditions of fastening and symmetry:

$$w(0) = 0, \quad \frac{dw(b)}{\mathrm{d}x} = 0. \tag{18.8}$$

Hence,

$$C_2 = 0, \quad C_1 = -\frac{1}{EJ}\int_0^b M\,\mathrm{d}x. \tag{18.9}$$

Knowing σ_1, σ_2 longitudinal deformation at the central point of a cross-section of the arch can be found:

$$\varepsilon = \frac{\sigma_1 + \sigma_2}{E} = \frac{-N}{EA}. \tag{18.10}$$

On the other part

$$\varepsilon = \frac{\mathrm{d}u}{\mathrm{d}x} - \frac{w}{r}. \tag{18.11}$$

Here r is a radius of curvature of the arch. Therefore

$$u = \int_0^x \frac{w}{r}\mathrm{d}x + \int_0^x \varepsilon dx = \int_0^x \frac{w}{r}\mathrm{d}x - \frac{N}{EA}x. \tag{18.12}$$

To determine N we use the following symmetry condition

$$u(b) = 0. \tag{18.13}$$

Substituting here (18.12) we obtain the expression for N:

$$N = \frac{EA}{b}\int_0^b \frac{w}{r}\mathrm{d}x. \tag{18.14}$$

Thus, for the case of small displacements, the solution of the elastic task is known, which means that the stress field σ_1, σ_2 is known too. In the case of large displacements, as noted above, it is also possible to obtain an analytical solution, although more cumbersome (see in [10]).

Next, we move on to solving the task of finding the field of the speeds of creep flow deformations. Let us consider the case of the theory of flow, namely the simplest version, in the following form:

$$\dot{\xi}_1 = B\sigma_1|\sigma_1|^m, \quad \dot{\xi}_2 = B\sigma_2|\sigma_2|^m. \tag{18.15}$$

Here $\dot{\xi}_1, \dot{\xi}_2$ are the speeds of deformations at the central points of the cross-sections of the outer layers.

In the future, we will neglect an elastic part of the deformations. Then it is possible to connect $\dot{\xi}_1, \dot{\xi}_2$ with an additional deflection w^c, caused by a creep flow, by the following relation

$$\dot{\xi}_1 = -\frac{\mathrm{d}^2\dot{w}^c}{\mathrm{d}x^2}\frac{H}{2} + \dot{\xi}, \tag{18.16}$$

$$\dot{\xi}_2 = \frac{\mathrm{d}^2\dot{w}^c}{\mathrm{d}x^2}\frac{H}{2} + \dot{\xi}, \tag{18.17}$$

where $\dot{\xi}$ is a speed of deformation at the central point of the cross-section. From (18.16), (18.17) we find:

$$\frac{\mathrm{d}^2\dot{w}^c}{\mathrm{d}x^2} = \frac{\dot{\xi}_2 - \dot{\xi}_1}{H}, \quad \dot{\xi} = \dot{\xi}_1 + \dot{\xi}_2. \tag{18.18}$$

Therefore, we can express $\dot{w}^c$ and $\dot{\xi}$ in terms of stress. We get from (18.15):

$$\frac{\mathrm{d}^2\dot{w}^c}{\mathrm{d}x^2} = B(\sigma_2|\sigma_2|^m - \sigma_1|\sigma_1|^m)/H. \tag{18.19}$$

Then

$$\dot{w}^c = \frac{B}{H}\int\limits_0^x\left(\int\limits_0^x(\sigma_2|\sigma_2|^m - \sigma_1|\sigma_1|^m)dx\right)\mathrm{d}x + C_3x, \tag{18.20}$$

$$\xi_1 = B(\sigma_1|\sigma_1|^m + \sigma_2|\sigma_2|^m)/2. \tag{18.21}$$

The constant C_3 again found from the symmetry condition:

$$\frac{\partial\dot{w}^c(b)}{\partial x} = 0.$$

Then

$$C_3 = -\frac{B}{H}\int_0^b (\sigma_2|\sigma_2|^m - \sigma_1|\sigma_1|^m)\mathrm{d}x.$$

Let us express $\dot{\varepsilon}^c$ in terms of $\dot{u}^c$and $\dot{w}^c$

$$\dot{\varepsilon}^c = \frac{\mathrm{d}\dot{u}^c}{\mathrm{d}x} - \frac{\dot{w}^c}{r}, \tag{18.22}$$

$$\dot{u}^c = \int_0^x \dot{\xi}\mathrm{d}x + \int_0^x \frac{\dot{w}^c}{r}\mathrm{d}x + D. \tag{18.23}$$

The constant of integration D is found from the symmetry condition, too:

$$\dot{u}^c(b) = 0. \tag{18.24}$$

Since all the energy, spent on deformation of the creep flow, dissipates, and elastic deformations are much smaller than inelastic, then we can assume that in a new configuration of the body there are no residual stresses caused by deformations of the creep flow. Then a new configuration can be easily found by assuming

$$f^{\mathrm{new}}(x) = f(x) + \dot{w}^c(x)\Delta t, \tag{18.25}$$

$$\chi^{\mathrm{new}} = x + \dot{u}^c(x)\Delta t. \tag{18.26}$$

After this, it is necessary to approximate the obtained curve according to some nodal points.

Suppose that before the start of the process there were K nodes on the curve $f = f(x)$, whose coordinates are denoted by x_k, f_k. Then the new coordinates of the nodes will receive the values determined by the relations (18.25), (18.26):

$$f_k^{\mathrm{new}} = f_k + \dot{w}^c(x_k)\Delta t, \tag{18.27}$$

$$x_k^{\mathrm{new}} = x_k + \dot{u}^c(x_k)\Delta t. \tag{18.28}$$

From them the new functions $f^{\mathrm{new}}(x)$, $r^{\mathrm{new}}(x)$ can be found by approximation over a certain system of functions $\varphi_k(x)$

$$f^{\mathrm{new}}(x) = \sum_{k=1}^{K} f_k\varphi_k(x), \tag{18.29}$$

$$r^{\text{new}}(x) \cong \frac{d^2 f^{\text{new}}}{\mathrm{d}x^2}. \tag{18.30}$$

Let us immediately load the arch with the strength P and find elastic deformations and displacements, stresses σ_1, σ_2 and force N.

18.3 Stability Assessment

With some new configuration, at some point in time t_{cr} the arch can lose stability by clapping. To find this time, we apply the following algorithm. At each time step after determining the function $f^{\text{new}}(x)$ the critical strength P_{cr} can be found, for example, according to the simplified method [23]. At the moment of stability loss, the arch loses stability regarding the form depicted in Fig. 18.2. The critical force N_{cr} can be calculated approximately by the Euler formula:

$$N_{cr} = \frac{\pi^2 EJ}{(\mu l)^2} = \frac{\pi^2 EJ}{b^2}. \tag{18.31}$$

In the center of the arch at the point of inflexion the bending moment is equal to zero. Then the equilibrium equation of the left side of the arch will take the following form:

$$(P_{\text{cr}}/2)b - N_{\text{cr}} f^{\text{new}}(b) = 0. \tag{18.32}$$

From relation (18.32) we obtain the critical load

$$P_{\text{cr}} = \frac{2\pi^2 EJ}{b^3} f^{\text{new}}(b). \tag{18.33}$$

If at the given moment in time t we get $P_{\text{cr}} < P$, then it means that there has been a loss of stability, and this time can be taken as critical t_{cr}.

An arch can lose not only stability, but also strength, since the stresses σ_1, σ_2, as a rule, increase due to the flattening of the arch. Therefore, it is necessary to check not only its stability, but also the strength of the bearing layers. In our case, this is done quite simply by checking the conditions

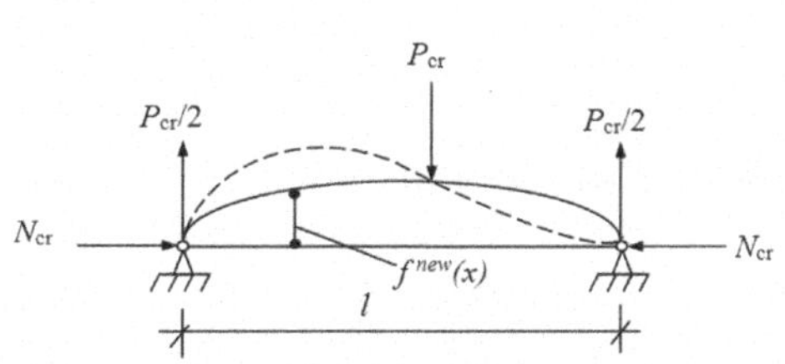

Fig. 18.2 At the moment of stability loss

$$|\sigma_1| \leq \sigma_{\text{cmp}}, \quad |\sigma_2| \leq \sigma_{\text{cmp}}, \quad \sigma_2 \leq \sigma_{\text{str}}.$$

where σ_{str}, σ_{cmp} are the strength limits of the material of the layers regarding stretching and compression. Besides, if there are restrictions on deflections, the operating time of the panel can be limited by the condition

$$w(b) \leq w_0,$$

where w_0 is an allowable deflection of the panel.

18.4 Calculation Example

Examples of numerical experiments conducted for various statements of the problems of creep and loss of stability of panels, rods, arches and shells can be found in [22–29].

Let us present the results of numerical experiments for a gentle circular arch, for which the following geometric parameters were used:

$$a/h = 10, \quad H/h = 10, \quad b/h = 100, \quad f/h = 25.$$

Figure 18.3 shows the deformed states of a gentle circular arch at various points in time. Figure 18.4 shows the values of the dimensionless critical load $\tilde{P}_{cr} = P_{cr} \cdot 10^4/(Eah)$ for the deformed state depending on the time. From a comparison with the current dimensionless load $\tilde{P}_{textcr} = P_{\text{cr}} \cdot 10^4/(Eah) = 30$ the critical time t_{cr} can be found, upon reaching which there is a stability loss in an elastic way. Figure 18.5 shows the maximum compressive stress $\tilde{\sigma_1} = \sigma_1 \cdot 10^5/(Ea^2h^2)$. In all these figures, some of the curves (green) represent solutions in the case when a fourth-degree polynomial is used to approximate the curve $y = f(x)$, and the others—for the case of use of a sixth-degree polynomial. We can see that the difference in time between the achievement of any values regarding voltage and critical loads is small (about 3%). The pictures of the arch deformation at different points in time are also quite close. Figure 18.6 shows the values of the parameters of the stress-strain state (referred to their values at the maximum number of time steps), depending on the number of time steps. We can also see that a degree of convergence is quite good. If we consider that the solution is approximate, then we can conclude that it is enough to use a not very small time step.

18.5 Conclusion

To solve creep flow tasks at large displacements a simplified Lagrange approach was used, according to which the geometry is recalculated only on the basis of deformations of creep flow. It is also assumed that the elastic deformations and

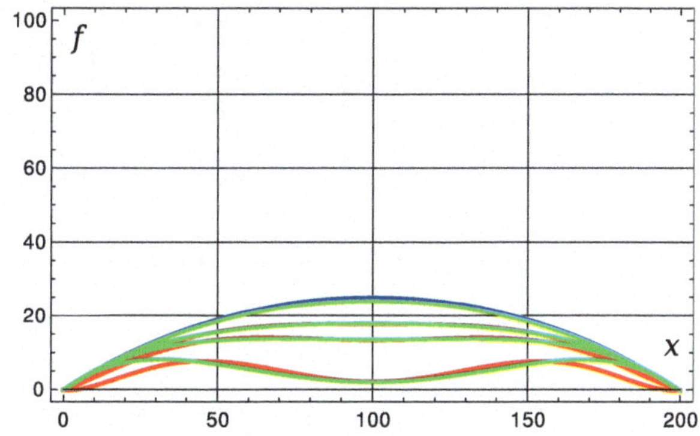

Fig. 18.3 Forms of a gentle initially circular arch at various points in time. The upper line is an original form, the green lines—in the case of approximating the curve $y = f(x)$ a fourth-degree polynomial, red lines—in the case of using a sixth-degree polynomial

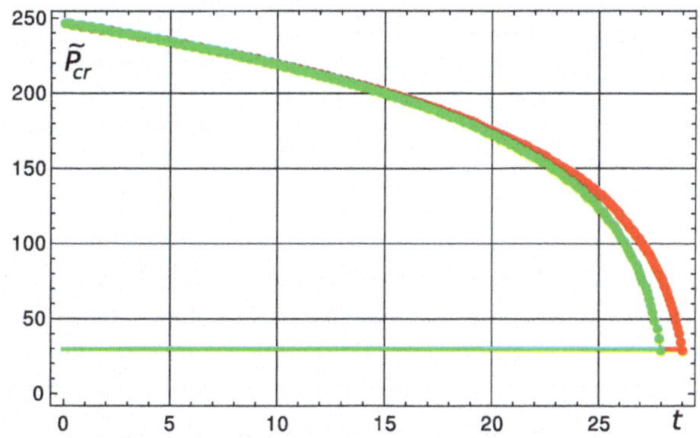

Fig. 18.4 Values of dimensionless critical load $\tilde{P}_{\text{cr}} = P_{\text{cr}} \times 10^4/(Eah)$ for a deformed arch depending on the time (lower green line is a value of the constant acting dimensionless load $\tilde{P} = P \times 10^4/(Eah) = 30$

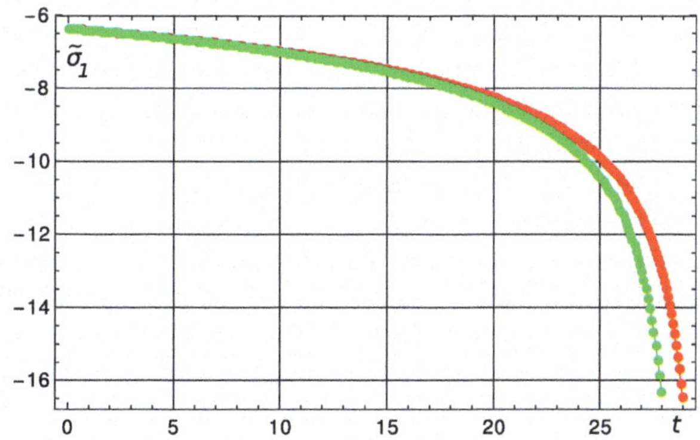

Fig. 18.5 Values of maximum compressive dimensionless stress $\tilde{\sigma_1} = \sigma_1 \times 10^5/(Ea^2h^2)$ depending on the time

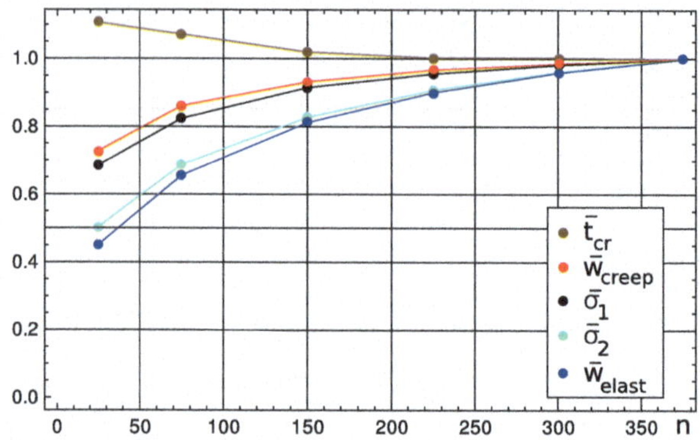

Fig. 18.6 Values of parameters of the stress-strain state (referred to their values at the maximum number of time steps) depending on the number of time steps (the curves from top to bottom: relative critical time $\bar{t}_{cr} = t_{cr}/t_{cr(n=350)}$, the lower one is a relative deflection in the center $\bar{w}_{creep} = w_{creep}/w_{creep(n=350)}$, caused by a creep flow, the third one from above is a relative maximum compressive stress $\bar{\sigma}_1 = \sigma_1/\sigma_{1(n=350)}$, the blue one is a relative maximum stretching stress $\bar{\sigma}_2 = \sigma_2/\sigma_{2(n=350)}$, the lowest is a relative elastic deflection in the center $\bar{w}_{elast} = w_{elast}/w_{elast(n=350)}$

displacements, caused by them, are small. This allows us to solve a linearly elastic task for a new geometry at every step. Then, based on the stresses, we have found, the speeds of creep flow deformations are calculated, with the help of which some small displacements caused by a creep flow of the material are determined over a short period of time. The proposed approach is especially convenient when using the finite element method. The analysis of the numerical experiments shows that the method used allows us to obtain not only a stress-strain state of the body, but also to evaluate the time after which a loss of its strength occurs. To assess stability of the body, it is necessary to apply some standard methods for determining a critical load for the elastic body, but in a new configuration.

The proposed approach is demonstrated on the task of bending and buckling of a flat three-layer arch, in which only the outer layers are bearing. A study of the calculation results at various values of the time step shows, that the method converges quickly and allows a satisfactory description of the structure behavior taking into account a creep flow even when using a large step. However, when approaching a critical state, it is better to reduce the time step. This can be done in proportion to how the elastic maximum displacements increase compared to the initial ones. It is expected that small quantities (in this work—tension stresses and elastic deflection) are more sensitive to the time step, but they are of little significance when calculating the strength and stiffness of the body under consideration.

In addition, the analysis of the results of numerical experiments has shown that an increase in the accuracy of approximation of the deformed body configuration leads to an increase in the degree of convergence of the SSS parameters of the body depending on the number of time steps.

Acknowledgements This work was supported by the Russian Science Foundation (project No. 19-19-00059, 19-79-10018 Sect. 2-3), the Russian Federal Property Fund (project No. 19-08-00349, Sect. 4).

References

1. Alfutov, N.A.: Foundations for Computation of Stability of Elastic Systems. Mashinostroienie, Moscow (1978). (in Russian)
2. Brovman, M.Ya.: On creep deformation during the longitudinal bending of beams. Bull. RAS. MTT, No. 4, pp. 121–129 (2014)
3. Gerard, G.A.: Greep buckling hypoteses. J. Aeronaut. Sci. **23**(9) (1956)
4. Grigolyuk, E.I., Lipovtsev, Yu.V.: Application of the variational principle in tasks of shell stability under creep conditions. Engine magazine "Mechanics of solids" No. 2 (1966) (in Russia)
5. Hoff, N.D.: Longitudinal Bending and Stability. IL, Moscow (1955)
6. Kachanov, L.M.: Creep Theory. Fizmatgiz (1960)
7. Kayumov, R.A.: Plastic deformation of fibrous composites and failure of fibrous composite structures. Mech. Compos. Mater. **29**(1), 60–65 (1993)
8. Kayumov, R.A.: Models of plastic flow of fibrous composites and destruction of structures from them. News of universities, Mathematics, No. 4, pp. 82–87 (1993)
9. Kayumov, R.A.: Postbuckling behavior of compressed rods in an elastic medium. Mech. solids. **52**(5), 575–580 (2017)
10. Kayumov, R.A., Tazyukov, B.F.: Stability of a curved thin elastic plate loaded with shear force. News of universities. Aircraft Technol. No. 4, pp. 12–15 (2001)
11. Kornishin, M.S., Mushtari, H.M.: Stability of an infinitely long flat cylindrical panel under the influence of normal uniform pressure. Bull. KFAN USSR Ser. Phys.-Math. Tech. Sci. **7**, 36–50 (1995)
12. Kurshin, L.M.: To the statement of a task about creep buckling of the shell. Rep. USSR Acad. Sci. **163**(1), 46–49 (1965)
13. Kurshin, L.M.: To the solution of tasks about stability of plates under the conditions of creep flow according to a quasistatic theory. PMTF, No. 5, pp. 154–158 (1962)
14. Lokoshchenko, A.M.: The buckling of a visco-elastic rod. PMTF, pp. 156–160 (1966)
15. Paimushin, V.N.: Problems of geometric non-linearity and stability in the mechanics of thin shells and rectilinear rods. J. Appl. Math. Mech. **71**(5), 772–805 (2007)
16. Paimushin, V.N., Kayumov, R.A., Kholmogorov, S.A.: Deformation features and models of $[\pm 45]_{2s}$ cross-ply fiber-reinforced plastics in tension. Mech. Compos. Mater. **55**(2), 141–154 (2019)
17. Pian, T.H.H.: Creep buckling of curved beam under lateral loading. In: Proceedings of 3rd US Congress Appl. Mech., NY, pp. 649–654 (1958)
18. Rabotnov, YuN: Creep Problems in Structural Members. North-Holland Publishing Company, Amsterdam (1969)
19. Rabotnov, Yu.N., Shesterikov, S.A.: Stability of rods and plates in a state of creep flow. PMM **21**(3) (1957)
20. Romanov, K.I.: Critical time in the theory of plate buckling under creep conditions. Mech. Solids, No. 5,pp. 114–120 (2004)
21. Romanov, K.I.: Plate and shell buckling by the Karman scheme under creep conditions. Mech. Solids **43**(1), 112–116 (2008)
22. Rosenblum, V.I.: Stability of a compressed rod in a state of creep. Engin. compos. book. XVIII, pp. 99–104 (1954)
23. Rzhanitsyn, A.R.: Stability of systems with creep properties. In: Creep and Long-Term Strength, pp. 207–219 (1963)

24. Rzhanitsyn, A.R.: Building Mechanics. Higher school, p. 439 (1991)
25. Shanley, F.R.: Weight-Strength Analysis of Aircraft Structures. McGraw Hill, New York (1952)
26. Shesterikov, S.A.: Dynamic criterion of stability during creep flow for rods. In: PMTF, No. 1 (1961)
27. Shesterikov, S.A.: Creep buckling. PMTF **25**(3) (1961)
28. Teregulov, I.G.: Bending and Stability of Thin Plates and Shells During Creep Flow. Nauka, p. 206 (1969)
29. Volmir, A.S.: Stability of Elastic Systems. Fizmatgiz, p. 880 (1963)

Chapter 19
Elliptic Boundary Layer in Shells of Revolution Under Normal Edge Shock Loading

Irina V. Kirillova and Leonid Yu. Kossovich

Abstract The present paper is aimed to construction the asymptotic theory of the elliptic boundary layer in shells of revolution under normal edge shock loading. Asymptotic equations of this boundary layer are derived in the small vicinity of the surface Rayleigh wave front. First of all the problem for normal edge shock loading is reduced to the problem for surface shock loading by constructing the particular solution, satisfying only the boundary conditions on the shell edge. Asymptotic solution of the governing equations for addition problem is obtained by using the Lourye symbolic approach and the front asymptotic near the surface Rayleigh wave front. The symbolic solution allows us to derive governing equations of the boundary layer. The behaviour of this boundary layer along the thickness is defined by elliptic equations and the boundary conditions on the faces are defined by hyperbolic equations, characterized wave motion on these faces. Considered component together with early constructed ones allows describe solution for transient waves in all parts of the phase plane.

Keywords Shell of revolution · Asymptotic theory · Shock loading · Rayleigh wave · Lourye symbolic approach · Boundary layer

19.1 Introduction

Asymptotic theory of non-stationary wave propagation in thin elastic shells of revolution at edge shock loading is described in [10, 18]. This theory is used the non-stationary separation scheme into components with different variability and dynamicity indices. There was considered two types of edge shock loading which, according to [23], refer to the cases of tangential longitudinal and bending longitudinal loading.

Irina V. Kirillova (✉) · Leonid Yu. Kossovich
Saratov State University, 83 Astrakhanskaya Street, Saratov 410012, Russian Federation
e-mail: president@info.sgu.ru

Leonid Yu. Kossovich
e-mail: nano-bio@info.sgu.ru

H. Altenbach et al. (eds.), *Multiscale Solid Mechanics*,
Advanced Structured Materials 141,
https://doi.org/10.1007/978-3-030-54928-2_19

Here, the non-stationary wave solutions are generated by means of the membrane and bending components (low-frequency approximations, corresponding to the components of a solution on the base of the Kirchhoff-Love shell theory), the Saint-Venant quasi-static boundary layer, quasi-plane and quasi-antiplane high-frequency, short wave components and two types of boundary layers: the boundary layer of hyperbolic type in the vicinity of dilatational and shear wave fronts and the boundary layer of parabolic type in the vicinity of the quasi-front, which is corresponded to the front of membrane waves on the base of Kirchhoff-Love theory. This research completely defines not only asymptotic theory of transient waves in elastic shells of revolution, but also gives the basis for investigation of transient waves in bodies of arbitrary form [5, 27, 29, 30] and shells with complicated properties of material (anisotropic, viscoelastic) [2–5, 28–30].

To complete construction of asymptotic nonstationary theory in third case of shock edge loading (NW type) there where investigated solutions in small vicinity of the Rayleigh surface wave front. Consideration of this SSS (stress-strain state) began with analyzing the non-stationary SSS in the case of Lamb problem for the semi-plane [11]. This result was used to construct the asymptotical components of solutions for transient waves in plates and shells. There was constructed approximate equations for elliptic type boundary layer in the vicinity of the Rayleigh wave front as for plate and for cylindrical shells at shock surface loading and for shock edge loading [19–21].

Equations for the elliptic boundary layer in shells of revolution at normal shock surface loading were constructed in [17]. This component together with the bending component on the basis of low-frequency approximation, the Saint-Venant quasi-static boundary layer, the quasi-plane high frequency short wave component and the hyperbolic boundary layer in the vicinity of the shear wave front allows construct solutions for transient waves in all parts of the phase plane. Such asymptotical representation completely corresponds to the description of wave strain transient processes analysed in [1, 23] by numerical methods.

Analysis of non-stationary SSS in a small vicinity of the Rayleigh surface wave front in the Lamb problem for the semi-plane allows to consider wave motion in different complicated constructions [6–9, 12–16, 24–26, 31].

Present paper is devoted to extend the approach of [17] on the case of transient waves in semi-infinite shells of revolution at shock edge loading. This paper is be organised as follows. The problem formulation is given in Sect. 19.2. Section 19.3 describes constructing of a particular solution, which is satisfied the boundary conditions only on the edge. With the help of this particular solution the problem for the semi-infinite shell of revolution is reduced to the problem for the infinite shell and it could be used methods [17] for describing the solution in a small vicinity of the Rayleigh wave fronts. Governing equations of the elliptic boundary layer are derived in Sect. 19.4.

19.2 Statement of the Problem

Consider a shell of revolution (see Fig. 19.1), where α is a parameter of the meridian arc on the midsurface, θ is the angular coordinate, z is the distance from the midsurface along the normal.

Denote stresses and displacements as σ_{ij} and v_i $(i, j = 1, 2, 3)$, consequently. Consider the following boundary conditions on the shell faces:

$$\sigma_{33} = \sigma_{13} = 0, \quad z = \pm h \tag{19.1}$$

and on the edge ($\alpha = 0$), which define axisymmetric case of the SSS:

$$\sigma_{13} = IH(t), \quad v_1 = 0, \quad \alpha = 0, \tag{19.2}$$

where h is the shell halfthickness, t is time, I is the amplitude of the load, $H(t)$ is Heaviside unit function. We consider the homogeneous initial conditions

$$v_i = \frac{\partial v_i}{\partial t} = 0 \quad (i = 1, 2, 3) \quad \text{at} \quad t = 0. \tag{19.3}$$

Now we delate the scale of the independent variables, setting

$$\xi = \frac{\alpha}{h}, \quad \zeta = \frac{z}{h}, \quad \tau = \frac{tc_2}{h}, \qquad c_2 = \sqrt{\frac{E}{2\rho(1+\nu)}}, \tag{19.4}$$

where E, ν, ρ are the Young's modulus, Poisson's ratio and mass density of the body; c_2 is the distortion wave speed. We assume also that differentiation with respect to the dimensionless variables ξ and τ does not change the asymptotic order of unknown quantities. Then the dimensionless equations in displacements can be written as follows

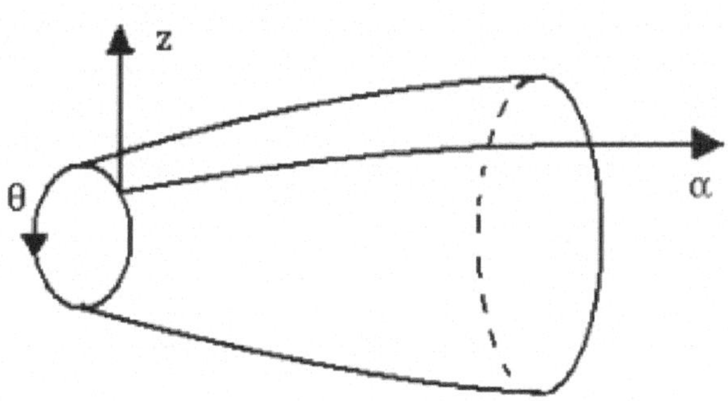

Fig. 19.1 Shell of revolution

$$
\begin{aligned}
&æ^{-2}\frac{\partial^2 v_1}{\partial\xi^2}+\frac{\partial^2 v_1}{\partial\zeta^2}+\frac{1}{1-2\nu}\frac{\partial^2 v_3}{\partial\xi\partial\zeta}-\frac{\partial^2 v_1}{\partial\tau^2}+\eta\frac{\zeta}{R_1}\left(-æ^{-2}\frac{\partial^2 v_1}{\partial\xi^2}+\frac{\partial^2 v_1}{\partial\zeta^2}-\frac{\partial^2 v_1}{\partial\tau^2}\right)\\
&\quad+\eta æ^{-2}\frac{B'}{B}\frac{\partial v_1}{\partial\xi}+\eta\left(\frac{1}{R_1}+\frac{1}{R_2}\right)\frac{\partial v_1}{\partial\zeta}+\eta\left(\frac{3-4\nu}{1-2\nu}\frac{1}{R_1}+\frac{1}{1-2\nu}\frac{1}{R_2}\right)\frac{\partial v_3}{\partial\xi}=0,\\
&\frac{1}{1-2\nu}\frac{\partial^2 v_1}{\partial\xi\partial\zeta}+\frac{\partial^2 v_3}{\partial\xi^2}+æ^{-2}\frac{\partial^2 v_3}{\partial\zeta^2}-\frac{\partial^2 v_3}{\partial\tau^2}+\eta\frac{\zeta}{R_1}\left(-\frac{\partial^2 v_3}{\partial\xi^2}+æ^{-2}\frac{\partial^2 v_3}{\partial\zeta^2}-\frac{\partial^2 v_3}{\partial\tau^2}\right)\\
&\quad-\eta\frac{3-4\nu}{1-2\nu}\frac{1}{R_1}\frac{\partial v_1}{\partial\xi}+\eta\frac{1}{1-2\nu}\frac{B'}{B}\frac{\partial v_1}{\partial\zeta}+\eta\frac{B'}{B}\frac{\partial v_3}{\partial\xi}+\eta æ^{-2}\left(\frac{1}{R_1}+\frac{1}{R_2}\right)\frac{\partial v_3}{\partial\zeta}=0,
\end{aligned}
\tag{19.5}
$$

and

$$
\begin{aligned}
\sigma_{11}=&\frac{E}{2(1+\nu)æ^2h}\left(\frac{\partial v_1}{\partial\xi}+\frac{\nu}{1-\nu}\frac{\partial v_3}{\partial\zeta}+\eta\left(-\frac{\zeta}{R_1}\frac{\partial v_1}{\partial\xi}+\frac{\nu}{1-\nu}\frac{B'}{B}v_1\right.\right.\\
&\left.\left.+\left(\frac{1}{R_1}+\frac{\nu}{1-\nu}\frac{1}{R_2}\right)v_3\right)\right),
\end{aligned}
\tag{19.6}
$$

$$
\begin{aligned}
\sigma_{33}=&\frac{E}{2(1+\nu)æ^2h}\left(\frac{\nu}{1-\nu}\frac{\partial v_1}{\partial\xi}+\frac{\partial v_3}{\partial\zeta}+\eta\left(-\frac{\nu}{1-\nu}\frac{\zeta}{R_1}\frac{\partial v_1}{\partial\zeta}\right.\right.\\
&\left.\left.+\frac{\nu}{1-\nu}\frac{B'}{B}v_1+\frac{\nu}{1-\nu}\left(\frac{1}{R_1}+\frac{1}{R_2}\right)v_3\right)\right),\\
\sigma_{13}=&\frac{E}{2(1+\nu)h}\left(\frac{\partial v_1}{\partial\zeta}+\frac{\partial v_3}{\partial\xi}-\eta\left(\frac{v_1}{R_1}+\frac{\zeta}{R_1}\frac{\partial v_3}{\partial\xi}\right)\right),
\end{aligned}
$$

where B is the distance from the axis of revolution, devided by R. R is the typical radius of curvature of the midsurface, R_i $(i=1,2)$ are the principal curvature radii of the midsurface.

Similarly to the case of surface loading [17] we consider our SSS as short-wave SSS and represent it as combination of symmetric and antisymmetric parts. Analysis [17] shows that the symmetric part of SSS is asymptotical main and governing equations can be written in the following form

$$
\begin{aligned}
&æ^{-2}\frac{\partial^2 v_1}{\partial\xi^2}+\frac{\partial^2 v_1}{\partial\zeta^2}+\frac{1}{1-2\nu}\frac{\partial^2 v_3}{\partial\xi\partial\zeta}-\frac{\partial^2 v_1}{\partial\tau^2}+\eta æ^{-2}\frac{B'}{B}\frac{\partial v_1}{\partial\xi}=0,\\
&\frac{1}{1-2\nu}\frac{\partial^2 v_1}{\partial\xi\partial\zeta}+\frac{\partial^2 v_3}{\partial\xi^2}+æ^{-2}\frac{\partial^2 v_3}{\partial\zeta^2}-\frac{\partial^2 v_3}{\partial\tau^2}+\eta\left(\frac{1}{1-2\nu}\frac{B'}{B}\frac{\partial v_1}{\partial\zeta}+\frac{B'}{B}\frac{\partial v_3}{\partial\zeta}\right)=0,
\end{aligned}
\tag{19.7}
$$

$$\sigma_{11} = \frac{E}{2(1+\nu)æ^2 h}\left(\frac{\partial v_1}{\partial \xi} + \frac{\nu}{1-\nu}\frac{\partial v_3}{\partial \zeta} + \eta\frac{\nu}{1-\nu}\frac{B'}{B}v_1\right),$$
$$\sigma_{33} = \frac{E}{2(1+\nu)æ^2 h}\left(\frac{\nu}{1-\nu}\frac{\partial v_1}{\partial \xi} + \frac{\partial v_3}{\partial \zeta} + \eta\frac{\nu}{1-\nu}\frac{B'}{B}v_1\right),$$
$$\sigma_{13} = \frac{E}{2(1+\nu)h}\left(\frac{\partial v_1}{\partial \zeta} + \frac{\partial v_3}{\partial \xi}\right). \tag{19.8}$$

19.3 Particular Solution

Construct particular solution of Eq. (19.7), which satisfy only boundary conditions (19.2) on the edge and denote it by the index "0". To do it we apply simplest asymptotic method of solution decomposition and represent the solution in the following form

$$v_1^{(0)} = 0,\ v_3^{(0)} = -I\sum_{n=1}^{\infty} F_n(\xi_0)(\tau_0 - \xi_0)^n H(\tau_0 - \xi_0), \tag{19.9}$$

where $F_n(\xi_0)$ are the sought for functions, dimensionless variables are defined as $\xi_0 = \alpha/R,\ \tau_0 = tc_2/R$ and R is a typical radius of curvature of the midsurface.

Substituting (19.9) into (19.7), we obtain the following equations for F_n

$$2\frac{dF_n}{d\xi_0} + \frac{B'}{B}F_n = \frac{1}{n}\left(\frac{d^2F_{n-1}}{d\xi_0^2} + \frac{B'}{B}\frac{dF_{n-1}}{d\xi_0}\right). \tag{19.10}$$

Solution of equation (19.10) has the form

$$F_n = \frac{C_n}{\sqrt{B}} + \frac{1}{2n\sqrt{B}}\int\limits_0^{\xi_0}\frac{1}{\sqrt{B}}\frac{d}{d\xi_0}\left(\frac{1}{B}\frac{dF_{n-1}}{d\xi_0}\right)d\xi_0, \tag{19.11}$$

where C_n are constants. Then taking into account boundary conditions (19.2) we obtain the particular solution as follows:

$$
\begin{aligned}
v_1^{(0)} &= 0,\ v_3^{(0)} = -\frac{2(1+\nu)hI}{E}\left[\sqrt{\frac{B(0)}{B}}(\tau_0-\xi_0)\right.\\
&\quad\left.+\sum_{n=2}^{\infty}\Phi_n(\xi_0)(\tau_0-\xi_0)^n\right]H(\tau_0-\xi_0),\\
\sigma_{11}^{(0)} &= \sigma_{33}^{(0)} = 0,\ \sigma_{13}^{(0)} = I\left[\sqrt{\frac{B(0)}{B}}+\sum_{n=2}^{\infty}\left(n\Phi_n(\xi_0)\right.\right.\\
&\quad\left.\left.-\Phi'_{n-1}(\xi_0)\right)(\tau_0-\xi_0)^{n-1}\right]H(\tau_0-\xi_0),\\
\Phi_1 &= \frac{1}{\sqrt{B}},\\
\Phi_n &= \frac{1}{2n\sqrt{B}}\int_0^{\xi_0}\frac{1}{\sqrt{B}}\frac{\mathrm{d}}{\mathrm{d}\xi_0}\left(\frac{1}{B}\frac{\mathrm{d}\Phi_{n-1}}{\mathrm{d}\xi_0}\right)\mathrm{d}\xi_0,\ n\geq 2.
\end{aligned}
\tag{19.12}
$$

Represent the solution of the initial problem in the form

$$
\mathrm{SSS} = \mathrm{SSS}^{(0)} + \mathrm{SSS}^{(1)} \tag{19.13}
$$

with the boundary conditions for $\mathrm{SSS}^{(1)}$ on the edge and surface

$$
\begin{aligned}
\frac{\partial v_3^{(1)}}{\partial\alpha} &= 0,\quad v_1^{(1)} = 0,\quad \alpha = 0,\\
\sigma_{33}^{(1)} &= 0,\quad \sigma_{13}^{(1)} = -S,\quad z = \pm h.
\end{aligned}
\tag{19.14}
$$

The expression for S defined as

$$
S = \pm I\left[\sqrt{\frac{B(0)}{B}}+\sum_{n=2}^{\infty}(n\Phi_n-\Phi'_{n-1}(\tau_0\mp\xi_0)^{n-1})\right]H(\tau_0\mp\xi_0), \tag{19.15}
$$

where the signs "$\mp$" and "$\pm$" correspond to the regions $\alpha > 0$ (top one) and $\alpha < 0$ (lower one).

Boundary conditions (19.14) show that the deformed edge section remains rectilinear and perpendicular to the axis ξ. Hence, if an infinite shell with a symmetric (with respect to the axis ζ) loading is taken, as shown in Fig. 19.2, deformation of each symmetric parts is equivalent to deformation of the considered semi-infinite shell. Therefore, it is possible to pass on to investigating the equivalent problem for the semi-infinite shell.

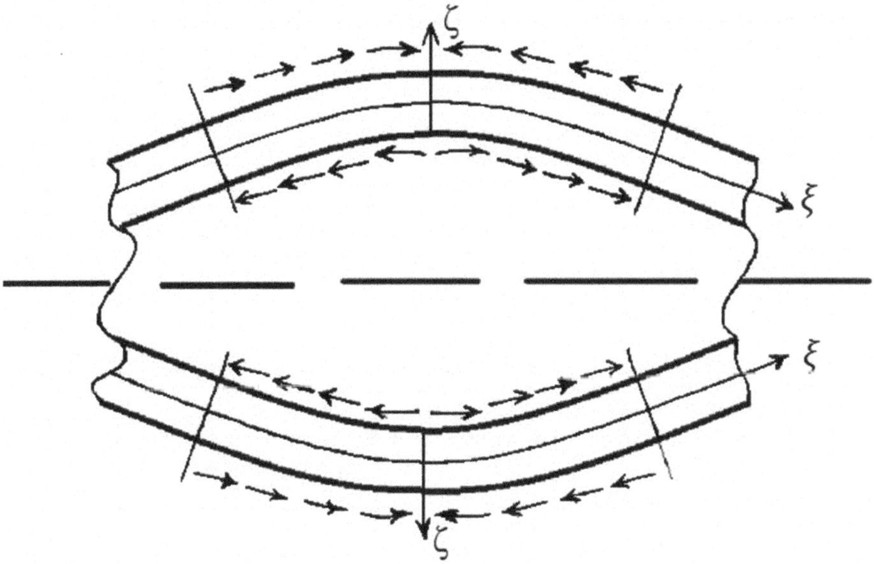

Fig. 19.2 Equivalent problem for semi-infinite shell

19.4 Elliptic Boundary Layer

Asymptotic solution of governing Eq. (19.7) can be obtained by introducing a symbolic notation wich was proposed in the case of a flat layer by Lurie [22]. Let's introduce operators $\partial_\xi = \partial/\partial\xi$, $\partial_\tau = \partial/\partial\tau$. According to the symbolic approach developed, Eq. (19.7) have to be treated as ordinary differential equations with respect to the independent variable ζ. In doing so the operators ∂_ξ and ∂_τ are considered as algebraic quantities of the order of unity. Then solutions to considered equations can be written as follows

$$\begin{aligned} v_1 &= \partial_\xi e^{-i(1-\zeta)\alpha_1} D_1 + \partial_\xi e^{-i(1+\zeta)\alpha_1} D_2 + \beta_1 e^{-i(1-\zeta)\beta_1} D_3 + \beta_1 e^{-i(1+\zeta)\beta_1} D_4, \\ v_3 &= i\alpha_1 e^{-i(1-\zeta)\alpha_1} D_1 - i\alpha_1 e^{-i(1+\zeta)\alpha_1} D_2 + i\left(\partial_\xi + \eta\frac{B'}{B}\right) e^{-i(1-\zeta)\beta_1} D_3 \\ &\quad - i\left(\partial_\xi + \eta\frac{B'}{B}\right) e^{-i(1+\zeta)\beta_1} D_4, \end{aligned} \tag{19.16}$$

where η is the small parameter ($\eta = h/R$), D_m ($m = 1, 2, 3, 4$) are arbitrary parameters, α_1 and β_1 are defined by the expressions

$$\alpha_1^2 = \partial_\xi^2 - æ^2\partial_\tau^2 + \eta\frac{B'}{B}\partial_\xi, \quad \beta_1^2 = \partial_\xi^2 - \partial_\tau^2 + \eta\frac{B'}{B}, \tag{19.17}$$

where $æ^2 = (1 - 2\nu)/(2 - 2\nu)$.

Substituting operator solution (19.16) into the boundary conditions on the faces, we obtain the following system of algebraic equations for D_m:

$$\gamma_1^2 D_1 + \gamma_1^2 e^{i2\alpha_1} D_2 + \left(\partial_\xi + \eta\frac{B'}{B}\right)\beta_1 D_3 + \left(\partial_\xi + \eta\frac{B'}{B}\right)\beta_1 e^{i2\beta_1} D_4 = 0,$$
$$\gamma_1^2 e^{i2\alpha_1} D_1 + \gamma_1^2 D_2 + \left(\partial_\xi + \eta\frac{B'}{B}\right)\beta_1 e^{i2\beta_1} D_3 + \left(\partial_\xi + \eta\frac{B'}{B}\right)\beta_1 D_4 = 0,$$
$$i\partial_\xi\alpha_1 D_1 - i\partial_\xi\alpha_1 e^{i2\alpha_1} D_2 + i\gamma_1^2 D_3 - i\gamma_1^2 e^{i2\beta_1} D_4 = -S,$$
$$i\partial_\xi\alpha_1 e^{i2\alpha_1} D_1 - i\partial_\xi\alpha_1 D_2 + i\gamma_1^2 e^{i2\beta_1} D_3 - i\gamma_1^2 D_4 = -S, \qquad (19.18)$$

where $\gamma_1^2 = \partial_\xi^2 - \partial_\tau^2/2 + \eta B'/B$. Equations (19.18) can be simplified by neglecting terms with multipliers $e^{i2\alpha_1},\ e^{i2\beta_1}$ [17].

Consider big values of the time, when $\tau \gg 1$, i.e. wave fronts propagate distances much more shell thickness: $\tau = O(T),\ T \gg 1$. Introduce for the case $\xi > 0$ new small parameter ε and variables $y,\ \tau_0$:

$$y = \frac{\xi - æ_R\tau}{\varepsilon}, \quad \tau_0 = \varepsilon\tau, \quad \varepsilon = \frac{1}{T}, \qquad (19.19)$$

where $æ_R = c_R/c_2$, c_R is the velocity of the surface Rayleigh waves. Also introduce new operators $\partial_y,\ \partial_\tau$:

$$\partial_y = \frac{\partial}{\partial y}, \quad \partial_{\tau_0} = \frac{\partial}{\partial \tau_0}. \qquad (19.20)$$

The asymptotically simplified form of Eq. (19.18) in new parameters can be written in the form

$$[g\partial_y^2 - \varepsilon æ_R\partial_y\partial_{\tau_0} - \eta\frac{B'}{B}\partial_y]D_m + [b\partial_y^2 - \varepsilon\frac{æ_R}{b}\partial_y\partial_{\tau_0} - \eta\left(b + \frac{1}{2b}\right)\frac{B'}{B}\partial_y]D_{m+2} = 0,$$
$$(\varepsilon 2æ_R\partial_y\partial_{\tau_0} + \eta k_c\frac{B'}{B}\partial_y)i\partial_y D_{m+2} = \mp\frac{æ_R B_\omega}{g}\partial_y S, \quad m = 1, 2 \qquad (19.21)$$

where $a = \sqrt{1 - æ^2æ_R^2}$, $b = \sqrt{1 - æ_R^2/2}$, $g = 1 - æ_R^2/2$, $k_c = \left(\frac{1}{b^2} + \frac{1}{a^2} + \frac{ab}{g^2} - \frac{4}{g}\right)/\left(\frac{1}{b^2} + \frac{æ^2}{a^2} - \frac{2}{g}\right)$, $B_\omega = 2\left[\frac{æ_R}{1-æ_R^2} + \frac{ææ_R}{1-æ^2æ_R^2} - \frac{4æ_R}{2-æ_R^2}\right]^{-1}$.

Asymptotical representation of displacements and stresses in new operators $\partial_y,\ \partial_{\tau_0}$ have the following form

$$v_1 = -\partial_y e^{ia(1+\zeta)\partial_y} D_1 - \partial_y e^{-ia(1+\zeta)\partial_y} D_2 - b\partial_y e^{-ib(1+\zeta)\partial_y} D_3 - b\partial_y e^{-ib(1-\zeta)\partial_y} D_4,$$
$$v_3 = -ia\partial_y e^{-ia(1+\zeta)\partial_y} D_1 + ia\partial_y e^{-ia(1-\zeta)\partial_y} D_2 - i\partial_y e^{-ib(1+\zeta)\partial_y} D_3$$
$$+ i\partial_y e^{-ib(1-\zeta)\partial_y} D_4, \qquad (19.22)$$

$$\sigma_{33} = -\frac{E}{(1+\nu)h}[g\partial_y^2 e^{-ia(1+\zeta)\partial_y} D_1 + g\partial_y^2 e^{-ia(1-\zeta)\partial_y} D_2$$
$$+ b\partial_y^2 e^{-ib(1+\zeta)\partial_y} D_3 + b\partial_y^2 e^{-ib(1-\zeta)\partial_y} D_4],$$
$$\sigma_{13} = \frac{E}{(1+\nu)h}[L_1 e^{-ia(1+\zeta)\partial_y} D_1 - L_1 e^{-ia(1-\zeta)\partial_y} D_2$$
$$+ L_2 e^{-ib(1+\zeta)\partial_y} D_3 - L_2 e^{-ib(1-\zeta)\partial_y} D_4],$$

where operators L_1, L_2 are defined by the expressions

$$L_1 = a\partial_y^2 - \varepsilon\frac{æ^2 æ_R}{a}\partial_y\partial_{\tau_0} - \eta\frac{1}{2a}\frac{B'}{B}\partial_y,$$
$$L_2 = g\partial_y^2 - \varepsilon æ_R \partial_y\partial_{\tau_0} - \eta\frac{B'}{B}\partial_y.$$

Introduce the dilatation and shear Lame potentials, φ and ψ. We shall divide them into parts, denoting with the index "1"—negative direction of the disturbance propagation, and the positive one with the index "2"

$$\varphi = \varphi_1 + \varphi_2, \quad \psi = \psi_1 + \psi_2. \tag{19.23}$$

Operator representation of these potentials is defined by the following expressions

$$\varphi_m = e^{-ia(1\pm\zeta)\partial_\xi} D_m, \quad \psi_m = e^{-ib(1\pm\zeta)\partial_\xi} D_{m+2}, \quad m = 1,\ 2. \tag{19.24}$$

These potential functions satisfies the elliptic equations

$$\left(1 - \frac{c_R^2}{c_1^2}\right)\frac{\partial^2\varphi_m}{\partial\alpha^2} + \frac{\partial^2\varphi_m}{\partial z^2} = 0,$$
$$\left(1 - \frac{c_R^2}{c_2^2}\right)\frac{\partial^2\psi_m}{\partial\alpha^2} + \frac{\partial^2\psi_m}{\partial z^2} = 0 \tag{19.25}$$

under the following boundary conditions

$$\frac{\partial^2\chi_1}{\partial\alpha^2} - \frac{1}{c_R^2}\frac{\partial^2\chi_1}{\partial t^2} + \frac{k_c}{æ_R^2}\frac{B'}{B}\frac{\partial\chi_1}{\partial\alpha} = \frac{B_\omega}{g æ_R}\frac{\partial S}{\partial\alpha},\ z = -h,$$
$$\frac{\partial^2\chi_2}{\partial\alpha^2} - \frac{1}{c_R^2}\frac{\partial^2\chi_2}{\partial t^2} + \frac{k_c}{æ_R^2}\frac{B'}{B}\frac{\partial\chi_2}{\partial\alpha} = -\frac{B_\omega}{g æ_R}\frac{\partial S}{\partial\alpha},\ z = h,$$
$$\left(g + \frac{æ_R^2}{2}\right)\frac{\partial^2\varphi_m}{\partial\alpha^2} - \frac{æ_R^2}{2c_R^2}\frac{\partial^2\varphi_m}{\partial t^2} + \frac{B'}{B}\frac{\partial\varphi_m}{\partial\alpha}$$
$$+ \left(b + \frac{æ_R^2}{2b}\right)\frac{\partial^2\psi_m}{\partial\alpha^2} - \frac{æ_R^2}{2bc_R^2}\frac{\partial^2\psi_m}{\partial t^2} + \left(b + \frac{1}{2b}\right)\frac{B'}{B}\frac{\partial\psi_m}{\partial\alpha} = 0,$$
$$z = \mp h, \tag{19.26}$$

where

$$\chi_1 = \frac{h}{b}\frac{\partial \psi_1}{\partial z}, \quad \chi_2 = -\frac{h}{b}\frac{\partial \psi_2}{\partial z},$$

$$B_\omega = 2\left[\frac{æ_R}{1 - æ_R^2} + \frac{ææ_R}{1 - æ^2æ_R^2} - \frac{4æ_R}{2 - æ_R^2}\right]^{-1}.$$

Displacements and stresses are described throw potential functions by the expressions

$$v_1 = h\left(\frac{\partial \varphi_1}{\partial \alpha} + \frac{\partial \varphi_2}{\partial \alpha} + b\frac{\partial \psi_1}{\partial \alpha} + b\frac{\partial \psi_2}{\partial \alpha}\right),$$

$$v_3 = h\left(\frac{\partial \varphi_1}{\partial z} + \frac{\partial \varphi_2}{\partial z} + \frac{1}{b}\frac{\partial \psi_1}{\partial z} + \frac{1}{b}\frac{\partial \psi_2}{\partial z}\right),$$

$$\sigma_{33} = -\frac{Eh}{1+\nu}\left(g\frac{\partial^2 \varphi_1}{\partial \alpha^2} + g\frac{\partial^2 \varphi_2}{\partial \alpha^2} + b\frac{\partial^2 \psi_1}{\partial \alpha^2} + b\frac{\partial^2 \psi_2}{\partial \alpha^2}\right),$$

$$\sigma_{13} = \frac{Eh}{1+\nu}\left(\frac{1}{a^2}\frac{\partial^2 \varphi_1}{\partial \alpha \partial z} + \frac{æ^2æ_R^2}{a^2 c_R}\frac{\partial^2 \varphi_1}{\partial z \partial t} + \frac{1}{2a^2}\frac{B'}{B}\frac{\partial \varphi_1}{\partial z}\right.$$
$$+ \frac{1}{a^2}\frac{\partial^2 \varphi_2}{\partial \alpha \partial z} + \frac{æ^2æ_R^2}{a^2 c_R}\frac{\partial^2 \varphi_2}{\partial z \partial t} + \frac{1}{2a^2}\frac{B'}{B}\frac{\partial \varphi_2}{\partial z}$$
$$+ \frac{g + æ_R^2}{b}\frac{\partial^2 \psi_1}{\partial \alpha \partial z} + \frac{æ_R^2}{b c_R}\frac{\partial^2 \psi_1}{\partial z \partial t} + \frac{1}{b}\frac{B'}{B}\frac{\partial \psi_1}{\partial z}$$
$$\left. + \frac{g + æ_R^2}{b}\frac{\partial^2 \psi_2}{\partial \alpha \partial z} + \frac{æ_R^2}{b c_R}\frac{\partial^2 \psi_2}{\partial z \partial t} + \frac{1}{b}\frac{B'}{B}\frac{\partial \psi_2}{\partial z}\right. . \quad (19.27)$$

In the case $\xi < 0$ governing equations for our elliptic boundary layer coincide with Eqs. (19.25)–(19.27) with the exception of expression for the stress σ_{13} which is described in this case as

$$\sigma_{13} = \frac{Eh}{1+\nu}\left(\frac{1}{a^2}\frac{\partial^2 \varphi_1}{\partial \alpha \partial z} - \frac{æ^2æ_R^2}{a^2 c_R}\frac{\partial^2 \varphi_1}{\partial z \partial t} + \frac{1}{2a^2}\frac{B'}{B}\frac{\partial \varphi_1}{\partial z}\right.$$
$$+ \frac{1}{a^2}\frac{\partial^2 \varphi_2}{\partial \alpha \partial z} - \frac{æ^2æ_R^2}{a^2 c_R}\frac{\partial^2 \varphi_2}{\partial z \partial t} + \frac{1}{2a^2}\frac{B'}{B}\frac{\partial \varphi_2}{\partial z}$$
$$+ \frac{g + æ_R^2}{b}\frac{\partial^2 \psi_1}{\partial \alpha \partial z} - \frac{æ_R^2}{b c_R}\frac{\partial^2 \psi_1}{\partial z \partial t} + \frac{1}{b}\frac{B'}{B}\frac{\partial \psi_1}{\partial z}$$
$$\left. + \frac{g + æ_R^2}{b}\frac{\partial^2 \psi_2}{\partial \alpha \partial z} - \frac{æ_R^2}{b c_R}\frac{\partial^2 \psi_2}{\partial z \partial t} + \frac{1}{b}\frac{B'}{B}\frac{\partial \psi_2}{\partial z}\right). \quad (19.28)$$

References

1. Ainola, L.Ya., Nigul, U.K.: Wave strain processes in elastic plates and shells. Izv. ANESSR **14**(1), 3–63 (1965)
2. Anofrikova, N.S.: Investigation of boundary layer in vicinity of dilatation wave front in viscoelastic cylindrical shells. Math. Mech. **2**, 148–151 (2000)
3. Anofrikova, N.S.: Long wave approximations of three dimensional viscoelastic theory equations. In: Proceeding of International Conference. Voronezh University Press, pp. 10–15 (2010)
4. Anofrikova, N.S., Gureeva, E.V.: Nonstationary bending waves in viscoelastic cylindrical shell. Math. Mech. **6**, 183–186 (2004)
5. Bescrovniy, A.S., Anofrikova, N.S.: Construction of low frequency long wave approximations of three dimensional viscoelastic theory equations for the case of two layered plates. Math. Mech. **19**, 114–117 (2017)
6. Ege, N., Erbas, B., Prikazchikov, D.A.: On the 3D Rayleigh wave field on an elastic half-space subject to tangential surface loads. ZAMM **95**(12), 1558–1565 (2015)
7. Ege, N., Erbas, B., Chorozoglou, A., Kaplunov, J., Prikazchikov, D.A.: On surface wave fields arising in soil-structure interaction problems. In: X International Conference on Structural Dynamics (EURODYN2017), vol. 95, no. 12, pp. 2366–2371 (2017)
8. Erbas, B., Kaplunov, J., Prikazchikov, D.: The Rayleigh wave field in mixed problems for a half-plane. IMA J. Appl. Math. **79**(5), 1078–1086 (2013)
9. Erbas, B., Kaplunov, J., Palsu, M.: A composite hyperbolic equation for plate extension. Mech. Res. Commun. **99**, 64–67 (2019)
10. Kaplunov, J.D., Kossovich, LYu., Nolde, E.V.: Dynamics of thin walled elastic bodies. Academic Press, San Diego (1998)
11. Kaplunov, YuD, Kossovich, LYu.: Asymptotic model of Rayleigh waves in the far-field zone in an elastic half-plane. Doklsady Phys. **49**(4), 234–236 (2004)
12. Kaplunov, J., Zakharov, A., Prikazchikov, D.: Explicit models for elastic and piezoelastic surface waves. IMA J. Appl. Math. **71**, 768–782 (2006)
13. Kaplunov, J., Nolde, E., Prikazchikov, D.A.: A revisit to the moving load problem using an asymptotic model for the Rayleigh wave. Wave Motion **47**(7), 440–451 (2010)
14. Kaplunov, J., Prikazchikov, D.: Asymptotic theory for Rayleigh and Rayleigh-Type Waves. Adv. Appl. Mech. **50**, 1–106 (2017)
15. Kaplunov, J., Prikazchikov, D., Sultanova, L.: Rayleigh-type waves on a coated elastic half-space with a clamped surface. Philos. Trans. R. Soc. **A377** (2019)
16. Khajieva, L.A., Prikazchikov, D.A., Prikazchikov, L.A.: Hyperbolic-elliptic model for surface wave in a pre-stressed incompressible elastic half-space. Mech. Res. Commun. **92**, 49–53 (2018)
17. Kirillova, I.V., Kossovich, L.Yu.: Refined equations of elliptic boundary layer in shells of revolution under normal shock surface loading. Mathematics **50**, 68–73 (2017)
18. Kirillova, I.V., Kossovich, L.Yu.: Asymptotic theory of waves in thin walled shells at shock edge loading of tangential, bending and normal types. In: Proceedings of the XI Russian Conference on Fundamental Problems of Theoretical and Applied Mechanics. Kazan University Press, pp. 2008–2015 (2015)
19. Kovalev, V.A., Taranov, O.V.: The separation of non-stationary SSS of cylindrical shells under normal type edge loading. In: Morozov, N.F. (ed.) Mixed Problems of Mechanics of Solid. Proceedings of the V Russian Conference with Intern. Participation, pp. 191–193. Saratov University Press (2005)
20. Kushekkaliev, A.N.: Solution of the problems about the propagation of waves in the transversal-isotropic cylindrical shell under the normal loading. Contin. Mech. **14**, 106–115 (2002)
21. Kushekkaliev, A.N.: Rayleigh wave in the semi-infinite plate under the transverse normal loading. The Mechanics of the Deformed Media, pp. 66–73. Saratov University Press (2004)
22. Lurie, A.I.: Spatial Problems in Theory of Elasticity. Gostekhizdat, Moscow (1955)
23. Nigul, U.: Regions of effective application of the methods of 3D and 2D of analysis of transient of stress waves in shells and plates. Int. J. Solids Struct. **54**, 607–627 (1969)

24. Nobili, A., Prikazchikov, D.A.: Explicit formulation for the Rayleigh wave field induced by surface stress in an orthorhombic half-plane. Eur. J. Mech. A/Solids **70**, 86–94 (2018)
25. Prikazchikov, D.A.: Rayleigh waves of arbitrary profile in anisotropic media. Mech. Res. Commun. **50**, 83–86 (2013)
26. Prikazchikov, D.A., Rogerson, G.A.: On surface wave propagation in incompressible, transversely isotropic, pre-stressed elastic half-space. Int. J. Eng. Sci. **42**(10), 967–986 (2004)
27. Shevtsova, YuV: Boundary layer in the vicinity of quasi-front in circle cylindrical shell. Math. Mech. **2**, 180–182 (2000)
28. Shevtsova, Yu.V.: Boundary layer in the vicinity of quasi-front in transversal isotropic cylindrical shell. In: On the Strength of the Elements of Constructions under the Action of Loads and Working Media, pp. 114–117. Saratov Technical Univ. Press (2000)
29. Shevtsova, Yu.V., Anofrikova, N.S.: Asymptotical approximations of the three dimensional exudations of elasticity and viscoelasticity for the case of two-layered plates. In: Actual Problems of Continuum Mechanics. Proceedings of the Second International Conference. Dilizan, vol. 1, pp. 86–90 (2010)
30. Shevtsova, YuV, Anofrikova, N.S.: Low frequency long wave approximations of three dimensional equations of viscoelasticity theory for the case of twolayered plates. Math. Mech. **12**, 126–130 (2010)
31. Wootton, P.T., Kaplunov, J., Colquitt, D.J.: An asymptotic hyperbolic-elliptic model for flexural-seismic metasurfaces. Proc. R. Soc. A Math. Phys. Eng. Sci. **475**(2227), 20190079 (2019)

Chapter 20
Non-stationary Dynamic Problem for Layered Viscoelastic Cylinder

Ekaterina A. Korovaytseva and Sergey G. Pshenichnov

Abstract A problem of non-stationary waves propagation in cross section of a hollow layered linearly viscoelastic cylinder at its axisymmetric loading is considered. On the contact surfaces between the layers, the continuity conditions for the displacement and stress vectors are accepted. The solution is constructed using the integral Laplace transform in time with subsequent reversal. The solution in the originals is presented in an integral form convenient for numerical implementation. It allows studying the process of wave propagation in a multilayer cylinder with an arbitrary number of viscoelastic coaxial layers without any simplifying assumptions concerning the existence of a connection between the hereditary kernels. The results of studies performed on the basis of the constructed solution with specific initial data are presented. These results illustrate the influence of contacting materials physical-mechanical parameters difference on the transient wave process in a layered cylindrical body.

Keywords Dynamics of viscoelastic bodies · Piecewise homogeneous solids · Wave processes · Layered cylinder

20.1 Introduction

Investigation of non-stationary wave processes in solids consisting of homogeneous linearly viscoelastic components, using analytical and numerical-analytical methods, is actual. Various achievements in this field are represented, for example, in the works [4–8, 12, 13]. Over the last years, one of the main trends here has been carrying out research using methods of boundary integral equations and boundary elements [1, 2]. With the help of these methods, representatives of the corresponding scientific

E. A. Korovaytseva · S. G. Pshenichnov (✉)
Institute of Mechanics, Lomonosov Moscow State University, Michurinsky Prospect, 1, Moscow 119192, Russia
e-mail: serp56@yandex.ru

E. A. Korovaytseva
e-mail: katrell@mail.ru

H. Altenbach et al. (eds.), *Multiscale Solid Mechanics*, Advanced Structured Materials 141,
https://doi.org/10.1007/978-3-030-54928-2_20

school carried out a number of works concerning dynamics of piecewise homogeneous viscoelastic media [3, 9]. Along with this within the framework of this theme quite a number of questions requiring investigations remain. First of all, this regards the investigation of influence of material hereditary properties on transient wave processes in solids with arbitrary number of homogeneous components' boundaries in the absence of certain relationship between hereditary kernels.

Previously, the author of this work constructed solutions of a number of non-stationary dynamic problems for multilayer solids with plane-parallel, spherical and cylindrical boundaries of viscoelastic homogeneous layers [10]. Hereditary kernels were chosen in the form of finite sum of exponents. Besides, recently, the work [11] which touched upon some general questions concerning constructing solutions of dynamic problems for piecewise homogeneous linearly viscoelastic bodies using integral Laplace transform with the following inversion has been published. Statements concerning properties of the solutions in transforms that simplify originals construction were formulated in the work. Statements concerning relationship between points of branching and poles of problem solution in transforms with the spectrum of the corresponding problem of considered multicomponent body free oscillations were proved as well. At certain conditions, a relationship between the initial non-stationary dynamic problem and the static problem of elasticity theory, in which long-term moduli play the role of elastic constants, was pointed out. Various forms of solution representation in originals were discussed.

In the present work, results of transient wave processes in layered cylindrical solid at its axisymmetric loading investigations are represented. The results are obtained with the help of numerical realization of the corresponding initial-boundary problem solution. This solution, compared with the constructed one in the work [10], is more convenient for calculations. Besides, the variant of its integral form represented here is suitable for hereditary kernels of a wider variety. In the work [11], similar problem was used for demonstration of a special test; however, its solution was not represented there. Below, we shall build on theoretical results of the work [11] when necessary.

20.2 Dynamic Problem for Layered Cylinder Statement

Let us consider dynamic problem for infinitely long hollow circular cylinder: $R_0 \leq R \leq R_N$, $-\pi < \theta \leq \pi$, $-\infty < z < +\infty$, (R, θ, z are cylindrical coordinates), consisting of N coaxial isotropic homogeneous linearly viscoelastic layers. We consider that at the boundaries $R = R_m$, ($m = 1, 2, 3, \ldots, N-1$) between neighboring layers conditions of displacement vector and stress vector continuity are fulfilled. At the initial time, the cylinder is at rest and undeformed. Its external surface is free, and the internal surface at the moment $t = 0$ is subjected to influence of uniform distributed load $Q(t)$, which provides condition of plane deformation and axial symmetry (Fig. 20.1).

We introduce the following dimensionless variables:

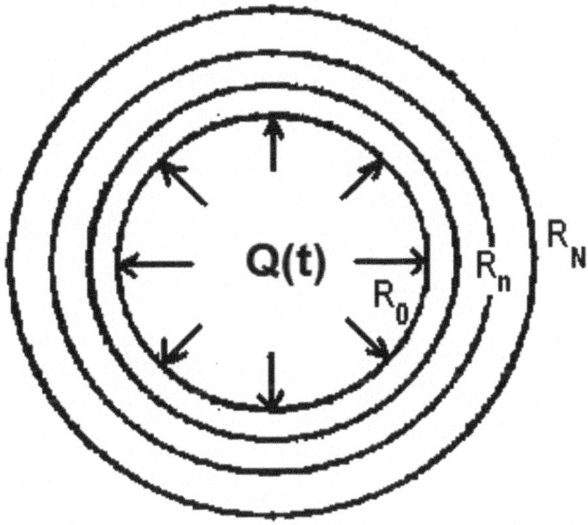

Fig. 20.1 Cross section and loading scheme of N-layered cylinder

$$r = R/R_N,\quad \tau = t/t_0,\quad r_0 = R_0/R_N,\quad r_m = R_m/R_N\quad (m = 1, 2, \ldots, N-1),$$

$$u^{(n)}(r,\tau) = \mathrm{v}^{(n)}(R,t)/R_N,$$

$$\sigma_{rr}^{(n)}(r,\tau) = P_{RR}^{(n)}(R,t)/2G_0^{(n)},\quad \sigma_{\theta\theta}^{(n)}(r,\tau) = P_{\theta\theta}^{(n)}(R,t)/2G_0^{(n)},$$

$$\gamma_s^{(n)}(\tau) = t_0 T_s^{(n)}(t),\quad \gamma_v^{(n)}(\tau) = t_0 T_v^{(n)}(t),\quad n = 1, 2, \ldots, N,$$

$$Q_0\psi(\tau) = Q(t)/2G_0^{(1)},\quad \alpha_n = c_N/c_n,\quad w_n = (1-\nu_0^{(n)})/(1-2\nu_0^{(n)}),$$

where $t_0 = R_N/c_N$, Q_0 is dimensionless constant; $\mathrm{v}^{(n)}(R,t)$, $P_{RR}^{(n)}(R,t)$, $P_{\theta\theta}^{(n)}(R,t)$ are radial displacement and radial and circular stresses; $T_s^{(n)}(t)$, $T_v^{(n)}(t)$, $G_0^{(n)}$, $\nu_0^{(n)}$ are volume and shear relaxation kernels and instantaneous shear modulus and Poisson's ratio of the nth layer material; c_n is the speed of longitudinal elastic waves in the nth layer.

We shall suppose that for all the layers' material conditions of creep restriction are fulfilled.

Mathematical problem statement for the layered cylinder in dimensionless form includes dynamics equations (where stresses are expressed through displacements):

$$\left(1-\hat{a}_1^{(n)}\right)\frac{\partial}{\partial r}\left[\frac{\partial u^{(n)}(r,\tau)}{\partial r} + \frac{u^{(n)}(r,\tau)}{r}\right] - a_n^2\frac{\partial^2 u^{(n)}(r,\tau)}{\partial \tau^2} = 0,\quad n = 1, 2, \ldots, N, \tag{20.1}$$

boundary conditions

$$\sigma_{rr}^{(1)}(r_0, \tau) = -Q_0\psi(\tau), \quad \sigma_{rr}^{(N)}(1, \tau) = 0, \quad \tau > 0 \tag{20.2}$$

conditions of contact between the layers ($m = 1, 2, \ldots, N-1$):

$$u^{(m)}(r_m, \tau) = u^{(m+1)}(r_m, \tau), \quad G_0^{(m)}\sigma_{rr}^{(m)}(r_m, \tau) = G_0^{(m+1)}\sigma_{rr}^{(m+1)}(r_m, \tau), \tag{20.3}$$

and initial conditions

$$u^{(n)}(r, 0) = 0, \quad \frac{\partial u^{(n)}}{\partial \tau}(r, 0) = 0, \tag{20.4}$$

here

$$\sigma_{rr}^{(n)}(r, \tau) = w_n\left(1 - \hat{d}_1^{(n)}\right)\frac{\partial u^{(n)}(r, \tau)}{\partial r} + (w_n - 1)\left(1 - \hat{d}_2^{(n)}\right)\frac{u^{(n)}(r, \tau)}{r}, \tag{20.5}$$

$$\sigma_{\theta\theta}^{(n)}(r, \tau) = w_n\left(1 - \hat{d}_1^{(n)}\right)\frac{u^{(n)}(r, \tau)}{r} + (w_n - 1)\left(1 - \hat{d}_2^{(n)}\right)\frac{\partial u^{(n)}(r, \tau)}{\partial r},$$

where operators $\hat{d}_1^{(n)}$, $\hat{d}_2^{(n)}$ and corresponding functions $d_1^{(n)}$, $d_2^{(n)}$ have the form

$$\hat{d}_j^{(n)}\xi(\tau) = \int_0^\tau d_j^{(n)}(\tau - \chi)\xi(\chi)d\chi, \quad j = 1, 2,$$

$$d_1^{(n)}(\tau) = \frac{1}{3(1 - \nu_0^{(n)})}\left[\left(1 + \nu_0^{(n)}\right)\gamma_V^{(n)}(\tau) + 2\left(1 - 2\nu_0^{(n)}\right)\gamma_s^{(n)}(\tau)\right],$$

$$d_2^{(n)}(\tau) = \frac{1}{3\nu_0^{(n)}}\left[\left(1 + \nu_0^{(n)}\right)\gamma_V^{(n)}(\tau) - \left(1 - 2\nu_0^{(n)}\right)\gamma_s^{(n)}(\tau)\right], \quad n = 1, 2, \ldots, N.$$

20.3 Solution Construction

Applying integral time Laplace transform to Eqs. (20.1) and relations (20.2), (20.3), (20.5), and taking into account homogeneous initial conditions (20.4), we shall obtain dynamics equations in transforms

$$\left[1 - D_1^{(n)}(s)\right]\frac{\partial}{\partial r}\left[\frac{\partial U^{(n)}(r, s)}{\partial r} + \frac{U^{(n)}(r, s)}{r}\right] - \alpha_n^2 s^2 U^{(n)}(r, s) = 0, \tag{20.6}$$

boundary conditions

$$S_{rr}^{(1)}(r_0, s) = -Q_0\Psi(s), \quad S_{rr}^{(N)}(1, s) = 0, \tag{20.7}$$

and conditions of contact between the layers ($m = 1, 2, \ldots, N-1$):

$$U^{(m)}(r_m, s) = U^{(m+1)}(r_m, s), \quad G_0^{(m)} S_{rr}^{(m)}(r_m, s) = G_0^{(m+1)} S_{rr}^{(m+1)}(r_m, s), \tag{20.8}$$

along with this

$$S_{rr}^{(n)}(r, s) = w_n\left[1 - D_1^{(n)}(s)\right]\frac{\partial\, U^{(n)}(r, s)}{\partial\, r} + (w_n - 1)\left[1 - D_2^{(n)}(s)\right]\frac{U^{(n)}(r, s)}{r}, \tag{20.9}$$

$$S_{\theta\theta}^{(n)}(r, s) = w_n\left[1 - D_1^{(n)}(s)\right]\frac{U^{(n)}(r, s)}{r} + (w_n - 1)\left[1 - D_2^{(n)}(s)\right]\frac{\partial U^{(n)}(r, s)}{\partial r},$$

where $U^{(n)}(r, s)$, $S_{rr}^{(n)}(r, s)$, $S_{\theta\theta}^{(n)}(r, s)$, $\Psi(s)$, $D_1^{(n)}(s)$, $D_2^{(n)}(s)$, $s \in C$.

are correspondingly transforms of functions
$u^{(n)}(r, \tau), \sigma_{rr}^{(n)}(r, \tau), \sigma_{\theta\theta}^{(n)}(r, \tau), \psi(\tau), d_1^{(n)}(\tau), d_2^{(n)}(\tau)$.

Problem solution in transforms (20.6)–(20.9) has the form ($n = 1, 2, \ldots, N$)

$$U^{(n)}(r, s) = Q_0\Psi(s)\frac{q_n(s)}{\alpha_1\beta_1(s)Z(s)}\left[I_1(y_n)\,\kappa_1^n(s) + K_1(y_n)\,\kappa_2^n(s)\right], \tag{20.10}$$

where

$$y_n(r, s) = r\alpha_n\beta_n(s), \quad \beta_n(s) = \frac{s}{\sqrt{1 - D_1^{(n)}(s)}}, \quad n = 1, 2, \ldots, N,$$

functions $\kappa_1^n(s), \kappa_2^n(s)$ are determined from recurrence relations, in which the variable with a smaller number is expressed through the variable with a greater number:

$$\begin{pmatrix} \kappa_1^m \\ \kappa_2^m \end{pmatrix} = \begin{pmatrix} z_{11}^m & z_{12}^m \\ z_{21}^m & z_{22}^m \end{pmatrix}\begin{pmatrix} \kappa_1^{m+1} \\ \kappa_2^{m+1} \end{pmatrix}, \quad m = N-1, N-2, \ldots, 2, 1,$$

here

$$\begin{aligned}
\kappa_1^N &= -\left[w_N(1 - D_1^{(N)})K_1'(\beta_N) + (w_N - 1)(1 - D_2^{(N)})\tfrac{K_1(\beta_N)}{\beta_N}\right],\\
\kappa_2^N &= w_N(1 - D_1^{(N)})I_1'(\beta_N) + (w_N - 1)(1 - D_2^{(N)})\tfrac{I_1(\beta_N)}{\beta_N},\\
z_{11}^m &= I_1(b_2^m)\,\psi_{12}^m - K_1(b_1^m)\,\psi_{21}^m,\\
z_{12}^m &= -[K_1(b_1^m)\,\psi_{22}^m - K_1(b_2^m)\,\psi_{12}^m],\\
z_{21}^m &= -[I_1(b_2^m)\,\psi_{11}^m - I_1(b_1^m)\,\psi_{21}^m],
\end{aligned}$$

$z_{22}^m = I_1(b_1^m)\,\psi_{22}^m - K_1(b_2^m)\,\psi_{11}^m,$
and also

$$\psi_{11}^m = \frac{G_0^{(m)}\beta_m\alpha_m}{G_0^{(m+1)}\beta_{m+1}\alpha_{m+1}}\left[w_m(1-D_1^{(m)})I_1'(b_1^m) + (w_m-1)(1-D_2^{(m)})\frac{I_1(b_1^m)}{b_1^m}\right],$$

$$\psi_{12}^m = \frac{G_0^{(m)}\beta_m\alpha_m}{G_0^{(m+1)}\beta_{m+1}\alpha_{m+1}}\left[w_m(1-D_1^{(m)})K_1'(b_1^m) + (w_m-1)(1-D_2^{(m)})\frac{K_1(b_1^m)}{b_1^m}\right],$$

$$\psi_{21}^m = w_{m+1}\left(1-D_1^{(m+1)}\right)I_1'(b_2^m) + (w_{m+1}-1)\left(1-D_2^{(m+1)}\right)\frac{I_1(b_2^m)}{b_2^m},$$

$$\psi_{22}^m = w_{m+1}\left(1-D_1^{(m+1)}\right)K_1'(b_2^m) + (w_{m+1}-1)\left(1-D_2^{(m+1)}\right)\frac{K_1(b_2^m)}{b_2^m},$$

$$b_1^m = r_m\alpha_m\beta_m,\quad b_2^m = r_m\alpha_{m+1}\beta_{m+1},\quad m = 1, 2, \ldots, N-1,$$

$$Z = -\left(T_1\kappa_1^{(1)} + T_2\kappa_2^{(1)}\right),$$

$$T_1 = w_1(1-D_1^{(1)})I_1'(b_2^0) + (w_1-1)(1-D_2^{(1)})\frac{I_1(b_2^0)}{b_2^0},$$

$$T_2 = w_1(1-D_1^{(1)})K_1'(b_2^0) + (w_1-1)(1-D_2^{(1)})\frac{K_1(b_2^0)}{b_2^0},\quad b_2^0 = r_0\alpha_1\beta_1,$$

I_1, K_1 are imaginary argument Bessel function and Macdonald function of the first index, and I_1', K_1' are their derivatives of corresponding argument, and also

$$I_1'(y) = I_0(y) - I_1(y)/y,\quad K_1'(y) = -K_0(y) - K_1(y)/y.$$

Values of q_n are determined from the relations

$$q_1 \equiv 1,\quad q_n = \delta_1\delta_2, \ldots, \delta_{n-1},\quad n = 2, 3, \ldots N,$$

$$\delta_m = -\frac{G_0^{(m)}}{G_0^{(m+1)}r_m\beta_{m+1}\alpha_{m+1}}w_m\left(1-D_1^{(m)}\right),\quad m = 1, 2, \ldots, N-1$$

We shall note that in the expressions mentioned above the values α_n, $G_0^{(n)}$, r_m, r_0, w_n are constants. Other values are functions of complex argument s, which is omitted here. Transforms of radial and circular stresses in the layers are obtained using expression (20.10) and relations (20.9).

Below, we shall consider the case when the following condition is met for the function of external load

$$\lim_{\tau\to\infty} \psi(\tau) = 1, \tag{20.11}$$

from which it follows $\lim_{s\to 0} s\Psi(s) = 1$.

If $\Psi(s)$, $D_1^{(n)}(s)$, $D_2^{(n)}(s)$ do not have branching points (e.g., if $\Psi(s) = 1/s$ is Heaviside function transform, and relaxation kernels $\gamma_s^{(n)}(\tau)$, $\gamma_v^{(n)}(\tau)$ are regular and can be represented as finite sums of exponents), then transforms $U^{(n)}(r, s)$, determined by Formula (20.10), do not have branching points in complex plain, despite the fact that they are expressed through the functions which have branching points. This becomes clear after a number of cumbersome computations. At the same time, it is not necessary to carry out these computations as this fact follows from a general statement concerning solution in transforms branching points for the problems of considered class [11]. In this case, after asymptotic investigation of the transform near the poles set limit points, the solution in original is written in the form of infinite sum of residues [10].

In the case of singular kernels, the set of functions (20.10) branching points is a conjugation of sets of functions $\Psi(s)$, $D_1^{(n)}(s)$, $D_2^{(n)}(s)$ branching points. Then, for obtaining the originals of displacements, it is convenient to use formulae

$$u^{(n)}(r,\tau) = \frac{1}{2}u_0^{(n)}(r) + \frac{1}{\pi}\int_0^\infty \mathrm{Re}[U^{(n)}(r, i\omega)e^{i\omega\tau}]d\omega \tag{20.12}$$

and for the stresses originals—analogous relations

$$\sigma_{rr}^{(n)}(r,\tau) = \frac{1}{2}\sigma_{0rr}^{(n)}(r) + \frac{1}{\pi}\int_0^\infty \mathrm{Re}[S_{rr}^{(n)}(r, i\omega)e^{i\omega\tau}]d\omega, \tag{20.13}$$

$$\sigma_{\theta\theta}^{(n)}(r,\tau) = \frac{1}{2}\sigma_{0\theta\theta}^{(n)}(r) + \frac{1}{\pi}\int_0^\infty \mathrm{Re}[S_{\theta\theta}^{(n)}(r, i\omega)e^{i\omega\tau}]d\omega,$$

where $u_0^{(n)}(r)$, $\sigma_{0rr}^{(n)}(r)$, $\sigma_{0\theta\theta}^{(n)}(r)$ are displacement and stresses which are the solution of plane axisymmetric static problem of elasticity theory for the considered layered cylinder. Then, elastic properties of the layers in static problem are determined by long-term moduli, corresponding to viscoelastic kernels of dynamic problem. As condition (20.11) is met for the function of external load $\psi(\tau)$, integrands in (20.12), (20.13) do not have singularity at $\omega \to 0$.

It is important that Formulae (20.12), (20.13) are correct in a special case, when the external load is described by Heaviside function $\psi(\tau) = h(\tau)$. Then, using the corresponding convolution, it is easy to obtain solution at other $\psi(\tau)$, not necessarily meeting condition (20.11). Formulae (20.12), (20.13) can be used both for regular and singular hereditary kernels, but then at least one of the layers must possess hereditary properties. When all the layers are linearly elastic, Formulae (20.12), (20.13) are not applicable; however, in this case, the originals can be obtained easily in the form of infinite sum of residues.

Let us represent expressions for the functions $u_0^{(n)}(r)$. Corresponding static problem statement for these functions contains equations

$$\frac{d}{dr}\left[\frac{du_0^{(n)}(r)}{dr}+\frac{u_0^{(n)}(r)}{r}\right]=0, \quad r_0\le r\le 1, \quad (n=1,2,\dots,N),$$

boundary conditions:

$$\sigma_{0rr}^{(1)}(r_0)=-Q_0 \quad \sigma_{0rr}^{(N)}(1)=0,$$

and layers contact conditions ($m=1,2,\dots,N-1$):

$$u_0^{(n)}(r_m)=u_0^{(m+1)}(r_m), \quad G_0^{(m)}\sigma_{0rr}^{(m)}(r_m)=G_0^{(m+1)}\sigma_{0rr}^{(m+1)}(r_m),$$

here

$$\sigma_{0rr}^{(n)}(r)=w_n[1-D_{01}^{(n)}]\frac{du_0^{(n)}(r)}{dr}+(w_n-1)[1-D_{02}^{(n)}]\frac{u_0^{(n)}(r)}{r},$$

$$\sigma_{0\theta\theta}^{(n)}(r)=w_n[1-D_{01}^{(n)}]\frac{u_0^{(n)}(r)}{r}+(w_n-1)[1-D_{02}^{(n)}]\frac{du_0^{(n)}(r)}{dr},$$

where $D_{01}^{(n)}=D_1^{(n)}(0)$, $D_{02}^{(n)}=D_2^{(n)}(0)$.

The solution of this problem has the form

$$u_0^{(n)}(r)=Q_0\frac{q_{0n}}{\alpha_1\beta_{01}Z_0}\left[r\frac{\alpha_n}{2}\beta_{0n}\kappa_{01}^n+\frac{1}{r}\frac{1}{\alpha_n\beta_{0n}}\kappa_{02}^n\right], \quad n=1,2,\dots,N,$$

where

$$\beta_{0n}(s)=\frac{1}{\sqrt{1-D_{01}^{(n)}}}, \quad n=1,2,\dots,N,$$

constants κ_{01}^n, κ_{02}^n are determined from recurrence relations, in which the variable with a smaller number is expressed through the variable with a greater number:

$$\begin{pmatrix}\kappa_{01}^m\\ \kappa_{02}^m\end{pmatrix}=\begin{pmatrix}z_{11}^{0m} & z_{12}^{0m}\\ z_{21}^{0m} & z_{22}^{0m}\end{pmatrix}\begin{pmatrix}\kappa_{01}^{m+1}\\ \kappa_{02}^{m+1}\end{pmatrix}, \quad m=N-1,N-2,\dots,2,1,$$

here

$\kappa_{01}^N=\frac{1}{(\beta_{0N})^2}\left[w_N(1-D_{01}^{(N)})-(w_N-1)(1-D_{02}^{(N)})\right]$,

$\kappa_{02}^N = \frac{1}{2}\left[w_N(1 - D_{01}^{(N)}) + (w_N - 1)(1 - D_{02}^{(N)})\right],$
$z_{11}^{0m} = \frac{b_{02}^m}{2}\psi_{12}^{0m} - \frac{1}{b_{01}^m}\psi_{21}^{0m},$
$z_{12}^{0m} = \frac{1}{b_{02}^m}\psi_{12}^{0m} - \frac{1}{b_{01}^m}\psi_{22}^{0m},$
$z_{21}^{0m} = \frac{1}{2}[b_{01}^m\,\psi_{21}^{0m} - b_{02}^m\,\psi_{11}^{0m}],$
$z_{22}^{0m} = \frac{b_{01}^m}{2}\psi_{22}^{0m} - \frac{1}{b_{02}^m}\psi_{11}^{0m},$
and also
$\psi_{11}^{0m} = \frac{G_0^{(m)}\beta_{0m}\alpha_m}{2G_0^{(m+1)}\beta_{0m+1}\alpha_{m+1}}[w_m(1 - D_{01}^{(m)}) + (w_m - 1)(1 - D_{02}^{(m)})],$
$\psi_{12}^{0m} = \frac{G_0^{(m)}\beta_{0m}\alpha_m}{G_0^{(m+1)}\beta_{0m+1}\alpha_{m+1}} \cdot \frac{1}{(b_{01}^m)^2}[(w_m - 1)(1 - D_{02}^{(m)}) - w_m(1 - D_{01}^{(m)})],$
$\psi_{21}^{0m} = \frac{1}{2}[w_{m+1}(1 - D_{01}^{(m+1)}) + (w_{m+1} - 1)(1 - D_{02}^{(m+1)})],$
$\psi_{22}^{0m} = \frac{1}{(b_{02}^m)^2}[(w_{m+1} - 1)(1 - D_{02}^{(m+1)}) - w_{m+1}(1 - D_{01}^{(m+1)})],$
$b_{01}^m = r_m\alpha_m\beta_{0m},$
$b_{02}^m = r_m\alpha_{m+1}\beta_{0m+1},$
$m = 1, 2, \ldots, N - 1,$
$Z_0 = -(T_{01}\kappa_{01}^{(1)} + T_{02}\kappa_{02}^{(1)}),$
$T_{01} = \frac{1}{2}[w_1(1 - D_{01}^{(1)}) + (w_1 - 1)(1 - D_{02}^{(1)})],$
$T_{02} = \frac{1}{(b_{02}^0)^2}[(w_1 - 1)(1 - D_{02}^{(1)}) - w_1(1 - D_{01}^{(1)})],$
$b_{02}^0 = r_0\alpha_1\beta_{01},$
Values q_{0n} are determined from the relations:

$$q_{01} = 1, \quad q_{0n} = \delta_{01}\delta_{02}, \ldots, \delta_{0n-1}, \quad n = 2, 3, \ldots, N,$$

$$\delta_{0m} = -\frac{G_0^{(m)}w_m}{G_0^{(m+1)}r_m\alpha_{m+1}\beta_{0m+1}}(1 - D_{01}^{(m)}), \quad m = 1, 2, \ldots, N - 1$$

We shall omit here expressions for $\sigma_{rr}^{0(n)}(r)$, $\sigma_{\theta\theta}^{0(n)}(r)$.

20.4 Calculation Results

Let us represent some characteristic results, obtained with the help of calculations carried out by Formulae (20.12), (20.13), in graphical form of τ-dependencies of values, $s_r^{(n)} = G_0^{(n)}\sigma_{rr}^{(n)}(r, \tau)/(G_0^{(1)}Q_0)$, $s_\theta^{(n)} = G_0^{(n)}\sigma_{\theta\theta}^{(n)}(r, \tau)/(G_0^{(1)}Q_0)$ at fixed points r for two-layered cylinder ($N = 2, n = 1, 2$). Plots in Figs. 20.2, 20.3, 20.4 and 20.5 correspond to calculations for the following initial data: $r_0 = 0.5, r_1 = 0.7$,

$$G_0^{(2)}/G_0^{(1)} = 1/8, \quad \nu_0^{(1)} = 0.3, \quad \nu_0^{(2)} = 0.33, \quad \alpha_2 = c_2/c_1 = 0.64$$

(internal layer is much stiffer),

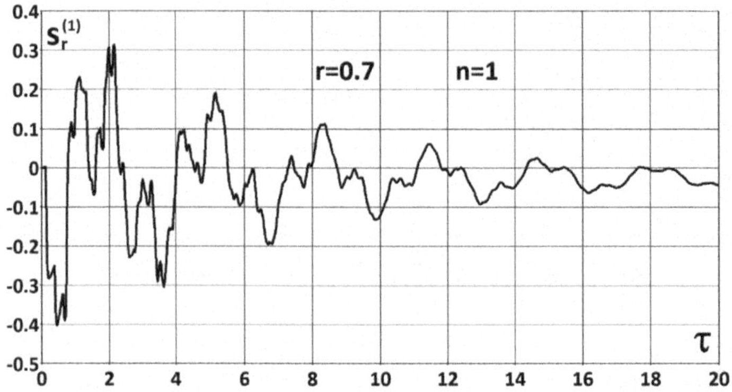

Fig. 20.2 Time dependence of $s_r^{(1)}$ at $r = 0.7$

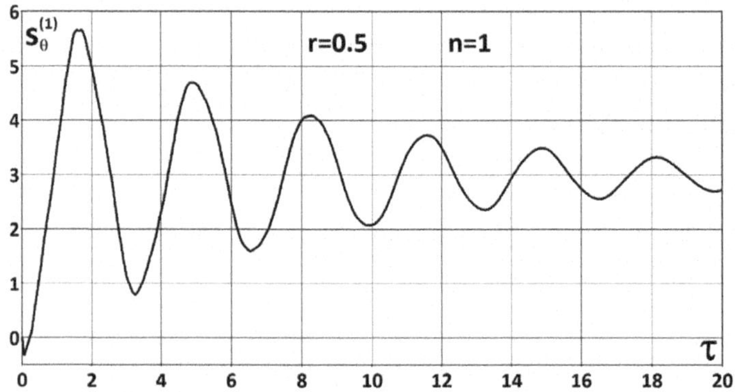

Fig. 20.3 Time dependence of $s_\theta^{(1)}$ at $r = 0.5$

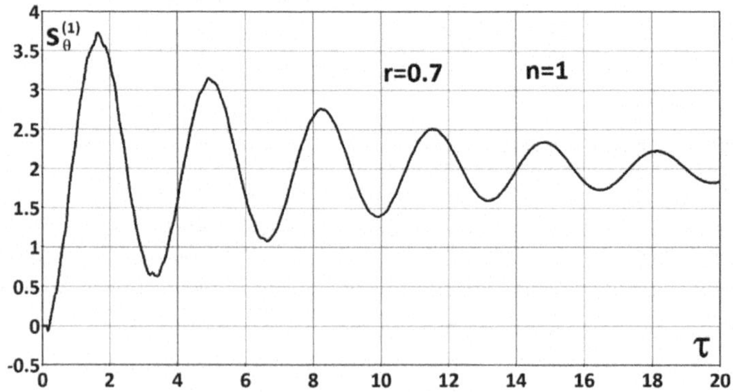

Fig. 20.4 Time dependence of $s_\theta^{(1)}$ at $r = 0.7$

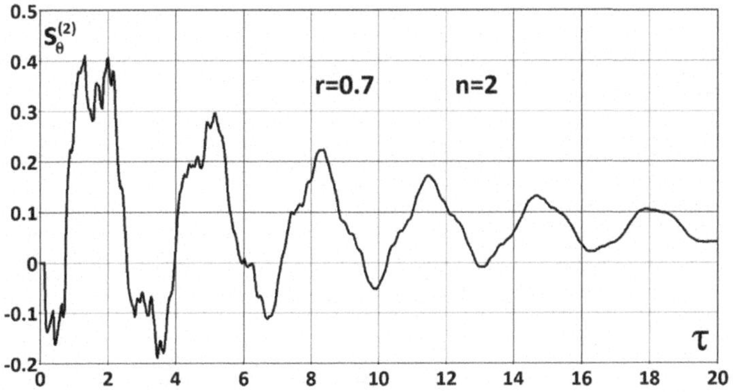

Fig. 20.5 Time dependence of $s_\theta^{(2)}$ at $r = 0.7$

$$\gamma_v^{(1)}(\tau) = \gamma_v^{(2)}(\tau) \equiv 0, \quad \gamma_s^{(1)}(\tau) = 0.3e^{-\tau}\tau^{-0.2}, \quad \gamma_s^{(2)}(\tau) = 0.5e^{-0.9\tau}\tau^{-0.4};$$

time dependence of external load has the form of "smooth step" $\psi(\tau) = 1 - e^{-50\tau}$ $(\tau > 0)$.

Hereinafter, negative stresses are compressional. From Figs. 20.3 and 20.4, it is obvious that circular stresses in a stiffer internal layer practically do not react to excitations brought by the waves reflected from the boundary of media, as well as from free or loaded boundary. At the same time, at decaying amplitude of circular stress, its time change is close to periodic one with a period corresponding to the first eigenfrequency of free oscillations of considered cylinder in the absence of external load (it is checked by the calculations). Radial stress, as well as circular stress on the contact surface from the side of a softer layer, at the initial stage of transient process, is sensitive to perturbations brought from the free or loaded boundary of the cylinder. At the same time, their time change is also close to periodic one with decaying amplitude.

Figures 20.6, 20.7, 20.8 and 20.9 represent the plots obtained for that same values of r_0, r_1 and load $\psi(\tau)$, as the plots in Figs. 20.2, 20.3, 20.4 and 20.5, but for another distribution of layers material properties. Thick lines correspond to the case

$$G_0^{(2)}/G_0^{(1)} = 8, \quad \nu_0^{(1)} = 0.33, \quad \nu_0^{(2)} = 0.3, \quad \alpha_2 = c_2/c_1 = 1.57,$$

$$\gamma_v^{(1)}(\tau) = \gamma_v^{(2)}(\tau) \equiv 0, \quad \gamma_s^{(1)}(\tau) = 0.5e^{-0.9\tau}\tau^{-0.4}, \quad \gamma_s^{(2)}(\tau) = 0.3e^{-\tau}\tau^{-0.2},$$

i.e., layers materials changed their places and now the external layer became stiffer. Thin lines refer to the results for a homogeneous cylinder, when at calculations, it was taken that:

$$r_0 = 0.5, \quad \nu_0^{(1)} = \nu_0^{(2)} = 0.3, \quad G_0^{(2)}/G_0^{(1)} = 1, \quad \alpha_2 = c_2/c_1 = 1,$$

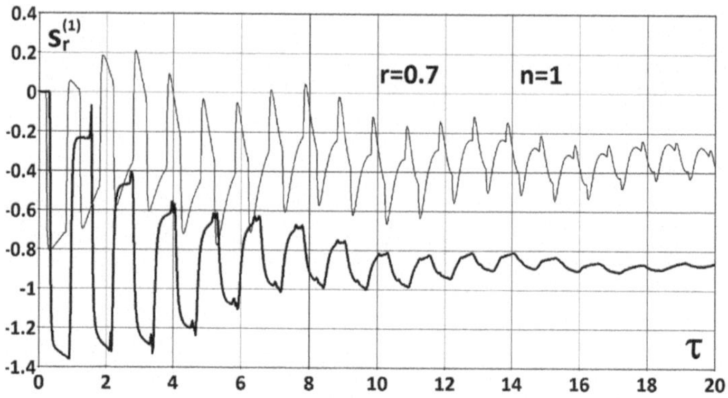

Fig. 20.6 Time dependence of $s_r^{(1)}$ at $r = 0.7$

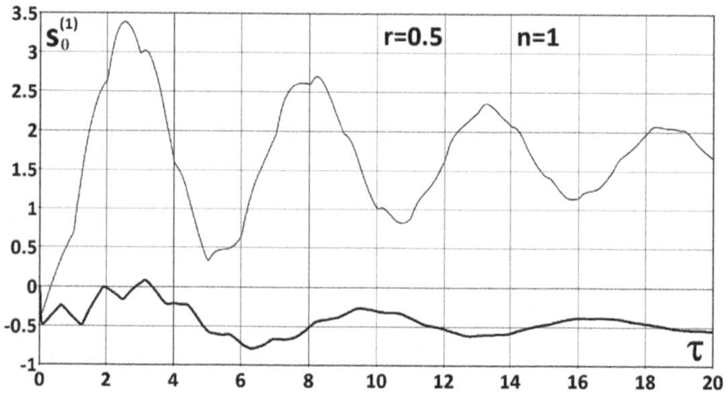

Fig. 20.7 Time dependence of $s_\theta^{(1)}$ at $r = 0.5$

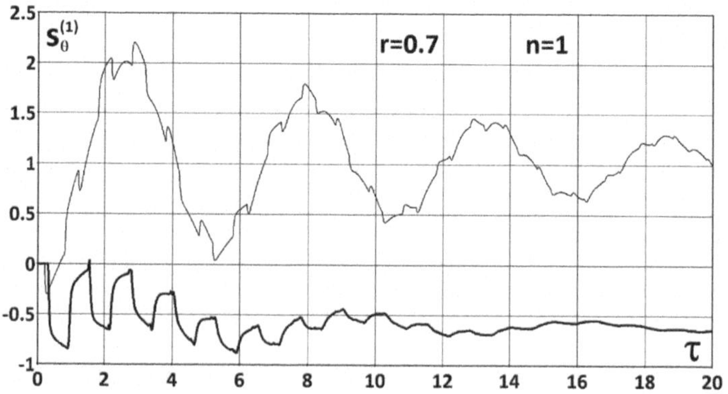

Fig. 20.8 Time dependence of $s_\theta^{(1)}$ at $r = 0.7$

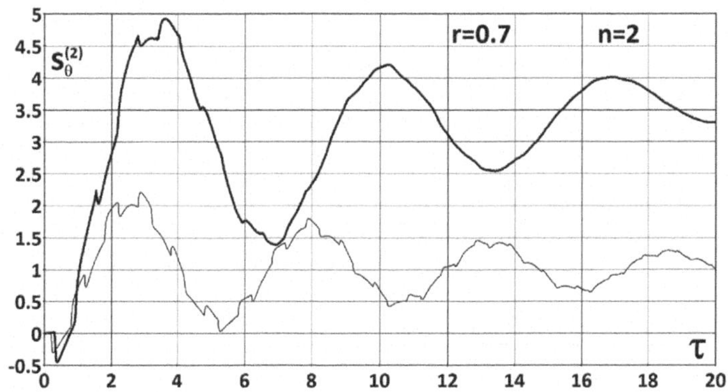

Fig. 20.9 Time dependence of $s_\theta^{(2)}$ at $r = 0.7$

$$\gamma_v^{(1)}(\tau) = \gamma_v^{(2)}(\tau) \equiv 0, \quad \gamma_s^{(1)}(\tau) = \gamma_s^{(2)}(\tau) = 0.3e^{-\tau}\tau^{-0.2}, \quad \psi(\tau) = 1 - e^{-50\tau}.$$

Now on layers contact boundary, we can observe less sensitivity of radial stress to the perturbations from cylinder boundaries. Besides, we should mention that compared to homogeneous material, layers existence drastically changes the character of wave process, so that compressional circular stresses appear in a softer internal layer because of internal pressure action on a two-layered cylinder.

20.5 Conclusion

The problem solution in an integral form (20.12), (20.13) is convenient for numerical realization. It allows investigating the process of wave propagation in multilayer cylinders with arbitrary number of viscoelastic coaxial layers without any simplifying assumptions concerning relation between hereditary kernels. Note that the more hereditary properties are manifested in the layers, the more effective application of Formulas (20.12), (20.13) is. But if all the layers are linear elastic, these formulae are not applicable, but then the solution can be obtained easily by calculating residues.

Calculation results represented above illustrate the influence of two contacting materials physical-mechanical parameters difference on transient wave process in cylindrical solid. The results obtained can be used not only directly for the purpose of wave propagation in layered cylindrical structure elements investigation, but also for testing various numerical algorithms developed for dynamics of piecewise homogeneous solids investigation.

The work was supported by RFBR (project No. 18-08-00471-a) and RFBR and Moscow City Government (project No. 19-38-70005 mol_a_mos).

References

1. Bazhenov, V.G., Belov, A.A., Igumnov, L.A.: Boundary-element modeling of piecewise homogeneous media and structures dynamics. N. Novgorod, 180 p. (2009)
2. Bazhenov, V.G., Igumnov, L.A.: Methods of boundary integral equations and boundary elements in solution of three-dimensional dynamic elasticity theory with conjugated fields problems. M.: Physmathlit, 252 p. (2008)
3. Belov, A.A.: Boundary-element analysis of composite viscoelastic bodies dynamics. Problemy Prochnosti I Plastichnosti: Mezhvuzovskij Sbornik. N. Novgorod: NNSU. issue 70, 164–170 (2008)
4. Hosseini, H.S., Bakhshi, K.H.: Dynamic behavior of multi-layered viscoelastic nanobeam system embedded in a viscoelastic medium with a moving nanoparticle. J. Mech. **33**(5), 559–575 (2017)
5. Lee, H.S.: Viscowave—a new solution for viscoelastic wave propagation of layered structures subjected to an impact load. Int. J. Pavement Eng. **15**(6), 542–557 (2014)
6. Kozlov, V.I., Kucher N.K.: Dynamic behavior of multilayer cylindrical structures with transient loads. Strength Mater. **12**(5), 639–648 (1980). [in Russian]
7. Lokshin, A.A.: The head wave at the boundary of two hereditary-elastic half-spaces. The case of a linear source. J. Appl. Math. Mech. **58**(1), 171–176 (1994)
8. Nuriev, B.R.: Impact on a viscoelastic layered composite. Izv. Akad. Nauk AzSSR. Ser. Fiz-Tekh. Mat. Nauk **4**, 35–41 (1985). [in Russian]
9. Petrov, A.N., Ermolaev, M.D.: Calculation of composite viscoelastic solids dynamics using boundary element method. Mechanics of deformed solid. Vestnik Nizhegorodskogo universiteta n.a. N.I. Lobachevskii. **4**(4), 1694–1696 (2011)
10. Pshenichnov, S.G.: Analytic solution of one-dimensional problems in the dynamics of piecewise-homogeneous viscous-elastic bodies. Izv. AN SSSR. Mekhanika Tverdogo Tela. **26**(1), 89–96 (1991)
11. Pshenichnov, S.G.: Dynamic linear viscoelasticity problems for piecewise homogeneous bodies. Mech. Solids **51**(1), 65–74 (2016)
12. Sabodash, P.F.: Propagation of longitudinal viscoelastic waves in a three-layer medium. Polymer Mech. **7**(1), 124–128 (1971). [in Russian]
13. Songnan, L., Ping, G.: Dynamic response of layered viscoelastic half-space and its application to dynamic foundation problems. Hubnan Daxue Xuebao. J. Hunan Univ. **20**(1), 57–64 (1993)

Chapter 21
Evaluating Contact Stresses for an Impactor Penetrating Soil, Accounting for Dynamic Compressibility, Internal Friction, and Initial Strength of the Soil

Vasily L. Kotov, Vladimir V. L. Balandin, Andrey K. Lomunov, and Leonid A. Igumnov

Abstract Formulas for evaluating contact stresses for an impactor penetrating soil, accounting for friction, are obtained. In analyzing dynamic deformation of soil, its compressibility, shear resistance and initial strength are taken into account. Impact compressibility of the medium is described based on Hugoniot's linear adiabat. Plastic deformation obeys the Mohr-Coulomb yield criterion with a constraint on a maximal value of tangential stresses according to Tresca's criterion. An earlier obtained analytical solution of the one-dimensional problem of a cavity expanding at a constant velocity from a point in a half-space occupied by a plastic medium is used. A formula for determining critical pressure, minimal pressure required for the formation of a cavity, accounting for internal pressure in the framework of Mohr-Coulomb yield criterion, is also applied, which is a generalization of a known solution for an elastic ideally plastic medium with Tresca's criterion. The obtained evaluations of resistance to an impactor penetrating soil are based on a quadratic relation between pressure normal to the surface of the impactor and impact velocity. Finite expressions for coefficients of the trinomial approximation as a function of experimentally determined physical-mechanical parameters of the soil—shock adiabatic and yield strength-pressure relations—have been determined for the first time. The

V. L. Kotov (✉)
National Research Lobachevsky State University of Nizhny Novgorod, 23 Gagarin Avenue, Nizhny Novgorod, Russian Federation 603950
e-mail: vkotov@inbox.ru

V. V. L. Balandin · A. K. Lomunov · L. A. Igumnov
Research Institute for Mechanics, National Research Lobachevsky State University of Nizhny Novgorod, 23 Gagarin Avenue, Building 6, Nizhny Novgorod, Russian Federation 603950
e-mail: rustydog2007@yandex.ru

A. K. Lomunov
e-mail: lomunov@mech.unn.ru

L. A. Igumnov
e-mail: igumnov@mech.unn.ru

H. Altenbach et al. (eds.), *Multiscale Solid Mechanics*,
Advanced Structured Materials 141,
https://doi.org/10.1007/978-3-030-54928-2_21

derived formulae have been verified by comparing them with already known experimental data on the penetration of a steel conical impactor into frozen sandy soil. It is shown that the difference between the numerical and experimental results is within 15%.

Keywords Soil · Strength · Internal friction · Compressibility · Spherical cavity expansion · Mohr-Coulomb-Tresca plasticity condition · Penetration · Conical impactor

21.1 Introduction

In the theory of high-velocity impact, special attention is paid to determining the parameters of contact interaction of the bodies. Practical experience shows [1–5] that a solution of the problem of spherical or cylindrical cavity expansion is a good approximation of pressure on the contact interface between a rigid impactor and a resisting medium. The cavity expansion problem is a classical problem of mechanics of deformable bodies, the analysis of which was the subject of the works by R. Hill, R. Bishop, H. Hopkins, A. Ya. Sagomonyan, N. V. Zvolinskiy et al. There exist many formulations of this problem, differing in the way the material compressibility is accounted for—a linearly compressible medium [6], an incompressible medium or a medium of limited ultimate strain [7, 8]; in the assumed yield criterion—Tresca's [6, 9], Mohr-Coulomb's [8], Mohr-Coulomb's with Tresca's constraint [7]. In [10–13], approximations of pressure on the cavity wall as a function of its expansion velocity were obtained for varying mechanical characteristics of the material, and the obtained solutions were parametrically analyzed. For an incompressible elastoplastic medium, analytical solutions for various kinds of yield criterion were obtained.

To determine stress and velocity fields in the region of plastic deformation adjacent to the cavity, an effective algorithm [6, 7] of numerical analysis was developed, that makes it possible to obtain a solution of the problem with an accuracy sufficient for practical application. Numerical solutions of the spherical cavity expansion problem and their application to evaluating contact stresses and forces resisting penetration of rigid bodies into soils, concretes and metals are given in [6–19]. In [20, 21], an analytical solution of the problem was obtained, based on the assumption of incompressibility of the medium behind the shockwave front. Dynamic compressibility of a medium is characterized by the shock adiabat, and shear resistance is defined by the Mohr-Coulomb plasticity condition. Based on the introduced analytical solution, a methodology for evaluating forces resisting penetration of a rigid body into a soft soil was developed [22]. Relations for the maximal force resisting penetration of a rigid sphere and an impactor with a conical head into dry and water-saturated sand were obtained [23]. Comparison of the analytical, numerical and experimental evaluations of forces resisting penetration demonstrates their good qualitative and quantitative agreement for impact velocities in the range of 50–400 m/s.

Papers [24–26] give both experimental and theoretical data testifying to the limited nature of yield strength of soils under high pressures. In [24] it is shown that a nonlinear relation between yield strength and pressure, taking into account scatter of the data and measuring inaccuracies, can be represented by a two-part broken line—a linear relation for small pressures, as it is assumed in the Mohr-Coulomb plasticity condition, and a constraint on the value of yield strength for high pressures, according to Tresca's condition. In this connection, problems of penetration of rigid bodies into soft soils can be effectively analyzed using an analytical solution of the problem of spherical cavity expansion in a medium with the Mohr-Coulomb-Tresca plasticity condition [18, 23].

21.2 The Model of Cavity Expansion in a Soil Medium with the Mohr-Coulomb-Tresca Plasticity Condition

A one-dimensional self-similar model of a spherical cavity expanding from a point in an infinite elastoplastic medium at constant velocity V is considered. A mathematical model of dynamic deformation of the elastoplastic medium is defined by a system of differential equations expressing laws of continuity and change of momentum, which is written in spherical Euler coordinates r, ϕ, φ.

$$\rho\left(\frac{\partial \upsilon}{\partial r}+2\frac{\upsilon}{r}\right)=-\left(\frac{\partial \rho}{\partial t}+\upsilon\frac{\partial \rho}{\partial r}\right),\ \frac{\partial \sigma_r}{\partial r}+2\frac{(\sigma_r-\sigma_\phi)}{r}=-\rho\left(\frac{\partial \upsilon}{\partial t}+\upsilon\frac{\partial \upsilon}{\partial r}\right)$$

Experimentally determined finite relations close the system of partial differential equations

$$\sigma_r = f_1(\theta), \sigma_r - \sigma_\varphi = f_2(\theta) \tag{21.1}$$

where υ is velocity of particles of the medium, σ_r and $\sigma_\phi = \sigma_\varphi$ are radial, circumferential and meridional tensor components of Cauchy stresses (assumed positive in compression) $\theta = 1 - \rho_0/\rho$ is volumetric strain, ρ_0 and ρ are initial and current densities of the medium, functions f_1 and f_2 define the equations of state and the plasticity condition of the medium.

Experiments [27–29] show that dynamic compressibility of many of the media are characterized by Hugoniot's shock adiabat in the form of linear relation:

$$U_s = C_0 + s u_p, \tag{21.2}$$

Correlating plane shockwave velocity U_s and velocity of particles behind the wave front u_p. Here, C_0 corresponds to the sonic velocity in the medium under zero pressure, and s is constant.

Use of relation (21.2) and the Renkin-Hugoniot conditions yields a correlation between stress σ_r and volumetric strain θ, which will have the form $f_1(\theta) \equiv \rho_0 C_0^2 \theta (1 - s\theta)^{-2}$, where constant s characterizes ultimate compressibility of the medium.

A solution of the problem in the plastic yield region defined by radii $r = Vt$ and $r = ct$ is obtained in the following assumptions:

- the medium is assumed to be rigid-plastic (elastic deformation of a soft soil medium is neglected), i.e., the plastic yield region is adjacent to the region of the unperturbed medium, where $\sigma_r = \upsilon = 0$;
- the solution is a plastic shockwave (SW) propagating over the unperturbed half-space;
- density ρ_s behind the shockwave front is assumed constant, i.e., incompressibility behind the shockwave front is assumed, the value of ρ_s is a function of cavity expansion velocity;
- propagation velocity of the plastic wave front c is equal to the propagation velocity of the plane shockwave front, defined by Hugoniot's linear adiabat (21.2), i.e., it is assumed that $U_s \equiv c$.

Such assumptions are considered to be justified for soft soils possessing relatively low cohesion and high porosity.

Using self-similar substitution $\xi = r/ct$, the system of partial differential equations is reduced to a system of ordinary differential equations (ODE). The boundary-value problem for a system of two first-order ODE's for dimensionless velocity $U = \upsilon/c$ and dimensional stress $S = \sigma_r / \rho_s c^2$, taking into account the above assumptions, will assume the following form:

$$U' + 2\frac{U}{\xi} = 0, \; U(\xi = \varepsilon) = \varepsilon, \; U(\xi = 1) = \theta_s$$

$$S' + 2\frac{\tilde{f}_2}{\xi} = (\xi - U)U', \; S(\xi = 1) = \theta_s - \theta_s^2,$$

where: $\varepsilon = V / c$ is value of the dimensionless coordinate, corresponding to the cavity boundary, $\tilde{f}_2 = f_2 / \rho_s c^2$ is dimensionless function in the elasticity condition, the volumetric strain along the SW front will assume the value $\theta_s = \left(1 - C_0 / c\right) / s$, where the stroke designates differentiation for ξ. Equation system (21.3)–(21.7), beside dimensionless velocity U and dimensionless stress S, includes unknown parameter c—propagation velocity of the plastic shockwave front.

The plasticity function for the Mohr-Coulomb condition, with the account of the constraint on the maximal value of Tresca's yield strength, has the following form:

$$f_2 \equiv \begin{cases} \tau_0 + \mu\sigma_r, & 0 < \sigma_r \le \sigma_M, \\ \tau_M, & \sigma_r > \sigma_M \end{cases}$$

where $\sigma_M = (\tau_M - \tau_0) / \mu$.

An analytical solution of the problem of a spherical cavity expanding in a rigid-plastic medium with the Mohr-Coulomb-Tresca plasticity condition in the assumption of incompressibility behind the shockwave front was obtained in [18]. According to the solution of [18], the stress along the cavity boundary will be defined by equation

$$\sigma(V)=\begin{cases}\frac{\tau_0}{\mu}\left(\varepsilon^{-2\mu}-1\right)+\frac{\rho_0 V^2}{1-\varepsilon^3}\left(\frac{3}{(\mu-2)(2\mu-1)}+\frac{2\mu+1}{2\mu-1}\varepsilon^{1-2\mu}-\frac{\mu-1}{\mu-2}\varepsilon^{4-2\mu}\right), & 0<V<V_0;\\ \sigma_M+\tau_M\ln\left(\frac{\xi_M}{\varepsilon}\right)^2+\frac{\rho_0 V^2}{1-\varepsilon^3}\left(\frac{3}{2}-\frac{2\varepsilon}{\xi_M}+\frac{\varepsilon^4}{2\xi_M^4}\right), & V_0\le V\le V_M,\\ -2\tau_M\ln\varepsilon+\frac{\rho_0 V^2}{1-\varepsilon^3}\left(\frac{3}{2}-\varepsilon-\frac{\varepsilon^4}{2}\right), & V>V_M.\end{cases} \tag{21.3}$$

where $\xi=\xi_M$ is value of the dimensionless coordinate, for which equality $\sigma_r=\sigma_M$ holds, V_0 and V_M are cavity expansion velocities defining the upper and lower limits, respectively, of the scope of use of the Mohr-Coulomb and Tresca's condition. To determine the plastic wave front velocity, an approximate formula is used:

$$c=\sqrt[3]{s}V+C_0/3 \tag{21.4}$$

Cavity expansion velocities V_M and V_0 are determined as follows. It follows from the relations on the SW for $\xi=1$ that $\sigma_r=\rho_0 cv=\rho_0 c^2\theta$. The values of SW velocity, volumetric strain and ε, corresponding to $V=V_M$, will be designated as c_M, θ_M and ε_M. If c_M is assumed to be defined by formula (21.4), then

$$c_M=\sqrt[3]{s}V_M+C_0/3,\ \theta_M=\varepsilon_M^3=\frac{V_M^3}{c_M^3},\ \sigma_M=\rho_0 c_M^2\theta_M=\rho_0\frac{V_M^3}{c_M}$$

and V_M will be determined from the following cubic equation:

$$\left(\frac{\tau_M-\tau_0}{\mu}\right)\left(\sqrt[3]{s}V_M+C_0/3\right)=\rho_0 V_M^3 \tag{21.5a}$$

Similarly, the value of cavity expansion velocity, for which equation $\sigma_r(\xi=\varepsilon)=\sigma_M$ holds, will be designated by V_0. To determine V_0, it is required to solve nonlinear equation

$$\begin{aligned}\frac{\tau_M-\tau_0}{\mu}&=\frac{\tau_0}{\mu}\left(1-\varepsilon_0^{-2\mu}\right)+\frac{\rho_0 V_0^2}{1-\varepsilon_0^3}\left(\frac{3}{(\mu-2)(2\mu-1)}\right.\\&\left.+\frac{2\mu+1}{2\mu-1}\cdot\varepsilon_0^{1-2\mu}-\frac{\mu-1}{\mu-2}\cdot\varepsilon_0^{4-2\mu}\right),\end{aligned} \tag{21.5b}$$

where $\varepsilon_0=V_0/c_0$, $c_0=\sqrt[3]{s}V_0+C_0/3$.

According to the widely used model of local interaction (LIM), expression (21.3) for the pressure on the cavity boundary is adopted as contact pressure in the impactor/resisting medium interaction. A projection of the motion velocity vector of the impactor on the normal to an element on the surface of the body is taken as cavity expansion velocity. Such an approach was used to evaluate forces resisting penetration of spherical and conical impactors into dry and wet sandy soils in the penetration velocity range of 50–450 m/s, providing good agreement with the experimental data [23].

21.3 Approximating the Stress-Velocity Relation

Analysis of expression (21.4) shows that, for zero velocity, the value of $\varepsilon = V/c$ also becomes equal to zero. At the same time, the stress on the cavity boundary (21.3) will tend to infinity. Thus, for low velocities, expression (21.3) cannot be used to evaluate stress.

Earlier, the issue of evaluating a minimal stress on the cavity boundary, required for its expansion (critical pressure), was considered. The Cauchy problem was obtained, the solution of which determines critical pressure in a linearly compressible elastic ideally plastic medium with Tresca's [12] and Mohr-Coulomb [17] plasticity conditions.

$$\frac{d\sigma_r}{d\xi} + 2\frac{\tau_0 + \mu\sigma_r}{\xi} = 0,\ \varepsilon \leq \xi \leq 1,\ \sigma_r(\xi = 1) = 2\tau_0/3$$

The solution of the Cauchy problem is function $\sigma_r(\xi) = \frac{2}{3}\tau_0\xi^{-2\mu} + \tau_0\left(\frac{\xi^{-2\mu}-1}{\mu}\right)$, which, for $\xi = \varepsilon$, defines critical pressure.

In the present paper, it is proposed to use this critical pressure value as the value of stress for zero velocity

$$\sigma_C = \frac{2}{3}\tau_0\varepsilon_C^{-2\mu} + \tau_0\left(\frac{\varepsilon_C^{-2\mu} - 1}{\mu}\right),\ \varepsilon_C = \sqrt[3]{\frac{3(1-\nu)\tau_0}{E}} \quad (21.6a)$$

where E is Young modulus, ν is Poisson's coefficient. The value of critical pressure for $\mu = 0$ can be found by passing to the limit $\mu \to 0$ in (21. 6a)

$$\sigma_C = \frac{2}{3}\tau_0\left(1 + \ln\varepsilon_C^{-3}\right) \quad (21.6b)$$

In [17], it is shown that the approximation inaccuracy of formula (21.6) is within 6% for the values of the internal friction coefficient changing all over the admissible range of $0 < \mu < 0.75$ and the initial value of yield strength changing by three orders of magnitude in the range of $0.01 < \tau_0 < 10$ MPa.

When the velocity tends to infinity, the value of $\varepsilon < \varepsilon_s$ remains limited, $\varepsilon_s = s^{-1/3}$. Analysis of Eq. (21.3) shows that the stress on the cavity boundary for expansion velocities $V > C_0$ will satisfy inequality

$$\rho_0 V^2 < \sigma < 1.1\rho_0 V^2 \tag{21.7}$$

for the variation of the quantity characterizing ultimate compressibility of the soil in a sufficiently wide range of $1 < s < 4$. Thus, using (21.7), the coefficient with the squared velocity in the trinomial quadratic stress-velocity relation can be determined. Taking into account (21.6), the following expression, for example, is obtained:

$$\sigma(V) = \sigma_C + 1.05\rho_0 V^2 \tag{21.8a}$$

The constructed approximation of the normal stress-velocity relation of the form of (21.8) approximates accurately enough stresses on the cavity boundary, both for low and high velocities. For better agreement with (21.3), in the middle velocity range, approximation (21.8) can be completed with a linear member. The coefficient with a linear-velocity member can be determined using the least squares method in the velocity range of interest. The coefficient can also be chosen based on the condition of equality of the values of stresses for $V = V_M$. It should be kept in mind that, for $V > V_M$, stresses over the entire range of plastic yield satisfy Tresca's criterion.

Thus, the resulting form of the sought trinomial quadratic approximation of the stress-velocity relation will be

$$\sigma(V) = \gamma + \beta V + \alpha V^2 \tag{21.8b}$$

$$\gamma = \frac{2}{3}\tau_0 \varepsilon_C^{-2\mu} + \tau_0\left(\frac{\varepsilon_C^{-2\mu} - 1}{\mu}\right),\ \beta = -\frac{2\tau_M \ln \varepsilon_M}{V_M},$$

$$\alpha = \frac{\rho_0}{1 - \varepsilon_M^3}\left(\frac{3}{2} - \varepsilon_M - \frac{\varepsilon_M^4}{2}\right)$$

where ε_C is determined according to Eq. (21.6), $\varepsilon_M = V_M \big/ \left(\sqrt[3]{s}V_M + C_0/3\right)$, V_M is determined by solving cubic Eq. (21.5a), using Cardano formulae, and parameter β is assumed to be large or equal to zero.

21.4 Results of Numerical Verification

Numerical implementation of relations (21.8) is demonstrated, using an example of a soil medium with the shock adiabat having parameters $C_0 = 1700$ m/s, $s = 3.4$. The parameters of the Mohr-Coulomb-Tresca plasticity condition will vary. The function in the yield criterion will be appropriately represented as a function of pressure

$$f_2(p) \equiv \begin{cases} \tau_0 + kp, & 0 < p \le p_M, \\ \tau_M, & p > p_M \end{cases}$$

where $k = \mu / (1 - 2\mu/3)$ is internal friction coefficient.

Two possible choices of yield strength will be considered. In the first choice, the maximal value of yield strength varies, $\tau_M = 20$, 35 and 50 MPa, with fixed values of $\tau_0 = 3$ MPa and $k = 0.3$. In the second choice, $\tau_0 = 5$ MPa and $\tau_M = 50$ MPa are fixed, while coefficient k assumes values 0.3, 0.45 and 0.6. These relations are presented in Fig. 21.1.

Figure 21.2 shows the values of critical stress (a stress required for the cavity to expand at a velocity close to 0) as a function of the initial value of yield strength (a) and as a function of the internal friction coefficient (b). The exact solution was obtained using the algorithm [6, 7] as a result of numerically solving [17] a boundary-value problem for a system of two first-order ordinary differential equations according to the fourth-order-accuracy Runge-Kutta method. The approximate solution was obtained using formulae (21.6) and differs from the exact one by not more than 2–3%.

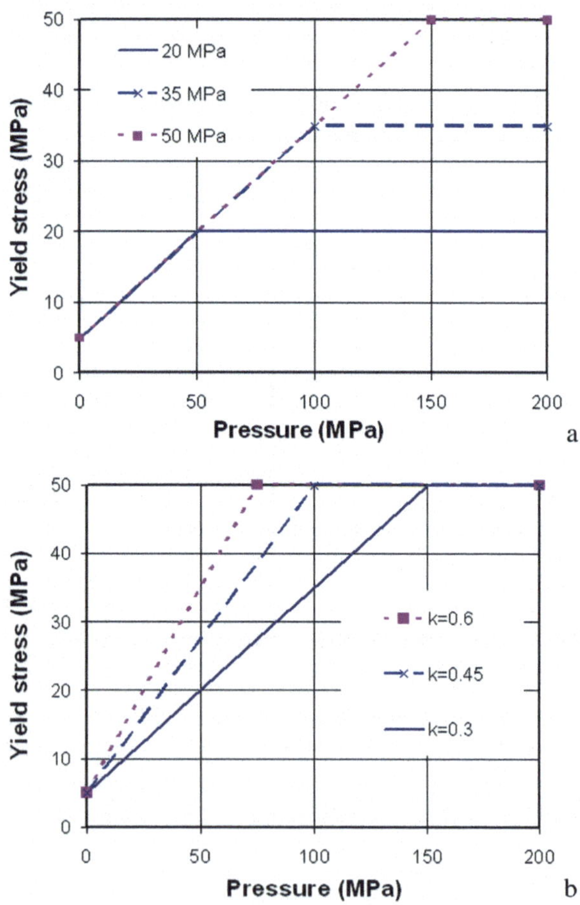

Fig. 21.1 Versions of yield strength/Mohr-Coulomb-Tresca pressure relations: varied maximal value of yield strength (**a**) and internal friction coefficient (**b**)

Fig. 21.2 Critical stress on the cavity wall as a function of initial value of yield strength τ_0 **a** and as a function of internal friction coefficient k **b** an exact solution (solid line) and an approximation using formulae (21.6)

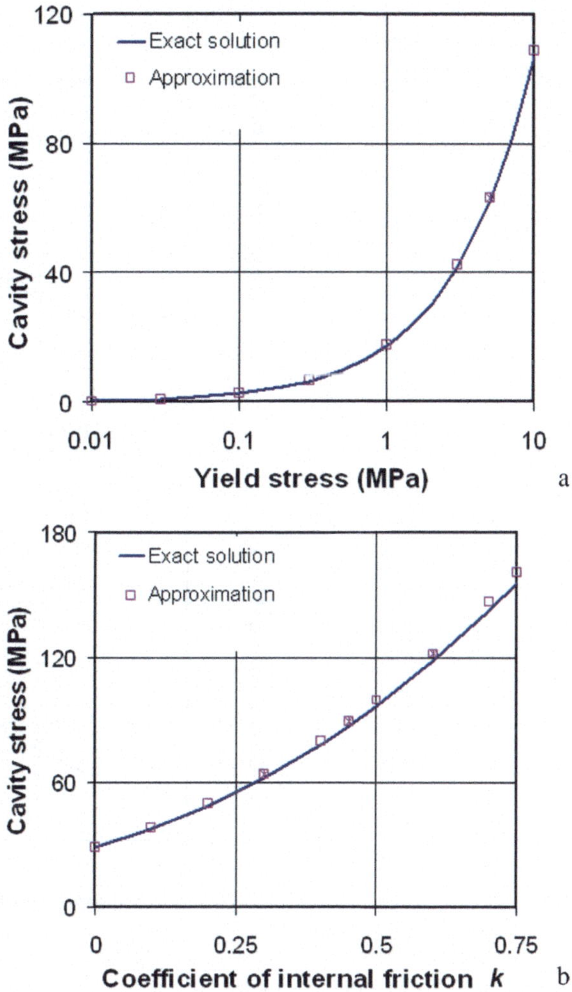

For the first choice of yield strength (Fig. 21.1a), the obtained value of critical pressure is $\sigma_C = 63.5$ MPa (shown by a cross in Fig. 21.2a). The values of stress are $\sigma_M = 80$, 120 and 180 MPa, with the constraints on the velocity range $V_0 = 97.4$, 144 and 194 m/s, $V_M = 310$, 408 and 481 m/s.

For the second choice of yield strength (Fig. 21.1b), the obtained values of critical pressure are $\sigma_C = 63.5$, 89.4 and 122 MPa for $k = 0.3$, 0.45 and 0.6, respectively (shown by crosses in Fig. 21.2b). The values of stress are $\sigma_M = 180$, 130 and 117 MPa, with the constraints on the velocity range $V_0 = 194$, 130 and 87.5 m/s, $V_M = 481, 422$ and 387 m/s.

Figure 21.3 shows the diagrams of stresses on the cavity wall as a function of its expansion velocity, as obtained by using the solution of the problem of a spherical cavity expansion in a medium with the Mohr-Coulomb plasticity condition (the

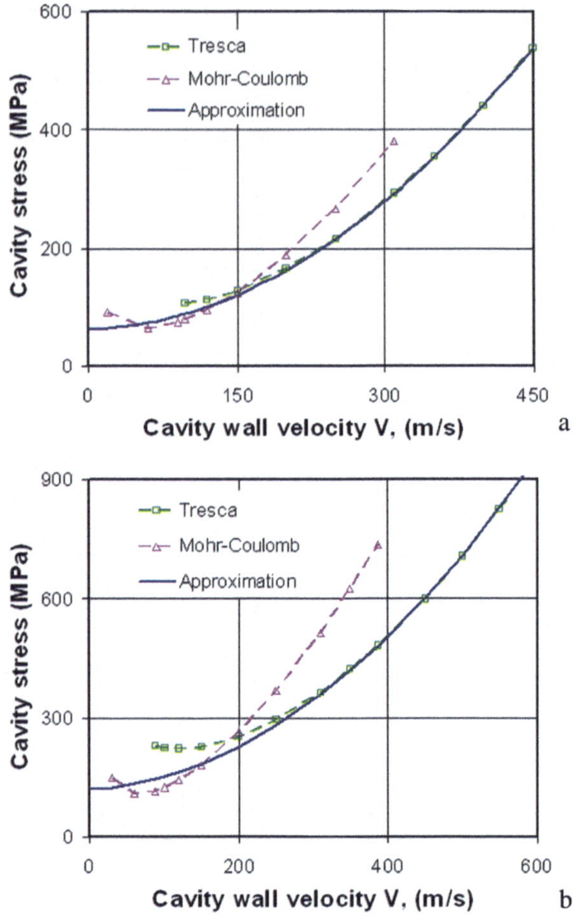

Fig. 21.3 Stress on the cavity wall as a function of its expansion velocity: $k = 0.3$, $\tau_M = 20$ MPa (**a**) and $k = 0.6$, $\tau_M = 50$ MPa (**b**)

dashed line with triangles), with Tresca's plasticity condition for a fixed value of yield strength, equal to τ_M (the dashed line with squares), the solid line corresponding to the quadratic approximation according to formula (8). Good agreement of the approximation results is observed for the value of cavity expansion velocity $V > V_M$. The computations show that the maximal relative inaccuracy of approximation (21.8) in the velocity range of $V_M < V < 10V_M$ is within 2–3%.

It was earlier shown [23] that LIM based on the analytical solution of the spherical cavity expansion problem in the assumption of incompressibility of the soil behind the shockwave front satisfactorily describes the forces resisting penetration of cylindrical impactors into compacted dry and water-saturates sand soils in a general case with the account of the limited dependence of yield strength on pressure. However, in a limited impact velocity range, satisfactory results can be obtained using simpler models. For instance, in [23], good agreement was demonstrated between the experimental data and numerical results for dry sand, using the Mohr-Coulomb yield criterion for impact velocities from 100 to 400 m/s.

In modeling penetration resistance of water-saturated soil, the model using the Mohr-Coulomb yield criterion satisfactorily describes the force as a function of velocity only for impact velocities not exceeding 120 m/s. For higher impact velocities, the general model is required, that accounts for the limited dependence of yield strength on pressure according to the Mohr-Coulomb-Tresca yield criterion. At the same time, completely neglecting shear properties of water-saturated soils leads to inaccuracies in evaluating the impact resistance force several times as big as that of the experiment. Thus, nonlinear shear resistance of dry and water-saturated soil must be taken into account in analyzing impact and penetration problems.

However, cohesion or initial strength of both dry and water-saturated soils is a negligibly small quantity. The applicability of the LIM to determining maximal stress resisting penetration of a conical impactor into a soil medium, accounting for its initial strength, will be assessed by analyzing a spherical cavity expansion problem. According to the LIM, the force resisting penetration at constant velocity u_0 of a cone with the apex angle of 2η, is defined as follows

$$F = (\sigma + \tau ctg\eta)S,$$

where σ is normal stress, and τ is tangential stress acting on the lateral surface of the conical impactor, S is cone basis area.

Nominal stress is determined based on formula (21.8b), where $V = u_0 \sin\eta$ is projection of the velocity vector onto a normal to the cone surface. Tangential stresses for $V < V_M$ are determined based on Coulomb's law of dry friction $\tau = k_f\sigma$. For higher impact velocities, $V \geq V_M$, the maximal tangential stress is assumed to be limited by the value of

$$\tau = k_f\left(\gamma + \beta V_M + \alpha V_M^2\right).$$

Thus, the relation between the force resisting penetration of a steel conical impactor into frozen soil and the velocity will be

$$F(V) = \begin{cases} \left(1 + k_f ctg\eta\right)\left(\gamma + \beta V + \alpha V^2\right)S, & V < V_M, \\ \left(\left(\gamma + \beta V + \alpha V^2\right) + k_f ctg\eta\left(\gamma + \beta V_M + \alpha V_M^2\right)\right)S, & V \geq V_M \end{cases} \quad (21.9)$$

In [30], the results of studying the processes of impact and penetration of a steel conical impactor into frozen sandy soil, based on the inverse experiment methodology, using the pressure bar technique, are published. Diagrams of maximal values of the force resisting penetration of cones with the base diameters of 10, 12 and 19.8 mm into soil as a function of impact velocities in the range of 100–400 m/s are presented. The condition of the sandy soil specimens before freezing at a temperature of −18 °C is characterized as fully water-saturated. As a result of a series

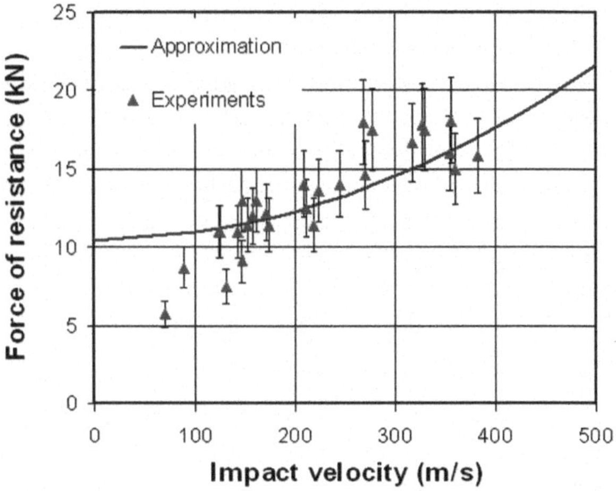

Fig. 21.4 Maximal value of the force resisting penetration of a cone with the apex angle of 30° into frozen sand: experiment and approximation

of numerical analyses, parameters of Grigoryan's soil medium model were determined, that provide good agreement between the results of axisymmetric numerical computations and experimental data.

In the present paper, similar parameters of the initial elastic portion for frozen sand are used: Young's modulus $E = 21$ GPa, Poisson's coefficient $v = 1/3$ and the parameters of the shock adiabat: $C_0 = 1700$ m/s, $s = 3.4$. To approximate the nonlinear relation between yield strength and pressure, used in the numerical analyses of [30], the following values of the parameters of the Mohr-Coulomb-Tresca yield criterion (cohesion, internal friction and maximal yield strength) are used: $\tau_0 = 16$ MPa, $k = 0.15$, $\tau_M = 32.5$ MPa, the friction coefficient being equal to $k_f = 0.1$.

Figure 21.4 presents maximal values of the force resisting penetration of a 10 mm-dia steel conical impactor with the apex angle of $\eta = 30$ degrees into frozen soil, as obtained in the inverse experiment [30] and based on approximation (21.8), (21.9). The vertical lines show a relative error of 15%. Good agreement between the experimental and simulated results is observed in the impact velocity range of 100–400 m/s.

21.5 Conclusion

The problem of a spherical cavity expanding at a constant velocity from a point in an infinite elastoplastic medium with the Mohr-Coulomb-Tresca plasticity condition has been analyzed. A relation between the stress on the cavity boundary and its expansion velocity has been obtained in the assumption if incompressibility of the medium behind the shockwave front.

Finite expressions for coefficients of the 'stress-velocity' trinomial approximation have been obtained for the first time as a function of the experimentally determined physical-mechanical parameters of the soil (coefficients of the stress adiabat and the yield strength-pressure relation).

The obtained formulae have been verified by comparing them with the already known experimental data on penetration of steel conic impactors into frozen sandy soil. It is shown that disagreement between the computational and experimental results is within 15%.

Acknowledgements This work was supported by a grant from the Government of the Russian Federation (contract No. 14.Y26.31.0031).

References

1. Aptukov V.N., Murzakaev R.T., Fonarev A.V. Prikladnaya teoriya pronikaniya [Applied Penetration Theory]. Moscow. Nauka Publ. (1992). 105 p. (In Russian)
2. Ben-Dor G., Dubinsky A., Elperin T. Ballistic Impact: Recent Advances in Analytical Modeling of Plate Penetration Dynamics–A Review. Applied Mechanics Reviews. 2005. V. 58. P. 355–371
3. Rakhmatulin K.A., Sagomonian A.I., Alekseev N.A.: Voprosy dinamiki gruntov [Ground Dynamics Issues]. Moscow. MGU Publ. (1964). 239p. (In Russian)
4. Sagomonian A.Ia. Pronikaniye [Penetration]. Moscow. MGU Publ. (1974). 299p. (In Russian)
5. Vysokoskorostnoye vzaimodeystviye tel [High-Speed Interaction of Bodies]. ed. V.M. Fomin. Novosibirsk. SB RAS Publ. (1999). 600p. (In Russian)
6. Forrestal, M.J., Longcope, D.B.: Closed-form solution for forces on conical-nosed penetrators into geological targets with constant shear strength. Mech. Mater. **1**, 285–295 (1982)
7. Forrestal, M.J., Luk, V.K.: Penetration into soil targets. Int. J. Imp. Eng. **12**(3), 427–444 (1992)
8. Forrestal, M.J., Norwood, F.R., Longcope, D.B.: Penetration into targets described by locked hydrostats and shear strength. Int. J. of Solids and Structures **17**(9), 915–924 (1981)
9. Forrestal, M.J., Luk, V.K.: Dynamic spherical cavity-expansion in a compressible elastic-plastic solid. J. Appl. Mech. **55**(2), 275–279 (1988)
10. Masri, R., Durban, D.: Cylindrical cavity expansion in compressible Mises and Tresca solids. European J. of Mechanics A/Solids. **26**, 712–727 (2007)
11. Masri, R., Durban, D.: Deep penetration analysis with dynamic cylindrical cavitation fields. Int J Imp. Eng **36**, 830–841 (2008)
12. Rosenberg, Z., Dekel, E.: A numerical study of the cavity expansion process and its application to long-rod penetration mechanics. Int. J. Imp. Eng. **35**(3), 147–154 (2008)
13. Rosenberg, Z., Dekel, E.: Analytical solution of the spherical cavity expansion process Int. J. Imp. Eng. **36**, 193–198 (2009)
14. Shi, C., Wang, M., Li, J., Li, M.: A model of depth calculation for projectile penetration into dry sand and comparison with experiments. Int. J. Imp. Eng. **73**, 112–122 (2014). https://doi.org/10.1016/j.ijimpeng.2014.06.010
15. Shi, C., Wang, M., Zhang, K., Cheng, Y., Zhang, X.: Semi-analytical model for rigid and erosive long rods penetration into sand with consideration of compressibility. Int. J. Imp. Eng. **83**, 1–10 (2015)
16. Kong, X.Z., Wu, H., Fang, Q., Peng, Y.: Rigid and eroding projectile penetration into concrete targets based on an extended dynamic cavity expansion model. Int. J Imp Eng **100**, 13–22 (2017)

17. Kotov, V.L.: Approksimatsiya napryazheniy v okrestnosti polosti, rasshiryayushcheysya s postoyannoy skorostyu v srede s usloviem plastichnosti Mora-Kulona [Approximating stresses in the vicinity of a cavity expanding at a constant velocity in a medium with the Mohr-Coulomb plasticity condition]. Problemy prochnosti i plastichnosti [Problems of Strength and Plasticity]. **81**(2), 177–190 (2019). (In Russian)
18. Kotov, V.L., Timofeev, D.B.: Analyzing the spherical cavity expansion problem in a medium with Mohr-Coulomb-Tresca's plasticity condition. Problemy prochnosti i plastichnosti [Problems of Strength and Plasticity]. **81**(3), 292–304 (2019). (In Russian)
19. Sun, Q., Sun, Y., Liu, Y., Li, R., Zhao, Y.: Numerical analysis of the trajectory stability and penetration ability of different lateral-abnormal projectiles for non-normal penetration into soil based on modified integrated force law method. Int. J. Imp. Eng. **103**, 159–168 (2017)
20. Kotov, V.L., Linnik, E.Y., Tarasova, A.A., Makarova, A.A.: Analiz priblizhennykh resheniy zadachi o rasshirenii sfericheskoy polosti v gruntovoy srede [The analysis of approximate solutions for a spherical cavity expansion in a soil medium problems]. Problemy prochnosti i plastichnosti [Problems of Strength and Plasticity]. **73**, 58–63 (2011). (In Russian)
21. Linnik, E.Y., Kotov, V.L., Tarasova, A.A., Gonik, E.G.: The solution of the problem of the expansion of a spherical cavity in a soil medium assuming incompressibility beyond the shock front. Problems of Strength and Plasticity. **74**, 49–58 (2012). [In Russian]
22. Kotov VL, Balandin VV, Bragov AM, Linnik EYu, and Balandin VV. Using a local-interaction model to determine the resistance to penetration of projectiles into sandy soil. Journal of Applied Mechanics and Technical Physics. 2013; 54(4): 612–621
23. Bragov, A.M., Balandin, V.V., Igumnov, L.A., Kotov, V.L., Krushka, L., Lomunov, A.K.: Impact and penetration of cylindrical bodies into dry and water-saturated sand. Int. J. Impact Eng **122**, 197–208 (2018)
24. Bazhenov, V.G., Bragov, A.M., Kotov, V.L.: Experimental-Theoretical Study of the Penetration of Rigid Projectiles and Identification of Soil Properties. J. Appl. Mech. Tech. Phys. **50**(6), 1011–1019 (2009)
25. Bazhenov, V.G., Kotov, V.L.: Method for identifying elastoplastic properties of ground media by penetration of impactors. Mech. Solids **43**(4), 678–686 (2008)
26. Grigorian, S.S.K.: Resheniyu zadachi o podzemnom vzryve v myagkikh gruntakh [To the solution of the problem of an underground explosion in soft soils]. Prikladnaya matematika i mekhanika [J. Appl. Math. Mech.] **28**(6), 1070–1082 (1964). (In Russian)
27. Balandin, V.V., Bragov, A.M., Igumnov, L.A., Konstantinov, AYu., Kotov, V.L., Lomunov, A.K.: Dynamic deformation of soft soil media: experimental studies and mathematical modeling. Mech. Solids **50**(3), 286–293 (2015)
28. Bragov, A.M., Lomunov, A.K., Sergeichev, I.V., Proud, W., Tsembelis, K., Church, P.: Tech. Phys. Lett. **31**(6), 530–531 (2005)
29. Prowd, W.G., Chapman, D.J., Williamson, D.M. et al.: The dynamic compaction of sand and related porous systems. CP955, Shock Compression of Condensed Matter-2007. 1403–1408 (2007)
30. Bragov, A.M., Balandin, V.V., Kotov, V.L., Linnik, E.Y.: Experimental study of the impact and penetration of a cone in frozen sand. J. Appl. Mech. Tech. Phys. **59**(3), 482–490 (2018)

Chapter 22
A Heterogeneous Medium Model and Its Application in a Target Perforation Problems

Alexander E. Kraus, Evgeny I. Kraus, and Ivan I. Shabalin

Abstract The cermet composite is considered as a mixture of solid materials with a given volume concentration and joint deformation of the components. The REACTOR software package developed at the Institute of Theoretical and Applied Mechanics of the Siberian Branch of the Russian Academy of Sciences (ITAM SB RAS) is used for the modeling. Various approaches to the prediction of elastic–plastic characteristics, such as elastic moduli and yield strength, describing the behavior of heterogeneous materials, are considered. It is shown that the mixture rule, which approximates the value of elastic sound velocities from the volume concentration of the components, proved itself well for the calculation of elastic moduli, while the yield strength must be calculated through the mass concentration. The proposed model of heterogeneous medium has been applied to solving a number of problems on the resistance of heterogeneous and gradient barriers to impact by a hardened striker. For barriers from various combinations of mild steel and Al_2O_3 ceramics at equal volume concentrations, ballistic curves of their resistance to impact of a solid steel core are constructed. It is shown that a layered barrier, consisting of equal in volume plates of Al_2O_3 ceramics and steel, is more resistant to impact by a solid core in comparison with other cermet barriers. In a numerical experiment, it has been found out that heterogeneous and gradient barriers have an advantage over a two-layer barrier (B_4C ceramics + steel) in the range of encounter velocities exceeding 650 m/s. A series of 2D and 3D calculations for modeling the perforation of heterogeneous barriers has been performed. It is shown that the resistance of heterogeneous barriers to perforation by a solid striker is approximately at the same level. The

A. E. Kraus (✉) · E. I. Kraus (✉) · I. I. Shabalin
Khristianovich Institute of Theoretical and Applied Mechanics SB RAS, 630090 Novosibirsk, Russia
e-mail: akraus@itam.nsc.ru

E. I. Kraus
e-mail: kraus@itam.nsc.ru

I. I. Shabalin
e-mail: shabalin@itam.nsc.ru

H. Altenbach et al. (eds.), *Multiscale Solid Mechanics*,
Advanced Structured Materials 141,
https://doi.org/10.1007/978-3-030-54928-2_22

heterogeneous model of ice has made it possible to explain the anomalous dependence of the depth of the cavity in ice on the speed of impact with a steel cylinder, which was discovered.

Keywords Heterogeneous medium model · Elastic moduli · Yield strength · Cermet composite · The REACTOR software package · Gradient barrier · Shock adiabat

22.1 Introduction

Experimental work on the practical creation of technologies for the manufacture of heterogeneous media with desired properties often outperforms the methods for predicting the properties themselves. Therefore, there is a need in parallel with conducting experiments to create mathematical models to minimize costs in experiments and achieve the desired results. One such material is a heterogeneous cermet composite based on nickel and boron carbide. So, at the ITAM SB RAS, using the method of cold gas-dynamic spraying, a metal matrix with the addition of dispersed particles of boron carbide is obtained. The laser energy treatment of this substrate gives a well-mixed cermet composite [1]. The purpose of this work is to forecast the mechanical parameters of the obtained cermet composite by numerical simulation of the propagation of elastoplastic waves in it.

22.2 Heterogeneous Media Model

The cermet composite, consisting of a metal matrix and ceramic inclusions, has the following geometric representation in discrete form, for example, for the two-dimensional case. Let the composite region be divided by a difference grid, the cells of which fill the space without gaps and overlaps. Then, a given volume of ceramic inclusions is randomly distributed over the matrix volume (see Fig. 22.1). Using this algorithm, it is possible to build various forms of ceramic inclusions in the metal matrix. The examples are shown in Fig. 22.2.

Each triangular cell has its own physical and mechanical properties. When switching from cell to cell, properties can change abruptly. At the cell boundaries, the conditions for the joint movement of the cermet components are satisfied. Inside the cells, the calculations are carried out in accordance with the explicit difference scheme.

To evaluate the bulk modulus of the cermet composite, the additive mixture model can be used [2–4]. We assume that the size of the composite grains is small enough to substantially distort the front and amplitude of the compression wave propagating through the cermet at a speed D. The specific volume of the mixture

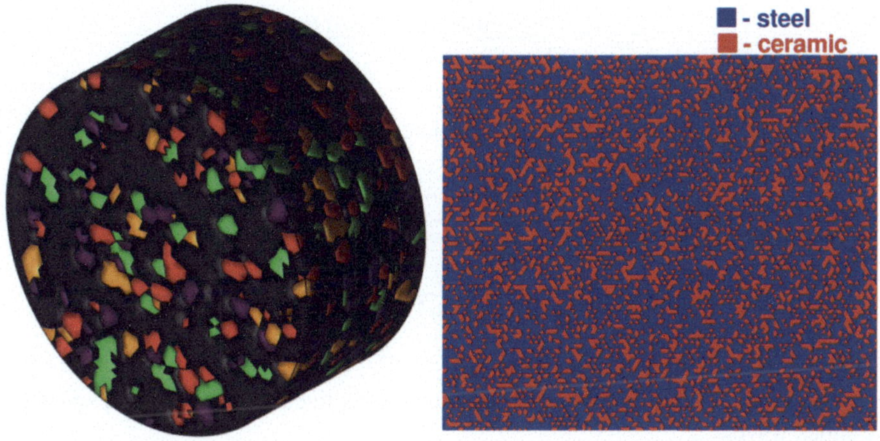

Fig. 22.1 Geometric representation of the 3D cermet composite is on the left. The cross section of the sample is on the right, and the volumetric content of ceramics is 35%

Fig. 22.2 Ceramic inclusions: acute-angled ones on the left rounded ones on the right. The volume concentration of ceramics is 35%

$$V_{\mathrm{mix}}(P) = \sum_{j=1}^{N} \alpha_j V_j(P), \quad \alpha_j = \frac{m_j}{\sum_{j=1}^{N} m_j}$$

where V_j is the specific volume of the j-th component, P is the pressure, α is the mass concentration of the component, m_j is the mass of the jth component, N is the number of components in the mixture.

Following [5], we write the pressure in the mixture through the parameters of the linear dependence of the velocity of the shock wave on the mass velocity, which has the form:

$$P_{\text{mix}} = \frac{\rho_{0\,\text{mix}} C_{0\,\text{mix}}^2 (1 - \rho_{0\,\text{mix}}/\rho_{\text{mix}})}{[1 - S_{\text{mix}}(1 - \rho_{0\,\text{mix}}/\rho_{\text{mix}})]^2},$$

while the parameters for the mixture are related as

$$D_{\text{mix}} = C_{0\,\text{mix}} + S_{\text{mix}} U_{\text{mix}},$$

$$\frac{1}{(C_{0\,\text{mix}} \rho_{0\,\text{mix}})^2} = \sum_{j=1}^{N} \frac{\alpha}{(C_{0\,j} \rho_{0\,j})^2}, \quad \frac{1}{S_{\text{mix}} \rho_{\text{mix}}} = \sum_{j=1}^{N} \frac{\alpha}{\rho_{0j} S_j},$$

where $C_{0\,\text{mix}}$, $c_{0\,j}$ and S_{mix}, S_j are the parameters in linear relation for the speed of the shock wave.

We calculate the velocity of the body waves as

$$C_{B\,\text{mix}}^2 = \frac{\partial P_{\text{mix}}}{\partial \rho_{\text{mix}}} = \frac{\rho_{0\,\text{mix}}^2 C_{0\,\text{mix}}^2 [1 + S_{\text{mix}}(1 - \rho_{0\,\text{mix}}/\rho_{\text{mix}})]}{\rho_{\text{mix}}^2 [1 - S_{\text{mix}}(1 - \rho_{0\,\text{mix}}/\rho_{\text{mix}})]^2}.$$

Passing to the limit in expression $C_{B\,\text{mix}}^2$ at $\rho_{\text{mix}} \to \rho_{0\,\text{mix}}$, we obtain that $C_{B\,\text{mix}}^2 = C_{0\,\text{mix}}^2$. This shows that the velocity of the body waves is determined by $C_{0\,\text{mix}}^2$, i.e., by the accuracy of the low-speed part of the shock adiabat, where the shock wave of small amplitude is a wave of elastic–plastic compression.

The rapid development of the production of composite and other heterogeneous materials has necessitated a theoretical understanding of the nature of such materials and methods for predicting the physical and mechanical properties from the properties of the components of their constituents [6–8]. On the basis of mathematical processing of numerous experimental data and theoretical constructions, empirical rules of mixture have been developed. Let a compression wave propagate along a sample of materials mixture. The propagation time of the wave at distance L consists of its propagation times in materials 1 and 2,

$$t = \frac{L}{C} = \frac{x}{C_2} + \frac{L - x}{C_1}.$$

Using the volumetric content of the second material $\beta = x/L$ with a constant cross-sectional area of the sample, we obtain

$$\frac{1}{C} = \sum_{j=1}^{N} \frac{\beta_j}{C_j}, \quad \beta_j = \frac{A_j}{\sum_{j=1}^{N} A_j},$$

where A_j is the volume of the jth component, β is the volume concentration of the component, N is the number of components in the mixture.

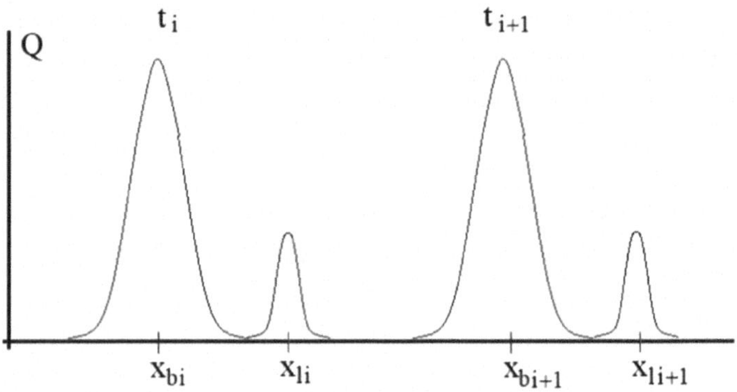

Fig. 22.3 Behavior of artificial viscosity at the front of elastic precursor and bulk compression wave at two time moments

Given the volume concentration of component materials in the cermet, we find the necessary sound speeds. Knowing the velocities of elastic and body waves, we determine the elastic moduli K_{mix} and G_{mix} of the cermet:

$$K_{\text{mix}} = C_{B\,\text{mix}}^2 \rho_{0\,\text{mix}}, \quad G_{\text{mix}} = \frac{3\left(C_{L\,\text{mix}}^2 - C_{B\,\text{mix}}^2\right)\rho_{0\,\text{mix}}}{4}.$$

Using the REACTOR software package [9], a one-dimensional shock wave can be used to calculate the elastic moduli of a cermet composite. Let the composite block, placed in the rigid sidewalls, colliding with hard surface. An elastoplastic compression wave propagates from the rigid wall into the composite. In an explicit difference scheme, artificial viscosity Q is used to smooth the gaps. From the change of the artificial viscosity, one can evaluate the magnitude of the gradients at the front of the compression waves. The elastic precursor propagates with speed C_L of the elastic longitudinal wave. Next, a plastic wave moves with speed C_B of the body waves (see Fig. 22.3).

Knowing the sequence of time intervals $\{t_i\}$ and the set of coordinates $\{x_{li}, x_{bi}\}$ of the front of the elastic and body waves, we find the average values of velocities C_L and C_B:

$$C_{B\,i} = \frac{x_{Bi+1} - x_{Bi}}{t_{i+1} - t_i}, \quad C_{B\,\text{mix}} = \frac{\sum_{i=1}^{n} C_{B\,i}}{n},$$

$$C_{L\,i} = \frac{x_{L\,i+1} - x_{L\,i}}{t_{i+1} - t_i}, \quad C_{L\,\text{mix}} = \frac{\sum_{i=1}^{n} C_{L\,i}}{n},$$

where n is the number of intervals.

The REACTOR software package uses the Mises form yield strength. For a one-dimensional elastoplastic wave, $S_{xy} = 0$; then, through the known values of the deviator component of the stress tensor behind the shock wave front, we find the yield strength of the composite [10]:

$$Y_0 = \sqrt{3(S_{xx}^2 + 2S_{yy}^2)/2}.$$

Thus, we can write the explicit formula for calculating yield strength Y_0 of the mixture through the corresponding values of the composite components and the mass concentration:

$$Y_{0\,\mathrm{mix}} = \sum_{j=1}^{N} \alpha_j Y_{0\,j},$$

where $Y_{0\,j}$ are the yield strengths of the components of the mixtures.

22.3 Ballistic Performances of Heterogeneous Plates

The model of heterogeneous medium described above allows one to construct heterogeneous and gradient plates and to compare their ballistic resistance with the classical two-layer cermet plate. A gradient model of cermet barrier with direct distribution of ceramics concentration over the volume, i.e., an obstacle in which the ceramics concentration decreases linearly from 100% in the front layers to 0% in the back layers of the barrier, is shown in Fig. 22.4a. Gradient barrier with inverse distribution of ceramics concentration over the volume, i.e., an obstacle in which the steel concentration of 100% in the front layers linearly decreases to 0%, being replaced in volume by ceramics, is shown in Fig. 22.4b. A model of heterogeneous cermet barrier constructed by the Gaussian distribution of ceramic clusters over the volume of steel is presented in Fig. 22.4c. A model of classical two-layer cermet plate, is shown in Fig. 22.4d.

For numerical simulation of interaction of solid projectiles with heterogeneous and gradient barriers, the REACTOR software package was used [9]. The cermet components were mild steel AISI1015 and ceramics Al_2O_3, and solid steel 2P was used for the projectile. The impact speed was 750 m/s. The thickness of the barriers varied from 1 to 3 cm. Based on the results of a series of calculations, it was shown that the layered barrier, consisting of equal in volume Al_2O_3 ceramic and steel plates, was more resistant to impact compared to other cermet barriers (see Fig. 22.5). The second by resistance to a solid core impact is the gradient barrier with direct distribution of Al_2O_3 ceramics concentration over the volume. The gradient barrier with inverse distribution of Al_2O_3 ceramics concentration over the volume and the heterogeneous barrier with uniform distribution of Al_2O_3 ceramics concentration over the volume have approximately the same resistance [11].

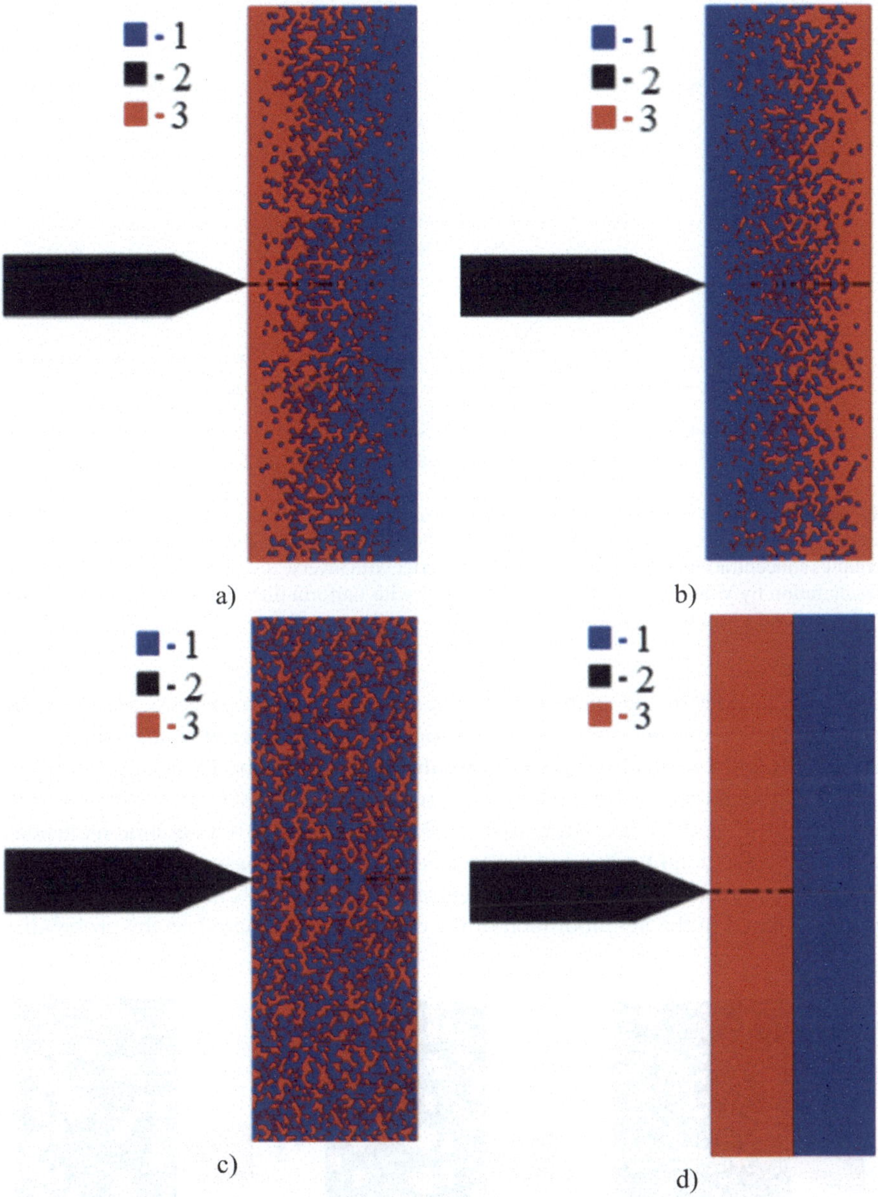

Fig. 22.4 Gradient, heterogeneous, and layered barriers of mild steel and ceramics: 1—mild steel, 2—high-strength hardened steel 2P, 3—ceramics

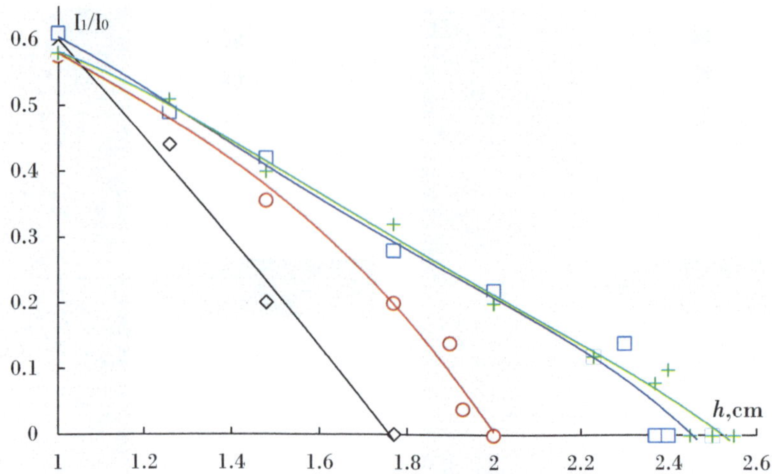

Fig. 22.5 Ballistic curve: dependence of the reduced residual momentum of the core on the thickness of the barrier. ◇—layered barrier; ○—gradient barrier with direct distribution of Al_2O_3 ceramics concentration by volume; +—gradient barrier with inverse distribution of Al_2O_3 ceramics concentration by volume; □—heterogeneous barrier with uniform distribution of Al_2O_3 ceramics concentration by volume

Let us consider the dependence of the resistance of heterogeneous plates on the concentration of Al_2O_3 ceramics with various shapes of ceramic inclusions. The REACTOR software package provides for the possibility of building a cermet material by creating three types of models of heterogeneous material matrix, which we will call model I, II, and III, according to Fig. 22.6: model I—cluster ceramic inclusions combined in the neighborhood of the faces of the element, based on the probability distribution, according to a given concentration; model II—cluster ceramic inclusions combined in the neighborhood of the element nodes, based on the probability

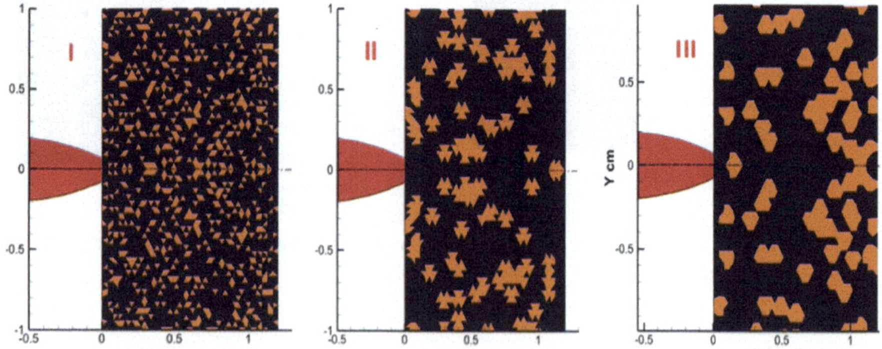

Fig. 22.6 Examples of heterogeneous materials with various shapes of inclusions, the roman numerals correspond to the model

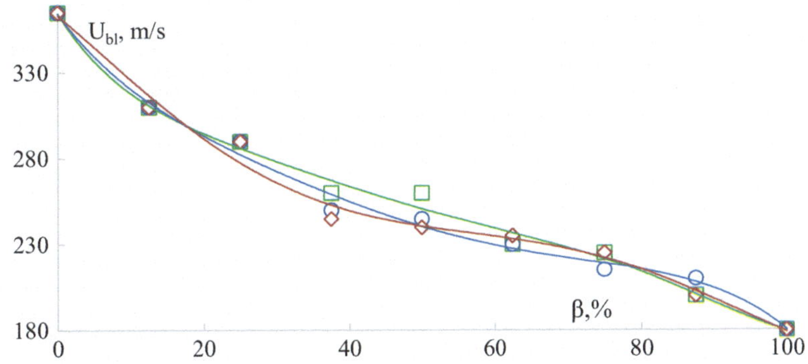

Fig. 22.7 Dependence of the limiting ballistic velocity (U_{bl}) of heterogeneous plates on volume fraction ($\beta\%$) of ceramics with different shapes of inclusions. □—model I; ○—model II; ◇—model III

distribution, according to a given concentration; model III—single inclusions of ceramic material based on a probabilistic normal distribution, according to a given concentration.

A series of calculations of the plates shock loading was made. The collision speeds were in the range of 100–600 m/s. The thickness of the heterogeneous plates is $L =$ 1.2 cm. The projectile with length of 2.2 cm and diameter of 0.4 cm was made in the form of a core with a truncation of the lively head part made of 2P-hardened steel.

Based on a series of calculations for heterogeneous plates with various shapes of inclusions, the dependence of the limiting ballistic velocity (the speed, at which the plate of a given thickness ceases to withstand the load of the projectile) on the volume fraction of ceramics in the plate was constructed (see Fig. 22.7). It is shown that, regardless of the shape of the ceramic grain included in the steel matrix, the limiting ballistic velocity decreases with increase in the volume fraction of ceramics.

Let B_4C ceramics be used in the cermet plates. A series of calculations was made for penetration of plates with the same weight and size parameters in the range of meeting speeds of 400–1000 m/s. The calculation results are summarized in the form of ballistic curves (see Fig. 22.8), which are constructed by the least squares method according to the scattered values of the residual velocities. The scattering is due to a random distribution of the materials over the barrier volume. At meeting speeds of the projectile with the obstacle of less than 600 m/s, the two-layer barrier has higher armor resistance; however, at speeds of more than 670 m/s, the advantage passes to heterogeneous and gradient barriers [12].

We studied the resistance of barriers constructed from heterogeneous plates as follows: a monolithic plate, a plate of two layers in contact, two plates with spacing between them of 1.2 and 2.4 cm (see Fig. 22.9, examples of two configurations). The calculations were performed for two cases: In the first case, the yield strength of the

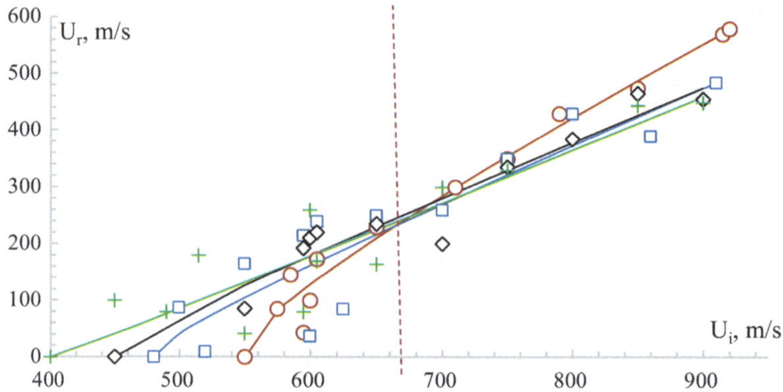

Fig. 22.8 Ballistic curves of heterogeneous, gradient, and layered barriers. U_i is the impact velocity, U_r is the residual velocity of the projectile: ○—layered barrier; ◊—gradient barrier with direct distribution of B_4C ceramics concentration by volume; +—gradient barrier with inverse distribution of B_4C ceramics concentration by volume; □—heterogeneous barrier with uniform distribution of B_4C ceramics concentration by volume

Fig. 22.9 On the left, the barrier of heterogeneous plates in contact; on the right, the plates spaced 2.4 cm apart

projectile was increased to $Y = 5$ GPa (rigid projectile), and in the second case, $Y =$ 1.9 GPa (deformable projectile).

The results of model calculations revealed rather interesting trends in the resistance of heterogeneous barriers [13]. The limiting ballistic velocity for both projectiles is the same and is approximately equal to $U_{bl} = 275$ m/s. However, the behaviors of the residual velocity are different. For the rigid projectile, all the barrier configurations demonstrate practically the same resistance with increase in the impact velocity (see Fig. 22.10a). In the case of deformable projectile, the ballistic curves diverge in a fan style with increase in the impact velocity, and the calculation results have a significant spread in the residual velocity (see Fig. 22.10b). This is due to the

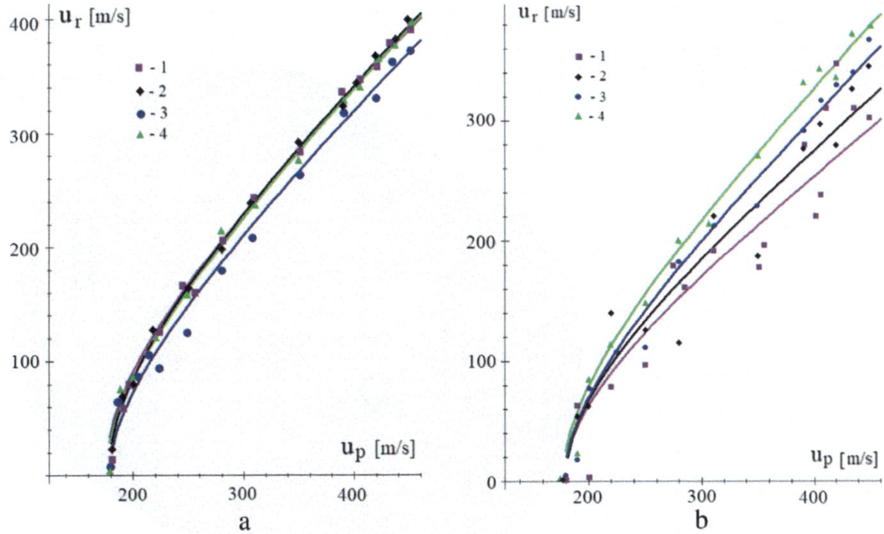

Fig. 22.10 Ballistic curves of perforation of heterogeneous barriers: **a** a rigid projectile, **b** a deformable projectile. 1—monolithic plate, 2—two-layer one with layers contact, 3—two-layer one with layers spacing of 2.4 cm, 4—two-layer one with layers spacing of 1.2 cm

random distribution of ceramic particles over the volume of the metal, because for a deformable projectile, the presence of ceramics in the front layers of the barrier, especially the monolithic one, leads to deformation of the head part and an increase in the cross section, i.e., to increase in the resistance. Thin heterogeneous plates deform the head of the projectile to a lesser extent; therefore, the scatter of the residual velocity is much lower.

22.4 Perforation of Ice Coverings

At the ITAM SB RAS, a series of experiments was carried out on a single-stage gas facility on penetration of steel cylinders into massive ice targets. The steel cylinders have diameter $d = 1.5$ cm, height $h = 1.5$ cm, and mass $m \approx 20$ g. The ice targets have diameter $d_t = 26.2$ cm and height $h_t = 35.0$ cm. The experimental results showed an abnormal dependence of depth of the ice caverns on the speed of the incident cylinder (see Table 22.1) [14].

Table 22.1 Results of experiments on the penetration of a steel cylinder into ice barriers

The impact velocity, *U* m/s	191	275	285
The cavern depth, *L* cm	11.0	11.5	10.5

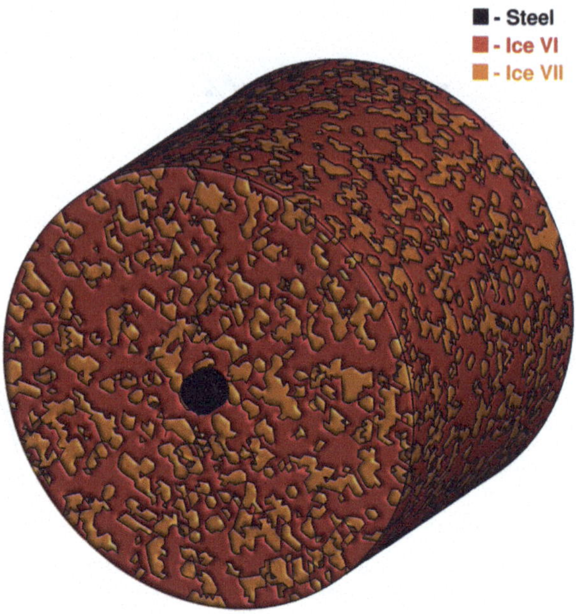

Fig. 22.11 Statement of the problem of a steel cylinder collision with a heterogeneous ice target

The study of the dynamic properties of ice shows the complexity of modeling in a wide range of loading speeds [15, 16], where shock adiabats of ice for different ranges of mass velocities appearing behind the front of the shock wave are given. We assume that for the mass velocities in the range of 200–1000 m/s, the ice behind the front of the shock wave will be a mixture of different solid phases (see Fig. 22.11). This is also confirmed by experimental studies [17], where a curve of volume compressibility of ice is constructed, which has a characteristic shelf in the range $0.15 < \Delta V/V < 0.45$ of the change in ice volume. The numerical values of the strength parameters of ice, for different authors, vary within the same order. The most suitable values are given in [18].

The process of cavity formation in a massive ice target during the impact of a steel cylinder was simulated in both axial two-dimensional and three-dimensional settings. When the projectile was introduced into the surface layers of the target, the ice was destroyed, and its fragments formed on the periphery of the projectile were carried outside the cavity. Fragments of ice in front of the steel cylinder were compacted and moved with it to the complete stop.

Figure 22.12 shows the results of numerical solution of the two-dimensional and three-dimensional problems on the penetration of a steel cylinder into a barrier from ice phase VII. In the calculations for ice phase VII, we used the shock adiabat with parameters $C_0 = 1340$ m/s, $S = 1.4$ [4]. At impact velocities $U = 275$, 285 m/s, the cavity depths obtained in the calculations and experiments are in good agreement, while at $U = 191$ m/s, they differ significantly. Using the parameters of the shock adiabat for ice phase VI $C_0 = 388$ m/s, $S = 2.61$ [4], the crater depth was 10.3 cm at the impact velocity $U = 191$ m/s, which is in a good agreement with experimental

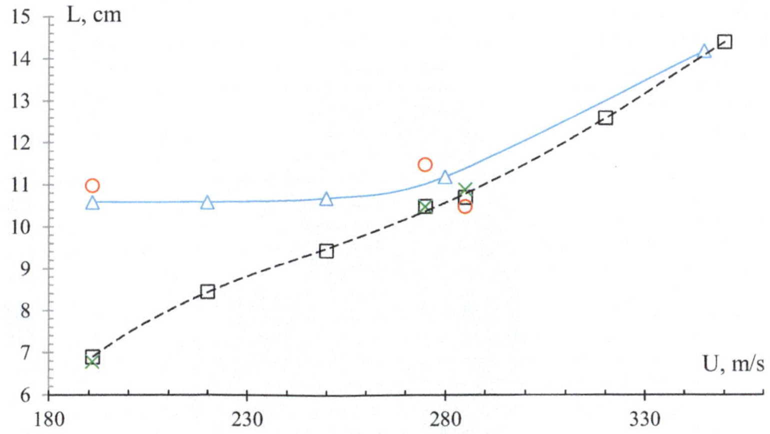

Fig. 22.12 Dependence of the cavity depth L on impact velocity U: Legend: ○—experiment; homogeneous ice model: □—2D simulation, ×—3D simulation; heterogeneous ice model: △—a mixture of ice VI and VII; dashed black line is the calculated dependence $L(U)$ for ice phase VII, solid one-calculated dependence for a mixture of ice phases VI and VII

data. The use of the heterogeneous ice model made it possible to reproduce with sufficient accuracy the experimental dependence of the crater depth on the collision velocity. The volume concentration of the ice phase VII linearly increases from its initial value $\alpha_{\mathrm{VII}} = 0$ at $U = 190$ m/s to the final value $\alpha_{\mathrm{VII}} = 1$ at $U = 300$ m/s.

Consider the problem of breaking through the ice cover. Let a steel ring with diameter of 1.92 m and thickness of 0.1 m fly with initial speed of 2000 m/s onto an ice barrier with thickness of 1 m. The high-speed ring forms an annular crater in the ice barrier at the initial stage of interaction. The unloading of the shock wave accelerates the central part of the ice plate, while the sprouting annular cavity separates it from the resting barrier (see Fig. 22.13). The diameter of the knocked out hole is 2.3 m.

Figure 22.14 shows a diagram of the distribution of ice fragments outside the knockout hole. Most of the fragments (about 98%) have a size from 0.2 to 2.5 cm, i.e., the central body is almost completely destroyed. On the periphery of the hole, there are larger fragments, but with a velocity vector directed outside the hole. To ensure the removal of the entire mass of the central body, the steel ring impact at an angle is recommended.

22.5 Conclusion

1. The additive model of the mixture is applicable only when there is reliable accuracy of the shock adiabat in the low-velocity region;
2. The rule of mixtures based on the processing of numerous experimental data is well suited for calculating the elastic moduli K and G.

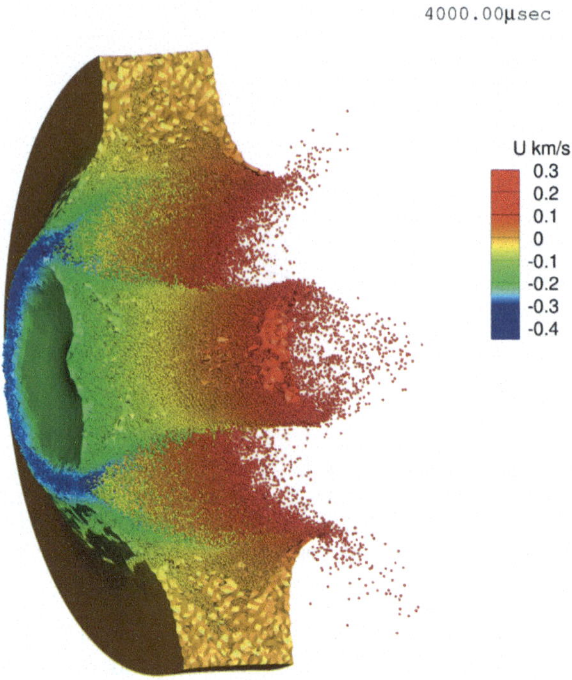

Fig. 22.13 Removal of the central part from the ice plate by a high-speed steel ring

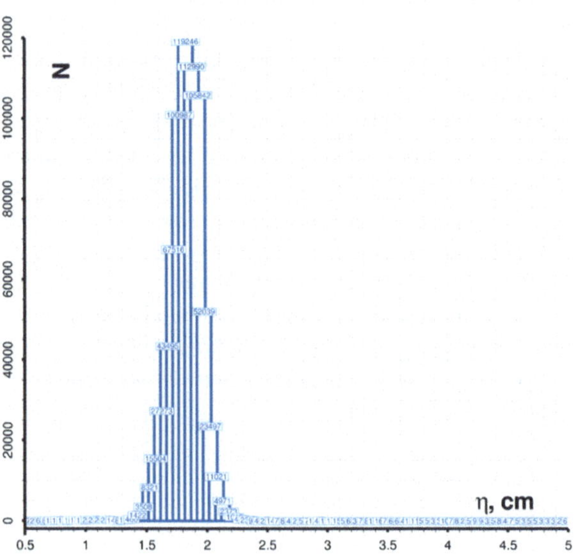

Fig. 22.14 Distribution of ice fragments in the space behind the barrier. Legend: N is the number of fragments, η is the average linear size of fragments

3. To predict the yield strength of the cermet composite, approximation by mass concentration is better suited.
4. Based on direct numerical simulation of the processes of interaction between solid projectiles and barriers, it is shown that a layered barrier consisting of equal in volume Al_2O_3 ceramic and steel plates is more resistant to impact compared to other cermet barriers. The second by resistance to a solid core impact is a gradient barrier with direct distribution of Al_2O_3 ceramics concentration over the volume. A gradient barrier with inverse distribution of Al_2O_3 ceramics concentration over the volume and a heterogeneous barrier with uniform distribution of Al_2O_3 ceramics concentration over the volume have approximately the same resistance.
5. It was found out in a numerical experiment that heterogeneous and gradient barriers had an advantage over a two-layer barrier (B_4C ceramics + steel) in the range of meeting velocities exceeding 650 m/s.
6. Model calculations of the perforation of heterogeneous barriers based on metal-ceramics showed that for a rigid projectile, all the configurations of barriers had the same resistance over a wide range of meeting speeds. Then, as for a deformable striker, the ballistic curves diverge in a fan style with increase in the speed of the meeting.
7. The heterogeneous model of ice made it possible to explain the anomalous dependence of the depth of the ice cavity on the speed of impact by a steel cylinder, which was discovered experimentally at the ITAM SB RAS.

Acknowledgements The study was conducted within the framework of the basic part of state task of the Khristianovich Institute of Theoretical and Applied Mechanics SB RAS (GR No. AAAA-A17-117030610136-3).

References

1. Fomin, V.M., et al.: Creation of heterogeneous materials on the basis of B4C and Ni powders by the method of cold spraying with subsequent layer-by-layer laser treatment. J. Appl. Mech. Tech. Phys. **58**, 947–955 (2017)
2. Bakanova, A.A., et al.: Compliance with the additivity rule for a number of alloys under shock compression. J. Appl. Mech. Tech. Phys. **13**, 903–907 (1972)
3. Dremin, A.N., Karpukhin, I.A.: Method of determining the shock adiabats for disperse materials. Zh. Prikl. Mekh. i Tekh. Fiz. **1**, 184–188 (1960)
4. Nikolaevskii, V.N.: Hydrodynamic analysis of shock adiabats of heterogeneous mixtures of substances. J. Appl. Mech. Tech. Phys. **10**, 406–411 (1969)
5. Kraus, E.I., et al.: Accounting for electronic components in the equation of state in the calculation of shock waves in a mixture of metals. PNRPU Mech. Bull. **9**, 78–84 (2001)
6. Bakhvalov, N., Panasenko, G.: Homogenisation: Averaging Processes in Periodic Media. Springer, Netherlands, Dordrecht (1989)
7. Duvall, G.E., Taylor, S.M.: Shock Parameters in a Two Component Mixture. J. Compos. Mater. **5**, 130–139 (1971)
8. Hashin, Z.: The Elastic Moduli of Heterogeneous Materials. J. Appl. Mech. **29**, 143–150 (1962)

9. Kraus, E.I., Shabalin, I.I.: Reactor2D: A tool for simulation of shock deformation. AIP Conf. Proc. **1770**, 030092 (2016)
10. Kraus, E.I., et al.: Numerical analysis of wave propagation in a cermet composite. AIP Conf. Proc. **1893**, 030130 (2017)
11. Kraus, E.I., Shabalin, I.I.: Modeling of penetration of gradient metal-ceramic barriers with finite thickness. AIP Conf. Proc. **2027**, 030164 (2018)
12. Kraus, E.I., Shabalin, I.I.: Comparison of ballistic resistance of layered and gradient barriers to impact. AIP Conf. Proc. **2125**, 030065 (2019)
13. Kraus, E., Shabalin, I.: Comparison of the resistance of monolithic and layered heterogeneous barriers to penetration by rigid and deformable strikers with an ogival nose based on numerical simulation. EPJ Web Conf. **221**, 01022 (2019)
14. Kraus, E.I., et al.: Penetration of Steel Projectiles through Finite-Thickness Ice Targets. J. Appl. Mech. Tech. Phys. **60**, 526–532 (2019)
15. Stewart, S.T., Ahrens, T.J.: Shock Hugoniot of H_2O ice. Geophys. Res. Lett. **30**, 6 (2003)
16. Stewart, S.T., Ahrens, T.J.: Shock properties of H_2O ice. J. Geophys. Res. Planets. **110**, E3 (2005)
17. Balandin, V.V., et al.: Numerical modeling of shock interaction of elastic cylinder with ice. Probl. Strength Plast. **79**, 93–103 (2017)
18. Lobanov, V.A.: Modeling of the interaction of ice with structures. Vestn. nauchno-tekhnicheskogo Razvit. **10**, 31–39 (2011)

Chapter 23
Panel Flutter Under Conditions of Local Strong Viscous-Inviscid Interaction

Igor I. Lipatov and Van Khiem Pham

23.1 Formulation of the Problem

We consider the flow around a flat plate by a supersonic flow of viscous gas. The Cartesian coordinate system is connected with the plate, the OX axis is directed along the plate surface, the OY axis is normal to the surface. It is assumed that at a distance l_0 from the leading edge is a flexible portion having a length l. The following notation is introduced for coordinates measured along the plate surface and normal to it, time, components of the velocity vector, density, pressure, total enthalpy, viscosity coefficient: $l_0 x$, $l_0 y$, $l_0 t/u_\infty$, $u_\infty u$ $u_\infty v$, $\rho_\infty \rho$, $\rho_\infty u_\infty^2 p$ $u_\infty^2 H/2$, $\mu_\infty \mu$, respectively. The index ∞ refers to the dimensional parameters of the unperturbed incident flow. It is assumed that the Reynolds number is large, but does not exceed the critical value, so that the laminar flow regime is maintained.

Depending on the ratio of geometric parameters and the number Re, various flow regimes are possible. The displacement thickness of the boundary layer changes under the influence of a pressure perturbation $\Delta p \ll 1$ and the main contribution to this change being made by the near-wall region at low speeds [3]. With a nonlinear effect on the near-wall flow, changes in velocity can be estimated as $\Delta u \sim u \sim \Delta p^{1/2}$. This estimate is true if the influence of the viscosity forces is, to a first approximation, insignificant.

I. I. Lipatov (✉)
Central Aerohydrodynamic Institute (TsAGI), Zhukovsky, 1 Zhukovsky Street, Moscow, Russian Federation
e-mail: igor_lipatov@mail.ru

V. K. Pham
Moscow Institute of Physics and Technology, Dolgoprudny, Moscow, Russian Federation
e-mail: van.fam@phystech.edu

H. Altenbach et al. (eds.), *Multiscale Solid Mechanics*,
Advanced Structured Materials 141,
https://doi.org/10.1007/978-3-030-54928-2_23

With nonlinear changes in velocity, the thickness of the near-wall region of the shear flow also changes in the main term, which follows from the condition of conservation of flow. Then from the estimate for the longitudinal velocity $y \sim \varepsilon u$, $\varepsilon = \mathrm{Re}^{-1/2}$, the estimate for the change in the displacement thickness follows $\Delta y \sim \varepsilon \Delta p^{1/2}$. Moreover, it is significant that the main part of the boundary layer with finite velocities makes a significantly smaller contribution to the total change in the displacement thickness $\Delta\delta \sim \varepsilon \Delta p$, since the velocity changes here are linear at small pressure amplitudes.

If the perturbations introduced into the external supersonic flow have the same order of the initial pressure perturbation $\Delta p \sim \varepsilon \Delta p^{1/2}/\Delta x$, then the estimate for the pressure value directly determines the length of the perturbed flow region $\Delta x \sim \varepsilon/\Delta p^{1/2}$. It follows that for all small pressure perturbations, the length of the perturbed flow region exceeds the thickness of the boundary layer. This makes it possible to use the Akkeret's formula to determine the induced pressure perturbation. For the condition of local strong interaction, it is assumed that the length of the perturbed region of the flow is much less than the characteristic length of the streamlined body l_0.

The assumption of the effect of viscosity in the region of nonlinear changes of velocity leads to well-known estimates of the theory of free interaction [3]. Below we consider a regime for which the effect of viscosity is insignificant in the first approximation. With small perturbations, such this regime is realized if the pressure amplitude satisfies the inequality $\varepsilon^{1/2} \ll \Delta p \ll 1$.

Under these conditions, 4 characteristic regions can be distinguished in the field of the perturbed flow. The first region contains trickles of an inviscid supersonic flow; the characteristic transverse dimension of this region is determined by its length and the slope of the characteristics, then for finite Mach numbers $y_1 \sim \varepsilon/\Delta p^{1/2}$.

Region 2 is the main part of the boundary layer. At the bottom of this region is a region of nonlinear perturbations of the longitudinal velocity (region 3), in which the influence of viscosity is insignificant in the first approximation. To take into account the effect of viscosity, it is necessary to introduce into consideration region 4, the transverse size of which is estimated from the condition of the balance of viscosity and inertia forces $y_4 \sim \varepsilon^{3/2}/\Delta p^{1/2}$. It should be noted that the possibility of the existence of the proposed flow structure depends on the existence of a continuous flow in the local boundary layer (region 4).

The solution in region 3 can be written in the following form, based on the above estimates

$$x = x_3 / \rho_w^{1/2} a^{1/2} \beta \Delta p^{1/2} \tag{23.1}$$

$$y = \rho_w^{-1/2} a^{-1} \Delta p^{1/2} y_3 \tag{23.2}$$

$$t = a^{-1} \beta^{-1} \Delta p^{-1} t_3 \tag{23.3}$$

$$u(x, y, t, \varepsilon) = \rho_w^{-1/2} \Delta p^{1/2} u_3(x_3, y_3, t_3) + \dots \tag{23.4}$$

$$v(x, y, t, \varepsilon, \Delta p) = \rho_w^{1/2} a^{-1/2} \beta \Delta p^{3/2} v_3(x_3, y_3, t_3) + \dots \tag{23.5}$$

$$p(x, y, t, \varepsilon, \Delta p) = 1/\left(\gamma M_\infty^2\right) + \Delta p p_3(x_3, t_3) + \dots \tag{23.6}$$

$$\rho = \rho_w + \dots \tag{23.7}$$

where parameter a is determined from the solution for the flow in the unperturbed boundary layer $a = \partial u / \partial y_w$. It should be noted that this parameter is equal $O\left(\text{Re}^{-1/2}\right)$ in order of magnitude. Substitution of expressions (23.1–23.7) into the system of Navier-Stokes equations and marginal transition $\text{Re} \to \infty,\ \Delta p \to \infty$ lead to a system of equations

$$\frac{\partial u_3}{\partial t_3} + u_3 \frac{\partial u_3}{\partial x_3} + v_3 \frac{\partial u_3}{\partial y_3} + \frac{\partial p_3}{\partial x_3} = 0 \tag{23.8}$$

$$\frac{\partial u_3}{\partial x_3} + \frac{\partial v_3}{\partial y_3} = 0 \tag{23.9}$$

$$\frac{\partial p_3}{\partial y_3} = 0 \tag{23.10}$$

with boundary conditions

$$v_3 = 0,\ u_3 = y_3,\ x \to -\infty$$

The solution can be found in the form $u_3 = y_3 + A_3(x_3,\ t_3)$
Then we can transform Eqs. (23.8)–(23.10) to the form

$$\frac{\partial A}{\partial t} + A \frac{\partial A}{\partial x} + \frac{\partial p}{\partial x} = 0 \tag{23.11}$$

where the subscript “3” is omitted.

The physical meaning of the function A is a change in the thickness of the boundary layer, taken with the opposite sign. In the external flow it induces a pressure perturbation $\Delta p = -\frac{\partial A}{\partial x}$.

We assume that either small periods of time are considered, or the separation is suppressed in one way or another. Under conditions when a part of the boundary layer is located above the flexible surface, the total change in the displacement thickness will be determined by the change in the thickness of the boundary layer and the surface deformation w. Then the system of equations for two regions has the form

$$\Delta p = -\frac{\partial A}{\partial x} + \frac{\partial w}{\partial x} \tag{23.12}$$

$$\frac{\partial A}{\partial t} + A\frac{\partial A}{\partial x} + v_w + \frac{\partial p}{\partial x} = 0 \tag{23.13}$$

where v_w is the vertical speed of the plate elements.

An equation describing the deformation of a flexible section [6, 7] should be added to this system in order to finally obtain a closed system of equations

$$D\frac{\partial^4 w}{\partial x^4} - M\frac{\partial^2 w}{\partial x^2} + \frac{\partial^2 w}{\partial t^2} + p(x,t) = 0 \tag{23.14}$$

where w is the deflection of the plate, $D = Eh^3/(12(1-\nu^2))$ is the cylindrical stiffness of the plate; $M = M_0 + M_x$, M_0-plate tension, $M_x = \frac{Eh}{2l}\int_{-l/2}^{l/2}\left(\frac{\partial w}{\partial x}\right)^2 \mathrm{d}x$—tension due to plate deflection, E—Young's modulus, h—plate thickness, ν—Poisson's ratio.

Equation (23.14) is a formula of the Von Karman's theory. In addition, the equation for the kinematic connection of parameters in region 3 and the plate

$$\frac{\partial w}{\partial t} = v_w - A\frac{\partial w}{\partial x} \tag{23.15}$$

From (23.12–23.15), a system of equations was obtained for the deflection of a plate and a change in the thickness of the boundary layer

$$\begin{cases} D\frac{\partial^4 w}{\partial x^4} - M\frac{\partial^2 w}{\partial x^2} + \frac{\partial^2 w}{\partial t^2} - \frac{\partial A}{\partial x} + \frac{\partial w}{\partial x} = 0 \\ \frac{\partial A}{\partial t} + A\frac{\partial A}{\partial x} - \frac{\partial^2 A}{\partial x^2} + \frac{\partial^2 w}{\partial x^2} + \frac{\partial w}{\partial t} + A\frac{\partial w}{\partial x} = 0 \end{cases} \tag{23.16}$$

With boundary conditions

$$\begin{gathered} w(x, t=0) = g_1(x);\ w(x \geq l/2, x \leq -l/2,\ t) \equiv 0; \\ \frac{\partial^m w(x = \pm l/2, t)}{\partial x^m} = 0;\ \frac{\partial w(x,\ t=0)}{\partial t} = g_2(x); \\ A(x, t=0) = g_3(x),\ \frac{\partial A(x = \pm L/2, t)}{\partial x} = 0 \end{gathered} \tag{23.17}$$

The region $-l/2 \leq x \leq l/2$ corresponds to the flexible section of the plate, and the region $-L/2 \leq x \leq -L/2$ corresponds to the perturbed region of the flow.

The case $m = 1$ corresponds to the boundary condition of clamped panel, and the case $m = 2$ corresponds to the boundary condition of simply supported panel.

In this paper, we will solve the system of Eqs. (23.16) with boundary conditions (23.17) for dimensionless parameters $D,\ M$.

The nonlinear term in the Burgers equation $A\frac{\partial A}{\partial x}$ characterizes the process of energy transfer between long waves and short waves, the nonlinear term $A\frac{\partial w}{\partial x}$ corresponds to the interaction of the plate with the viscous gas flow.

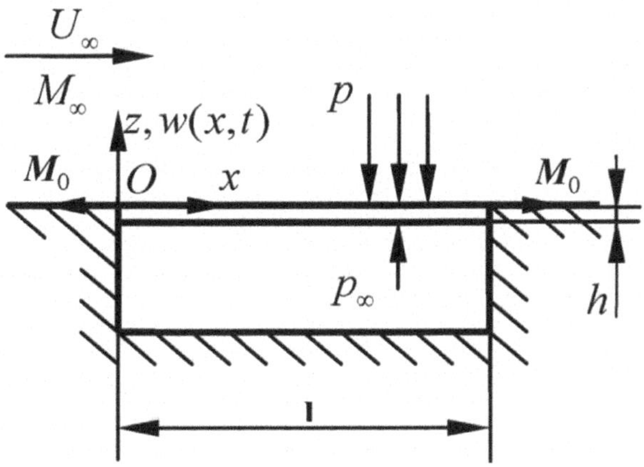

Fig. 23.1 The scheme of elastic panel with unperturbed incident flow

In Fig. 23.1. a scheme of elastic panel with unperturbed incident flow is given, on the one side, a supersonic flow of viscous gas with speed U_∞, and on the other side, the gas is at rest; U_∞, M_∞—speed and Mach number of the unperturbed incident flow, p_∞—pressure of the unperturbed incident flow, p—pressure due to aerodynamic forces, M_0—plate tension, l—plate length, h—plate thickness, $w(x, t)$—plate deflection.

23.2 Linear Stability Analysis of the System of Equations

After linearizing the system of Eqs. (23.16), we obtain a simpler system of equations

$$\begin{cases} D\frac{\partial^4 w}{\partial x^4} - M\frac{\partial^2 w}{\partial x^2} + \frac{\partial^2 w}{\partial t^2} - \frac{\partial A}{\partial x} + \frac{\partial w}{\partial x} = 0 \\ \frac{\partial A}{\partial t} - \frac{\partial^2 A}{\partial x^2} + \frac{\partial^2 w}{\partial x^2} + \frac{\partial w}{\partial t} = 0 \end{cases} \quad (23.18)$$

Using the spectral method [1, 4, 8], we can reduce the system of linear partial differential Eqs. (23.18) to the system of ordinary differential equations.

The system of Eqs. (23.18) can be represented as

$$\begin{cases} DD_{4x}w - MD_{2x}w + \frac{\partial^2 w}{\partial t^2} - D_{1x}A + D_{1x}w = 0 \\ \frac{\partial A}{\partial t} - D_{2x}A + D_{2x}w + \frac{\partial w}{\partial t} = 0 \end{cases}$$

$$\begin{cases} \frac{\partial w}{\partial t} = u \\ \frac{\partial u}{\partial t} = -(DD_{4x} - MD_{2x} + D_{1x})w + D_{1x}A \\ \frac{\partial A}{\partial t} = -D_{2x}w - u + D_{2x}A \end{cases}$$

$$\frac{\partial}{\partial t}\begin{pmatrix} w \\ u \\ A \end{pmatrix}_{3n\times n} = \begin{bmatrix} Z & I & Z \\ D_w & Z & D_{1x} \\ -D_{2x} & -I & D_{2x} \end{bmatrix}_{3n\times 3n} \begin{pmatrix} w \\ u \\ A \end{pmatrix}_{3n\times n} \tag{23.19}$$

where D_{1x}, D_{2x}, D_{4x} are the differential matrices of the first, second and fourth orders of size $n \times n$, Z is the zero matrix of size $n \times n$, I is the identity matrix of size $n \times n$, $D_w = -(DD_{4x} - MD_{2x} + D_{1x})$

In [4], formulas for the determination of differential matrices of the first and second orders are shown

If n is an even number

$$D_{1x,kj} = \begin{cases} 0 & k = j \\ \frac{1}{2}(-1)^{k-j}\cot\frac{(k-j)h}{2} & k \neq j \end{cases}$$

$$D_{2x,kj} = \begin{cases} -\frac{\pi^2}{3h^2} - \frac{1}{6} & k = j \\ -(-1)^{k-j}\frac{1}{2}\sin^{-2}\frac{(k-j)h}{2} & k \neq j \end{cases}$$

$$h = \frac{2\pi}{n}$$

If n is an odd number

$$D_{1x,kj} = \begin{cases} 0 & k = j \\ \frac{1}{2}(-1)^{k-j}\sin^{-1}\frac{(k-j)h}{2} & k \neq j \end{cases}$$

$$D_{2x,kj} = \begin{cases} -\frac{\pi^2}{3h^2} - \frac{1}{12} & k = j \\ -(-1)^{k-j}\frac{1}{2}\sin^{-1}\frac{(k-j)h}{2}\cot\frac{(k-j)h}{2} & k \neq j \end{cases}$$

$$D_{4x} = (D_{1x})^4$$

$$h = \frac{2\pi}{n}$$

the fourth-order differential matrix is calculated by the method shown in [8]. In this paper the number of nodes $n = 64$, when the number of nodes is doubled, the calculation results remain practically unchanged (Table 23.1).

Table 23.1 The eigenvalues of the matrix in (23.19) for $D = 1;\ 2$

$D = 1,\ M = 1$		$i\omega$	$D = 2,\ M = 1$		$i\omega$
k	1	**0.42 ± 1.69i**	k	1	**0.39 ± 1.96i**
	2	0.25 ± 4.68i		2	0.23 ± 6.15i
	3	0.17 ± 9.64i		3	0.15 ± 13.18i
	4	0.12 ± 16.61i		4	0.117 ± 23.06i
	5	**0.10 ± 25.6i**		5	**0.094 ± 35.77i**
	10	**0.05 ± 100.55i**		10	**0.047 ± 141.7i**

Using the spectral method, we can find unstable modes with the highest growth increment for each integer wave number. We can say that there are always growing modes for all integer wave numbers. This reasoning is also valid for all positive real wave numbers $k \geq 1$. It is clear that when using spectral methods, traveling waves are represented as sums of harmonics whose wave numbers are integers.

$$w = \sum_{k_j=-n/2}^{n/2} w_j e^{-i(k_j x - \omega t)}$$

$$A = \sum_{k_j=-n/2}^{n/2} A_j e^{-i(k_j x - \omega t)}$$

n is an even number.

In Figs. 23.2 and 23.3, with an increase in the wave number, the maximum growth increment decreases linearly, and the frequency increases according to a quadratic law. Numerical results showed that the maximum growth increments at small wavenumbers are greater than the maximum growth increments at large wavenumbers. Therefore, unstable modes of long waves grow faster than unstable modes of short waves.

23.3 Solution of the Problem in Nonlinear Mode

The system of Eq. (23.16) with boundary conditions (23.17) are solved using the finite difference method of the third and fourth order of accuracy, built on a uniform grid, and the second order Runge-Kutta method. The following functions and values are selected

$$L = 2;\ l = 1;\ g_1(x) = 0\ \ \forall x;\ g_2(x) = 0\ \forall x;$$
$$g_3(x) = A_0 \exp\left(-25x^2\right);\ A_0 = 1;\ m = 1;\ D = 10^{-3};\ M_0 = 1$$

As the cylindrical stiffness of the plate D decreases, its oscillation becomes more complex due to the nonlinearity of the system. This is explained by the fact that, as the cylindrical stiffness D decreases, the term in the fourth-order derivative becomes less important, and the nonlinear term for tension in (23.16) in the second-order derivative plays the main role.

Using the Bubnov-Galerkin method, we can reduce the plate vibration equation (Von Karman's equation) to a system of ordinary differential equations. Let us expand the solution for deflection of the plate according to the eigenmodes of oscillation of the plate $w_j(x)$ with unknown amplitudes $A_j(t)$

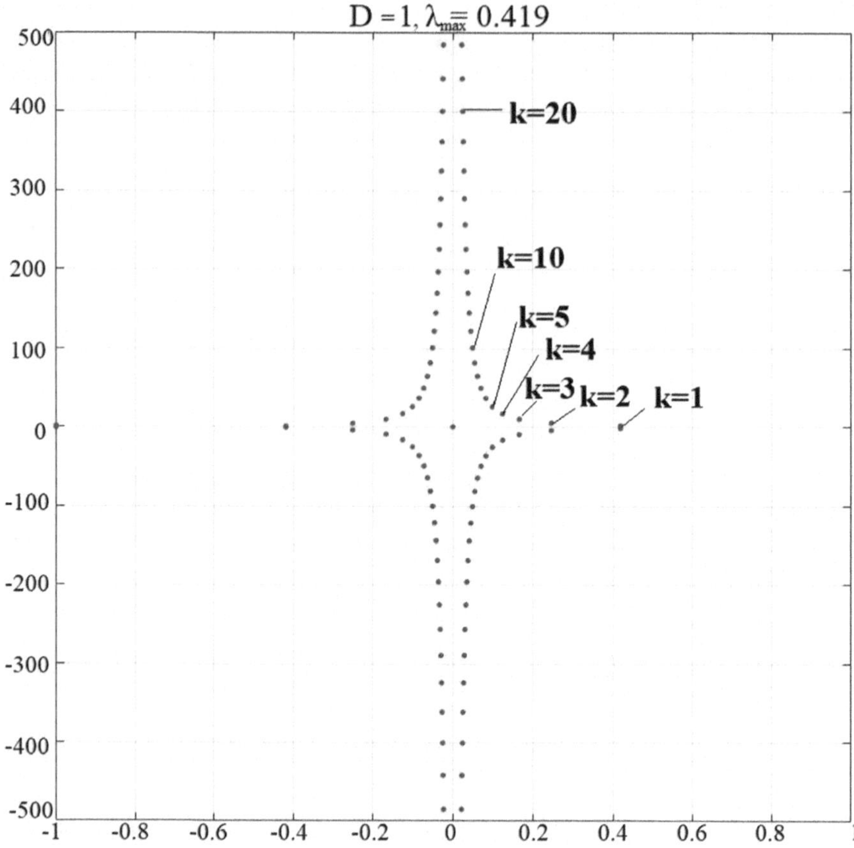

Fig. 23.2 The eigenvalues of the matrix in (23.19) in the complex plane for $D = 1$

$$w(x,\ t) = \sum_{j=1}^{\infty} w_j(x) A_j(t)$$

We substitute the last expression for the deflection of the plate into Eq. (23.14) in vacuum

$$D\frac{\partial^4 w}{\partial x^4} - (M_0 + M_x)\frac{\partial^2 w}{\partial x^2} + \frac{\partial^2 w}{\partial t^2} = 0 \tag{23.20}$$

where $M_x = \frac{6D}{l}\int_{-l/2}^{l/2}\left(\frac{\partial w}{\partial x}\right)^2 \mathrm{d}x$.

Multiply (23.20) by the eigenfunction $w_n(x)$ and integrate the resulting product in x from $-l/2$ to $l/2$. Due to the orthonormality of the eigenfunctions, we obtain the following equation for the nth amplitude [6]

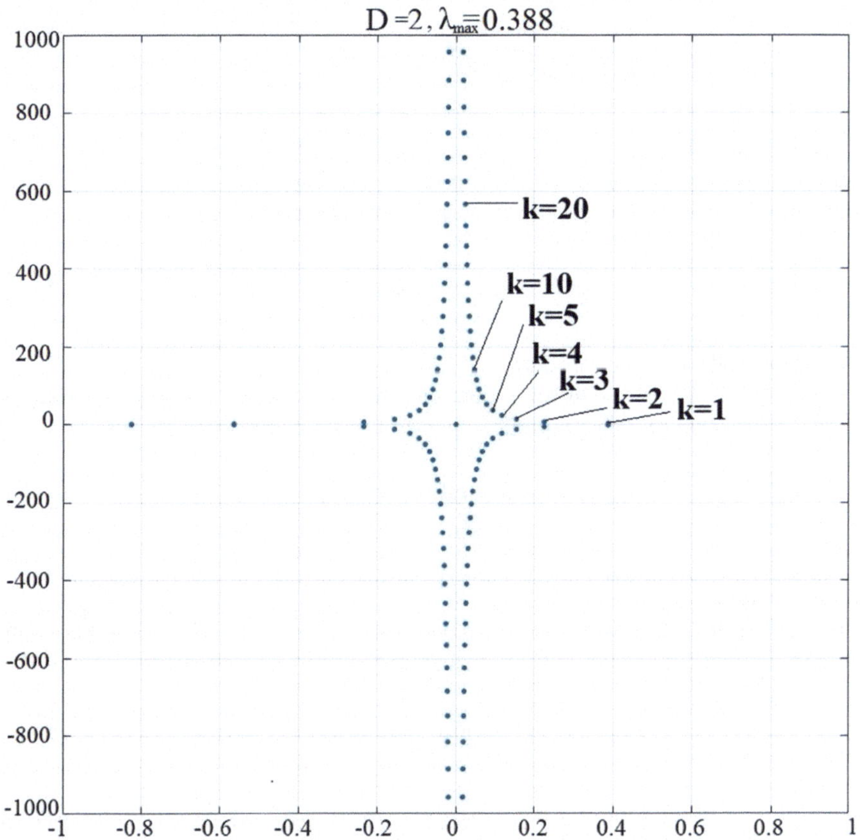

Fig. 23.3 The eigenvalues of the matrix in (23.19) in the complex plane for $D = 2$

$$\frac{\partial^2 A_n}{\partial t^2} + \omega_{0n}^2 A_n + 12D \sum_{m,k,j=1}^{\infty} a_{mk} a_{jn} A_m A_k A_j = 0 \tag{23.21}$$

where

$$\int \left(\frac{\partial^4 w_j}{\partial x^4} - M_0 \frac{\partial^2 w_j}{\partial x^2} \right) w_n dx = \omega_{0n}^2 \delta_{nj}$$

$$\int_{-l/2}^{l/2} w_j w_n dx = \delta_{nj}$$

δ_{nj} is Kronecker symbol

$$a_{jn} = \frac{1}{\sqrt{2l}} \int\limits_{-l/2}^{l/2} \frac{\partial w_j}{\partial x} \frac{\partial w_n}{\partial x} \mathrm{d}x = \frac{1}{\sqrt{2l}} \int\limits_{-l/2}^{l/2} \frac{\partial^2 w_j}{\partial x^2} w_n \mathrm{d}x$$

It is believed that $|A_n| \ll |A_1| \; \forall n > 1$ and $a_{mn} \ll a_{11} \; \forall m, n > 1$, then in the equation for amplitude A_1 we can drop the terms that contain amplitudes with an index above 1. Hence, the equation for A_1 is written in the form

$$\frac{\partial^2 A_1}{\partial t^2} + \omega_{01}^2 A_1 + K a_{11}^2 A_1^3 = 0 \tag{23.22}$$

where $K = 12D$.

Equation (23.22) is a Duffing equation, we look for an approximate solution (23.22) in the form of an asymptotic expansion of Poincare type [2]

$$\begin{aligned} A_1 &= A_{10} + \varepsilon A_{11} + \varepsilon^2 A_{12} + \ldots \\ \omega_{01}^2 &= \omega^2 + \varepsilon c_1 + \varepsilon^2 c_2 + \ldots \end{aligned} \tag{23.23}$$

where $\varepsilon = K a_{11}^2$.

By substituting the expansion (23.23) into the Duffing Eq. (23.22) and equating the coefficients at the same powers, we obtain the following problems for A_{10}, A_{11}

$$\ddot{A}_{10} + \omega^2 A_{10} = 0; \; A_{10}(0) = a_0, \; \dot{A}_{10} = 0 \tag{23.24}$$

$$\ddot{A}_{11} + \omega^2 A_{11} + c_1 A_{10} + A_{10}^3 = 0; \; A_{11}(0) = 0, \; \dot{A}_{11} = 0 \tag{23.25}$$

Solution of Eq. (23.24) $A_{10} = a_0 \cos(\omega t)$.

By substituting the resulting solution (23.24) into Eq. (23.25) and using the trigonometric identity $\cos(3x) = 4\cos^3(x) - 3\cos(x)$, we obtain

$$\ddot{A}_{11} + \omega^2 A_{11} + a_0\left(c_1 + 3a_0^2/4\right)\cos(\omega t) + a_0^3 \cos(3\omega t)/4 = 0$$

We discard the secular term $a_0\left(c_1 + 3a_0^2/4\right)\cos(\omega t)$ therefore $c_1 = -3a_0^2/4$. For the first approximation

$$\omega_{01}^2 = \omega^2 + \varepsilon c_1$$

$$\omega = \sqrt{\omega_{01}^2 - 3\varepsilon a_0^2/4}$$

From here we get the solution of the Duffing equation in the first approximation

$$A_1 = a_0 \cos(\omega t) + \frac{a_0^3}{32}\cos(3\omega t)$$

where $\omega = \sqrt{\omega_{01}^2 - \frac{3\varepsilon a_0^2}{4}}$, $\varepsilon = 12Da_{11}^2$.

Hence the first mode contains two harmonics with frequencies $\omega/2\pi$, $3\omega/2\pi$. In addition, it also follows from [6] that an interaction was observed between the first two main modes of plate vibration. This confirms the reliability of the calculation of the oscillation of the plate, as shown below.

Using the Singular Value Decomposition (SVD) [5], it was revealed that there are 8 main modes (in Fig. 23.4 and Table 23.2), and these modes contain more than 99% of the vibrational energy of the plate.

In Fig. 23.5 and Table 23.3 the ratio of the harmonics frequencies of the first mode $f_{11} : f_{12} \approx 1 : 3$ due to cubic nonlinearity and resonance $f_{22} : f_{11} : f_{21} : f_{23} : f_{12} : f_{24} \approx 1 : 3 : 5 : 7 : 9 : 11$ due to the interaction between the first two main modes are found.

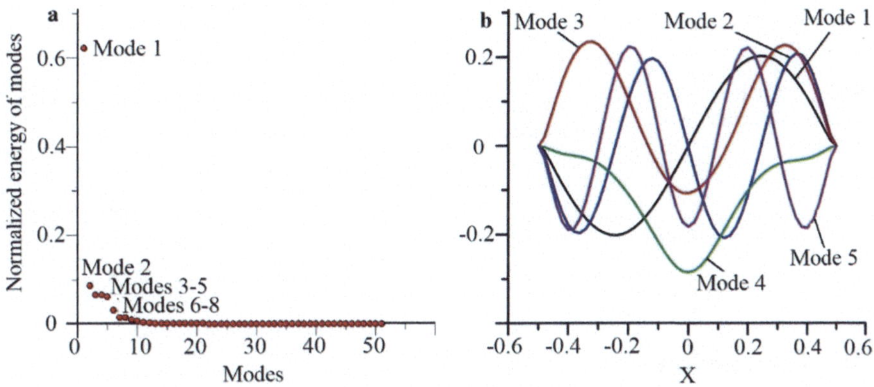

Fig. 23.4 Normalized energy of modes (**a**) and 5 first normal modes (**b**)

Table 23.2 Normalized energy of modes

Mode number	Normalized energy of modes
1	0.628
2	0.090
3	0.074
4	0.068
5	0.061
6	0.030
7	0.019
8	0.006

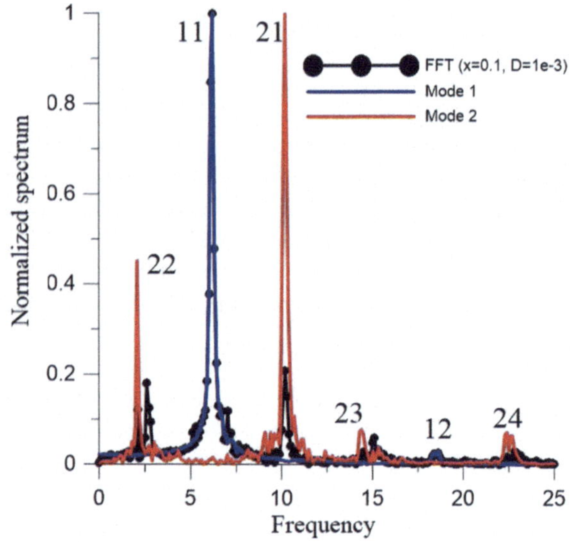

Fig. 23.5 Normalized vibrational spectra of plate for $x = 0.1$ and its first two main modes

Table 23.3 Frequency of harmonics of the first two main modes

Frequency of harmonics of the first mode		Frequency of harmonics of the second mode	
f_{11}	6.2	f_{21}	10.4
f_{12}	18.6	f_{22}	2.1
		f_{23}	14.6
		f_{24}	22.8

23.4 Conclusion

It was found that there are unstable linear modes for all positive wave numbers under certain conditions for the cylindrical stiffness and tension of the plate. In this case, unstable modes of long waves grow faster than unstable modes of short waves.

In the nonlinear mode, the ratio of the harmonics frequencies of the first mode $f_{11} : f_{12} \approx 1 : 3$ due to cubic nonlinearity and the ratio of the harmonics frequencies $f_{22} : f_{11} : f_{21} : f_{23} : f_{12} : f_{24} \approx 1 : 3 : 5 : 7 : 9 : 11$ due to the interaction between the first two main modes are found.

References

1. Canuto, C., Hussaini, M.Y., Quarteroni, A., Zang, T.A.: Spectral Methods in Fluid Dynamics, 567 p. Springer, Berlin (1988)

2. He, J.H.: Some asymptotic methods for strongly nonlinear equations. Int. J. Mod. Phys. B. **20**, 1141–1199 (2006)
3. Neiland, V.Y., Bogolepov, V.V., Dudin, G.N., Lipatov, I.I.: Asymptotic Theory of Supersonic Viscous Gas Flows, 456 p. Fizmatlit, Moscow (2003). (In Russian)
4. Trefethen, L.N.: Spectral methods in MATLAB, 181 p. SIAM, Philadelphia (2000)
5. Trefethen, L.N.; Bau III, David Numerical Linear Algebra, 361 p. SIAM, Philadelphia (1997)
6. Vedeneev, V.V.: Nonlinear high-frequency flutter of a plate. Fluid Dyn. **5**, 858–868 (2007)
7. Vedeneev, V.V.: Panel flutter at low supersonic speeds. J. Fluids Struct. **29**, 79–96 (2012)
8. Weideman, J.A.C., Reddy, S.C.: A MATLAB differentiation matrix suite. ACM Trans. Math. Softw. **26**(4), 465–519 (2000)

Chapter 24
Investigation of the Influence of Operational Loading Regimes on the Service Life of Nuclear Power Plants

Nikolay A. Makhutov, Mikhail M. Gadenin, Sergey V. Maslov, Dmitry O. Reznikov, Sergey N. Pichkov, and Vladimir A. Panov

Abstract The paper presents the results of design and experimental studies of strength and service life of NPP components subjected to multi-frequency cyclic loading regimes. Experimental studies were carried out on laboratory specimens of low-alloy austenitic reactor steels for a wide range of variation of frequencies and amplitudes of cyclic stresses, as well as on models during bench life tests and on the equipment of the reactor primary circuit during commissioning and during the initial period of operation. In the general case, a comprehensive design and experimental analysis of the initial and residual life of the equipment of nuclear power plants are based on an analytical assessment of the conditions for the accumulation of operational damage under various operating loading conditions, with accounting for the corresponding constitutive laws and the kinetics of the mechanical properties of materials, as well as on the study of conditions for transition to limit states

N. A. Makhutov · M. M. Gadenin · S. V. Maslov · D. O. Reznikov · S. N. Pichkov ·
V. A. Panov (✉)
Mechanical Engineering Research Institute of the RAS, 101990 4 Maly Kharitonievsky lane, Moscow, Russia
e-mail: vapanov@okbm.nnov.ru

N. A. Makhutov
e-mail: kei51@mail.ru

M. M. Gadenin
e-mail: safety@imash.ru

S. V. Maslov
e-mail: maslovsv@inbox.ru

D. O. Reznikov
e-mail: mibsts@mail.ru

S. N. Pichkov
e-mail: okbm@okbm.nnov.ru

N. A. Makhutov · M. M. Gadenin · S. V. Maslov · D. O. Reznikov
Joint Stock Company Afrikantov OKB Mechanical Engineering, 603074 Burnakovsky proezd, 15, Nizhny Novgorod, Russia

H. Altenbach et al. (eds.), *Multiscale Solid Mechanics*,
Advanced Structured Materials 141,
https://doi.org/10.1007/978-3-030-54928-2_24

using criteria of strength, crack resistance and damage tolerance. The approaches to assessment of the reduction of fatigue durability due to combined action of low- and high-frequency harmonics of loading are described. A model for estimation of fatigue damage of structural components under two-frequency loading is presented which considers damage accumulation as a summation of damages due to static loading plus damages due to low- and high-frequency loading. The obtained results present the basis for the estimation of the durability of NPP components at two-frequency loading.

Keywords Service life of NPP · Multi-frequency cyclic loading · Accumulation of damage · Mathematical modeling · Experimental data on durability

24.1 Introduction

The age of nuclear energy began in 1954 when the first nuclear power plant with channel-type 5 MW reactor was put into service in Obninsk (USSR-Russia). Since then, in the world's leading countries (USSR-Russia, USA, Great Britain, France, etc.), a full spectrum of new nuclear type of energy generation has appeared. By 2019, there are ten nuclear power stations with 35 power units operating in the Russian Federation. The total capacity of these NPP equals to 29 GW.

The studies of strength and service life of nuclear power plants carried out in Russia and abroad form the scientific basis for development of

(1) regulatory documents on design;
(2) methods for assessment of reliability, damage tolerance and conditions of initiation and propagation of emergencies according to risk criteria;
(3) new principles, technologies and tools for ensuring safe operation of nuclear power plants.

The decision made by Russian (USSR) regulatory bodies to carry out a special research program to back up standards for design of nuclear reactors was of particular importance for the country [15, 16]. Such standards have also been developed in the USA [1]. Subsequently, the IAEA developed the relevant international regulatory documents [7]. Russian national standards for the design of nuclear power stations [15, 16] include sections focused on assessment of strength and service life as well as on the extension of the term of their safe operation.

For many years, Joint Stock Company Afrikantov OKB Mechanical Engineering and Mechanical Engineering Research Institute of the Russian Academy of Sciences have been conducting systematic computation and experimental studies of strength and service life of load carrying components of NPP equipment with water-cooled water-moderated energy reactor (WWER) and BN (sodium-cooled fast breeder) reactors. Academicians F. M. Mitenkov and K. V. Frolov considered these studies of fundamental importance in terms of the laws of elastoplastic cyclic deformation, damage accumulation and fracture under complex multi-frequency regimes of thermomechanical loading. These regimes are related to low-frequency cycles of start-ups, shutdowns, power control and actuation of emergency protection systems superimposed on high-frequency vibration cycles as well as pulses of pressure and temperature.

Experimental studies were carried out on

(i) laboratory specimens of low-alloy austenitic reactor steels for a wide variety of the frequencies and amplitudes of stresses;
(ii) standard models during bench service life tests and
(iii) on the equipment of the first circuit of WWER and BN reactors at the stage of starting-up and adjustment and during the initial operation period.

On this basis, the assessments of the effect of two-frequency and multi-frequency loading on the strength of NPP equipment were carried out. The corrections of the standard checking calculations were made. The theory and criteria of service life and reliability have recently been included in traditional analysis of the working capacity of NPP equipment [8, 9, 11–14]. The main findings of the joint design and experimental studies in this field are presented below.

24.2 Modes of Operational Loading of NPP Equipment

An analysis of the operational loading conditions of reactor components (Fig. 24.1a) shows that along with the traditional representation of cyclic loading with sinusoidal or triangular cycles (Fig. 24.1b) or schematization of loading by known methods, for example, the rainflow cycle counting method (Fig. 24.1c), in some cases, it is possible to describe complex loading regimes as polyharmonic processes of multi-frequency cyclic impacts which, in turn, may after the exclusion of harmonics with small amplitudes be converted to a two-frequency process (Fig. 24.1d), characterized by the superposition of the low- and high-frequency loadings. Such loading conditions are typical for power units. Here, low-frequency loading corresponds to stresses caused by start-ups and shutdowns of the units, and high-frequency loading corresponds to vibrations of hydrodynamic and aerodynamic origin. In addition, high-frequency (in a relative sense) variable stresses in these conditions can be caused by the change in the power of the units or by the control of the operating mode at a specific level (Fig. 24.1e).

The two-frequency loading mode can be characterized by four load parameters, namely, the high cycle stress amplitudes and frequency, and the low-cycle stress amplitude and frequency. These parameters depend on the specific operating conditions of the power unit [11–14] that are described in Table 24.1.

Due to the relative similarity of the general mechanisms of damage accumulation and service life exhaustion, two types of calculations are provided for the design and operation stages of typical NPP components (with ensuring the initial and residual strength, service life and safety) [8–16].

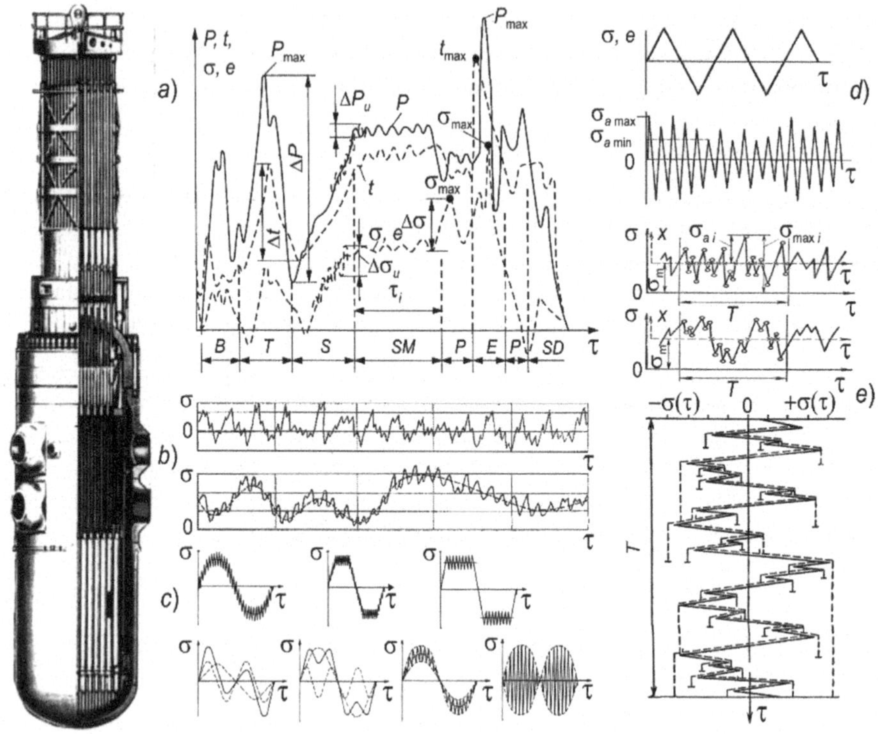

Fig. 24.1 Schematization of operating modes of loading

The total operating time τ of the analyzed components may change from $10^1 \div 10^2$ s to $10^8 \div 10^9$ s. These components are subjected to various loading regimes with the following ranges of the number of loading cycles *N*:

- 10^0–10^1 for extreme cycles (start-ups, tests, emergency shutdowns and emergency situations);
- 10^2–10^3 for operational mode cycles (approach to operating mode, power regulation, actuation of protection systems);
- 10^4–10^5 for operational adjustment cycles (technological cycles, regulation);
- 10^6–10^8 for operational technological cycles (technologies for maintaining rotor rotation modes, pressure changes);
- 10^9–10^{12} for operational vibration cycles (vibration, pulsation of temperature and pressure).

Figure 24.2 shows a detailed diagram of the change in the design loading parameters of highly loaded components over time τ. These parameters at the certain moments of time reach their minimum and maximum values that determine the ranges and amplitude values of pressure of the working medium p—$p_{\max}$, p_a, Δp; temperature t—$t_{\max}$, $t_{\min}$, t_a, Δt; nominal and local stresses σ—$\sigma_{\max}$, $\sigma_{\min}$, σ_a or

Table 24.1 Parameters of operating modes of the power plant

No.	Loading mode	Number of cycles
I Warm-up and cooling cycles		
1	Planned warm-up from the room temperature to the nominal one	100–500
2	Planned cooling from the nominal temperature to the room one	100–500
II Energy (operational) cycles		
1	Change in power from 0 to 100% with the rate of 5% per min	0–15,000
2	Change in power from 100% to 0 at the rate of 5% per min	0-15,000
3	Power change from 50% to 100% at the rate of 15% per min	2000–15,000
4	Power change from 100 to 50% at the rate of 15% per min	2000–15,000
5	Stepwise increase in power by 10% in the range 0–100%	0–2000
6	Stepwise reduction of power by 10% in the range 0–100%	0–2000
7	Stepwise reduction of power by 50% in the range of 50–100%	0–200
8	Pressure fluctuations near the stationary level in the range from 0.03 to 0.07 MPa, +2.5 °C	More than 300,000
III Ensuring safe operation		
1	Shutdowns due to malfunctions of the reactor	200–400
2	Shutdowns due to turbine malfunctions	0–40
3	Hydrostatic tests	5–40
4	Leak tests	5–300
5	Functioning of the safety valve	0–200
6	Operation of the cooling system	10

strains e—$e_{\max}$, $e_{\min}$, e_a. This diagram also shows the possible change in these basic parameters at various stages of operation of the analyzed components: building-up (*B*), start-up (*S*) and hydrotesting (HT), stationary modes (SM), power change (PC), emergency modes (E), actuation of protection system (*P*), vibration (*V*), shutdowns (SD).

Changes in pressure Δp, temperatures Δt and the impact of vibrational loads cause the presence of high-frequency stress amplitudes σ_{au}, which, together with the basic loads from the main operating modes create two-frequency and multi-frequency loading regimes with frequency ratios in the range $f_a/f_{\text{au}} = 10^1$–10^5.

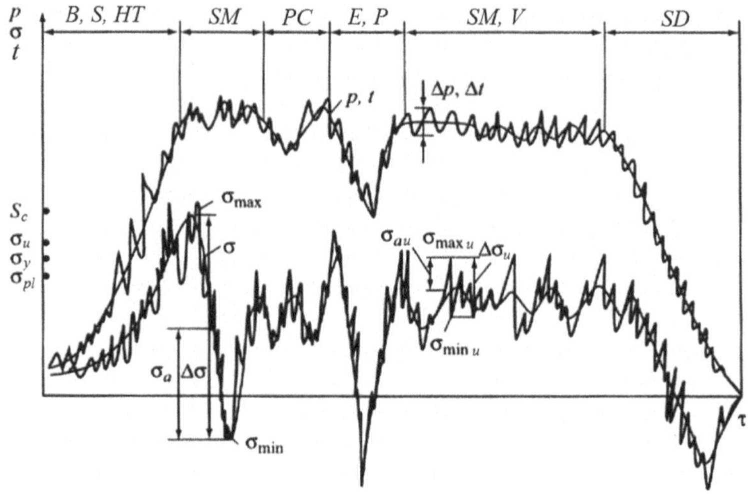

Fig. 24.2 Diagram of changes in operational parameters of loading and stress

24.3 Mathematical Modeling of Damage Caused by the Two-Frequency Loading Mode

When forecasting the initial cyclic design life, a diagram of cyclic strength can be constructed with accounting for the initial characteristics of the mechanical properties of structural materials and their changes at various stages of the life cycle [4, 8–14]. The same diagram is also used to assess the accumulated cyclic damage, as a rule, using the linear law of summing the component damage, which in turn varies nonlinearly depending on the parameters and operating conditions [2–6, 8, 14].

The assessment of durability under two-frequency loading conditions can be carried out by various methods [2, 3, 5, 6, 8, 9, 11–14, 17], one of them being a method for assessing the constituents of damage from individual components of such loading mode (low frequency, high frequency and static components) with accounting for the ratio of frequencies, amplitudes and variable asymmetries of the acting stresses [2, 5, 6]. Another method consists in direct conduction of two-frequency tests at a given range of load levels under the required parameters of two harmonics. This method uses empirical relationships to determine the reduction in durability under these conditions as compared to the durability under the equivalent single-frequency loading [3, 5].

The degree of damage caused by the high-frequency component under two-frequency loading is affected by both the ratio of the amplitudes of acting stresses σ_2/σ_1 and the ratio of their frequencies f_2/f_1. In the region of small values of f_2/f_1, a change in this ratio within one order of magnitude does not cause a significant change in durability because in this case, the constituent of damage from the high-frequency component varies insignificantly. With an increase in σ_2, as well as with an extension of the frequency range, the ratio of the degrees of damage from both components changes, which entails a significant change in durability [2, 3, 5, 6, 8, 17].

In the general case, the resistance to low-cycle fracture is described by the equation of the low-cycle fatigue curve in the form [8, 9]:

$$e_a = \frac{1}{2\cdot(4N)^{m_p^t}}\ln\frac{100}{100-\psi_c^{t\tau}} + \frac{1}{(4N)^{m_e^t}}\cdot\frac{S_c^{t\tau}}{E^t} \tag{24.1}$$

where all the characteristics of the mechanical properties of the material and the parameters included in Eq. (24.1) are defined as quantities that depend on temperature t and loading time τ.

The total damage d_Σ accumulated under low-cycle loading may be represented as summation of fatigue $d_f^{t\tau}$ and quasistatic $d_s^{t\tau}$ damages: $d_\Sigma = d_f^{t\tau} + d_s^{t\tau}$ [8],

where

$$d_f^{t\tau} = F[e_a(k), t, \tau]\, d_s^{t\tau} = F[e_p^{(k)}(k), t, \tau] \tag{24.2}$$

Damage accumulation under stress controlled two-frequency low-cycle loading can also be described on the basis of linear summation of the additional component of fatigue damage d_f^{II} from superimposed high-frequency amplitude of stress σ_2 or strain e_{a2} with variable asymmetry, with components of fatigue d_f^I and quasistatic d_s damage included in expressions (24.2). Then, the total damage under two-frequency loading mode can be expressed as [2, 5, 6]:

$$d_\Sigma = d_f^I + d_f^{II} + d_s \tag{24.3}$$

or taking into account expressions (24.1)–(24.3), it can be rewritten as

$$d_\Sigma = \int_0^{N_T} \frac{1}{N[e_a(n), t, \tau]}\mathrm{d}n + \int_0^{N_T^*} \frac{1}{N^*[e_{a2}, t, \tau, r(n^*)]}\mathrm{d}n^* + \int_0^{N_T} \frac{e_n}{\varepsilon_{t\tau}}\mathrm{d}n = 1.0 \tag{24.4}$$

Estimation of the service life under these conditions can be performed on the basis of expression (24.4), provided that the total damage d_Σ reaches a critical value equal to 1.0.

It is known from the experience of operation, tests and calculations that under the conditions of two-frequency loading, especially under extreme loading conditions, the value of durability (service life) decreases from that observed in case of single-frequency loading which is equivalent in terms of maximum loads. This decrease becomes more substantial when the ratio of high frequency and total strains goes up as well as the ratio of the corresponding loading frequencies (Fig. 24.3).

Along with the described above method for estimation of durability under two-frequency loading that is based on the criterion of summation damages which are

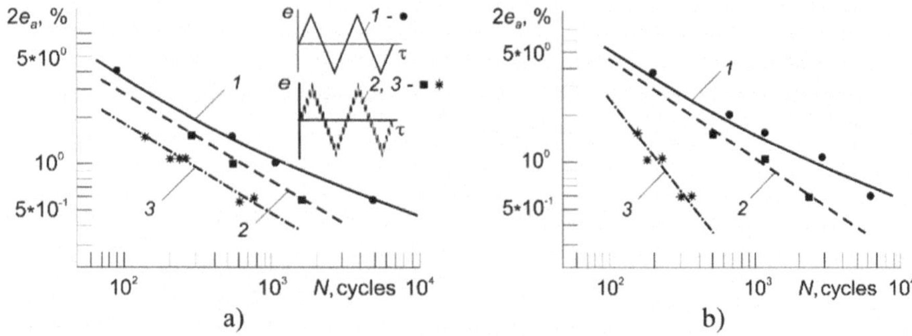

Fig. 24.3 Low-cycle fatigue curves for steel 12Cr2MFA (**a**) and steel Cr18N10T (**b**) under single-frequency (1) and two-frequency (2 and 3) low-cycle loading at the frequency ratio $f_2/f_1 = 1500$ and temperature $t = 300$ °C

expressed in terms of the deformation characteristics according to expressions (24.1)–(24.4), the effect of reduction of the two-frequency durability N_2 from the value of the equivalent (in terms of the total load) single-frequency durability N_1 can be approximately estimated by the ratio of frequencies and amplitudes of the high-frequency and low-frequency components of the loading process in the form [3, 5, 6]:

$$N_1 = N_2(f_2/f_1)^{k_N(e_{a2}/e_a)} \ or \ N_1 = N_2(f_2/f_1)^{k_N(\sigma^*_{a2}/\sigma^*_a)} \quad (24.5)$$

where e_{a2} and σ^*_{a2} as well as e_a and σ^*_a are the amplitude values of high-frequency strains and the corresponding pseudoelastic stresses ($\sigma^*_a = e_a E$, where E is the elastic modulus), as well as their total values (along with low frequency), respectively. In a certain range of ratios of amplitudes and frequencies of the superimposed harmonics, the decrease in durability calculated by expressions (24.5) also satisfactorily agrees with experimental data described in the standard for power equipment design [16].

Expression (24.5) assumes a linear relationship between the logarithm of the ratio N_2/N_1 and the ratio e_{a2}/e_a (σ^*_{a2}/σ^*_a) depicted by a straight line in semilogarithmic coordinates. Figure 24.4 presents theoretical lines constructed using expressions (24.5) and the corresponding experimental data on the durability under single-frequency and two-frequency loading modes. One can see that the straight lines corresponding to the tested specimens made of 12Cr2MFA and Cr18N10T steels (solid lines) under the strain-controlled regime of two-frequency loading with different ratios of amplitudes of high-frequency and total low-frequency deformations (pseudoelastic stresses) satisfactorily correspond to the results of experiments. In this case, the value of the durability reduction factor k_N in expressions (24.5) for steel 12Cr2MFA is $k_N = 1.2$ (line 1 in Fig. 24.4a and triangular dots), and for steel Cr18N10T – $k_N = 1.9$ (line 1 on Fig. 24.4b and square dots).

These values of the factor k_N are determined on the basis of assumptions about the average values of the corresponding experimental durability. If the indicated values

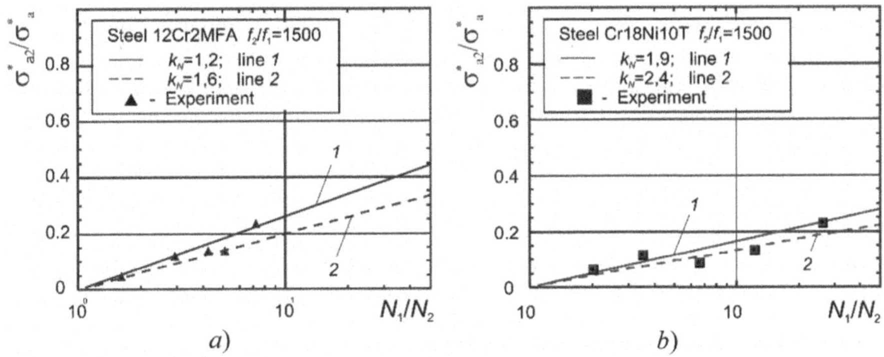

Fig. 24.4 Dependences of the change in durability for various ratios of low- and high-frequency amplitudes of pseudoelastic stresses in case of a two-frequency loading mode for steels 12Cr2MFA (**a**) and Cr18N10T (**b**)

of k_N are taken from the points that are out of the general set, i.e., with safety factor against durability—dashed lines in Fig. 24.4a–b, then the values of the durability decrease factor k_N will be equal to 1.6 for 12Cr2MFA steel (line 2 in Fig. 24.4a) and equal to 2.4 for Cr18N10T steel (line 2 in Fig. 24.4b). These values of the durability decrease factor k_N under two-frequency loading were used in the corresponding calculations [5, 12, 15].

It should be noted that expression (24.5) fits well to experimental data only for relatively small values of the ratio e_{a2}/e_a that do not exceed $0.2 \div 0.3$. Otherwise, the calculated values of the durability decrease turn out to be overestimated. This requires that appropriate limitations on the use of expressions (24.5) are introduced. Such limitations may consist of the following conditions:

- the ratio of the amplitude of the pseudoelastic high-frequency stresses σ^*_{a2} to the amplitude of the corresponding total low-frequency stresses σ^*_a is in the range of $0 < \sigma^*_{a2}/\sigma^*_a < 0.3$;
- the absolute value of the maximum and minimum stresses during loading with superposition of high-frequency stresses does not exceed the value of $(0.2 \times 10^{-2} E^t + \sigma^t_{0,2})$
- the ratio of the frequencies of the applied vibrational and main cyclic loading f_2/f_1 does not exceed 5×10^5;
- the number of cycles with the amplitude of the superimposed high-frequency stresses σ^*_{a2} according to Fig. 24.1d exceeds 10 ($f_2/f_1 > 10$).

The presented above assessments were performed for two steels with different cyclic properties (cyclically softened steel 12Cr2MFA and cyclically hardened steel Cr18N10T) at certain ratios of variable amplitudes σ^*_{a2}/σ^*_a (e_{a2}/e_a) and fixed frequencies $f_2/f_1 = 1500$ of harmonics according to Fig. 24.1d at a temperature $t =$ 300 °C.

In the general case, a comprehensive analytical and experimental analysis of the initial and residual service life of NPP equipment is based primarily

(i) on assessing the conditions of the damage accumulation under various loading regimes, with accounting for the corresponding constitutive equations,
(ii) and also on studying the conditions of transition to limit states with accounting for the kinetics of mechanical properties of structural materials, criteria of strength, crack resistance and damage tolerance.

The described above procedures for assessment of damage accumulation are implemented using a set of criteria expressions, design equations and design characteristics used to analyze and determine the normal and limit states of engineering facilities [4, 8–14]. The set of criteria expressions includes the following:

- To evaluate static and long-term strength

$$F_Q\{\sigma, e, t, \tau\} = F_Q\left\{f_1\left(\frac{\sigma_y^\tau}{n_y}, \frac{\sigma_u^\tau}{n_u}, \frac{\sigma_l^\tau}{n_\sigma}, \frac{e_c^\tau}{n_e}, \frac{\tau_c}{n_\tau}\right), f_2(m)\right\} \tag{24.6}$$

where F_Q is the functional characterizing the dependence of stresses on the set of force actions Q; σ, e are stresses and strains acting in time τ at temperature t; f_1 is the functional dependence; σ_y^τ, σ_u^τ and σ_l^τ are the yield strengths, ultimate strengths and long-term strengths of the material for the given time of deformation τ, respectively; e_c^τ is the value of critical (upon failure) strain at a given time τ; n_y, n_u, n_σ, n_e, n_τ are safety factors against yielding, fracture, stress, strain and time; f_2 is the functional (in most cases exponential) dependence of the strain hardening exponent m in the elastoplastic region of deformation [8].

- To estimate the service life using the parameters of the number of cycles N and time τ:

$$F_T\{\sigma, e, N, \tau\} = \left\{f_1\left(\frac{\sigma_a}{n_\sigma}, \frac{e_a}{n_e}, \frac{N_p}{n_N}\right) f_2(\sigma_y, \psi_c, m_p, m_e)\right\} \tag{24.7}$$

where F_T is the functional characterizing the dependence of the service life on the amplitudes of stresses σ_a, strains e_a, number of cycles to failure N_p and corresponding safety factors, as well as the available plasticity of the material ψ_c (reduction in cross-sectional area) and the power exponent in the equation of the fatigue curve of type (24.1) for the plastic m_p and elastic components m_e of cyclic strain e_a [8–14].

24.4 Conclusions

An analysis of the design and experimental data on durability under one-frequency and two-frequency loading conditions to assess the influence of the amplitude–frequency parameters of these modes on cyclic durability shows that the superposition of the high-frequency component of cyclic deformation on the main process of low-cycle loading leads to a decrease in cyclic durability. The higher the level of ratios of amplitudes and frequencies of the superimposed harmonic processes, the more substantial is the reduction of durability. As a first approximation, the estimation of the durability reduction under these conditions can be made using both the criterion expression (24.4) for summation of damages and empirical expressions (24.5), with accounting for the certain above noted limitations regarding the ratios of the frequencies and amplitudes of low- and high-frequency harmonics.

References

1. Boiler and Pressure Vessel Code—2010 Edition. ASME, NY (2011)
2. Gadenin, M.M.: Specific features of the development of deformations and damage accumulation under two-frequency low-cycle loading and elevated temperatures. J. Machinovedenie (Soviet Machine Science) **1**, 69–77 (1976)
3. Gadenin, M.M.: Change in resistance to low-cycle deformation and fracture due to the two-frequency loading process. In: Problems of Reliability and Durability of Structural Components in Mechanical Engineering and the Construction industry. Abstracts of the XIII Scientific and Technical Conference, pp. 5–6. NTO, Sverdlovsk (1978)
4. Gadenin, M.M.: Characteristics of mechanical properties of materials in studies of conditions of attainment of marginal stated. J. Inorg. Mater. (2013)
5. Gadenin, M.M.: Study of the effect of strain amplitude ratio at two-frequency cyclic loading. J. Inorg. Mater. (2019)
6. Gadenin, M.M.: Study on damaging and fatigue life of constructions under single- and two-frequency loading modes based on deformational and energy approaches. J. Inorg. Mater. (2018)
7. IAEA Safety Standards: Safety of Nuclear Power Plants: Design. Specific Safety Requirements. International Atomic Energy Agency—IAEA. No. SSR-2/1. IAEA, Vienna (2012)
8. Makhutov, N.A.: Strength and Safety. Basic and Applied Research. Nauka Publ, Novosibirsk (2008)
9. Makhutov, N.A.: Safety and Risks: System Research and Development. Nauka Publ, Novosibirsk (2017)
10. Makhutov, N.A., Gadenin, M.M.: Basic research on the safe operation of man-made facilities based on criteria of strength, service life and risk. J. Eng. Autom. Probl. **3**, 7–24 (2018)
11. Makhutov, N.A., Dragunov, YuG, Frolov, K.V., et al.: Dynamics and Strength of Water-Moderated Power Reactors (in series "Research on Stress and Strength of Nuclear Reactors"). Nauka Publ, Moscow (2004)
12. Makhutov, N.A., Frolov, K.V., Dragunov, YuG, et al.: Problems of Strength and Safety of Water-Moderated Power Reactors (in series "Research on Stress and Strength of Nuclear Reactors"). Nauka Publ, Moscow (2008)
13. Makhutov, N.A., Frolov, K.V., Dragunov, YuG, et al.: Maintenance of Operating Life and Survivability of Water-Moderated Power Reactors (in series "Research on Stress and Strength of Nuclear Reactors"). Nauka Publ, Moscow (2009)
14. Mitenkov, F.M., Kaydalov, V.B., Korotkikh, YuG, Panov, V.A., Pichkov, S.N.: Methods for Substantiation of the NPP Service Life. Mashinostrieniye Publ, Moscow (2007)
15. Standards for Strength Calculating of Equipment and Pipelines of Nuclear Power Plants. Energoatomizdat Publ., Moscow (1989)

16. Standards for Strength Calculating of Reactors Components: Steam Generators, Vessels and Pipelines of Nuclear Power Plants. Experimental and Research Nuclear Reactors and Installations. Metallurgy Publ, Moscow (1973)
17. Locati, L.: In: Colloqufum on Fatigue, Stockholm-1955, p. 160. Springer-Verlag, Berlin, Gottingen, Heidelberg (1956)

Chapter 25
On the Theory of Rocking Structures in Seismically Active Zones

Vladimir S. Metrikin, Leonid A. Igumnov, and Andrei V. Metrikine

Abstract Vibrations of a rectangular block on a rigid, horizontally vibrating plane are analyzed in the framework of the theory of the controlled dynamical systems (CDS). The friction force between the block and the supporting plane is assumed to be sufficiently high for the block not to slide along the plane. Under this assumption the block may only rotate about its vertices that are in contact with the supporting plane. Energy loss occurs due to impact interactions of the block and the supporting plane. The impacts are modelled using the angular velocity restitution coefficient. Using qualitative CDS methods the controllability regions for the block to return to any arbitrarily small vicinity of point (0, 0), corresponding to zero values of the angular velocity are derived. Dimensions of a safe zone of the controllability region in the phase space of the CDS are also determined. It is shown that the safe zone decreases with the increase of the maximum value of acceleration of the plane. The controllability regions are shown for different geometrical dimensions of the block and the upper limit of the supporting plane acceleration.

Keywords Mathematical model · Point mapping · Stability · Bifurcation diagrams · Chaos

---

V. S. Metrikin (✉) · L. A. Igumnov
Research Institute for Mechanics, National Research Lobachevsky State University of Nizhny Novgorod, 23 Gagarin Avenue, Building 6, Nizhny Novgorod 603950, Russian Federation
e-mail: v.s.metrikin@mail.ru

L. A. Igumnov
e-mail: igumnov@mech.unn.ru

A. V. Metrikine
Faculty of Civil Engineering and Geosciences, Delft University of Technology, Building 23 Stevinweg 1, 2628 CN Delft, The Netherlands
e-mail: a.metrikine@tudelft.nl

H. Altenbach et al. (eds.), *Multiscale Solid Mechanics*,
Advanced Structured Materials 141,
https://doi.org/10.1007/978-3-030-54928-2_25

25.1 Introduction

A rigid rectangular block, dynamics of which is analyzed in the present paper, can be used as a model of an ancient civil structure (arches, buildings made of large stones, etc). Alternatively, it can model separate objects such as a railway wagon or a plane under the action of earthquakes. The characteristic feature of the model under consideration is that it is not attached to a foundation and can exhibit rocking motion about the contact points with the supporting structure. The issue of stabilization of such and similar structures is of importance for seismically active zones [1–5]. Such systems are also of interest for studying because some structures that, at first sight, seem to be unstable (e.g., separate stone poles or elevated water reservoirs), sometimes survive earthquakes better than apparently more stable structures. For instance, Fig. 25.1 in [3] shows electrical facilities at Sylmar converter station destroyed during the 1971 San Fernando earthquake, and Fig. 25.2 depicts a train overturned in the 1906 Point Reyes earthquake, San Francisco. Investigations into the rocking of a solid block on a horizontal base were presented in [3] where accelerations of the supporting plane were assumed to be of finite duration and represented by a block function or a single-period sine function. Expressions for the critical acceleration magnitude leading to the overturning of the rectangular block were obtained. The excitations considered in [3] are unrealistic, which make the fundamental results obtained in [3] to be of purely theoretical interest [6–8]. The work by Housner [3] was followed by a large number of studies on the complex dynamics of one of the simplest artificial structures—a rectangular block on a vibrating surface [2, 6–20]. The rocking response of the block under the effect of a steady-state harmonic load was studied in detail in [17], where 'hazardous' and 'safe' zones in the phase space were constructed and analytical methods were developed for determining periodic and subharmonic regimes of the system motion. The latter study was continued in [11], were a further explanation of the mathematical structure of the problem was offered, using the notions of the orbit stability and Poincare's cross-sectional surfaces. The latter study gave a thorough insight into the rocking response to a cosine and sine acceleration pulse. These two trigonometric pulses are physically realizable and, in a number of cases, resemble non-parallel and faulty components of motions registered in the vicinity of the source

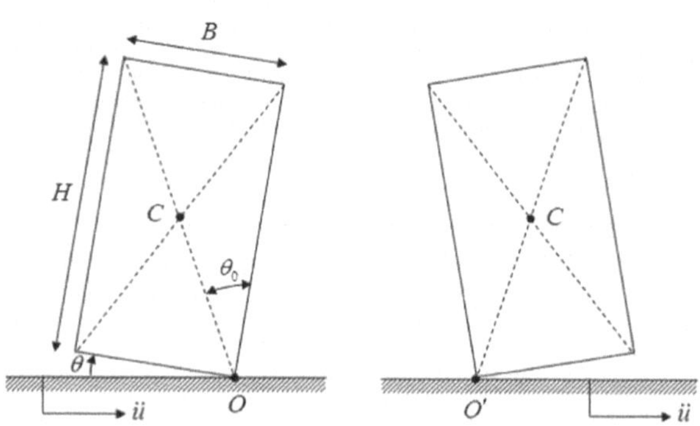

Fig. 25.1 System considered (shown schematically)

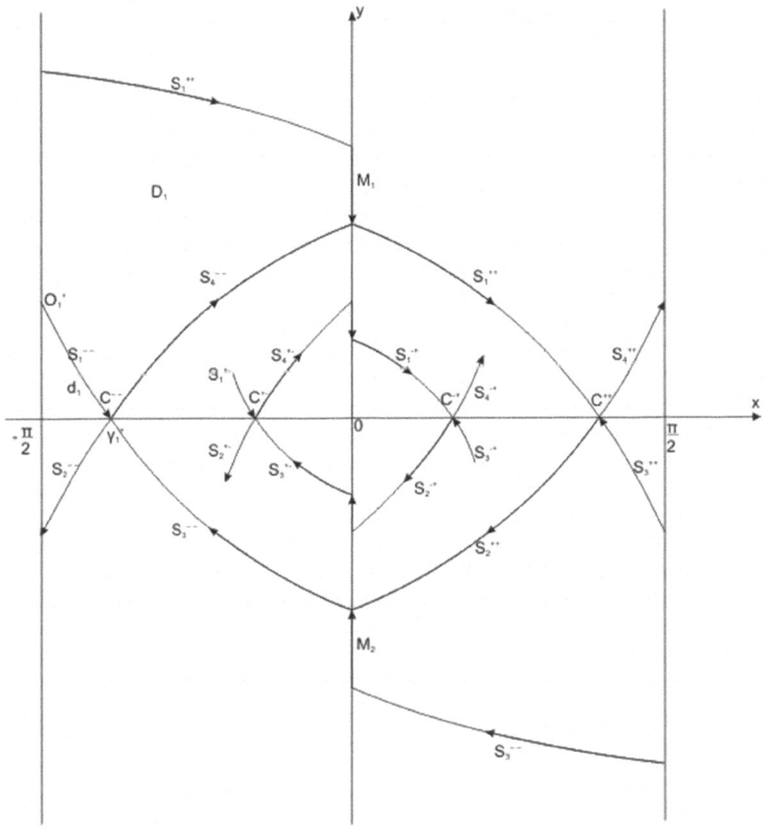

Fig. 25.2 Controllability region

of strong earthquakes [6, 7]. Finally, there are quite recent papers [21] and [22] in which the block response to particular acceleration signatures is studied.

In the present paper the CDS methods are employed to determine the controllability regions to an arbitrarily small vicinity of the point (0, 0) corresponding to undisturbed position of the block (the body-at-rest position). The meaning of the controllability region in the phase space is that being at a certain moment within this region the block will necessarily return to the rest position for any earthquake signature with a prescribed maximum absolute value of the supporting plane acceleration. The dimensions of the safe zone of the controllability region in the phase space of the CDS are also given. It is shown that the safe zone decreases with the increase of the maximum value of the supporting plane acceleration. Various shapes of the controllability regions are shown for different geometrical dimensions of the block and the upper limits of the absolute value of the supporting plane acceleration $\ddot{U}(t)$.

25.2 Mathematical Model

Vibrations of a rectangular rigid block positioned on a smooth horizontal surface (Fig. 25.1) are considered.

The surface is assumed to vibrate strictly horizontally. The friction force between the surface and the block is assumed to be sufficiently high for the block not to slide. Under the assumptions made, the block can only perform rotational–vibrational motions relative to two corner points, O and O', belonging to the lower side of the block. Vibrations of the block can be unambiguously characterized by the angle θ between the surface and the lower side of the block. The transition from roattion about one vertex of the block to rotation about the other one occurs by way of impact that is considered to have negligible duratoin. The energy lost in the impact is taken into account by introducing an angular velocity restitution coefficient, $0 < r \leq 1$, so that $\dot{\theta}^{+} = r\dot{\theta}^{-}$, where $\dot{\theta}^{-}$ is the angular velocity of the block prior to the impact, and $\dot{\theta}^{+}$ is the angular velocity of the block after the impact.

Equations of motion of the block in the time intervals between the impacts can be written in the form of the balance of angular momentum. In the frame of reference moving together with the vibrating surface, the only force acting on the block is gravity. Thus, equations of motion in this reference system can be written as follows:

$$
\begin{aligned}
J\frac{\mathrm{d}^2\theta}{\mathrm{d}t^2} &= -mg\,R\,\sin(\alpha - \theta), \quad \theta > 0 \\
J\frac{\mathrm{d}^2\theta}{\mathrm{d}t^2} &= +mg\,R\,\sin(\alpha + \theta), \quad \theta < 0,
\end{aligned}
\tag{25.1}
$$

equating the angular inertia of the block to the gravitational moment.

In the above equations, $\alpha = \theta_0$ is the angle between the lateral side of the block and the diagonal, and $2R$ is length of the diagonal. The mass moment of inertia J of the block relative to points O and O' is expressed as

$$
j = \frac{4m}{3}R^2
$$

To change to a fixed coordinate system, account must be taken of the force of inertia resulting from the motion of the surface. This force acts on the center of mass of block C, is directed contrary to the acceleration of the surface and is equal to $m\ddot{u}$. Keeping in mind the moment of this force, the equations can be rewritten, taking into account impact interactions of the block, as follows:

$$
\begin{aligned}
&J\frac{\mathrm{d}^2\theta}{\mathrm{d}t^2} = -mg\,R\,\sin(\alpha - \theta) - mR\frac{d^2u}{dt^2}\cos(\alpha - \theta), \quad \theta > 0 \\
&J\frac{\mathrm{d}^2\theta}{\mathrm{d}t^2} = +mg\,R\,\sin(\alpha + \theta) - mR\frac{d^2u}{dt^2}\cos(\alpha + \theta), \quad \theta < 0 \\
&\dot{\theta}^{+} = r\dot{\theta}^{-}, \quad \theta = 0
\end{aligned}
\tag{25.2}
$$

The first equation in system (25.2) describes vibratory motions of the block in the vicinity of point O. The second equation of system (25.2) describes vibratory motions of the block in the vicinity of point O'. The third equation of the system describes impact on the block.

Introducing dimensionless time $\tau = t\sqrt{3g/2\sqrt{B^2 + H^2}}$ and a new variable $x = \theta$, Eq. (25.2) can be rewritten in the following form:

$$\begin{cases} \dot{x} = y \\ \dot{y} = \begin{cases} \sin(x+\alpha) - u(\tau)\cos(x+\alpha), -\pi/2 < x < 0 \\ \sin(x-\alpha) - u(\tau)\cos(x-\alpha), 0 < x < \pi/2 \end{cases} \end{cases} \quad (25.3)$$

$$y^+ = ry^-, \quad x = 0 \quad (25.4)$$

where $u(\tau)$ is an arbitrary dimensionless piecewise continuous function with the values belonging to interval $[-U, +U]$, y^-, y^+ are velocities of the block before and after the impact, respectively, $0 < r < 1$ is the restitution coefficient characterizing loss of energy in the impact, $0 < \alpha < \pi/2$, and dots over the variables imply differentiation with respect to the dimensionless time τ.

25.3 Qualitative Study of the System

Phase space $\Phi(|x| < \pi/2, y, \tau)$ of system (25.3, 25.4) is three-dimensional in coordinates x, y, τ and truncated along phase variable x. System (25.3) defines two controllable standard dynamic systems (CDS) with Butenina's phase constraints, [1, 23], one in subspace $\Phi_1^-(-\pi/2 < x < 0, y, \tau)$ and another in subspace $\Phi_1^+(0 < x < \pi/2, y, \tau)$. A standard CDS defined in $\Phi_1^-(-\pi/2 < x < 0, y, \tau)$ will be designated by (CDS$^-$), and one defined in $\Phi_1^+(0 < x < \pi/2, y, \tau)$ by (CDS$^+$).

Butenina's contact curve, [1, 23], for (CDS$^-$) is defined by equation $F = -y\cos(x + \alpha) = 0$. Taking into consideration that, for (CDS$^-$), inequality $\cos(x + \alpha) > 0$ holds, the equation of contact curve will have the form $y = 0$. It is evident that $F > 0(< 0)$ if $y < 0(y > 0)$. A contact curve for (CDS$^+$) is evidently defined by the same equation as for (CDS$^-$).

In what follows, (CDS) with $U(t) = -U$ will be designated by (1^-), and (CDS$^+$) with $U(t) = U$ by (1^+). It is noteworthy that dynamic systems (1^-) and (1^+) are autonomous (CDS).

According to the general theory of controlled dynamic systems described in [1, 23], the following is true:

1. Trajectories of system (1^-) in field $F^-(y > 0, x \neq 0)$ of plane x, y, where $F < 0$, are intersected by other admissible trajectories of system (25.3) in the

negative direction (clockwise), and in field $F^{+}(y < 0, x \neq 0)$ in the positive direction ($F > 0$);
2. Trajectories of system (1^{+}) in field $F^{-}(y > 0, x \neq 0)$ are intersected by other admissible trajectories of system (25.3) in the positive direction, and in field $F^{+}(y < 0, x \neq 0)$ in the negative direction.

Thus, in fields $P_{-}(-\pi/2 < x < 0)$ and $P_{+}(0 < x < \pi/2)$, planes x, y of the trajectory of CDS (25.3) for $u(\tau) = -U\ \mathrm{sign}(y), y \neq 0$ and $u(\tau) = U\ \mathrm{sign}(y), y \neq 0$, extended across axis OX on continuity, are trajectories of a unilateral intersection. All the other admissible trajectories can intersect the above trajectories only in the negative (positive) direction.

If in CDS (25.3), it is assumed that $u(\tau) = \varphi(\tau), |\varphi(\tau)| \leq U$, where $\varphi(\tau)$ is a particular function of variable τ, a particular non-autonomous system of differential equation is obtained. Each of the trajectories of this non-autonomous system, passing across axis ($x = 0, y \neq 0$), has on this axis a discontinuity point in the form of a finite jump.

In view of abovementioned points (1 and 2), the following statements can be proved to be true:

25.3.1 *Statement 1*

Let

(1) Trajectory l of any particular system of CDS family (25.3) for $\tau = t$ has a point in half-band ($|x| < \pi/2, y > 0$) to the right of trajectory L^{-} of system (1^{-}) (to the left of trajectory L^{+} of system (1^{+}).
(2) For all $\tau, \tau \in [t, T]$, the arc of trajectory l lies in the abovementioned half-band.

Then, for all $\tau, \tau \in [t, T]$, the arc of trajectory l lies to the right of L^{-} (to the left of L^{+}).

25.3.2 *Statement 2*

Let

(1) Trajectory l of CDS (25.3) for $\tau = t$ has a point in half-band $\Pi^{-}(|x| < \pi/2, y < 0)$ to the left of trajectory L^{-} of system (1^{-}) (to the right of trajectory L^{+} of system (1^{+}),
(2) For all $\tau, \tau \in [t, T]$, the arc of trajectory l lies in half-band Π^{-}.

Then, for all $\tau, \tau \in [t, T]$, the arc of trajectory l in Π^{-} lies to the left of L^{-} (to the right of L^{+}).

The above cross-linked autonomous systems [9, 24] are then completed by system $[1^{-+}]([1^{+-}])$, for which in CDS (25.3) $u(t) = -U[+U]$, if $y > 0$; $u(t) = +U[-U]$, if $y < 0$, $|u(t)| \leq U$ if $y = 0$.

The above statements and introduced notations lead to the next statement.

25.3.3 Statement 3

Trajectories of cross-linked autonomous systems $[1^{-+}]$ and $[1^{+-}]$ are, for CDS (25.3), trajectories of unilateral intersection: In band $(|x| < \pi/2, x \neq 0)$, trajectories of system $[1^{-+}]([1^{+-}])$ are intersected by other admissible trajectories of CDS (25.3) in the negative (positive) direction.

25.3.4 Studying Special Points of the Supplementary Autonomous Systems

In CDS (25.3), $u(\tau)$ will be substituted for by scalar parameter μ, $|\mu| \leq U$. Then

$$\begin{cases} \dot{x} = y = P(x, y) \\ \dot{y} = \begin{cases} \sin(x + \alpha) - \mu\cos(x + \alpha), -\pi/2 < x < 0, = Q_1(x, y) \\ \sin(x - \alpha) - \mu\cos(x - \alpha), 0 < x < \pi/2, = Q_2(x, y) \end{cases} \end{cases} \quad (1\mu)$$

It follows from the first equation of system (1μ) that all the special points of system $(x \neq 0)$ lie along axis OX. If $x < 0$, then $x = -\alpha + \operatorname{arctg}\mu$; if $x > 0$, then $x = \alpha + \operatorname{arctg}\mu$.

The types of special points are defined, as is known [25], in terms of the roots of characteristic equation $\lambda^2 - (a + d)\lambda + \Delta = 0$, where

$$\begin{aligned} a &= P'_x(M_0) = 0, b = P'_y(M_0) = 1, c = Q'_x(M_0) \\ &= \begin{cases} \cos(x_0 + \alpha)(1 + \mu^2), x_0 < 0 \\ \cos(x_0 - \alpha)(1 + \mu^2), x_0 > 0 \end{cases}, \\ d &= Q'_y(M_0) = 0, \Delta = ad - bc = -c = \lambda_1\lambda_2 \end{aligned}$$

λ_1, λ_2 are roots of the characteristic equation. Thus, in each special point of system (1μ), on intervals $-\pi/2 - \alpha < x < 0$ and $0 < x < \pi/2 + \alpha$, $\Delta < 0$. It implies that the roots of the characteristic equation are real and have different signs, i.e., each special point of the system in question $(x \neq 0)$ is a saddle.

The saddle of system (1^-) (for which $u(t) = -U$), located on interval $(-\pi/2 - \alpha < x < 0)$ $[0 < x < \pi/2 + \alpha]$ will be designated by $C^{--}[C^{-+}]$.

The saddle of system (1^+), located on interval ($-\pi/2 - \alpha < x < 0$) [$0 < x < \pi/2 + \alpha$] will be designated by $C^{+-}[C^{++}]$.

The separatrix of each of the saddles, which tends to its saddle from the upper half-surface ($y > 0$), will be designated, respectively, by S_1^{--}, S_1^{-+}, S_1^{+-}, S_1^{++}. The remaining separatrices of each of the saddles are renumbered, moving over the related saddle in the counterclockwise direction.

25.3.5 Controllability Regions of CDS (1)

A controllability region into an arbitrarily small vicinity of point (0, 0) will be called O-controllability region and designated by $D(0)$. Consider the structure of region $D(0)$ as a function of U and parameter α.

Let $0 < \alpha < \pi/4$, $|u(t)| \leq U < \text{tg}\alpha$.

Then

Theorem 1 There exists region G_0 containing point (0, 0) and function $u(t)$ (admissible control) translating any point of this region, in finite time, to an arbitrarily small vicinity of point (0, 0) and retaining it in this vicinity for an arbitrarily long time.

Proof Construct region G_0 containing point (0, 0), the boundary of which is formed by:

(1) Curve Γ^{--} containing saddle C^{--} arcs of separatrices S_1^{--} and S_3^{--} located in the band $|x| < \pi/2$;
(2) Curve Γ^{++} containing saddle C^{++} and arcs of separatrices S_1^{++} and S_3^{++} located in the band $|x| < \pi/2$.

It is evident that the boundary of region G_0 in the band $|x| < \pi/2, x \neq 0$ is intersected by other admissible trajectories of CDS (25.3) only in the outside direction (see statement 3).

The trajectory of CDS (25.3), that, for $\tau = t_0$, has a point in band $|x| < \pi/2, x \neq 0$ beyond region G_0, for $\tau > t_0$ in a finite time will be outside the boundary of this band (beyond G_0 in band $|x| < \pi/2$ there are no points, in which $\dot{x}^2 + \dot{y}^2 = 0$). It is noted that the diagrams of the solutions of systems (1^-) and (1^+) (as it follows from Eq. (25.3)) are symmetric relative to point (0, 0). In particular, in band $|x| \leq \pi/2, x \neq 0$, points C^{--} and C^{++}, half-separatrices S_4^{--} and S_2^{++} and half-separatrices S_3^{--} and S_1^{++} are symmetric relative to point (0, 0); hence, pre-impact point of half-separatrix S_4^{--} and the post-impact point of separatrix S_1^{++} (point M_1) coincide; the pre-impact point of separatrix S_2^{++} and the post-impact point of separatrices S_3^{--} (point M_2) coincide. Saddles C^{--} and C^{++} and the abovementioned half-separatrices of these saddles are connected by points M_1 and M_2 into a closed-curve defining region D_0, containing point (0, 0).

Region D_0 belongs to controllability region $U(0, 0)$. In fact, for any point M_0, $M_0 \in D_0$ for

$$u(t) = \begin{cases} -U, x < 0 \\ +U, x > 0 \end{cases}$$

the half-trajectory of CDS (25.3) with its origin in point M_0 has limiting point (0, 0).

If arc γ_1^+ of system (1^+) is drawn through point C^{--} to the intersection with straight line, $x = -\pi/2$ in point O_1^+ (Fig. 25.2).

Arc γ_1^+ will divide the part of region G_0, lying outside D_0 in half-plane $y > 0$, into two parts. The part that has on its boundary half-separatrix S_4^{--} containing point M_1 will be designated by D_1, the other one by d_1. Arc $(\overset{\frown}{C^{--}M_1})$ of separatrix S_4^{--} and region D_1 belong to region $U(0)$, as, whichever point $\overset{\frown}{M}$ in region D_1 or on arc $(\overset{\frown}{C^{--}M_1})$ is taken, the half-trajectory of system (1^+) with its origin in point $\overset{\frown}{M}$, with increasing τ, transfers into region D_0. $D_0 \in U(0, 0)$; hence, $D_1 \cup (C^{--}M_1) \in U(0, 0)$.

Region d_1 and arc γ_1^+ belong to $U(0, 0)$, as, whichever point of this set is taken, the half-trajectory of system (1^+) with its origin in this point, with increasing τ, transfers into region D_1. But $D_1 \in U(0, 0)$, that is, $d_1 \cup \gamma_1^+ \in U(0, 0)$. For the part of region D_0, lying in half-band $\{0 < x < \pi/2, y < 0\}$, the proof is analogous.

25.3.6 Safe Zone of the Controllability Region

Let $|u(t)| < \mathrm{tg}\alpha, 0 < \alpha < \pi/4$. Then, each of the systems (1^-) and (1^+) has in band $|x| < \pi/2, x \neq 0$ two special saddle points (apart from point (0, 0)), one of which lies in the left half-plane, and the other in the right one.

Special points of system (1^-)—saddle $C^{--}(x < 0)$ and saddle $C^{-+}(x > 0)$

Special points of system (1^+)—saddle $C^{+-}(x < 0)$ and saddle $C^{++}(x > 0)$.

Separatrix S_1^{-+} of saddle C^{-+}, that has a discontinuity point on axis $Oy(y > 0)$, intersects separatrix S_1^{+-} of saddle C^{+-} in point N_1^*.

Separatrix S_3^{+-} of saddle C^{+-}, that has a discontinuity point on axis $Oy(y < 0)$, intersects separatrix S_3^{-+} of saddle C^{-+} in point N_3^*.

Saddles C^{+-}, C^{-+} and half-separatrices of these saddles with their origins in points N_1^* and N_3^*, respectively, define in region $U(0, 0)$ region $B(0)$ containing point (0, 0). Admissible trajectories of CDS (25.3), differing from the considered 'boundary' trajectory, can intersect the boundary of this region only in the 'inside' direction.

$\overline{B}(0)$ (the closure of region $B(0)$) is the safe zone of the controllability region for the assigned constraints on controlling (Fig. 25.3).

The part of region $U(0, 0)$ lying outside $\overline{B}(0)$ is a zone of risk. However, for any point K of the zone of risk, there exists control $(u = u_k(t))$ transferring the representing point of system (25.3) from the zone of risk to the safe zone. For $U = \mathrm{tg}\alpha$, the safe zone disappears in point (0, 0). If $|U| > \mathrm{tg}\alpha$, then the entire controllability region with punctured point (0, 0) is a zone of risk (Fig. 25.3).

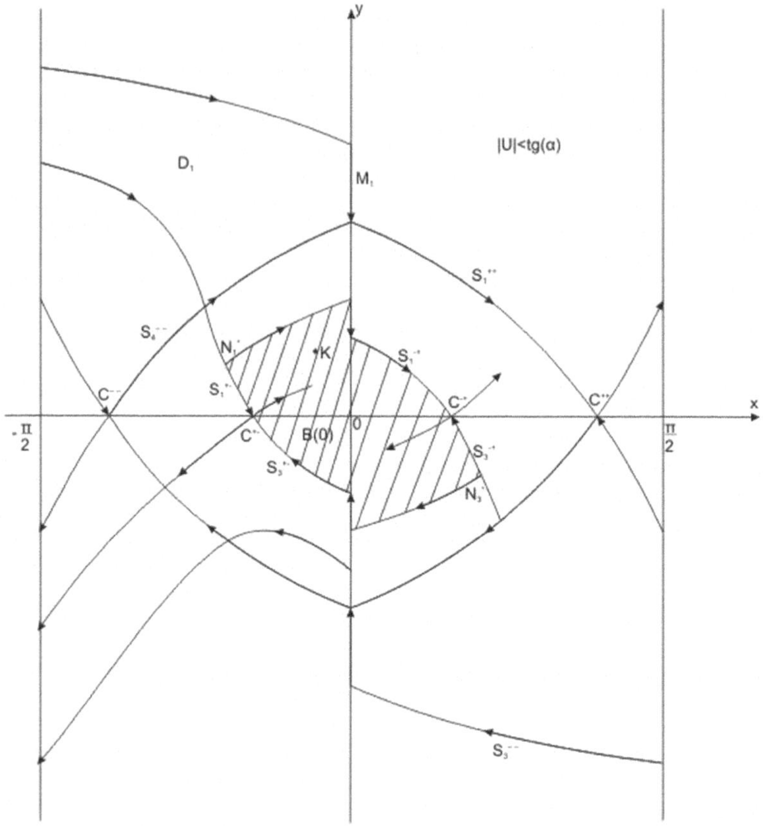

Fig. 25.3 Phase portrait for $|u(t)| < \text{tg}\alpha$, $0 < \alpha < \pi/4$

In Fig. 25.2, safe zone $\overline{B}(0)$ is hatched. Anything outside $\overline{B}(0)$ in region $\overline{U}(0)$ is a zone of risk. As was noted above, with increasing τ, only the trajectories of CDS (25.3) that have points outside $U(0, 0)$ a fortiori leave the limits of band $|x| < \pi/2$. Let $\text{tg}\alpha \leq |u(t)| < \text{ctg}\alpha$. The phase portrait of CDS (25.3) for this case is depicted in Fig. 25.4.

For any point of region $U(0, 0)$ (other than point (0, 0)), there exist phase trajectories coming outside this region and arriving in point (0, 0).

Let $|u(t)| = \text{ctg}\alpha$. Saddles C^{--} and C^{++} lie, respectively, in points $(-\pi/2, 0)$ and $(+\pi/2, 0)$. Region $U(0, 0)$ is defined in band $|x| < \pi/2, x \neq 0$ by the arcs of separatrices S_1^{++} and S_3^{--} (Fig. 25.5).

There is no safe zone.

If $|u(t)| > \text{ctg}\alpha$, saddles C^{--} and C^{++} lie, respectively, on intervals $(-\pi/2 - \alpha; -\pi/2)$ and $(\pi/2, \pi/2 + \alpha)$. Due to phase constraints $|x| < \pi/2, x \neq 0$, the boundary of the controllability region is defined by: (1) half-trajectory l_1^- of system

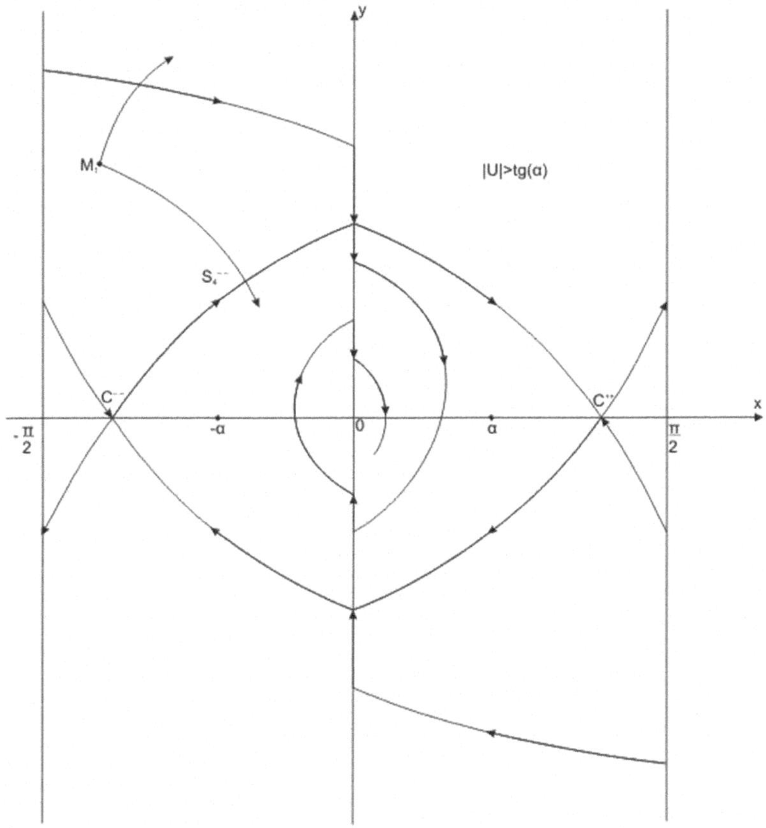

Fig. 25.4 Phase portrait for $\text{tg}\alpha \leq |u(t)| < \text{ctg}\alpha$

(1^{+}) with the origin in point and 2) half-trajectory l_2^- of system (1^-) with the origin in point $(-\pi/2, 0)$ (Fig. 25.6).

Thus, for the case of $0 < \alpha < \pi/4$, the controllability region of CDS (25.3) has been fully analyzed.

Let $\alpha = \pi/4$. In this case, $\text{tg}\alpha = \text{ctg}\alpha = 1$. For $U < 1$, the boundary of region $U(0, 0)$ includes saddles C^{--} and C^{++} and arcs of separatrices $S_1^{--}, S_3^{--}, S_1^{++}, S_3^{++}$ lying in band $|x| < \pi/2$. In the controllability region, there exists safe zone $\overline{B}(0)$ (see the case of $0 < \alpha < \pi/4$).

For $U = 1$, the safe zone of the controllability region disappears in point (0, 0); the boundary of the controllability region is defined by saddles $C^{--}(-\pi/2, 0)$, $C^{++}(+\pi/2, 0)$ and the arcs of separatrices S_1^{++} and S_3^{--} lying in band $|x| < \pi/2, x \neq 0$.

For $U > 1$, the system has no special points in band $|x| < \pi/2, x \neq 0$. The boundary of region $U(0, 0)$ is defined by negative half-trajectories $l_1^-(+\pi/2, 0)$ of system (1^{+}) and $l_2^-(-\pi/2, 0)$ of system (1^{+}) (see Fig. 25.4).

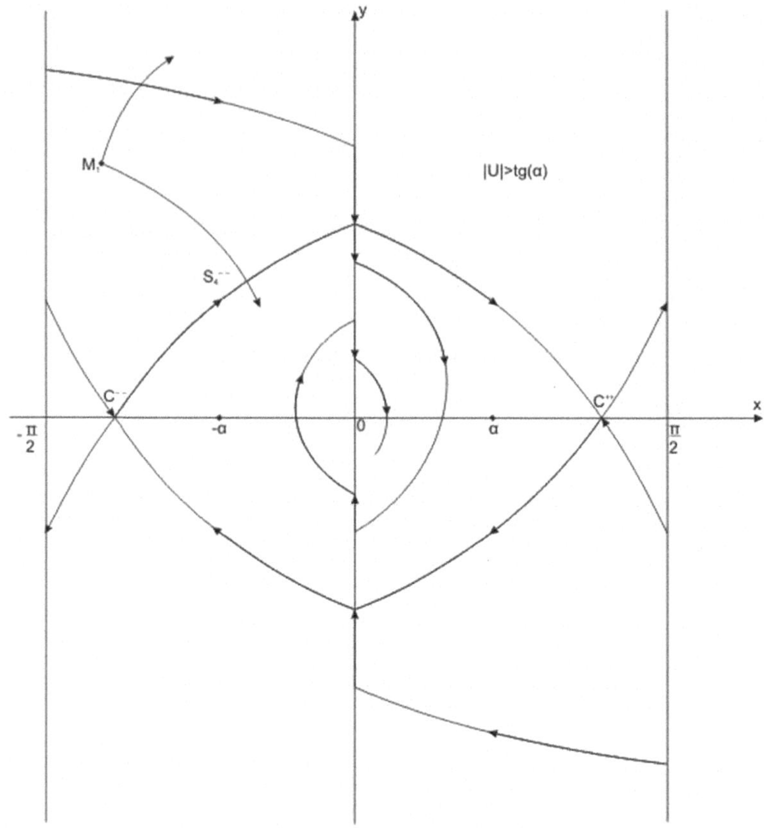

Fig. 25.5 Phase portrait for $|u(t)| = \text{ctg}\alpha$

If $\pi/4 < \alpha < \pi/2$, then saddles C^{--} and C^{++} lie, respectively, on intervals $(-\pi/2 - \alpha; -\pi/2)$ and $(\pi/2, \pi/2 + \alpha)$ of axis OX. Thus, due to phase constraints $|x| < \pi/2$, the boundary of controllability region $U(0, 0)$ is defined by: the arc of half-trajectory $l^-(+\pi/2, 0)$ of system (1^+) with the end in the intersection point with straight line $x = -\pi/2$, and the arc of half-trajectory $l_0^-(-\pi/2, 0)$ of system (1^-) with the end in the intersection point with straight line $x = \pi/2$ (see Fig. 25.6).

If $|u(t)| < \text{tg}\alpha$, then in region $U(0, 0)$ there exists safe zone $\overline{B}(0)$. If $|u(t)| \geq \text{tg}\alpha$, then the entire region $U(0, 0)$ (except for point $(0, 0)$) is a zone of risk.

Remark If $\alpha \to \pi/2$, then $\text{tg}\alpha \to \infty$. Hence, for any $u_0 > 0$ there will exist $\delta > 0$, such that, for $\alpha \in (\pi/2 - \delta, \pi/2)$, $\text{tg}\alpha > u_0$. It implies that, for $|u(t)| \leq u_0$, a safe zone in the controllability region exists for $\alpha \in (\pi/2 - \delta, \pi/2)$. With increasing u_0, the diameter of the controllability zone tends to 0.

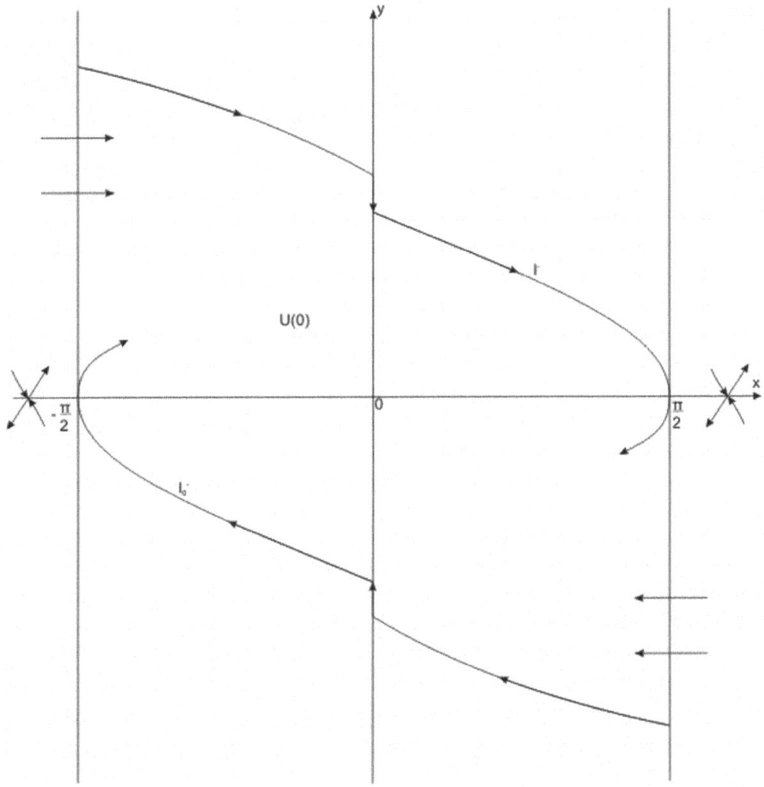

Fig. 25.6 Phase portrait for $|u(t)| > \mathrm{ctg}\alpha$

25.4 Main results

The main results of the present work are as follows:

- Using qualitative methods of CDS, controllability regions have been derived for a rectangular block rocking on a horizontal plane due to the acceleration of the latter. These regions are applicable to any dimensions of the block and time signatures of the acceleration of the supporting plane provided that the absolute value of the acceleration remains within specified bounds.
- The dimensions of the safe zone of the controllability region in the phase space of CDS are given.
- It has been proven that the safe zone decreases with the increase of the maximum value of the acceleration.

Acknowledgements The work was supported by a grant of the Russian Science Foundation (16-19-10237-P).

References

1. Butenina, N.N., Metrikin, V.S.: Primenenie metodov kachestvennoy teorii upravlyaemyh dinamicheskih sistem k issledovaniyu neavtonomhyh differentsyalnyh uravneniy. Nelineynaya dinamika **6**(1), 143–150 (2010)
2. Campillo, M., Gariel, J.C., Aki, K., Sanchez-Sesma, F.J.: Destructive strong ground motion in Mexico city: source, path and site effects during the great 1985 Michoagan earthquake. Bull. Seismol. Soc. Am. **79**(6), 1718–1735 (1989)
3. Housner, J.: The behavior or inverted pendulum structures during earthquakes. Bullet. Seismol. Soc. Am. **53**(2), 403–417 (1963)
4. Pena, F., Prieto, F., Lourenco, P.B., et al.: On the dynamics of rocking motions of single rigid-block structures. Earthquake Eng. Struct. Dyn. **36**, 2383–2399 (2007)
5. Prieto, F., Lourenco, P.B.: On the rocking behavior of rigid object. Meccanica **40**, 121–133 (2005)
6. Makris, N., Roussos, Y.: Rocking response and overturning of equipment under horizontal pulse-type motions. Report No. PEER98/05, Pacific Earthquake Engineering Research Center, University of California, Berkeley, Calif (1998)
7. Makris, N., Roussos, Y.: Rocking response of rigid blocks under near-source ground motions. Geotechnique **50**(3), 243–262 (2000)
8. Shi, B., Anooshehpoor, A., Zeng, Y., Brune, J.N.: Rocking and overturning of precariously balanced rocks by earthquake. Bull. Seismol. Soc. Am. **86**(5), 1364–1371 (1996)
9. Aslam, M., Scalise, D.T., Godden, W.G.: Earthquake rocking response of rigid bodies. J. Struct. Div. **106**(2), 377–392 (1980)
10. Heaton, T.H., Shi, B., Brune, J.N.: Estimates of the ground acceleration at point reyes station during the 1906 San Francisco earthquake. Bull. Seismol. Soc. Am. **89**(4), 843–853 (1999)
11. Hogan, S.J.: The many steady state responses of a rigid block J. Eng. Mech. 127(5), 473–483 (2001). Downloaded from ascelibrary.org by University of Florida on 02/04/18. Copyright ASCE. Journal of Engineering Mechanics 483 under harmonic forcing. Earthquake Eng. Struct. Dyn. **19**(7), 1057–1071
12. Housner, G.W.: The behaviour of inverted pendulum structures during earthquakes. Bull. Seismol. Soc. Am. **53**(2), 404–417 (1963)
13. Iwan, W.D., Chen, X.D.: Important near-field ground motion data from the landers earthquake. In: Proceedings of the 10th European Conference on Earthquake Engineering, Balkema, Rotterdam, The Netherlands (1994)
14. Jacobsen, L.S., Ayre, R.S.: Engineering Vibrations. McGrawHill, New York (1958)
15. Pompei, A., Scalia, A., Sumbatyan, M.A.: Dynamics of rigid block due to horizontal ground motion. J. Eng. Mech. **124**(7), 713–717 (1998)
16. Shenton, H.W. III.: Criteria for initiation of slide, rock, and slide-rock rigid-body modes. J. Eng. Mech. **122**(7), 690–693 (1996)
17. Spanos, P.D., Koh, A.S.: (1984) Rocking of rigid blocks due to harmonic shaking. J. Eng. Mech. **110**(11), 1627–1642 (1984)
18. Tso, W.K., Wong, C.M.: Steady state rocking response of rigid blocks part 1: analysis. Earthquake Eng. Struct. Dyn. **18**(1), 89–106 (1989)
19. Tso, W.K., Wong, C.M.: Steady state rocking response of rigid blocks part 2: experiment. Earthquake Eng. Struct. Dyn. **18**(1), 107–120 (1989)
20. Yim, C.S., Chopra, A.K., Penzien, J.: Rocking response of rigid blocks to earthquakes. Earthquake Eng. Struct. Dyn. **8**(6), 565–587 (1980)
21. Munitdyn, A., Munitdyna, M.: Oscillations of a rigid block on supported base. J. Vibroeng. Procedia **8**, 63–67 (2016)
22. Karapetyan, M.A.: Munitsyna oscillations of a rigid block on supported base. Automation Remote Control **76**(3), 394–404 (2015)
23. Butenina, N.N., Pavlyuchonok, Z.G., Savelyev, V.P.: Upravlyaemost dinamicheskih system. NNGY (1997)

24. Andreaus, U., Casini, P.: On the rocking-uplifting motion of a rigid block in free and forced motion: influence of sliding and bouncing. Acta Mech. **138**, 219–241 (1999)
25. Neimark, Yu.I.: The Point Mapping Method in the Theory of Nonlinear Oscillations, p. 472. Librokom, Moscow (2010)

Chapter 26
Variational Principle and Equations of Dynamics of Electromagnetoelastic Shells

Anatoly S. Okonechnikov, Marina Yu. Ryazantseva, and Dmitry V. Tarlakovskii

Abstract The boundary value problem of transient dynamics of a homogeneous thin anisotropic electromagnetoelastic shell is constructed with consideration for transverse normal strains, normal unit vector rotation, and variation of the electromagnetic field parameters along the normal vector. The piezoeffects, generalized Ohm's law, and Lorentz force are also taken into account. The earlier derived equations of mechanical motion and the boundary conditions at a given electromagnetic field are used. The well-known functional for a three-dimensional body is modified as applied to electromagnetic processes in anisotropic media. It is assumed that the vector and scalar potentials are interrelated, which makes it possible to obtain the wave equation for the former potential. The corresponding functional for a thin shell is constructed using the linear approximations in the normal coordinate of the potentials. The corresponding equations of electromagnetodynamics and the boundary conditions at a given displacement field are a necessary condition for the stationary state of this functional. It is shown that in the particular case of an isotropic medium, the total number of unknowns diminishes. In this case, the electromagnetic disturbances must be normally directed.

Keywords Shell · Anisotropy · Unsteady electromagnetoelasticity · Functional · Thickness integral characteristics · Variational equation · Necessary condition for extremum · Equation of motion

---

A. S. Okonechnikov (✉)
Moscow Aviation Institute, Moscow, Russia
e-mail: oas.mai@mail.ru

M. Yu. Ryazantseva
Institute of Mechanics Lomonosov Moscow State University, Moscow, Russia
e-mail: m-rznt@yandex.ru

D. V. Tarlakovskii
Institute of Mechanics Lomonosov Moscow State University,
Moscow Aviation Institute, Moscow, Russia
e-mail: tdvhome@mail.ru

H. Altenbach et al. (eds.), *Multiscale Solid Mechanics*,
Advanced Structured Materials 141,
https://doi.org/10.1007/978-3-030-54928-2_26

26.1 Introduction

There has been much research concerned with the problems of the three-dimensional interaction between electromagnetic fields and elastic media (see, e.g., [3, 5, 26, 27]). Among these studies, we will distinguish those used in theory of thin shells, namely devoted to variational principles and equations of electromagnetodynamics [18, 22] and electromagnetoelasticity [19, 20, 24, 29], including those dealing with discontinuous fields [1, 2].

The problems of electromagnetoelastic oscillations of three-dimensional bodies of particular geometries were solved, for example, in studies [6, 8, 11] and [13, 14]. The former group of studies considers free oscillations, while the latter group deals with transient processes.

At the same time, the problems concerned with electromagnetoelastic processes in thin-walled bodies are still poorly explored. Thus, in papers [9, 10], free oscillations of plates were investigated. In the fundamental study [4], the equations and boundary conditions for homogeneous and layered shells were constructed. For this purpose, the variational approach was used in combination with the method of indefinite factors. In [28], the equations of transient dynamics of a homogeneous thin elastic anisotropic shell with account for transverse normal strain and normal unit vector rotation are used for the same purpose. The electromagnetic part of the self-consistent problem is obtained directly as the linear approximation in the normal coordinate of the three-dimensional equations for an electromagnetic field. In this case, the piezoeffects, generalized Ohm's law, and Lorentz force are taken into account.

Below we put forward the most complete closed system of equations and the boundary conditions of the electromagnetoelasticity of a homogeneous thin shell. The electromagnetic field potentials are directly related, without invoking to indefinite factors.

26.2 Problem Formulation

We will consider a homogeneous anisotropic elastic shell of constant thickness h, whose material possesses the symmetry about its midsurface Π. As the equations most completely describing its dynamics (with account for the normal unit vector rotation and its transverse normal strain), we will use the initial and boundary value problems constructed in [21].

To take account for the electromagnetic field action on the shell, we will, first, represent the outer forces in the form of two terms (here and in what follows, the summation over Roman indices is from 1 to 2):

$$\mathbf{q} = \mathbf{q}_* + \mathbf{q}_e, \quad \mathbf{m} = \mathbf{m}_* + \mathbf{m}_e,$$
$$\mathbf{q}_* = q_*^i \pi_i + q_* \mathbf{n}, \quad \mathbf{m}_* = m_*^i \pi_i + m_* \mathbf{n}, \quad \mathbf{q}_e = q_e^i \pi_i + q_e \mathbf{n}, \quad \mathbf{m}_e = m_e^i \pi_i + m_e \mathbf{n}. \tag{26.1}$$

Here, the subscript "*e*" corresponds to the electromagnetic field action on the shell, while the other-in-nature forces are marked by the asterisks; **q** and **m** are the vectors of distributed pressure and moments; π_1 and π_2 are the basis of the surface Π given by the curvilinear coordinates ξ^1 and ξ^2 and the metric tensor g_{ij}; and **n** is the unit normal vector to the outward side of this surface, co-directional with the normal coordinate z.

The forces produced by the electromagnetic field are determined as follows:

$$q_e^i = \int\limits_{-h/2}^{h/2} F^i \mathrm{d}z, \; q_e = \int\limits_{-h/2}^{h/2} F_3 \mathrm{d}z, \; m_e^i = \int\limits_{-h/2}^{h/2} zF^i \mathrm{d}z, \; m_e = \int\limits_{-h/2}^{h/2} zF_3 \mathrm{d}z. \quad (26.2)$$

Here, under the integral signs, there are the coordinates of the Lorentz force which is given by formula [15–17, 22, 23, 26, 28] linearized with respect to the initial electromagnetic field (referred to with the subscript "0")

$$\mathbf{F} = F^i \pi_i + F_3 \mathbf{n} = \rho_{e0} \mathbf{E} + \rho_e \mathbf{E}_0 + c^{-1}([\mathbf{j}_0, \mathbf{B}] + [\mathbf{j}, \mathbf{B}_0]) \quad (26.3)$$

where $\mathbf{E} = E^i \pi_i + E_3 \mathbf{n}$ and $\mathbf{B} = B^i \pi_i + B_3 \mathbf{n}$ are the vectors of the electromagnetic field strength and magnetic induction; $\mathbf{j} = j^i \pi_i + j_3 \mathbf{n}$ is the current density; c is the speed of light; and ρ_e is the density of the charges.

Secondly, we use the generalized Hooke's law.

$$\begin{aligned} \sigma^{ij} &= C^{ijkl} \hat{\varepsilon}_{kl} + C^{ij33} \varepsilon_{33} - \frac{1}{4\pi} \left(\kappa^{ijk} E_k + \gamma^{ijk} H_k \right), \\ \sigma^{i3} &= 2C^{i3k3} \hat{\varepsilon}_{k3} - \frac{1}{4\pi} \left(\kappa^{i33} E_3 + \gamma^{i33} H_3 \right), \\ \sigma^{33} &= C^{33kl} \hat{\varepsilon}_{kl} + C^{3333} \varepsilon_{33} - \frac{1}{4\pi} \left(\kappa^{33k} E_k + \gamma^{33k} H_k \right), \end{aligned} \quad (26.4)$$

where σ^{ij}, σ^{i3}, σ^{33}, and $\hat{\varepsilon}^{ij}$, $\hat{\varepsilon}^{i3}$, ε^{33} are the components of the stress and strain tensors; H_k and H_3 are the coordinates of magnetic field strength vector $\mathbf{H}$; C^{ijkl}, C^{ij33}, C^{i3k3}, and C^{3333} are the components of the tensor of elastic constants; and κ^{ijk}, κ^{ij3}, κ^{i33}, and γ^{ijk}, γ^{ij3}, γ^{i33} are the components of the tensors of piezoelectric and piezomagnetic constants.

The tensors of the physical constants possess the symmetry about the midsurface:

$$C^{ijk3} = C^{i3kl} = C^{i333} = C^{33k3} = 0, \; \gamma^{ij3} = \gamma^{333} = 0, \; \kappa^{ij3} = \kappa^{333} = 0. \quad (26.5)$$

The deformations are related to the kinematic parameters of the shell as follows:

$$\begin{gathered}
\hat{\varepsilon}_{kl} = \varepsilon_{kl} + z\kappa_{kl}, \; \hat{\varepsilon}_{k3} = \frac{1}{2}\left[\theta_k + z\left(b_k^n\theta_n + \nabla_k\psi_3\right)\right], \; \varepsilon_{33} = \psi_3, \\
\varepsilon_{ij} = \frac{1}{2}\left(\alpha_{ij} + \alpha_{ji}\right), \; \kappa_{ij} = \frac{1}{2}\left(\beta_{ij} + \beta_{ji}\right), \alpha_{ij} = \nabla_i u_j - b_{ij}w, \\
\beta_{ij} = \nabla_i\psi_j - b_{ij}\psi_3 + b_i^k\alpha_{kj}, \; -\vartheta_i = \nabla_i w + b_i^k u_k, \; \theta_k = \psi_k - \vartheta_k,
\end{gathered} \tag{26.6}$$

where ε_{kl} and κ_{kl} are the components of the tensors of tangential deformation and variation in the curvature; u_j and w are the tangential and normal displacements; ψ_i are the angles of rotation of a normal unit vector, and ψ_3 is a corresponding normal strain; and b_{ij} are the components of the tensor of the curvature tensor of surface Π.

Here, we use the following approximation for the coordinates of the displacement vector $\mathbf{u} = \hat{u}^i\pi_i + u_3\mathbf{n}$

$$\begin{gathered}
\hat{u}_i\left(\xi^1, \xi^2, z\right) = u_i\left(\xi^1, \xi^2\right) + z\psi_i\left(\xi^1, \xi^2\right), \\
u_3\left(\xi^1, \xi^2, z\right) = w\left(\xi^1, \xi^2\right) + z\psi_3\left(\xi^1, \xi^2\right).
\end{gathered}$$

Equation (26.4) is necessary for constructing physical relations using the formulas [21]

$$\begin{gathered}
\hat{T}^{ij} = \int_{-h/2}^{h/2} \sigma^{ij}\mathrm{d}z, \; M^{ij} = \int_{-h/2}^{h/2} z\sigma^{ij}\mathrm{d}z, \\
\hat{Q}^i = \int_{-h/2}^{h/2} \sigma^{i3}\mathrm{d}z, \; \mu^i = \int_{-h/2}^{h/2} z\sigma^{i3}\mathrm{d}z, \; \hat{N} = \int_{-h/2}^{h/2} \sigma^{33}\mathrm{d}z
\end{gathered} \tag{26.7}$$

To obtain the explicit form of Eq. (26.7), the electromagnetic field characteristics should be of the same approximation order in the coordinate z with the parameters of the stress–strain state. As distinct from [28], this procedure will be performed below like in study [4].

26.3 Functional and Electromagnetic Field Equations for an Anisotropic Three-Dimensional Body

We will use the functional of the electromagnetic field [7, 15, 22, 23] modified as applied to the anisotropic three-dimensional body G.

$$H\left(\mathbf{A}, \varphi\right) = \int_{t_1}^{t_2} \left(W + A_G + A_S\right)\mathrm{d}t, \; t_2 > t_1, \tag{26.8}$$

where

$$W = \frac{1}{8\pi}\int_G [(\mathbf{E}, \mathbf{D}) - (\mathbf{H}, \mathbf{B})]\,\mathrm{d}\mathbf{x},$$
$$A_G = \int_G \left[\frac{1}{c}(\mathbf{j}, \mathbf{A}) - \rho_e\varphi\right] d\mathbf{x},\ \ A_S = \frac{1}{4\pi}\int_{\partial G} [P\varphi - (\mathbf{P}, \mathbf{A})]\,\mathrm{d}S. \tag{26.9}$$

Here, t is time; $\mathbf{H} = H^i\pi_i + H_3\mathbf{n}$ and $\mathbf{D} = D^i\pi_i + D_3\mathbf{n}$ are the vectors of the magnetic field strength and electric induction; $\mathbf{A} = A^i\pi_i + A_3\mathbf{n}$ and φ are the vector and scalar potentials; and $\mathbf{P}$ and P are the disturbances pre-assigned at the disturbance boundary ∂G. We note that in the three-dimensional version, the basis may be arbitrary but for the uniformity of presentation we will everywhere use the basis given by Eq. (26.1).

The potentials in Eq. (26.9) are determined by the formulas:

$$\mathbf{B} = \mathrm{rot}\mathbf{A},\ \ \mathbf{E} + c^{-1}\dot{\mathbf{A}} = -\mathrm{grad}\varphi \tag{26.10}$$

while the inductions and strengths are related by the physical law.

$$\begin{aligned} &D^i = e^{ij}E_j + e^{i3}E_3 + D^i_\varepsilon,\ \ D_3 = e^{3j}E_j + e^{33}E_3 + D_{3\varepsilon},\\ &B^i = \mu^{ij}H_j + \mu^{i3}H_3 + B^i_\varepsilon,\ \ B_3 = \mu^{3j}H_j + \mu^{33}H_3 + B_{3\varepsilon},\\ &D^i_\varepsilon = \kappa^{ijk}\hat{\varepsilon}_{jk} + \kappa^{i33}\varepsilon_{33},\ D_{3\varepsilon} = 2\kappa^{3j3}\hat{\varepsilon}_{j3},\\ &B^i_\varepsilon = \gamma^{ijk}\hat{\varepsilon}_{jk} + \gamma^{i33}\varepsilon_{33},\ B_{3\varepsilon} = 2\gamma^{3j3}\hat{\varepsilon}_{j3}. \end{aligned} \tag{26.11}$$

where e^{ij}, e^{i3}, e^{33}, and μ^{ij}, μ^{i3}, μ^{33} are the components of the tensors of dielectric and magnetic permeabilities.

The following statement can be fairly simply proved.

Statement 1. The necessary condition for an extremum of functional (26.8) under the conditions

$$\delta\mathbf{A}|_{t=t_1} = \delta\mathbf{A}|_{t=t_2} = \mathbf{0},\ \ \delta\varphi|_{t=t_1} = \delta\varphi|_{t=t_2} = 0$$

and at a fixed current density is represented by two Maxwell equations (here and in what follows, the dots mean time derivatives)

$$\mathrm{crot}\,\mathbf{H} = 4\pi\mathbf{j} + \dot{\mathbf{D}},\ \ \mathrm{div}\mathbf{D} = 4\pi\rho_e \tag{26.12}$$

together with the boundary conditions

$$[\mathbf{H}, \nu]|_{\Pi_H} = \mathbf{P},\ \ (\mathbf{D}, \nu)|_{\Pi_D} = P,\ \ \mathbf{A}|_{\Pi_A} = \mathbf{A}_0$$
$$\varphi|_{\Pi_\varphi} = \Phi,\ \ \partial G = \Pi_H \bigcup \Pi_A = \Pi_D \bigcup \Pi_\varphi,$$

where ν is the unit vector of the outward normal to ∂G , $\mathbf{A}_0$ and Φ are some given quantities. The surfaces Π_H and Π_A, as well as Π_D and Π_φ, intersect along a null-measure sets.

We note that the introduction of the vector potential by the first equation (26.10) leads to the identical satisfaction of two more Maxwell equations [25]

$$\text{rot}\,\mathbf{E} = -c^{-1}\dot{\mathbf{B}}, \ \text{div}\mathbf{B} = 0$$

The system of equations (26.10), (26.11), and (26.12) must be supplemented with one more physical relation, namely the generalized Ohm's law, whose linearized form is as follows [15, 23, 26–28]:

$$\begin{gathered} \mathbf{j} = \sigma\mathbf{E} + \mathbf{j}_v, \ \mathbf{j}_v = \sigma c^{-1}[\mathbf{v}, \mathbf{B}_0] + \rho_{e0}\mathbf{v}, \\ \mathbf{v} = v^i\pi_i + v_3\mathbf{n}, \ v^i = \dot{u}^i + z\dot{\psi}^i, \ v_3 = \dot{w} + z\dot{\psi}_3, \end{gathered} \tag{26.13}$$

where σ is the electrical conductivity coefficient.

Then, in the case of a given mechanical field, these relations form a system of 19 scalar equations in 20 unknowns, namely the coordinates of vectors $\mathbf{H}, \mathbf{B}, \mathbf{j}, \mathbf{D}, \mathbf{E}, \mathbf{A}$, and the functions ρ_e, φ. That is not closed which is due to the fact that the potentials are determined non-uniquely. Their relation for an anisotropic medium can be pre-assigned in the same fashion, as it was done in [22] for the isotropic case. For this purpose, we will use the inverse form of the physical relations following from Eq. (26.11)

$$H^i = 4\pi\beta^{ij}B_j + H^i_\varepsilon, \ H^i_\varepsilon = 4\pi q^{ijk}\hat{\varepsilon}_{jk} \tag{26.14}$$

where β^{ij} and q^{ijk} are the components of the tensors of magnetic and piezomagnetic compliance related to the components of the magnetic permeability tensors and the piezomagnetic constants, as follows (δ^i_k is the Kronecker symbol):

$$4\pi\mu^{ij}\beta_{jk} = \delta^i_k, \ q_{nkl} = -\gamma_{ikl}\beta^i_n = 0. \tag{26.15}$$

Here and in what follows in this section, all the Roman indices take the values 1, 2, and 3. The third-rank tensor q^{ijk} possesses the symmetry noted in Eq. (26.5).

Then, eliminating the coordinates of the vectors $\mathbf{E}$, $\mathbf{H}$, and $\mathbf{B}$ from Eqs. (26.10) and (26.11) and taking Eq.(26.14) into account, we obtain

$$\begin{aligned} & c^{-2}e^{ij}\ddot{A}_j + \nabla_m\left[4\pi\left(\varepsilon^{ijl}\beta_{lk}\varepsilon^{kmn} - \beta^{im}g^{jn}\right)\nabla_j A_n \ +4\pi\beta^{im}\text{div}\mathbf{A} + c^{-1}e^{im}\dot{\varphi}\right] \\ & = c^{-1}4\pi j^i + c^{-1}\dot{D}^i_\varepsilon - \varepsilon^{ijl}\nabla_j H_{l\varepsilon}, \end{aligned} \tag{26.16}$$

where ε^{ijl} are the components of the discriminant tensor [12].

Letting now

$$4\pi\beta^{im}\text{div}\mathbf{A} + c^{-1}e^{im}\dot{\varphi} = 0 \tag{26.17}$$

and convolving both parts of Eq. (26.16) with tensor $4\pi d_{iq}$, we arrive at hyperbolic equations in the coordinates of potential **A**

$$c^{-2}\ddot{A}^i = K^{ijmn}\nabla_j\nabla_m A_n + 4\pi d^i_k\left[c^{-1}4\pi j^k + c^{-1}\dot{D}^k_\varepsilon - \varepsilon^{kjl}\nabla_j H_{l\varepsilon}\right], \quad K^{ijmn} = 16\pi^2 d^i_q \beta_{lk}\left(g^{ql}g^{km}g^{jn} - \varepsilon^{qjl}\varepsilon^{kmn}\right). \tag{26.18}$$

Here, Eq. (26.15) is taken into account, together with the relation

$$4\pi e^{ij}d_{jk} = \delta^i_k \tag{26.19}$$

Precisely, Eq. (26.17) is the required form of the relation between the potentials. Using relations $4\pi e^{ij}d_{ij} = 3,\ 4\pi\mu^{ij}b_{ij} = 3$ following from Eq. (26.19), we can bring Eq. (26.17) into the form:

$$\mathrm{div}\mathbf{A} = -c^{-1}a\dot{\varphi},\ a = e^{im}\mu_{im}/3\ , \tag{26.20}$$

Now the system of Eqs. (26.10), (26.11), (26.13), and (26.20) is closed.

In the case of an anisotropic medium, an equation for the scalar potential similar to Eq. (26.18) cannot be constructed. However, the system consisting of Eq. (26.20) and the second equation (26.12) transformed with the use of Eqs. (26.10) and (26.11) can be regarded as its counterpart

$$-c^{-1}e^{ij}\nabla_i\dot{A}_j = e^{ij}\nabla_i\nabla_j\varphi + 4\pi\rho_e - \mathrm{div}\mathbf{D}_\varepsilon. \tag{26.21}$$

In the particular case of an isotropic medium, the physical relations (26.11) become considerably simpler

$$\mathbf{D} = \varepsilon\mathbf{E},\ \mathbf{B} = \mu_e\mathbf{H},\ e^{ij} = \varepsilon g^{ij},\ \mu^{ij} = \mu_e g^{ij}. \tag{26.22}$$

Hence, using Eqs. (26.15) and (26.19), we obtain that tensors β^{ij} and d^{ij} are also spherical.

$$4\pi\beta^{ij} = g^{ij}/\mu_e\ ,\ 4\pi d^{ij} = g^{ij}/\varepsilon\ .$$

Then tensor K^{ijmn} in Eq. (26.18) takes the form:

$$\varepsilon\mu_e K^{ijmn} = g^{im}g^{jn} - g_{lk}\varepsilon^{ijl}\varepsilon^{kmn} = g^{in}g^{jm}.$$

The validity of the last equality is established by direct verification. Equations (26.18) are transformed as follows:

$$c^{-2}\varepsilon\mu_e\ddot{A}^i = g^{jm}\nabla_j\nabla_m A^i + 4\pi\mu_e c^{-1}j^i,$$

which corresponds to the vector equation derived in [22]

$$\ddot{\mathbf{A}} = \frac{c^2}{\mu_e \varepsilon}\Delta\mathbf{A} + \frac{4\pi c}{\varepsilon}\dot{\mathbf{j}}.$$

In the isotropic version of the medium, Eqs. (26.20) and (26.21) with account for Eq. (26.22) take the form:

$$\mathrm{div}\mathbf{A} = -c^{-1}\varepsilon\mu_e\dot{\varphi}, \quad -c^{-1}\varepsilon\mathrm{div}\dot{\mathbf{A}} = \varepsilon\Delta\varphi + 4\pi\rho_e.$$

Hence, it follows the wave equation in the scalar potential

$$\ddot{\varphi} = \frac{c^2}{\varepsilon\mu_e}\Delta\varphi + \frac{4\pi c^2}{\varepsilon^2\mu_e}\rho_e,$$

which agrees with corresponding relation from [22].

26.4 Functional for the Electromagnetic Field of a Shell

To construct this functional, we will use, likewise in [21], functional (26.8) for the three-dimensional body G occupied by the shell. The body boundary $\partial G = \Pi_+ \bigcup \Pi_- \bigcup \Pi_b$, where Π_+ and Π_- are the outer and inner surfaces and Π_b is the lateral surface. Similarly to the displacement field, we will write the electromagnetic field potentials in the linear approximation in the coordinate z:

$$\begin{aligned}
&\mathbf{A}\left(\xi^1,\xi^2,z,t\right) = \tilde{\mathbf{A}} + \tilde{A}_3\mathbf{n}, \ \tilde{\mathbf{A}} = \tilde{A}^i\pi_i, \\
&\varphi\left(\xi^1,\xi^2,z,t\right) = \varphi_0\left(\xi^1,\xi^2,t\right) + z\varphi_1\left(\xi^1,\xi^2,t\right), \\
&\tilde{A}^i = A^i\left(\xi^1,\xi^2,t\right) + z\Psi^i\left(\xi^1,\xi^2,t\right), \ \tilde{A}_3 = A_3\left(\xi^1,\xi^2,t\right) + z\Psi_3\left(\xi^1,\xi^2,t\right).
\end{aligned} \tag{26.23}$$

In this case, likewise in [21], the covariant derivatives and the operators in Eqs. (26.10) and (26.12) take the form:

$$\begin{aligned}
&\nabla_i\tilde{A}_j = a_{ij} + z\gamma_{ij}, \ \nabla_i\tilde{A}_3 = t_i + \zeta_i z, \ \nabla_3\tilde{A}_i = \Psi_i, \ \nabla_3\tilde{A}_3 = \Psi_3, \\
&\mathrm{div}\mathbf{A} = A + \Psi_3 + z\gamma, \ \mathrm{rot}\mathbf{A} = \chi^{ij}\left[\left(t_j + \zeta_j z - \Psi_j\right)\pi_i + \left(\nabla_i A_j + zB_{ij}\right)\mathbf{n}\right], \\
&B_{ij} = \nabla_i\Psi_j + b_i^k\nabla_k A_j, \ \mathrm{grad}\varphi = \left(\nabla_i\varphi_0 + z\nabla_i\varphi_1\right)\pi^i + \varphi_1\mathbf{n}.
\end{aligned} \tag{26.24}$$

Here

$$\begin{aligned}
&a_{ij} = \nabla_i A_j - b_{ij}A_3, \ \gamma_{ij} = \nabla_i\Psi_j - b_{ij}\Psi_3 + b_i^k a_{kj}, \ t_i = \nabla_i A_3 + b_i^k A_k, \\
&\zeta_i = b_i^k\left(\Psi_k + t_k\right) + \nabla_i\Psi_3, \ B_{ij} = \nabla_i\Psi_j + b_i^k\nabla_k A_j, \\
&\gamma = \gamma_i^i = \nabla_i\Psi^i - 2H\Psi_3 + b^{ij}\nabla_i A_j - CA_3, \\
&A = a_i^i = \nabla_i A^i - 2HA_3, \ 2H = b_i^i, \ K = \det\left(b_j^i\right), \ C = 2\left(2H^2 - K\right),
\end{aligned} \tag{26.25}$$

where H and K are the mean and Gaussian curvatures of surface Π.

Using Eqs. (26.9) and (26.23)–(26.25), together with the procedure of the passage from the triple integrals over G and the surface integrals over Π_+ and Π_- to the integrals over midsurface Π and from the integral over Π_b to the curvilinear integral over $\Gamma = \partial\Pi$, which is standard for thin shells [21], we can write down the functional components in Eq. (26.9) in the form:

$$8\pi W = -\iint\limits_{\Pi} \{ \left(c^{-1}\dot{A}_i + \nabla_i\varphi_0\right) T^i + \left(c^{-1}\dot{\Psi}_i + \nabla_i\varphi_1\right) M^i + c^{-1}\dot{\Psi}_3 M^3 + \left(c^{-1}\dot{A}_3 + \varphi_1\right) T^3 + \chi^{ij}\left[\left(t_j - \Psi_j\right) H_i + \zeta_j N_i + H_3\nabla_i A_j + B_{ij}N_3\right]\} \, \mathrm{d}S; \tag{26.26}$$

$$A_G = \iint\limits_{\Pi} \left[c^{-1}\left(A_i J^i + A_3 J_3 + \Psi_i K^i + \Psi_3 K_3\right) - \left(\varphi_0 R_e + \varphi_1 S_e\right)\right] dS \tag{26.27}$$

$$4\pi A_S = \iint\limits_{\Pi} \left(r\varphi_0 + s\varphi_1 - r^i A_i - s^i\Psi_i - r_3 A_3 - s_3\Psi_3\right) \mathrm{d}S + \int\limits_{\Gamma} \left(P_{(0)}\varphi_0 + S_{(0)}\varphi_1 - P^i_{(0)}A_i - S^i_{(0)}\Psi_i - P^3_{(0)}A_3 - S^3_{(0)}\Psi_3\right) \mathrm{d}s. \tag{26.28}$$

Here

$$T^i = \int\limits_{-h/2}^{h/2} D^i \mathrm{d}z,\ T_3 = \int\limits_{-h/2}^{h/2} D_3 \mathrm{d}z,\ M^i = \int\limits_{-h/2}^{h/2} zD^i \mathrm{d}z,\ M_3 = \int\limits_{-h/2}^{h/2} zD_3 \mathrm{d}z,$$
$$H_k = \int\limits_{-h/2}^{h/2} H_k \mathrm{d}z,\ T_3 = \int\limits_{-h/2}^{h/2} H_3 \mathrm{d}z,\ N_k = \int\limits_{-h/2}^{h/2} zH_k \mathrm{d}z,\ N_3 = \int\limits_{-h/2}^{h/2} zH_3 \mathrm{d}z,$$
$$J^i = \int\limits_{-h/2}^{h/2} j^i \mathrm{d}z,\ J_3 = \int\limits_{-h/2}^{h/2} j_3 \mathrm{d}z,\ K^i = \int\limits_{-h/2}^{h/2} zj^i \mathrm{d}z,\ K_3 = \int\limits_{-h/2}^{h/2} zj_3 \mathrm{d}z,$$
$$R_e = \int\limits_{-h/2}^{h/2} \rho_e \mathrm{d}z,\ S_e = \int\limits_{-h/2}^{h/2} z\rho_e \mathrm{d}z;$$
$$r^i = r^i_+ + r^i_-,\ r^3 = r^3_+ + r^3_-,\ s^i = s^i_+ + s^i_-,\ s^3 = s^3_+ + s^3_-,$$
$$r = r_+ + r_-,\ s = s_+ + s_-,\ r^i_\pm = P^i\big|_{z=\pm h/2},\ r^3_\pm = P^3\big|_{z=\pm h/2},$$
$$s^i_\pm = \pm\tfrac{h}{2}r^i_\pm,\ s^3_\pm = \pm\tfrac{h}{2}r^3_\pm,\ r_\pm = P|_{z=\pm h/2},\ s_\pm = \pm\tfrac{h}{2}r_\pm; \tag{26.29}$$

$$P^i_{(0)} = \int\limits_{-h/2}^{h/2} P^i\big|_{\Pi_b} dz, \; S^i_{(0)} = \int\limits_{-h/2}^{h/2} zP^i\big|_{\Pi_b} dz, \; P_{3(0)} = \int\limits_{-h/2}^{h/2} P_3|_{\Pi_b} dz,$$
$$S_{3(0)} = \int\limits_{-h/2}^{h/2} zP_3|_{\Pi_b} dz, \, P_{(0)} = \int\limits_{-h/2}^{h/2} P|_{\Pi_b} dz, \; S_{(0)} = \int\limits_{-h/2}^{h/2} zP|_{\Pi_b} dz. \tag{26.30}$$

Thus, the functional of the electromagnetic field of the shell is determined by the equality following from Eq. (26.8)

$$H_o\left(\tilde{\mathbf{A}}, A_3, \tilde{\Psi}, \Psi_3, \varphi_0, \varphi_1\right) = \int\limits_{t_1}^{t_2} (W + A_G + A_S)\, dt, \;\; \tilde{\Psi} = \Psi^i \pi_i \tag{26.31}$$

into which formulas (26.26)–(26.28) should be substituted.

It is assumed that the physical relations following from Eqs. (26.6), (26.10), (26.11), (26.14), (26.24), and (26.29) are fulfilled

$$\begin{aligned}
&T^i = h\left[\kappa^{ijk}\varepsilon_{jk} + \kappa^{i33}\psi_3 - e^{ij}\left(c^{-1}\dot{A}_j + \nabla_j\varphi_0\right) - e^{i3}\varphi_1\right], \\
&T_3 = h\left[\kappa^{3k3}\theta_k - e^{3j}\left(c^{-1}\dot{A}_j + \nabla_j\varphi_0\right) - e^{33}\varphi_1\right], \; M^i = I\left[\kappa^{ijk}\kappa_{jk} - e^{ij}\nabla_j\varphi_1\right], \\
&M_3 = I\left[\kappa^{3k3}\left(b^n_k\theta_n + \nabla_k\psi_3\right) - e^{3j}\nabla_j\varphi_1\right], \\
&H^i = 4\pi h\left[\chi_{jm}\beta^{ij}\left(t^m - \Psi^m\right) + q^{ijm}\varepsilon_{jm} + q^{i33}\psi_3\right], \\
&H_3 = 4\pi h\left[\chi_{jm}\beta^{3j}\left(t^m - \Psi^m\right) + q^{3j3}\theta_j + q^{333}\psi_3\right], \\
&N^i = 4\pi I\left[\chi_{jm}\beta^{ij}\zeta^m + q^{ijm}\kappa_{jm} + q^{i33}\psi_3\right], \\
&N_3 = 4\pi I\left[\chi_{jm}\beta^{3j}\zeta^m + q^{3j3}\left(b^n_j\theta_n + \nabla_j\psi_3\right)\right].
\end{aligned} \tag{26.32}$$

26.5 Equations of Shell Electromagnetodynamics

Statement 2. The necessary condition for an extremum of functional (26.31) under the conditions

$$\begin{aligned}
&\delta\tilde{\mathbf{A}}\Big|_{t=t_1} = \delta\tilde{\mathbf{A}}\Big|_{t=t_2} = \mathbf{0}, \;\; \delta\tilde{\Psi}\Big|_{t=t_1} = \delta\tilde{\Psi}\Big|_{t=t_2} = \mathbf{0}, \\
&\delta\varphi_0|_{t=t_1} = \delta\varphi_0|_{t=t_2} = 0, \;\; \delta\varphi_1|_{t=t_1} = \delta\varphi_1|_{t=t_2} = 0
\end{aligned}$$

is represented by the equations

$$\begin{aligned}
&\chi^{il}\left(\nabla_i H_3 + b_i^k \nabla_k N_3\right) - \chi^{ij}\left(H_i b_j^l + N_i c_j^l\right) + c^{-1}\dot{T}^l + 4\pi c^{-1} J^l = r^l,\\
&\chi^{il}\nabla_i N_3 - \chi^{ij}\left(N_i b_j^l - H_i \delta_j^l\right) + c^{-1}\dot{M}^l + 4\pi c^{-1} K^l = s^l,\\
&\chi^{ij}\left(\nabla_j H_i + b_j^k \nabla_k N_i + c^{-1}\dot{T}_3\right) + 4\pi c^{-1} J_3 = r_3,\\
&\chi^{ij}\nabla_j N_i + c^{-1}\dot{M}_3 + 4\pi c^{-1} K_3 = s_3,\\
&\nabla_i T^i - 4\pi R_e + 4r = 0,\ \nabla_i M^i - T_3 - 4\pi S_e + s = 0
\end{aligned} \tag{26.33}$$

and the boundary conditions

$$\begin{aligned}
&\chi^{ij}\left(H_3 \nu_i + b_i^k \nu_k N_3\right)\big|_{\Gamma_H} = P^j_{(0)},\ \ \chi^{ij}\left(H_i \nu_j + b_j^k \nu_k N_i\right)\Big|_{\Gamma_H} = P_{3(0)},\\
&\chi^{ij}\nu_i N_3\big|_{\Gamma_H} = S^j_{(0)},\ \ \chi^{ij} N_i \nu_j\big|_{\Gamma_H} = S_{3(0)}\\
&\tilde{\mathbf{A}}\Big|_{\Gamma_A} = \mathbf{A}_0,\ \ A_3\big|_{\Gamma_A} = A_{30},\ \ \tilde{\Psi}\Big|_{\Gamma_A} = \Psi_0,\ \ \Psi_3\big|_{\Gamma_A} = \Psi_{30},\\
&\nu_i T^i\big|_{\Gamma_D} = -P_{(0)},\ \ \nu_i M^i\big|_{\Gamma_D} = -S_{(0)},\ \ \varphi_0\big|_{\Gamma_D} = \varphi_{00},\ \ \varphi_1\big|_{\Gamma_D} = \varphi_{10},
\end{aligned} \tag{26.34}$$

Here, $\nu = \upsilon_i \pi^i$ is the unit normal vector to the shell surface Π_b at $z = 0$ (on the intersection line $\Pi_b \bigcap \Pi = \Gamma$); δ_i^m is the Kronecker symbol; $c_j^k = b_j^m b_m^k$; curves Γ_H and Γ_A, as well as Γ_D and Γ_φ, intersect along a null-measure set; and $A_0,\ A_{30},\ \Psi_0, \Psi_{30}, \varphi_{00}$, and φ_{10} are some given quantities.

To prove this statement, we will use the variational equation for this functional, together with the following relations

$$\begin{aligned}
4\pi \int_{t_1}^{t_2} \delta W \mathrm{d}t = \int_{t_1}^{t_2} \mathrm{d}t \int_{\Pi} &\left\{\delta\varphi_0 \nabla_i T^i + \delta\varphi_1\left(\nabla_i M^i - T^3\right)\right.\\
&+ \left[\chi^{ij}\left(\nabla_j H_i + b_j^k \nabla_k N_i + c^{-1}\dot{T}^3\right)\right]\delta A_3 + \left(\chi^{ij}\nabla_j N_i + c^{-1}\dot{M}^3\right)\delta\Psi_3\\
&+ \left[\chi^{il}\left(\nabla_i H_3 + b_i^k \nabla_k N_3\right) - \chi^{ij}\left(H_i b_j^l + N_i c_j^l\right) + c^{-1}\dot{T}^l\right]\delta A_l\\
&+ \chi^{ij}\left[\left(H_i \nu_j + b_j^k \nu_k N_i\right)\delta A_3 + N_i \nu_j \delta\Psi_3\right.\\
&\left.\left. + \left(H_3 \nu_i + b_i^k \nu_k N_3\right)\delta A_j + N_3 \nu_i \delta\Psi_j\right]\right\} ds;
\end{aligned}$$

$$\delta A_G = \iint_{\Pi} \left[c^{-1}\left(J^i \delta A_i + J_3 \delta A_3 + K^i \delta\Psi_i + K_3 \delta\Psi_3\right) - \left(R_e \delta\varphi_0 + S\delta\varphi_{1_e}\right)\right] \mathrm{d}S;$$

$$4\pi\,\delta A_S = \iint\limits_{\Pi} \left(r\delta\varphi_0 + s\delta\varphi_1 - r^i\delta A_i - s^i\delta\Psi_i - r_3\delta A_3 - s_3\delta\Psi_3\right) \mathrm{d}S$$
$$+ \int\limits_{\Gamma} \left(P_{(0)}\delta\varphi_0 + S_{(0)}\delta\varphi_1 - P^i_{(0)}\delta A_i - S^i_{(0)}\delta\Psi_i - P^3_{(0)}\delta A_3 - S^3_{(0)}\delta\Psi_3\right) \mathrm{d}s;$$

The system of equations (26.33), together with the physical relations (26.32), contains 20 equations in 28 unknowns

$$A^k, A_3, \Psi^i, \Psi_3, T^k, T_3, H_i, H_3, N_i, N_3, M^k, M_3, J^k, J_3, K^k, K_3, R_e, S_e, \varphi_0, \varphi_1 \quad (26.35)$$

To obtain a system, which would be closed at a given displacement field, system (26.33) should be supplemented with equations following from Eqs. (26.20) and (26.23)–(26.25)

$$A + \Psi_3 + c^{-1}a\dot{\varphi}_0 = 0, \;\; \gamma + c^{-1}a\dot{\varphi}_1 = 0$$

and the relations following from the Ohm's law (26.13) and (26.29)

$$\begin{aligned}
J^i &= h\left\{\sigma\left[c^{-1}\chi^{ik}\left(\dot{u}_k B_{03} - \dot{w}B_{0k}\right) - c^{-1}\dot{A}^i - g^{ik}\nabla_k\varphi_0\right] + \rho_{e0}\dot{u}^i\right\},\\
J_3 &= h\left[\sigma\left(c^{-1}\chi^{ik}\dot{u}_i B_{0k} - c^{-1}\dot{A}_3 - \varphi_1\right) + \rho_{e0}\dot{w}\right],\\
K^i &= I\left\{\sigma\left[c^{-1}\chi^{ik}\left(\dot{\psi}_k B_{03} - \dot{\psi}_3 B_{0k}\right) - c^{-1}\dot{\Psi}^i - g^{ik}\nabla_k\varphi_1\right] + \rho_{e0}\dot{\psi}^i\right\},\\
K_3 &= I\left[\sigma c^{-1}\left(\chi^{ik}\dot{\psi}_{ii}B_{0k} - \dot{\Psi}_3\right) + \rho_{e0}\dot{\psi}_3\right].
\end{aligned} \quad (26.36)$$

26.6 Equations of Electromagnetoelasticity for an Anisotropic Shell

The system presented above is closed by the equations of shell motion [21] which must be supplemented with the equations for the force produced by the electromagnetic field and the expressions for internal forces and moments following from Eqs. (26.2)–(26.7). It is assumed that the initial electromagnetic field is independent of the coordinate z.

With account for Eqs. (26.10), (26.23), and (26.24), the first group of these relations takes the form:

$$\begin{aligned}
q^i_e = h&\left\{c^{-1}\left[j_{0k}\chi^{ik}\chi^{lj}\nabla_l A_j + j_{03}\left(t^i - \Psi^i\right)\right] - \rho_{e0}\left(c^{-1}\dot{A}^i + g^{ik}\nabla_k\varphi_0\right)\right\}\\
&+ R_e E^i_0 + c^{-1}\chi^{ik}\left(J_k B_{03} - J_3 B_{0k}\right),
\end{aligned}$$

$$\begin{aligned}
m^i_e = I&\left[c^{-1}\left(j_{0k}\chi^{ik}\chi^{lj}B_{lj} + j_{03}\zeta^i\right) - \rho_{e0}\left(c^{-1}\dot{\Psi}^i + g^{ik}\nabla_k\varphi_1\right)\right] + S_e E^i_0\\
&+ c^{-1}\chi^{ik}\left(K_k B_{03} - K_3 B_{0k}\right),
\end{aligned}$$

$$q_e = -h\left[c^{-1}j_{0i}\left(t^i - \Psi^i\right) + \rho_{e0}\left(\dot{A}_3 + \varphi_1\right)\right] + c^{-1}\chi^{ik}J_iB_{0k} + R_eE_{03},$$

$$m_e = -I\left(c^{-1}j_{0i}\zeta^i + \rho_{e0}\dot{\Psi}_3\right) + c^{-1}\chi^{ik}J_iB_{0k} + R_eE_{03}.$$

Here, we used the property of the discriminant tensor $\chi^{ik}\chi_{jk} = \delta^i_j$.

The second group of equations, with account for Equations (26.10) and (26.24), is written as follows:

$$\begin{aligned}
\hat{T}^{ij} &= h\left[C^{ijkl}\varepsilon_{kl} + C^{ij33}\psi_3 + \kappa^{ijk}\left(c^{-1}\dot{A}_k + \nabla_k\varphi_0\right)/4\pi\ \right] - \gamma^{ijk}H_k/4\pi,\\
M^{ij} &= I\left[C^{ijkl}\kappa_{kl} + \kappa^{ijk}\left(c^{-1}\dot{\Psi}_k + \nabla_k\varphi_1\right)/4\pi\ \right] - \gamma^{ijk}N_k/4\pi,\\
\hat{Q}^i &= h\left[C^{i3k3}\theta_k - \kappa^{i33}\left(c^{-1}\dot{A}_3 + \varphi_1\right)/4\pi\ \right] - \gamma^{i33}H_3/4\pi,\\
\mu^i &= Ih\left(C^{i3k3}\theta_k - \kappa^{i33}c^{-1}\dot{\Psi}_3/4\pi\ \right) - \gamma^{i33}N_3/4\pi,\\
\hat{N} &= h\left[C^{33kl}\varepsilon_{kl} + C^{3333}\psi_3 + \kappa^{33k}\left(c^{-1}\dot{A}_k + \nabla_k\varphi_0\right)/4\pi\ \right] - \gamma^{33k}H_k/4\pi.
\end{aligned} \tag{26.37}$$

26.7 Equations of Electromagnetoelasticity for an Isotropic Shell

If the shell material is isotropic, then Eqs. (26.32) and (26.37) can be considerably simplified at the expense of the symmetry property of the tensor of elastic constants [21] and the absence of piezoeffects [27]. In this case, Eq. (26.37) takes the form independent of the electromagnetic field, as noted in [21], while Eq. (26.32) with account for Eq. (26.22) is written as follows:

$$\begin{aligned}
&T^i = -h\varepsilon\left(c^{-1}\dot{A}^i + g^{ij}\nabla_j\varphi_0\right),\ H^i = h\chi_{jm}g^{ij}\left(t^m - \Psi^m\right)/\mu_e\,,\\
&M^i = -I\varepsilon g^{ij}\nabla_j\varphi_1, N^i = I\chi_{jm}g^{ij}\zeta^m/\mu_e\,,\ T_3 = -h\varepsilon\varphi_1,\\
&M_3 = 0,\ H_3 = 0,\ N_3 = 0.
\end{aligned} \tag{26.38}$$

This reduces the number of unknowns in Eq. (26.35) by three, brings Eq. (26.33) into a more compact form:

$$\begin{aligned}
c^{-1}\dot{T}^l + 4\pi c^{-1}J^l - \chi^{ij}\left(H_ib^l_j + N_ic^l_j\right) &= r^l,\\
c^{-1}\dot{M}^l + 4\pi c^{-1}K^l - \chi^{ij}\left(N_ib^l_j - H_i\delta^l_j\right) &= s^l,\\
\chi^{ij}\left(\nabla_jH_i + b^k_j\nabla_kN_i + c^{-1}\dot{T}_3\right) + 4\pi c^{-1}J_3 = r_3,\ & \chi^{ij}\nabla_jN_i + 4\pi c^{-1}K_3 = s_3,\\
\nabla_iT^i - 4\pi R_e + 4r = 0,\ & \nabla_iM^i - T_3 - 4\pi S_e + s = 0,
\end{aligned} \tag{26.39}$$

and, correspondingly, reduces the number of the boundary conditions (26.34).

$$\begin{gathered}\chi^{ij}\left(H_i\nu_j+b_j^k\nu_kN_i\right)\Big|_{\Gamma_H}=P_{3(0)},\ \ \chi^{ij}N_i\nu_j\big|_{\Gamma_H}=S_{3(0)},\\ \tilde{\mathbf{A}}\Big|_{\Gamma_A}=\mathbf{A}_0,\ \ A_3|_{\Gamma_A}=A_{30},\ \ \tilde{\Psi}\Big|_{\Gamma_A}=\Psi_0,\ \ \Psi_3|_{\Gamma_A}=\Psi_{30},\\ \nu_iT^i\big|_{\Gamma_D}=-P_{(0)},\ \ \nu_iM^i\big|_{\Gamma_D}=-S_{(0)},\ \ \varphi_0|_{\Gamma_D}=\varphi_{00},\ \ \varphi_1|_{\Gamma_D}=\varphi_{10}.\end{gathered}\tag{26.40}$$

Obviously, in this case, the following equalities must take place:

$$P^1_{(0)}=P^2_{(0)}=0,\ \ S^1_{(0)}=S^2_{(0)}=0,$$

In order for they be fulfilled, it is sufficient, by virtue of Eq. (26.30), to require that the electromagnetic disturbances on the lateral surface are direct normal to it:

$$\mathbf{P}|_{\Pi_b}=P_3|_{\Pi_b}\mathbf{n}$$

26.8 Conclusions

The most general boundary value problem on the transient dynamics of a homogeneous electromagnetoelastic shell is constructed. As a particular case, the passage to the isotropic shell is realized. In this case, Eq. (26.39) and boundary conditions (26.40) can be written using Eqs. (26.36) and (26.38) in the explicit form "in displacements", that is, in the kinematic parameters and potentials of the electromagnetic field.

Using the results obtained, we can further go over to different simplified versions of the equations neglecting the reduction, assuming the perpendicularity of the normal unit vector to the midsurface and the constancy of the electromagnetic field parameters with respect to the shell thickness, etc.

Acknowledgements This work was supported by the Russian Science Foundation (project No. 20-19-00217).

References

1. Altay, G., Dökmeci, M.C.: Fundamental variational equations of discontinuous thermopiezoelectric fields. Int. J. Eng.Sci. **34**, 769–782 (1996)
2. Altay, G., Dökmeci, M.C.: Fundamental equations of certain electromagneticacoustic discontinuous fields in variational form. Continuum Mech. Thermodyn. **16**, 53–71 (2004)
3. Altay, G., Dökmeci, M.C.: Variational principles for piezoelectric, thermopiezoelectric, and hygrothermopiezoelectric continua revisited. Mech. Adv. Mater. Struct. **14**(8), 549–562 (2007)

4. Altay, G., Dökmeci, M.C.: On the fundamental equations of electromagnetoelastic media in variational form with an application to shell-laminae equations. Int. J. Solids Struct. **47**, 466–492 (2010)
5. Bardzokas, D.I., Filshtinsky, M.L., Filshtinsky, L.A.: Mathematical Methods in Electro-Magneto-Elasticity. Springer, New York (2007)
6. Bhangale, R.K., Ganesan, N.: Free vibration studies of simply supported nonhomogeneous functionally graded magneto-electro-elastic finite cylindrical shells. J. Sound Vibr. **288**, 412–422 (2005)
7. Bredov, M.M., Rumyantsev, V.V., Toptygin, I.N: Klassicheskaya elektrodinamika: Uchebnoe posobie/Pod red. I.N. Toptygina. Nauka, Moscow (1992) [in Russian]
8. Buchanan, G.R.: Free vibration of an infinite magneto-electro-elastic cylinder. J. Sound Vibr. **268**, 413–426 (2003)
9. Chen, W.Q., Lee, K.Y., Ding, H.J.: On free vibration of non-homogeneous transversely isotropic magneto-electro-elastic plates. J. Sound Vibr. **279**, 237–251 (2005)
10. Chen, J.Y., Pan, E., Chen, H.L.: Wave propagation in magneto-electro-elastic multilayered plates. Int. J. Solids Struct. **44**, 1073–1085 (2007)
11. Dai, H.L., Fu, Y.M., Liu, T.X.: Electromagnetoelastic dynamic response of transversely isotropic piezoelectric hollow spheres in a uniform magnetic field. ASME J. Appl. Mech. **74**(1), 65–73 (2007)
12. Gorshkov, A.G., Rabinskii, L.N., Tarlakovskii, D.V.: Fundamentals of Tensor Analysis and Continuum Mechanics. Publishing house Nauka, Moscow (2000). [in Russian]
13. Hou, P.F., Leung, A.Y.T.: The transient responses of magneto-electro-elastic hollow cylinders. Smart Mater. Struct. **13**, 762–776 (2004)
14. Hou, P.F., Ding, H.J., Leung, A.Y.T.: The transient responses of a special nonhomogeneous magneto-electro-elastic hollow cylinder for axisymmetric plane strain problem. J. Sound Vibr. **291**, 19–47 (2006)
15. Il'yushin, A.A.: Mekhanika sploshnoj sredy. MSU publishing house, Moscow (1978). [in Russian]
16. Korotkina, M.R.: Electromagnetoelasticity. MSU Publishing House, Moscow (1988). [in Russian]
17. Kudryavtsev, B.A., Parton, V.Z.: Electromagnetoelasticity of Piezoelectric and Electrically Conductive Bodies. Nauka, Moscow (1988). [in Russian]
18. Landau, L.D., Lifshitz, E.M.: Electrodynamics of Continuous Media. Addison-Wesley, New York (1960)
19. Lee, P.C.Y.: A variational principle for the equations of piezoelectromagnetism in elastic dielectric crystals. J. Appl. Phys. **69**(11), 4770–4773 (1991)
20. Luo, E., Zhu, H., Yuan, L.: Unconventional Hamilton-type variational principles for electromagnetic elastodynamics. Sci. China Ser. G Phys. Mech. Astron. **49**(1), 119–128 (2006)
21. Mihajlova, EYu., Tarlakovskii, D.V., Fedotenkov, G.V.: A generalized linear model of dynamics of thin elastic shells. Uchenye Zapiski Kazanskogo Universiteta, Seriya Fiziko-Matematicheskie Nauki **160**(3), 561–577 (2018)
22. Morse P.M., Feshbach H.: Methods of Theoretical Physics. Series. International Series in Pure and Applied Physics, Part 1. McGraw-Hill Book Company (1953)
23. Nowacki, W.: Electromagnetic Effects in Solids. Panstwowe Wydawnictwo Naukowe, Warszawa (1983). [in Polish]
24. Qing, G.-H., Qui, J.-J., Liu, Y.-H.: Mixed H-R mixed variational principle for magneto-electroelastic bodies and state-vector equation. Appl. Math. Mech. **26**(7) (2005)
25. Sedov, L.I.: A Course in Continuum Mechanics, vol. I. Basic Equations and Analytical Techniques. Wolters-Noordhoff, Groningen (1971)
26. Tarlakovskii, D.V., Vestyak, V.A., Zemskov, A.V.: Dynamic processes in thermo-electro-magneto-elastic and thermo-elasto-diffusive media. Encyclopedia of Thermal Stresses, vol. 2, pp. 1064–1071. Springer, Dordrecht, Heidelberg, New York, London (2014)
27. Vestyak, V.A., Gachkevich, A.R., Musij, R.S., Tarlakovskij, D.V., Fedotenkov, G.V.: Dvumernye nestacionarnye volny v elektromagnitouprugih telah. FIZMATLIT, Moscow (2019). [in Russian]

28. Vestyak V.A., Tarlakovskii D.V.: The model of thin electromagnetoelastic shells dynamics. In: Proceedings of ICTAEM. Structural Integrity. Springer Nature Switzerland AG, pp. 254–258 (2019). https://doi.org/10.1007/978-3-030-21894-2
29. Yao, W.: Generalized variational principles of three-dimensional problems in magneto-electroelastic bodies. Chin. J. Comput. Mech. **20**(4), 487–489 (2003)

Chapter 27
Research of the Destruction of Ice Under Shock and Explosive Loads

Maxim Yu. Orlov and Yulia N. Orlova

Abstract A theoretical and experimental research of the behavior of ice destruction under shock and explosive loads was carried out. Full-scale underwater explosive experiments and laboratory impact experiments were performed. Especially to study the properties of ice with explosive loads, a mobile laboratory "Explosive Destruction of Natural Materials" was organized. The results of the full-scale experiment of the current year are given. Post-penetration analysis of destruction of three-layer ice targets with a low-velocity impact is presented. Briefly, the mathematical model of ice destruction under dynamic loads is described. The numerical method contains a new way for isolating discontinuity surfaces of materials. The impact of an ice cylinder on a rigid wall is modeled, which is considered as a quantitative test. The penetration of a metal container into thick ice is quantitatively described.

Keywords Ice destruction · Shock and explosive loads · Full-scale underwater explosive experiments · Rigid wall · Three-layer ice target · Projectile · Fracture

Acronym

EE	Emulsion explosives
RIAMM	Research Institute of Applied Mathematics and Mechanics
ScIcExe	Science Ice Exercise—US Research Scientific Program
UNDEX	Full-scale underwater explosive test
Mobilab	Mobile laboratory "Explosive Destruction of Natural Materials"
MDPMDS	Modern dynamic problems in the mechanics of a deformed solid

M. Yu. Orlov (✉)
Tomsk State University, 36, Lenin Avenue, Tomsk 634050, Russia
e-mail: orloff_m@mail.ru

Y. N. Orlova
Tomsk Polytechnic University, 30, Lenin Avenue, Tomsk 634050, Russia
e-mail: orlovaun@mail.ru

H. Altenbach et al. (eds.), *Multiscale Solid Mechanics*,
Advanced Structured Materials 141,
https://doi.org/10.1007/978-3-030-54928-2_27

DP Detonation product
DV Detonation velocity

27.1 Introduction

At present time, studies of the behavior of ice under shock and explosive loads are extremely important. This fact is widely known and explains the development of the northern territories, the need to increase the extraction of natural resources in the North, the fight against ice jams on Siberian rivers, the creation of protective structures against micrometeorites and some military applications. Recently, the need has arisen for the development of the Northern sea route and the extraction of natural resources, mainly gas on the shelf. Ice has been an object of constant research since the middle of the last century [2, 16]. The mechanical properties of ice under dynamic loads are investigated in [14, 26].

Therefore, it is necessary to understand the basic mechanisms and patterns of the process of ice destruction under dynamic loads. The main difficulty is that ice is the oldest natural material that is little studied in conditions of high-speed deformation. There are more than 15 types of ice, some of which are of extraterrestrial nature. A well-known fact is that the deformation of ice is accompanied by phase transitions, ice has unique plastic properties, and its strength depends on temperature [4, 25, 27]. Mathematical models of ice destruction are still being developed. For example, in work [5], a phenomenological model was developed for the destruction of ice upon impact. But the model is tied to ballistic experiments.

The situation is complicated by the fact that some experimental data do not agree with each other, which is most likely caused by the temperature of its formation. According to the analytical review of [22], experimental data on the explosive loading of ice are missing or have already become a bibliographical rarity.

The Research Institute of Applied Mathematics and Mechanics (hereinafter, RI AMM) conducts systematic research on the destruction of ice under shock and explosive loads. In Orlova [19], a mathematical model of ice deformation and fracture under dynamic loads was developed. In recently research [21], a numerical method was developed for solving tasks of ice destruction under detonation products (hereinafter, DP). The latest innovation can be considered as a mobile laboratory "Explosive Destruction of Natural Materials" organized on the basis of the RI AMM (hereinafter, mobilab). Currently, the mobilab is being developed as an alternative to the American research program ScIcExe.

27.2 Experimental

In the current paper, numerical and experimental research of the destruction of ice under explosive and shock loads are carried out. Experimental research deals with full-scale experiments and laboratory impact experiments. Snow-covered ice cover and needle ice cover were studied with mobilab. Snap-analysis of ice destruction under DP was obtained. Full-scale experiments were conducted with the help of KuzbasSpetsVzryv Ltd. Laboratory impact experiments were carried out on the basis of the RIAMM and Society for Practical Bullet Shooting.

27.2.1 Mobile Laboratory «Explosive Destruction of Natural Materials»

As mentioned earlier, ice is a poorly understood natural material. The state of art emphasizes the lack of experimental data on the behavior of ice under explosive loads. A few years ago, the mobile laboratory "Explosive Destruction of Natural Materials" was organized at the Tomsk State University. The first research objects were bare ice and natural limestone [23]. Mobilab's main goal is to deepen the experimental knowledge in the field of explosive destruction of ice. Currently, it has the status of an initial project and is developing as an alternative to the ScIcExe American research program [3]. Mobilab's permanent partners are the Ministry of Emergency Situations of Russia and the KuzbasSpetsVzryv Ltd.

In addition to ice, the traditional object of study was natural limestone. In previous research, the limestone was studied in detail under explosive loading. The shape of the explosive crater including the state of its edge and the morphology of destruction were obtained. Emulsion explosives (hereinafter, EE), granulate explosives and ammonite explosives and an explosive mixture (hereinafter, EM) of these components were used. EM maximum weight is 1000 kg of TNT. In current paper, only EE are considered.

By means of mobilab, snow-covered ice, bare ice, ice cover like needle ice and ice cover sandwich structures of the "Snow—Shuga—Ice" under explosive were studied. We use Sea Ice Nomenclatura of 1974 only. The subject of the study was the state of the ice cover after the explosion, including the diameter of the lane (or polynya), its shape, the state of the ice edge and the morphology of ice destruction. In Orlov et al. [20], it was possible to reveal the height and diameter of a cloud of ice fragments after blast.

27.2.2 Full-Scale Underwater Explosive Test

In this section, we focus on full-scale experiments. These experiments are called full-scale underwater explosive test (hereinafter, UNDEX). Similar experiments were in Cole [6], Zamyshlyaev and Yakovlev [29]. In this case, the blast was produced in water under ice without an air gap. Experiments are conducted during the last seven years on the river Tom. Four experimental sites were selected. At the first two sites, snow-covered ice and ice cover of the sandwich structure are formed every year. And in the last two sites, bare ice and "needle ice" are formed. This can be explained by the peculiarities of the formation of ice not by the river, including water current and temperature, wind, etc. In Glazyrin et al. [11], the behavior of these ice covers subjected to explosion is described.

The research object is snow-covered ice of medium thickness (ice thickness is approximately 70 cm). The approximate snow thickness is no more than 20 cm. The age of this ice is not more than 125 days. The research subject was the state of ice cover after the explosion, including the diameter of the lane (polynia) in ice, the state of the ice edge and the morphology of ice destruction.

In all cases, the explosion was performed in water under the ice. There was no air gap between explosives and ice. The explosive substance is EE (Emulast AS-FP-90). The explosive charge has a cylindrical shape and a mass of 4 kg (TNT equivalent 3.25 kg). Detonation velocity is 4600 m/s (hereinafter, DV). At the time of the blast, the explosive was located parallel to the ice cover. Water and air temperature were 4 °C. The depth of the water under the ice was approximately 5 m (hydroimpact was excluded). The river bottom was flat. The initiation point of explosion was at the top of EE charge. The total area of the experimental site was approximately 1000 m^2. The surface of the ice cover is smooth. There are no recurring polynya, crack and fracture of ice.

Figure 27.1 shows a polynya in a snow-covered ice subjected to 4-kg-EE blasting. Last year, such ice was identified as needle ice. This year the ice thickness was 65 cm. Due to ice fragments, the water inside the lane is almost invisible. Inside the lane, there

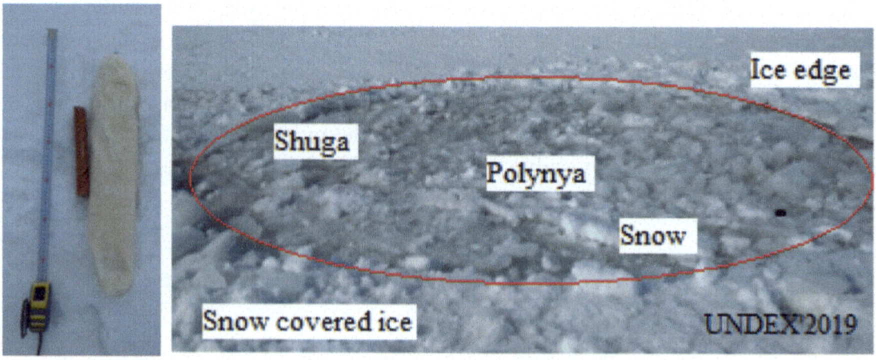

Fig. 27.1 One EE charge, polynya in a snow-covered ice. UNDEX's results of 2019

are fragments of ice of small size (≈15 cm). No large fragments of ice (≥50 cm) were found. The shape of the ice edge could not be identified. Most likely, it had a stepped shape. Near the edge of the ice, you can find small pieces of ice (≤ 10 cm). Initially, they were larger. It is possible that they are debris of medium or large fragments of ice. Radial cracks in the ice cover were also not detected. The approximate diameter of the lane was 500 cm. This fact is predictable and corresponds to last year's results [11].

27.2.3 Behavior of Ice Target Subjected to Normal Impact by Nosed Projectiles

This section outlines an experimental research of the destruction of ice under shock loads. Post-penetration analysis of low-velocity impact of a nosed projectile with single-layer and multilayer ice targets was performed. The low-velocity impact of ice cylinders with AU4G aluminum targets is considered in Combescure et al. [7]. Recently, several experiments were simulated. The subject of comparison was the morphology of the destruction of ice and the residual displacement of the aluminum plates. According to terminology from Babkin et al. [1], this experiment can be interpreted as a quantitative test.

In the current research, the impact resistance of a three-layer ice target was studied experimentally. In the first instance, the scientific interest is the features of the destruction of ice plates, including the penetration time, the morphology of ice destruction, etc. The research object is ice cylinder. Its dimensions are 10.5 × 4.5 cm. The cylinder was formed by freezing freshwater at a temperature of—24 °C and a freezing time of 48 h. Projectile is a well-known bullet 57-H-181C. The initial velocity is 315 m/s. Research subject is the condition of ice impacted to nosed projectile. In the previous experiment, the object of the research was one ice cylinder only.

Figure 27.2 illustrates one ice cylinder before impact and a three-layer ice target after impact. Due to the fact that the result was predictable, the results of the high-speed shooting of the ice destruction in this subsection are not presented. Ice fragments of various size, including small size (< 10 mm), are clearly visible in the photo. The projectile was not deformed. The results of this experiment almost completely repeat the results of the previous experiment. The presence of two additional ice cylinders in the composition of the target did not increase its impact resistance.

27.3 Model and Method

This section summarizes the mathematical model and method for numerical modeling ice destruction problems. A mathematical model of ice behavior during impact and explosion is based on a complex model of continuum mechanics. The numerical

Fig. 27.2 Ice sample and target of three-ice cylinder after impact. Scale bar represents 1 cm

method is based on the Lagrangian approach of the motion of continuous media and is a development of the numerical method of G. R. Johnson [12]. The method is developed by adding new calculation algorithms, include elements erosion algorithm, splitting nodes algorithm, etc. Numerical modeling was carried out on a non-commercial program package for solving modern dynamic tasks in the mechanics of a deformed solid (hereinafter, MDPMDS). According to the terminology of Gerasimov et al. [10], such tasks should be understood as it of penetration and penetration of impactors of complex geometry into structurally inhomogeneous targets.

27.3.1 Mathematical Model of Ice Destruction

Below is a mathematical model of ice destruction, which is based on a complex model of continuum mechanics. The model allows to describe the processes of deformation and destruction of ice under dynamic loads, including the formation of new free surfaces, the fragmentary destruction of materials. The governing equations are based on the fundamental laws of conservation of mass, momentum and energy. Ice is modeled by a single-phase medium without phase transitions with averaged strength characteristics. The material is assumed to be elastic–plastic, compressible, porous, taking into account the strength properties, shock-wave phenomena and joint formation of several types of fracture. The model was tested and detailed in Orlov [10, 19].

The elastic–plastic behavior of materials is described by the well-known Prandtl–Reis equations and von Mises yield condition. The model is not tied to any single equation of state. The equations of state of Walsh, Mi–Grüneisen, Tet, the Zhukov's wide-range equation, etc., were used. Pressure in detonation products (hereinafter,

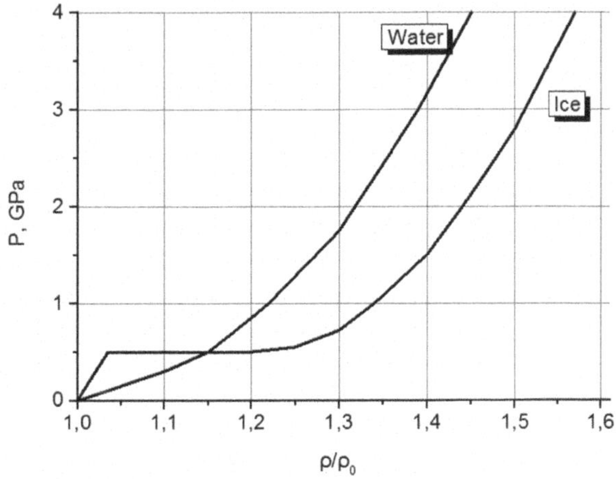

Fig. 27.3 Shock adiabat of ice and water [15]

DP) is described by Landau–Stanyukovich polytropic [18]. The shock adiabat of ice and water is shown in Fig. 27.3.

The model implemented the concept of destruction material from Tolkachev and Trushkov [28]. The concept is based on a well-known fact that dynamic destruction materials are accompanied by the formation of destruction like "spall" and destruction like "shift". There may be cases when one type of destruction prevails over another, for example, during an adiabatic shift. Therefore, when modeling fracture materials, both types must be considered. In the opinion of the authors, this will allow to model the processes of destruction most approximate to the real ones. The most important point is that it is possible to use different failure criteria of materials.

We used the shock adiabat of ice and water in the following form:

The equation of state of ice is as follows.

$$P(\rho) = \left(\frac{\rho}{\rho_0} - 1\right)\left(\frac{\rho}{\rho_0}\right)$$

where *P*—hydrostatic pressure, ρ—density, *B*—constant.

$B = 8.4$ GPa, $\rho_0 = 0.92$ g/cm^3.

The equation of state for water is a polynomial.

$$P(\rho) = 0.0225\left(\frac{\rho}{\rho_0} - 1\right) + 0.085\left(\frac{\rho}{\rho_0} - 1\right)$$

The pressure in the detonation products is described by the following formula.

$$P = \rho_0 D^2\left(\frac{\rho}{\rho_0}\right)\frac{1}{8}$$

where D is detonation's velocity.

27.3.2 Numerical Method and Test Calculations

In this subsection, we focus on the numerical method. The system of equations is solved in the two-dimensional axisymmetric statement on the basis of the Lagrangian approach to the description of the motion of continuous media. Well-known fact is that any Lagrangian method has serious problems with solving tasks of the deep penetration of projectiles of complex geometry into multilayer targets. For example, one is the penetrating of multilayer targets by nosed projectile. The problem is the overlap of the triangulated elements [9]. To overcome this lack, the algorithm erosion elements, algorithm splitting nodes, the algorithm for constructing the free surface were introduced. The last algorithm will be mentioned below when modeling the tasks of explosive destruction of the ice [22].

As said before, the numerical method contains a new way for isolating the surfaces of discontinuity of materials. A similar approach to modeling the perforation tasks developed in the work [24]. However, in the present approach, several ways of splitting nodes are possible. In this algorithm, it is not necessary to store any information in the nodes as in the work [17]. It allows us to use various failure criteria of solids. For the numerical solution of MCPMDS, this is of equal importance.

A software package has been developed for modeling of MCPMDS in the programming language C ++. The program package consists of a solver program and a viewer program. Until today, there are more than five versions of both programs. The following shows the capabilities of the latest version of the program package. In research [13], an attempt was made to simulate some tasks in the Euler's formulation using the original computer code.

27.3.3 Test Calculations

Before numerical simulation, test calculations were carried out. As a quantitative test, the impact of an ice cylinder on a rigid wall was simulated. This subsection uses terminology from Fomin et al. (1998). In the scientific literature, such a test is called the Taylor's test. Of course, the test results are predictable. The ice will be destroyed. The numerical results are also intended to demonstrate the capabilities of the software package and mathematical model.

The initial velocity cylinder was varied from 50 to 150 m/s. The diameter of the cylinder is 6.88 mm, and the height of one is 20.6 mm. A series of computational experiments consisted of five cases. The subject of research was the relative shortening, equal to the ratio of the final height of the cylinder to the original height of the cylinder.

Figure 27.4 shows ice at time 0 and 35 μs. In first case, the initial velocity is 125 m/s. Of all the cases considered, this was the penultimate. Due to the strong fragmentation of ice, the latter case is not given here. It can be seen that the ice cylinder is almost completely destroyed. Small ice pieces separated from the cylinder

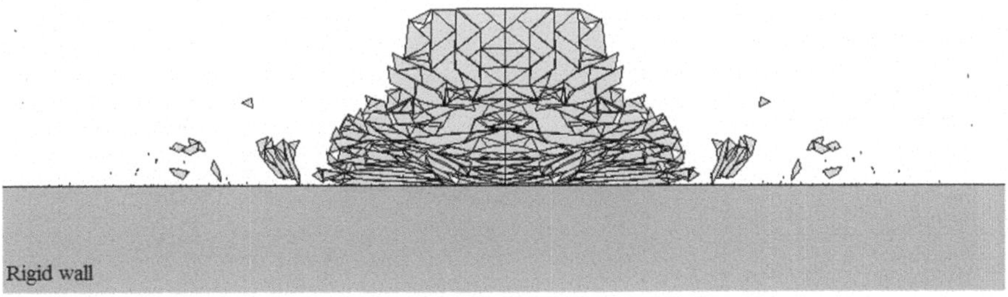

Fig. 27.4 Cross cection of ice cylinder—Rigid wall, $t = 40\ \mu s$

in the radial direction. The bottom part of the ice cylinder is cracked but still retained its original diameter. The height of the cylinder after the impact is difficult to measure, at this point in time, it was 6.2 mm. Obviously, after some time, the ice will be broken into small pieces. A similar picture of ice destruction is noted in Carney et al. [5].

Thus, an increase in the initial velocity leads to a decrease in the height of the cylinder after the impact. This fact is verified experimentally and corresponds to the physics of the process.

27.4 Deep Penetration of a Projectile into Thick Ice

This section presents simulation results. The deep penetration of metal container in thick ice in a subsonic range of initial velocities is modeled. Projectile is a metal container with inert filler. The walls of the metal container called a shell. The filler imitated explosive. The container diameter is 34 cm. The height of the container was 87.7 cm. The mass of the container along with the mass of the filler is 235 kg. Filler mass is 108.6 kg. The fill components are cement, sand, sawdust, freshwater and water glass. Strength characteristics of simulated materials are in Orlova [19].

Ice was a 400 cm barrier. In fact, such ice rarely meets on the Siberian rivers. Therefore, the object of study simulates sea ice. A series of numerical experiments consisted of four cases. The cases differed only in the initial velocity, which ranged from 150 m/s to 300 m/s. At the contact boundary of Shell – Filler and Shell – Ice, the slip condition was specified.

The subject of the research was the time of penetration (until the complete stop of the projectile), the diameter and depth of the impact crater in the ice, projectile's diameter after the impact, the gap between the filler and the shell of the projectile, etc. The time of birth of the first foci of ice destruction, their further evolution into ice were studied. This research does not address the problem of self-detonation.

Figure 27.5 shows the final stage of the penetrating for the last case. A fragment of the computational domain is shown. Outside this fragment, there were practically no destruction patterns. In front of the projectile, there is a zone of destroyed ice.

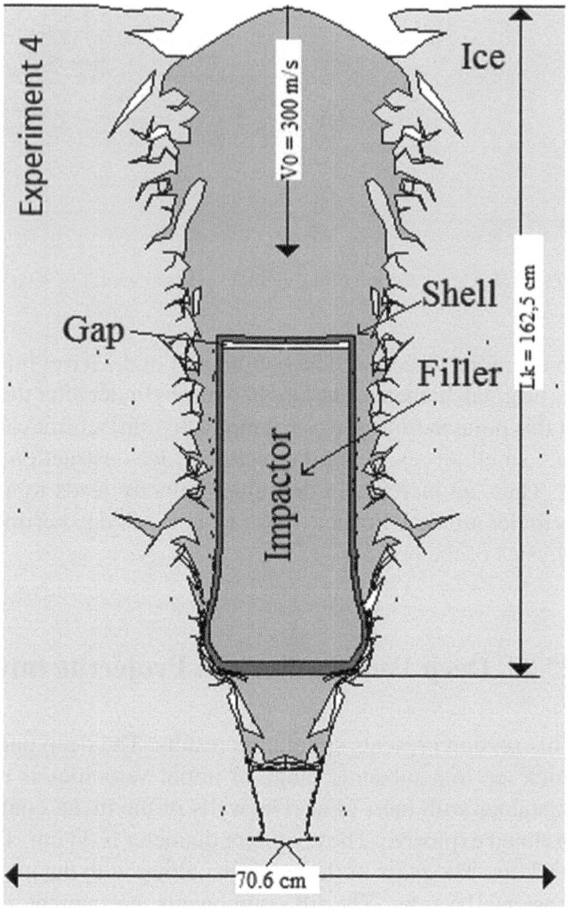

Fig. 27.5 Case 4, time 14,015 μs, cross section of Projectile—Ice

Therefore, it penetrated in the weakened ice. The impact crater is V-shaped. Penetration depth is practically almost twice the original diameter of the projectile. The front part of the container is deformed in the radial direction. In its bottom part, there is a gap between the shell and the filler. The pressure in the filler did not exceed 1 GPa.

Figure 27.6 shows the dependence of the axial velocity of the bottom of the shell on the time of penetration. Calculations are made at the control point on the axis of symmetry for case 1. The velocity curves for the other calculation cases were similar. The curve shows large-scale velocity oscillations. On the velocity curves for the bow of the shell and the filler, the oscillations are not found. At 14 ms, the velocity curve approaches zero.

The numerical results into ice are shown in Table 27.1, where D_{ice} is parameter of damage ice, L_k *is* depth of penetration of the projectile, d_l *is* diameter of the projectile after the penetration process, t_k is time of penetration, d_{d} is crater diameter, l_z is gap

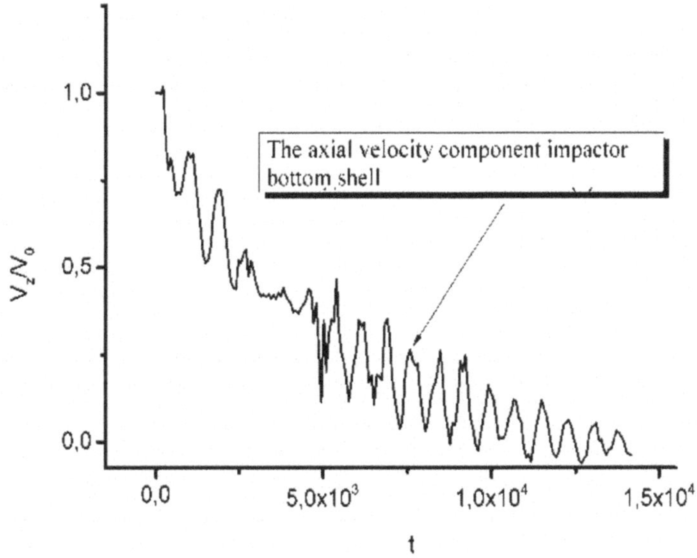

Fig. 27.6 Plot axial velocity shell's bottom versus penetration time

Table 27.1 Summary of the calculated parameters of the simulation of the projectile penetration into ice

Parameters	Case 1	Case 2	Case 3	Case 4
Projectile velocity V_0 (m/s)	150	200	250	300
Damage of ice D_{ice} (%)	1.120	1.946	3.285	4.254
Depth penetration L_k(cm)	56.11	94.14	128.7	162.5
Crater diameter d_d (cm)	55.52	57.08	58.32	61.9
Time of penetration t_k (μs)	13.042	14.510	17.270	20.15
Diameter after impact d_l (cm)	34.40	36.00	38.00	40.50
Gap l_z (cm)	0.25	0.25	1.15	1.50
Shortening ΔLShell/ΔLFiller	0.99/0.99	0.97/0.97	0.96/0.95	0.93/0.92

between the filler and shell, and $\Delta L_{Shell}/\Delta L_{Filler}$ is the relative shortening of the filling and shell. It should be noted that material damage has been proposed in Glazyrin et al. [11].

The table shows that with an increase in the initial velocity was an increase in all the parameters, except for the gap between the filler and the shell in cases 1 and 2. The volume of ice was a little damaged. Ice fracture zone formed, usually in the contact region "Projectile–Ice". No foci of destruction were found near the rear surface of ice barrier. It was found that the depth of penetration of the projectile in the range from 56.1 to 162.5 cm increases in proportion to its initial velocity. The time of the penetration of process exceeded the microsecond range and reached about

20 ms. Projectile's penetration was accompanied by a deformation of it (shell and filler) in the axial and radial directions. At the same time, d_{d} crater diameter ranges from 55.52 to 61.9 cm. It is found that at the first milliseconds gap is formed between the shell and filler

27.5 Conclusions

The experimental and numerical studies led to the following conclusions.

1. Based on the RI AMM, a mobilab *Explosive Destruction of Natural Materials* was organized. At the moment, its main goal is to snap-analysis of the behavior of ice under explosive loads, including the state of the ice edge, the morphology of the destruction and the diameter of the lane in the ice cover. In fact, the presence of mobilab seriously expands the research capabilities of the research team. This fact is of fundamental importance when studying the destruction of ice.
   At least four types of ice, including snow-covered ice, bare ice, sandwich structure ice and "needle ice", have been studied. The maximum mass of explosives was 10 kg of TNT equivalent. The diameter of the lane (polynya) varied from 200 cm to 800 cm. This year its diameter was 500 cm. There were differences in the morphology of destruction.
2. The results of laboratory experiments are presented. The research object was a three-layer ice target. Projectile is a well-known bullet 57-H-181C. There was no air gap between the ice cylinders. The initial velocity was subsonic and was 315 m/s. After the target was perforated, the projectile was not deformed. The ice target was completely destroyed. Among the fragments of ice, fragments of ice of medium size prevailed (sizes comparable to those of projectile). Two additional ice layers did not increase the impact resistance of the target. The results of the experiment turned out to be predictable and practically did not differ from the results of the experiment of the previous year.
3. Ice is described by one of the complex models of continuum mechanics based on fundamental conservation laws. The material is assumed to be elastic–plastic, compressible, porous, taking into account the strength properties, shock-wave phenomena and joint formation of several types of fracture. Ice is modeled by a single-phase medium without phase transitions with averaged strength characteristics. DP's pressure is described by Landau–Stanyukovich polytropic. The model implemented the concept of joint formation of fracture like spall and fracture like shear. This concept is based on a phenomenological approach to the description of destruction. Fragmentary destruction of the material is taken into account also.
4. The system of equations is solved in the two-dimensional axisymmetric statement based on the Lagrangian approach to the description of the motion of continuous media. The problem is the overlap of the triangulated elements. To overcome this lack, the algorithm of erosion elements, algorithm of splitting nodes and the

algorithm for constructing the free surface were introduced. According to terminology [8], the numerical method contains a new way for isolating discontinuity surfaces of materials, which does not impose serious restrictions on the solution of MCPMDS. One quantitative test was also given.

5. Quantitatively described projectile's penetration into the ice in the initial range of velocity below one of sound in air. A lot of parameters of the penetration for four cases are obtained. It was revealed that the loading process was accompanied by radial and axial deformation of the projectile and the formation of a conical (or V-shaped) crater into the ice. It was found that the depth of the crater increased in direct proportion to projectile's velocity. The volume of the destroyed ice was insignificant. Projectile's velocity including the bottom and the filler's bow changed hyperbolically. Hydrostatic pressure arising in ice can be categorized as low ($\leq$ 1GPa). In this paper, self-detonation was not considered.

Acknowledgements The reported study was funded by RFBR according to the research project № 19-08-01152.

References

1. Babkin, A., Veldanov, V., Gryaznov, E., Selivanov, V.: Means of destruction and ammunition. Publishing House of BMSTU, Moscow (2008)
2. Bogorodsky, V.V., Gavrilo, V.P.: Ice. Physical Properties of Modern Glaciology Methods. USSR, Leningrad (1980)
3. Bocharov, L.Y., Korchak, V.Y., Tuzhikov, E.Z.: US Scientific Research in the Arctic: Organizational Approach and Military. Arct. Ecol. Econ. **3**(19), 48–53 (2015)
4. Bragov, A., Igumnov, L., Konstantinov, A., Lomunov, A., Filippov, A.,Shmotin, Y., Didenko, R. and Krundaeva, A.: Investigation of strength properties of freshwater ice. In: EPJ Web Conferences (2015)
5. Carney, K.S., Benson, D.J., DuBois, P., Lee, R.: A phenomenological high strain rate model with failure for ice. Int. J. Solids Struct. **43**, 7820–7839 (2006)
6. Cole, R.: Underwater Explosions. Princeton University Press, Princeton (1948)
7. Combescure, A., Chuzel-Marmot, Y., Fabis, J.: Experimental study of high-velocity impact and fracture of ice. Int. J. Solids Struct. **48**(20), 2779–2790 (2011)
8. Fomin, V., Gulidov, A., Sapoznikov, A.: High Speed Body Interaction. Publishing House of the SB RAS, Novosibirsk (1999)
9. Flis, W.J.: Advanced algorithms for computer simulation of hypervelocity impact Int. J. Impact Eng. **5**(1–4), 269–275 (1987)
10. Gerasimov, A., Barashkov, V., Glazyrin, V., Konyaev, A., Orlov, M., Pashkov, S., Tolkachev, V., Trushkov, V., Khristenko, Yu.: Theoretical and Experimental Studies of High-SPEED Interaction of Bodies. Tomsk State University Press, Tomsk (2007)
11. Glazyrin, V., Orlov, M., Orlova, Yu.: Study destruction thick ice water substrate in blasting loads. Works Tomsk State Univ./Phys. Math. Ser. **282**, 329–334 (2012)
12. Holmquist, T., Johnson, G., Gerlach, C.: An improved computational constitutive model for glass. Phil. Trans. R. Soc. A (2017)
13. Konstantinov, A., Kochetkov, A., Krylov, S., Smirnov, I.: Simulation the dynamics of a composite cylindrical shell with a gas-permeable layer under the internal impulse. Materials Phys. Mech. **28**(1–2), 36–42 (2016)

14. Kraus, R.G., Senft, L.E., Stewart, S.T.: Impacts onto H2O ice: Scalin laws for melting, vaporization, excavation, and final crater size. Icarus **214**, 724–738 (2011)
15. Larson, D.: Shock wave of study of ice under uniaxial strain condition. Int. J. Glaciol. **30**(105), 235–240 (1984)
16. Maeno, N.: Science of Ice. USSR, Moscow (1988)
17. Nemirovich-Danchenko, M.: A model for the brittle hypoelastic medium: application to computation of deformations and failure in rock. Int. J. Phys. Mesomech. **1–2**, 101–108 (1998)
18. Orlenko, L.: Explosion Physics, (FizMatLit. Moscow **1**, 832 (2004)
19. Orlova, Y.N.: Comprehensive theoretical and experimental study of the behavior of ice under shock and explosive loads, Ph.D. Thesis, Tomsk State University (2014)
20. Orlov, M., Bogomolov, G., Orlova, Y.: Research of ice behavior at blasting emulsion. In: Gomes, J.F.S., Meguid, S. (eds.) Proceedings of the 7th International Conference on Mechanics and Materials in Design, pp. 1519–1520 (2017)
21. Orlov, M.Y., Orlova, Y., Bogomolov, G., Glazyrin, V.: Research of the behavior of ice on water under explosive loads. J. Phys. Conf. Ser. (2017)
22. Orlova, YuN, Orlov, MYu.: The study of the process of explosive loading of ice. Vestnik Tomskogo Gosudarstvennogo Universiteta, Matematika i Mekhanika. **38**, 81–89 (2015)
23. Orlov, M. Y., Orlova, Y., Tolkachev, V.: Mobile laboratory "Explosive destruction of natural materials": investigation of the behavior of ice and limestone under explosive loading. J. Phys.: Conf. Ser. (2015)
24. Olovsson, L., Limido, J., Lacome, J-L., Hanssen, A and Petit, J.: Modeling fragmentation with new high order finite element technology and node splitting. EPJ Web of Conferences, vol. 94 (2015)
25. Schulson, E., Duval, P.: Creep and Fracture of Ice. Cambridge University Press, (2009), p 400
26. Stewart, S.T., Ahrens, T.J.: Shock Hugoniot of H_2O ice. Geophys. Res. Letts 30 (2003). https://doi.org/10.1029/2002gl016789
27. Timco, G.W., Weeks, W.F.: A review of the engineering properties of sea ice. Cold Regions Sci. Technol. **60**, 107–129 (2010)
28. Tolkachev, V., Trushkov, V.: Mathematical modeling of shear and spall fractures during impact interaction of elastoplastic bodies. Russ. J. Phys. Chem. B **12**, 170–175 (1993)
29. Zamyshlyaev, B., Yakovlev, Yu.: Dynamic Loads during the Underwater Explosion Leningrad (1967)

Chapter 28
Experimental Investigations of Failure of Sandwich Specimens with Composite Facing Layers Under Four-Point Bending Conditions

Vitaly N. Paimushin, Vyacheslav A. Firsov, Sergey A. Kholmogorov, and Maksim V. Makarov

Abstract A series of experimental studies on the four-point bending of sandwich specimens with facing layers made of a unidirectional fiber-reinforced composite with specified geometric and physico-mechanical characteristics with varying geometric and physico-mechanical characteristics of the core was carried out. It is shown that when using a honeycomb core made of polymer paper, the implementation of the shear buckling mode in the facing layers is impossible, and the failure of the sample occurs due to the failure of the core. Specimens with stiffer cores are fail due to the failure of compressed facing layers in the vicinity of the loading roller at stresses comparable to the critical stresses of their buckling mode in a purely transverse-shear form. It was established that during the loading of the specimen, a significant increase in the mulitmodularity of the material is observed, apparently, mainly due to the implementation and continuous change of the internal micro- and mesoscale buckling modes of fibers and fiber bundles in areas with weakened physical and mechanical characteristics of the binder composite in compressed facing layer.

Keywords Sandwich specimen · Four-point bending · Honeycomb core · Rigid core · Failure · Shear buckling mode

V. N. Paimushin · V. A. Firsov (✉) · S. A. Kholmogorov · M. V. Makarov
Kazan National Research Technical University named after A.N. Tupolev,
10 K. Marks str., Kazan 420111, Russian Federation
e-mail: vafirsov@kai.ru

V. N. Paimushin
e-mail: vpajmushin@mail.ru

S. A. Kholmogorov
e-mail: hkazan@yandex.ru

M. V. Makarov
e-mail: makarovmaksim@mail.ru

V. N. Paimushin · M. V. Makarov
Kazan (Volga region) Federal University, 18 Kremlyovskaya str.,
Kazan 420008, Russian Federation

H. Altenbach et al. (eds.), *Multiscale Solid Mechanics*,
Advanced Structured Materials 141,
https://doi.org/10.1007/978-3-030-54928-2_28

28.1 Introduction

Sandwich structures are widely used in the shipbuilding and aerospace industries. As a rule, such constructions consist of two rigid outer layers, receiving tangential stresses, and a relatively low-rigid core, receiving transverse compression stresses and transverse shear stresses. The choice of materials for the outer layers and core depends on the specifics of the work of sandwich structures. In aerospace industry, as a rule, honeycomb based on light aluminum alloys or polymer paper is used. Such cores have the highest stiffness and strength characteristics with a low weight. In sandwich structures used in shipbuilding, porous metal cores is used. Such cores require high toughness and high ability to absorb impact energy. Both metals and composite materials based on glass and carbon plastics are used as materials for the facing layers.

The failure of sandwich structures can be caused by the implementation of different processes of deformation of their constituent elements under the loading [19, 21, 23, 24]: failure of the core due to the reaching of ultimate transverse shear stresses formed in them [3, 6]; fatigue fractures of the core [5, 23]; peeling of the facing layers from the core [1, 20]; reaching of ultimate transverse compressive stresses in the core [22]; buckling of the facing layers [2, 10].

In ultralight sandwich structures, the outer layers are usually made of carbon fiber-reinforced plastic. Unidirectional carbon plastics can have even greater tensile strength than high strength steels. Therefore, the main failure reason of such structures may be the buckling of the facing layers in one form or another under the conditions of compressive stresses. In particular, in the formation of such stresses, it is possible to realize the transverse-shear buckling mode. As is known [4, 7], when compressing specimens from unidirectional fiber-reinforced plastics along the fibers, the values of the transverse shear modulus can be taken as the ultimate compressive stresses. In a series of works [11, 15], it was shown that the theoretical identification of the shear buckling form of composite specimens under three-point bending conditions is possible using equations based on the simple kinematic model of S. P. Timoshenko. This work is a continuation of research [11, 15] and is devoted to experimental studies of the processes of deformation and failure of sandwich specimens with composite facing layers during four-point bending in order to identify the possibility of realizing a shear buckling mode in facing layers of a unidirectional fiber-reinforced plastic.

28.2 Bending of Sandwich Specimens with Honeycomb Core

For four-point bending, the sandwich specimens with honeycomb core made of paper ™Nomex with thickness $H = 5$ mm were manufactured. Dimension of the cell in average was $t = 4$ mm. Facing layers of the specimens were manufactured from carbon tape ELUR-P and cold curing epoxy XT-118. Total length of the specimens

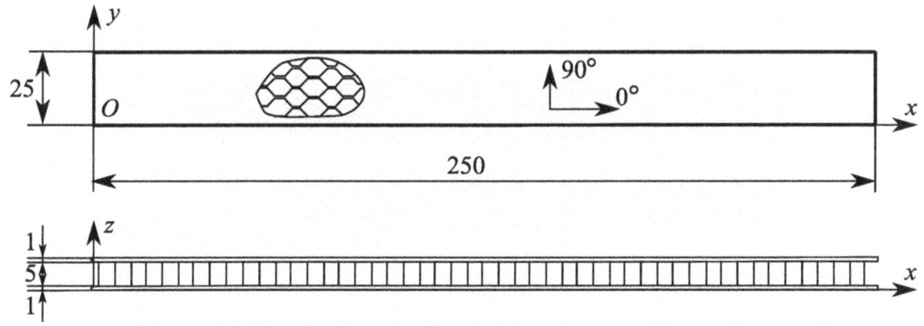

Fig. 28.1 Geometrical characteristics of the specimen

was $L = 210$ mm, and wide $b = 25$ mm. Lay-up direction of facing layers is 0° with respect to longitudinal axes of specimens Ox. Their average thickness is 1 mm and they consist of 8 laminas with a thickness of 0.12 mm. A sketch of the specimen with the basic geometric characteristics, the direction of the axes of the specimen, and the direction of the axes of orthotropy of the facing layers is shown in Fig. 28.1.

Specimens were cut from a flat plate made by vacuum molding. At the first stage of manufacturing, facing layers with lay-up sequence $[0]_8$ were formed. Laminas of carbon tape were molded layer-by-layer on a metal plate heated to $40°C$ and impregnated with an XT-118 cold-cured epoxy binder. Epoxy resin was mixed with a hardener in a proportion of 5:1. After the formation of the sequence of layers, a fabric was laid on top of the plate, absorbing excess binder and the package of layers was covered with a vacuum film, from which air was pumped out. The vacuum in the bag was held for 24 h, during which the binder completely polymerized. At the second stage, an XT-118 binder with a thickness of about 0.5 mm was applied to one of the surfaces of the finished packet of layers, and a honeycomb was placed and covered with the same packet of laminas on top. The package of the facing layers and the honeycomb core was again covered with a vacuum film and in a vacuum, the binder was polymerized within 24 h. After that, the plate was cut with a diamond cutting circular saw into the specimens. The view of the fabricated sandwich specimens is shown in Fig. 28.2.

Four-point bending tests were carried out on an Instron ElectroPuls E10000 servo-electric testing machine, on which a force transducer with a measuring range of ±10 kN is installed. The loading process was carried out in the stroke speed control mode. The speed of the stroke during the tests was constant and amounted to 5 mm/min. Bending fixtures were installed on the machine, which allow to change the distance between the support rollers. The supporting and loading rollers had a diameter of 10 mm. Equipment with loading rollers, the distance between which was 70 mm, was mounted on the specimen from above. The specimen installation diagram is shown in Fig. 28.3.

The loading curves of five specimens is shown in Fig. 28.4a. The deflection of the specimen during testing was recorded according to the stroke of the testing machine.

Fig. 28.2 Overview of the sandwich specimens

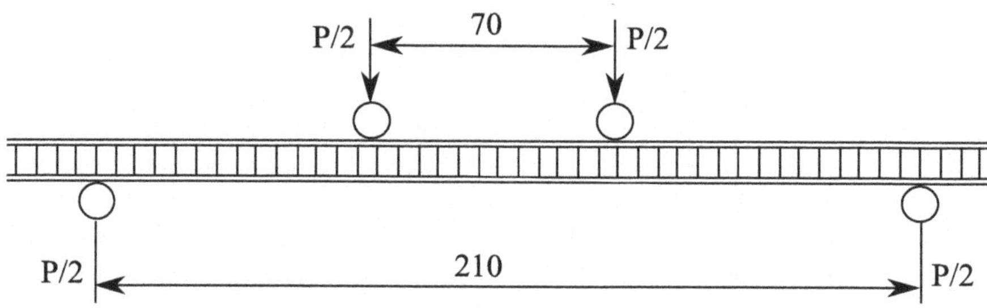

Fig. 28.3 Installation diagram of the sandwich specimen

In this regard, the deflection on the graphs reflects the deflection of the loading rollers. It can be seen from Fig. 28.4a that the maximum deflection for the tested specimen lies in the range of 3–4 mm, and the ultimate load $P_{\max}$ is about 0.3 kN.

Testing results and their statistical treatment are shown in Table 28.1. The magnitudes of ultimate stresses in the facing layers and ultimate shear stresses in the core are also shown in Table 28.1.

Normal stresses $\sigma_{\max}^{\text{bend}}$ calculated according to the formula

$$\sigma_{\max}^{\text{bend}} = M_z^{\max} / W_z, \quad M_z^{\max} = P_{\max} \cdot (l/2), \tag{28.1}$$

where value of modulus of section W_z determines by expression

$$W_z = [(b(H + 2h)^3 - bH^3)/12]/[(H + 2h)/2]. \tag{28.2}$$

Shear stresses were calculated by formula of Zuravskiy. The found values $\sigma_{\max}^{\text{bend}}$ and $\tau_{\max}$ are also given in Table 28.1 for each specimen. It can be seen that the average normal compressive stress $\sigma_{\max}^{\text{bend}} = 71.557$ MPa is much less than the ultimate compressive stress $\sigma_{\text{I}}^{-} = 529.286$ MPa found for the composite based on the ELUR-P carbon and binder XT-118 in [12]. Having studied the pattern of deformation of the

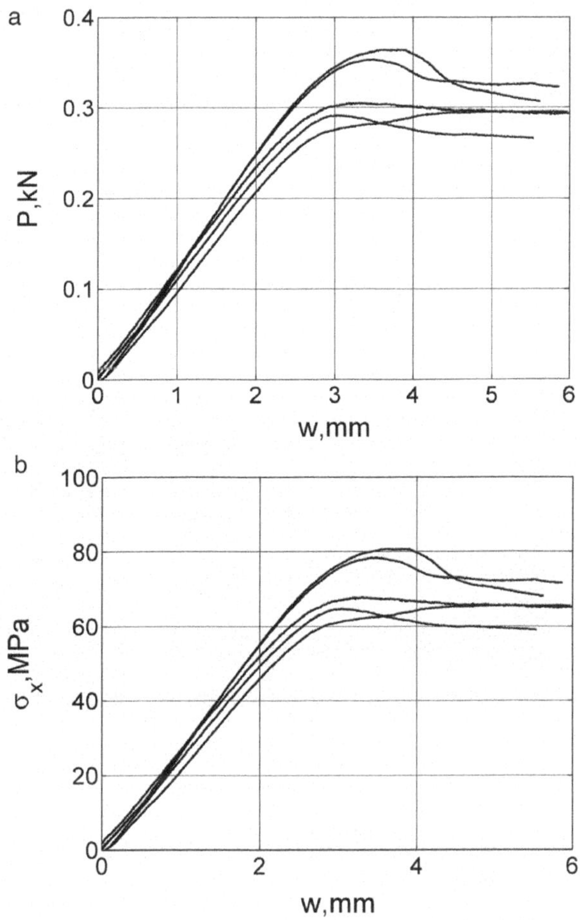

Fig. 28.4 Loading curves of the specimens with honeycomb core (**a** "load-deflection" curve, **b** "stress-deflection" curve)

specimens (Fig. 28.5) at the stages of the load drop on the diagram, it can be seen that the facing layers are not failed when the maximum load is reached.

It was established that failure of the specimens occurs due to the achievement of the shear stress in the core of the limiting value of $\tau_{xz}^{*} = 0.92$ MPa, which was established earlier in [16] on the basis of the corresponding experiments. The experimental average values $\tau_{xz}^{*} = 1.308$ MPa given in the table exceed the failure ones by 1.42 times. Based on this, we can conclude that for the specimens under study with the above geometric and physico-mechanical characteristics of the honeycomb core and the facing layers, failure is possible only due to the failure of the core, while the facing layers remain very underloaded.

Table 28.1 Testing results of the sandwich specimens

Specimen number	P_{max} (kN)	σ_{max}^{bend} (MPa)	τ_{max} (MPa)
1	−0.306	67.918	1.241
2	−0.292	64.786	1.184
3	−0.364	80.820	1.477
4	−0.354	78.501	1.435
5	−0.296	65.760	1.202
Average	−0.323	71.557	1.308
Standard deviation	0.034	7.529	0.138
Variation coefficient	−10.540	10.521	10.525

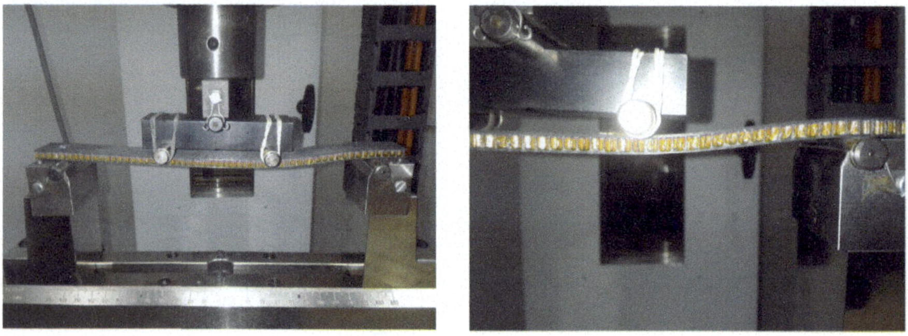

Fig. 28.5 Failure pattern of specimens with honeycomb core

28.3 Bending of Sandwich Specimens with Combined Rigid Core

Based on the analysis of the results of the four-point bending tests on specimens with a honeycomb core described above, wood (linden) was chosen as the core. Note that in order to realize the stresses of macroscale transverse-shear buckling mode in the facing layers, the compressive stresses in them must reach $\sigma_1^- = 529.286$ MPa. In this case, the core should not be failed due to the achievement of ultimate shear stresses. Ultimate shear stresses for different types of wood lie within $\tau_{xz}^* = 8 - 11$ MPa. Such ultimate stresses make it possible to form critical compressive stresses $\sigma_1^- = 529.286$ MPa in the facing layers without failure of the core. It is known [13] that in the region of support and loading rollers, when loading the specimen with cylindrical rollers, there is a concentration of shear stresses. In order to avoid failure of the core in the region of the support rollers, a fiberglass-based core was used, and the core in the working part of the specimen between the loading rollers was made of wood. A diagram of the specimen is shown in Fig. 28.6.

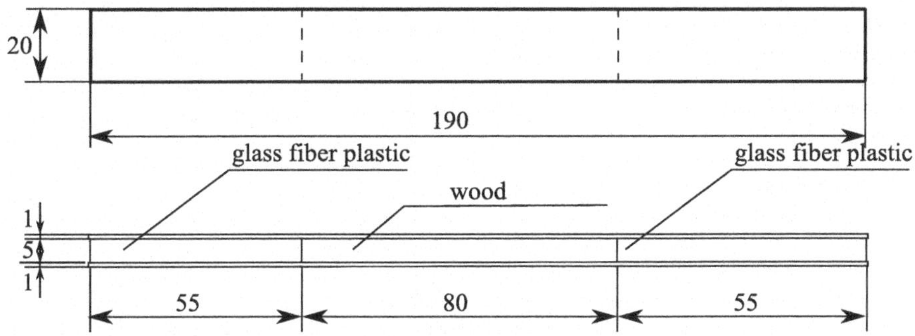

Fig. 28.6 Geometrical parameters of the specimen with combined core

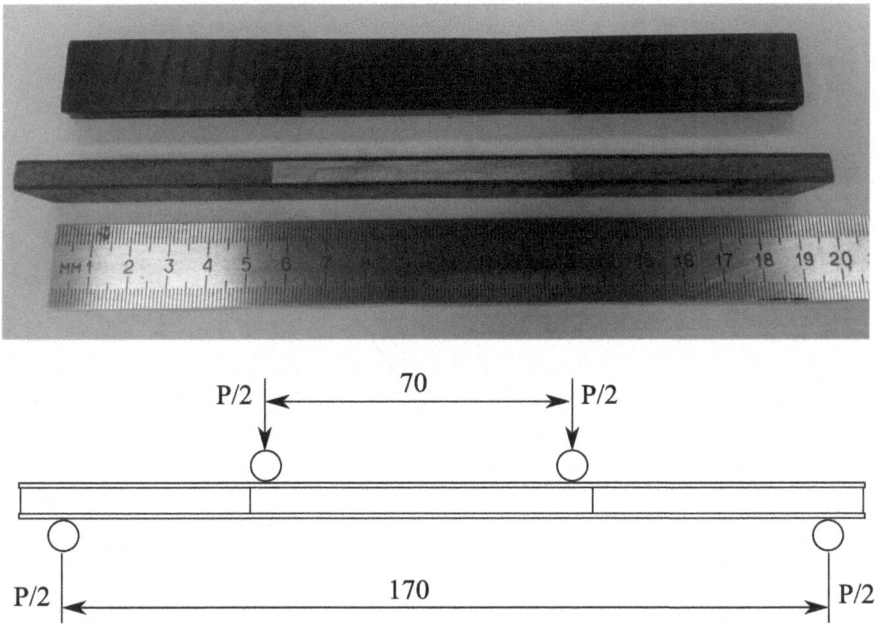

Fig. 28.7 Overall view of specimens with combined core and installation diagram

The manufacture of the specimens and four-point bending tests was carried out according to the same scheme as the specimens with honeycomb core. Only the center-to-center distance between the support rollers differed, which amounted to 170 mm (Fig. 28.7).

The "load-deflection" curves of the three specimens are shown in Fig. 28.8a, and the "stress-deflection" curves are shown in Fig. 28.8b. Axial normal stresses in the facing layers were determined by formulas (28.1) and (28.2).

The test results are summarized in Table 28.2. It can be seen from it that the average value of ultimate normal stresses is $\sigma_{max}^{bend} = 342.425$ MPa, and the shear stresses at

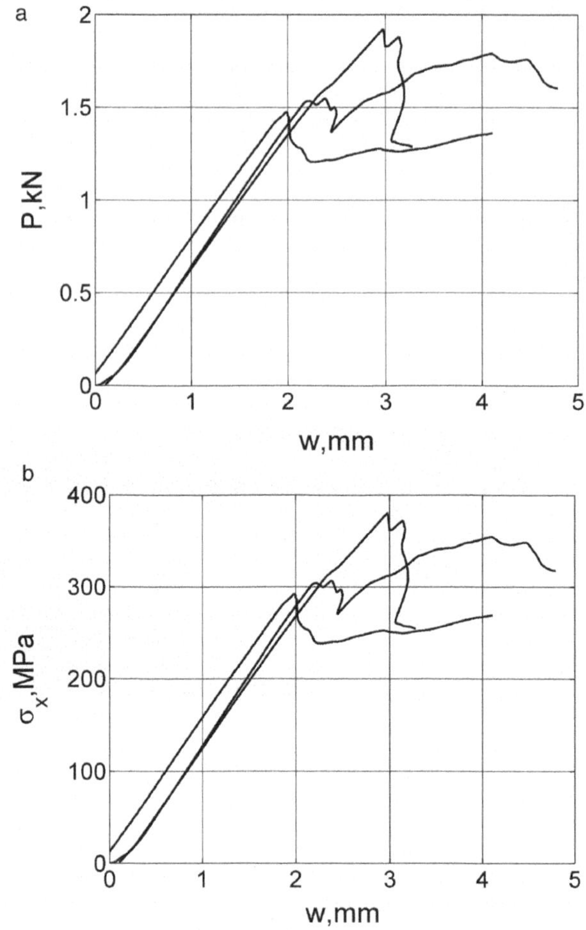

Fig. 28.8 Loading curves of the specimens with combined core (**a** "load-deflection" curve, **b** "stress-deflection" curve)

failure reach the value $\tau_{max} = 8.762$ MPa. Values σ_{max}^{bend} of normal stresses were much higher than values for bending specimens with honeycomb core. This value $\tau_{max} = 8.762$ MPa is almost failure for the used wood grade. Visual inspection of the upper facing layer of the specimens showed that failure in the upper (compressed) layer occurs at the contact area of the layer and the loading roller, while the core does not have any signs of failure.

28.4 Bending of Sandwich Specimens with Rigid Core

Tests of the specimens with a combined rigid core showed that the failure of the lower facing layer in the vicinity of the support rollers from the action of transverse normal stresses do not occur. In accordance with this, it is possible to refuse to use

Table 28.2 Testing results of the specimens with combined rigid core

Specimen number	P_{max} (kN)	σ_{max}^{bend} (MPa)	τ_{max} (MPa)
1	−1.477	292.436	7.482
2	−1.920	380.219	9.729
3	−1.791	354.620	9.074
Average	−1.729	342.425	8.762
Standard deviation	0.228	45.144	1.156
Variation coefficient	−13.175	13.184	13.189

a combined core in favor of using a rigid core (wood) along the entire length of the specimen. For testing, specimens were made with rigid core made of wood along the entire length of the specimens. The total length of the specimen was $L = 250$ mm, width $b = 25$ mm, core thickness $H = 5$ mm, nominal thickness of the facing layers $h = 1$ mm, the distance between the support rollers was 250 mm, and between the loading—70 mm.

The loading curves of the three specimens with a rigid core are shown in Fig. 28.9a, and the "stress-deflection" curves are shown in Fig. 28.9b.

Experimental results and evaluated by formulas (28.1) and (28.2), stresses summarized in Table 28.3.

The axial compressive stresses averaged over the results of tests of three specimens turned out to be equal to $\sigma_{max}^{bend} = 282.098$ MPa. This value turned out to be less than when bending specimens with a combined core ($\sigma_{max}^{bend} = 342.425$ MPa) and less than the value obtained by axial compression [12] ($\sigma_1^- = 529.286$ MPa). However, the shear stresses $\tau_{max} = 4.034$ MPa with such stiffness and geometric parameters were less than in experiments with a combined core. The values τ_{max} obtained by bending specimens with a uniform rigid core along the entire length exclude the failure of the core, which makes it possible to identify the causes of the facing layers failure.

28.5 Experimental Evaluation of Stresses in Facing Layers

As you know, honeycomb core belongs to the class of transversally flexible, which allows you to ignore the tangential stresses in the core when reducing the equations of the three-dimensional theory of elasticity to one or another two-dimensional model of sandwich shells or plates. The core used in the form of wood cannot be classified as transversally flexible, since the wood in the specimen is oriented with fibers along the Ox-axis (Fig. 28.1). The elastic modulus of wood along the fibers lies in the range $E_1 = 10$–12 GPa, while the elastic modulus of the unidirectional composite used in experiments is $E_1^+ = 104.73$ GPa. In this regard, the core in the form of fibrous wood can absorb part of the bending stresses, which is impossible for the honeycomb core. Therefore, in order to verify the validity of relations (28.1) and (28.2) when calculating normal stresses in the specimens of the class under study,

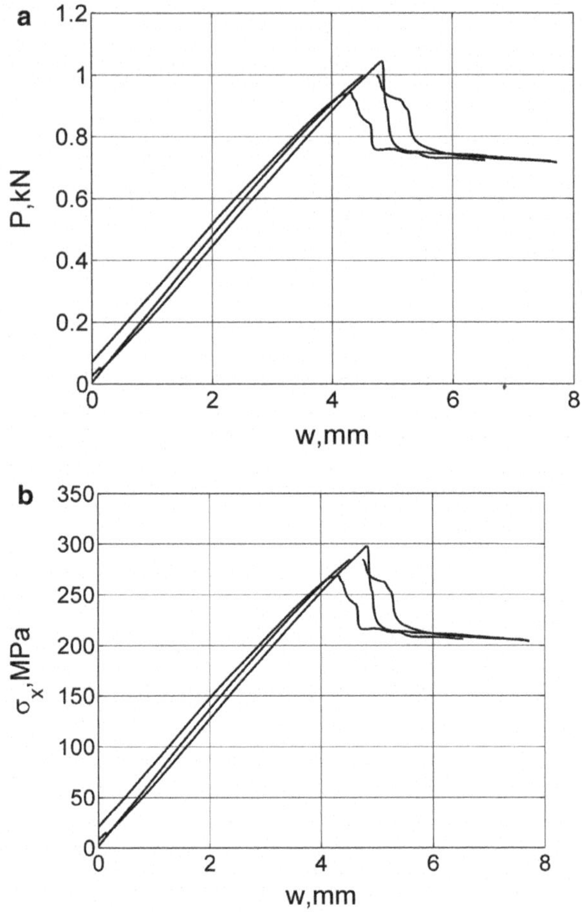

Fig. 28.9 Loading curves of the specimens with rigid core (**a** "load-deflection" curve, **b** stress-deflection" curve)

we experimentally measured strains in the facing layers during four-point bending. For this, BE 120-5AA (11-x) strain gauges were attached in the center of the facing layers of one of the specimen with a homogeneous core made of wood. To take readings, the National Instruments strain gauge station was used. A specimen with strain gauges installed in a fixture is shown in (Fig. 28.10).

The specimen was loaded upto failure with a speed of movement of the loading rollers of 5 mm/min. The loading curves in the axes "load-axial strain" are presented in Fig. 28.11a.

It can be seen that the ultimate strains of the compressed facing layer in the gauge area of the specimen reach a value $\epsilon_x^- = -2643\mu$, while the ultimate strains in the stretched facing layer are equal $\epsilon_x^+ = 1914\mu$. It can be seen that in the composite material under study, the elastic compressive characteristics are approximately to 28% lower than the tensile characteristics. In light of the results obtained in articles [12, 14, 17, 18], it can be assumed that, under the conditions of the formation of

Table 28.3 Testing results of the specimens with rigid core

Specimen number	P_{max} (kN)	σ_{max}^{bend} (MPa)	τ_{max} (MPa)
1	−1.043	292.436	4.230
2	−0.943	269.012	3.824
3	−0.999	284.847	4.049
Average	−0.995	282.098	4.034
Standard deviation	0.050	11.951	0.203
Variation coefficient	−5.037	4.237	5.042

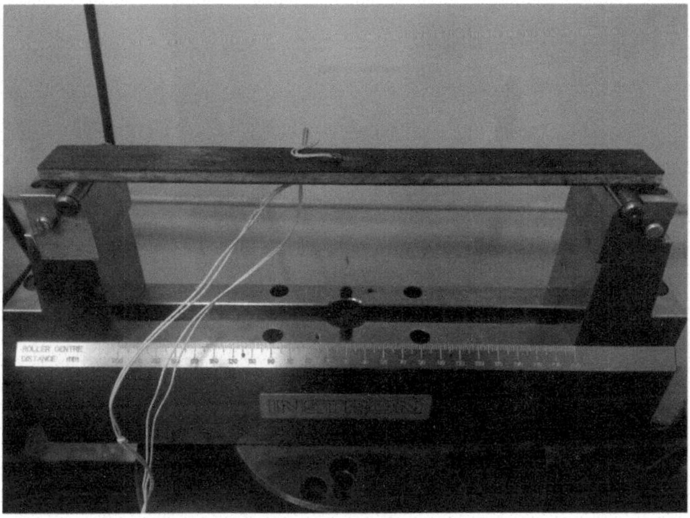

Fig. 28.10 Installation of the specimen with strain gauges

compressive stresses, noticeable structural changes occur in the upper facing layer of the specimen associated with the realization of internal micro- and mesoscale buckling form of structural elements of the composite with continuous their restructuring during loading. Using the elastic characteristics of the fibrous composite of the facing layers $E^+ = 104.73$ GPa and $E^- = 96.32$ GPa, found in [12] and the relations of Hooke's law (due to the almost linear dependence), it is possible to determine the axial normal stresses formed in the facing layers (Fig. 28.11b). In the compressed facing layer, they reach $\sigma_x^- = 280.69$ MPa, in the stretched $\sigma_x^+ = 202.37$ MPa. This significant difference in stresses is apparently explained by the fact that significant friction forces are formed at the contact areas of the support rollers and the lower facing layer, which prevent free stretching of the layer under conditions of four-point bending. However, the main reason for the nonfulfilment of equality $\sigma_x^+ = \sigma_x^-$ is that during loading of the specimen, there is a significant increase in the multimodularity of the material, mainly due, apparently, to the realization and continuous change of

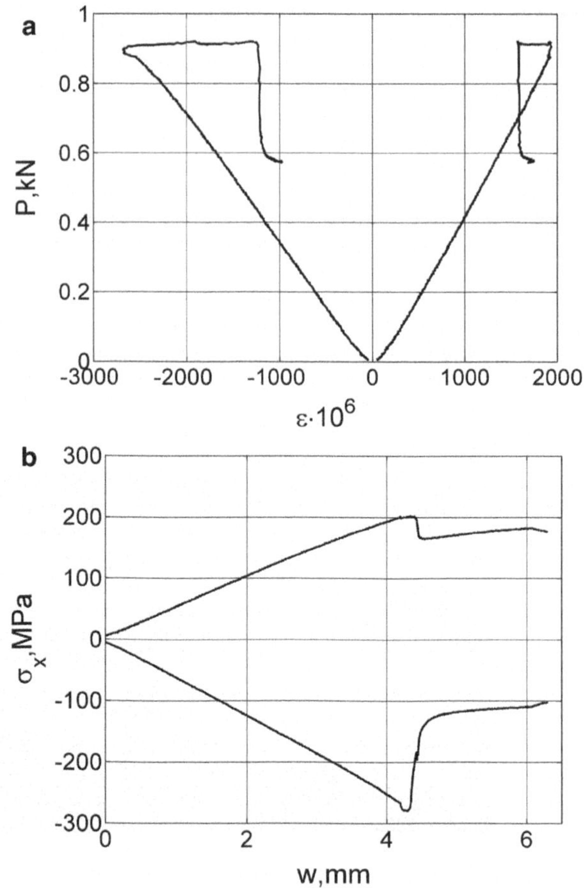

Fig. 28.11 Loading curves of the specimen according to the results of strain gauging

internal micro- and mesoscale buckling modes of fibers and fiber bundles in areas with weakened physical mechanical characteristics of the binder in a compressed facing layer [8, 9, 14, 17, 18]. The compiled equality, expressed through strains in the form $E_x^+\epsilon_x^+ = E_x^-\epsilon_x^-$, during loading can be fulfilled only if the secant modulus of elasticity E_x^- entering into it will significantly decrease during loading of the specimen while the initial value of the elastic modulus of $E^+ = 104.73$ GPa is retained.

28.6 Conclusion

Tests for four-point bending make it possible to exclude the formation of transverse shear stresses in the working part of the specimen and to realize a state of pure bending in it. As shown by the results of experimental studies, in a sandwich specimens, the failure of the facing layers of fiber-reinforced plastics occurs not in the working part of the specimen, but in the vicinity of the loading and support rollers when testing specimen with a slightly rigid core, and in the vicinity of the loading roller when testing specimens with a rigid core. The mathematical models of deformation known in the mechanics of sandwich structures, based on the description of the mechanics of deformation of the outer layers by the classical Kirhoff-Love model, do not have a sufficient degree of accuracy and meaningfulness for a theoretical study of the stress–strain state in the vicinity of loading and supporting rollers and identifying the types and causes of failure. In this regard, it is necessary to construct such refined versions of the theory of three-layer plates and shells that would take into account the nonuniform distribution of transverse shear stresses over the thickness of the facing layer, the physical nonlinearity of the behavior of the fiber-reinforced plastic under shear stresses, and also the developing (transforming) multimodularity of the fibrous composite material into conditions of tension and compression. To verify such refined deformation models, the results of the experimental studies described above can be used.

Acknowledgements This work was supported by Russian Science Foundation (Project No. 19-79-10018).

References

1. Alila, F., Fajoui, J., Gerard, R., Casari, P., Kchaou, M., Jacquemin, F.: Viscoelastic behaviour investigation and new developed laboratory slamming test on foam core sandwich. J. Sandw. Struct. Mater. (2018, in press)
2. Badriev, I.B., Makarov, M.V., Paimushin, V.N.: Solvability of physically and geomentrically nonlinear problem of the theory of sandwich plates with transversally-soft core. Russ. Math. **59**(10), 57–60 (2015)
3. Bahghai, J., Zhibin, L., Fangyun, L.: Failure mechanisms of sandwich beams subjected to three-point bending. Compos. Struct. **133**, 739–745 (2015)
4. Budiansky, B., Fleck, N.A.: Compressive failure of fibre composites. J. Mech. Phys. Solids **41**(1), 183–211 (1993)
5. Crupi, V., Epasto, G., Guglielmino, E.: Comparison of aluminium sandwiches for lightweight ship structures: honeycomb vs. foam. Mar. Struct. **30**, 74–96 (2013)
6. Fathi, A., Woff-Fabris, F., Altstadt, V., Gatzi, R.: An investigation of the flexural properties of balsa and polymer foam core sandwich structures: influence of core type and contour finishing options. J. Sandw. Struct. Mater. **15**(5), 487–508 (2013)
7. Jumahat, A., Soutis, C., Jones, F.R., Hodzic, A.: Fracture mechanisms and failure analysis of carbon fibre/toughened epoxy composites subjected to compressive loading. Compos. Struct. **92**(2), 295–305 (2010)
8. Paimushin, V.N., Gazizullin, R.K., Shishov, M.A.: Flat internal buckling modes of fibrous composite elements under tension and compression at the mini- and microscale. J. Appl. Mech. Tech. Phys. **60**(3), 548–559 (2019)
9. Paimushin, V.N., Gazizullin, R.K., Shishov, M.A.: Spatial buckling modes of a fiber (fiber bundle) of composite with a $[\pm 45_{2s}]$ stacking sequence under the tension and compression of test specimens. Mech. Compos. Mater. **55**(6), 743–760 (2019)

10. Pimushin, V.N., Makarov, M.V., Badriev, I.B., Kholmogorov, S.A.: Geometrically nonlinear strain and buckling analysis of sandwich plates and shells reinforced on their edge. Shell Structures: Theory and Applications—Proc. 11th Int. Conf., vol. 4, pp. 267–270 (2018)
11. Paimushin, V.N., Kholmogorov, S.A., Makarov, M.V., Tarlakovskii, D.V., Lukaszewicz, A.: Mechanics of fiber composites: forms of loss of stability and fracture of test specimens resulting from three-point bending tests. Z. Angew. Math. Mech. **99**(1), 1–25 (2018)
12. Paimushin, V.N., Kholmogorov, S.A.: Physical-mechanical properties of a fiber-reinforced composite based on an ELUR-P carbon tape and XT-118 binder. Mech. Compos. Mater. **54**(1), 2–12 (2018)
13. Paimushin, V.N., Kholmogorov, S.A., Makarov, M.V.: Numerical investigation of stress-strain state of layered composite specimen under three-point bending. In: Defor. Fail. Compos. Mater. Struct. Second Int. Conf., pp. 126–128 (2016) (in Russian)
14. Paimushin, V.N., Kholmogorov, S.A., Gazizullin, R.K.: Mechanics of unidirectional fiber-reinforced composites: buckling modes and failure under compression along fibers. Mech. Compos. Mater. **53**(6), 737–752 (2018)
15. Paimushin, V.N., Tarlakovskii, D.V., Kholmogorov, S.A.: On non-classical buckling mode and failure of composite laminated specimens under the three-point bending. Uch. Zap. Kaz. Univer. Seriya Fiz.-Mat. Nauki. **158**(3), 350–375 (2016)
16. Paimushin, V.N., Zakirov, I.M., Lukankin, S.A., Zakirov, I.I., Kholmogorov, S.A.: Average elastic and strength characteristics of a honeycomb core and a theoretical-experimental method of their determination. Mech. Compos. Mater. **48**(5), 511–524 (2012)
17. Paimushin, V.N., Polyakova, N.V., Kholmogorov, S.A., Shishov, M.S.: Non-uniformly scaled buckling modes of reinforcing elements in fiber reinforced plastic. Russ. Math. **9**, 89–95 (2017)
18. Paimushin, V.N., Polyakova, N.V., Kholmogorov, S.A., Shishov, M.A.: Buckling modes of structural elements of off-axis fiber-reinforced plastics. Mech. Compos. Mater. **54**(2), 133–144 (2018)
19. Petras, A., Sutcliffe, M.P.F.: Failure mode maps for honeycomb sandwich panels. Compos. Struct. **44**, 237–252 (2019)
20. Piovar, S., Kormanikova, E.: Sandwich beam in four-point bending test: experiment and numerical models. Adv. Mater. Res. **969**, 316–319 (2014)
21. Rupp, P., Elsner, P., Weidenmann, K.A.: Failure mode maps for four-point-bending of hybrid sandwich structures with carbon fiber reinforced plastic face sheets and aluminum foam cores manufactured by a polyurethane spraying process. J. Sandw. Struct. Mater. **21**(8), 2654–2679 (2019)
22. Russo, A., Zuccarello, B.: Experimental and numerical evaluation of the mechanical behaviour of GFRP sandwich panels. Compos. Struct. **81**, 575–586 (2007)
23. Shi, H., Liu, W., Fang, H.: Damage characteristics analysis of GFRP-Balsa sandwich beams under four-point fatigue bending. Compos. Part A: Appl. Sci. Manuf. **109**, 564–577 (2018)
24. Sokilinsky, V.S., Shen, H., Vaikhanski, L., Nutt, S.R.: Experimental and analytical study of nonlinear bending response of sandwich beams. Compos. Struct. **60**, 219–229 (2003)

Chapter 29
Sandwich Shells with Composite Facings and a Transversally Flexible Core: Refined Equations and Buckling Modes of Specimens Under Four-Point Bending Tests

Vitaly N. Paimushin, Ruslan K. Gazizullin, Natalya V. Polyakova, and Maksim A. Shishov

Abstract In order to develop the previously obtained results for the case of medium bending, a refined geometrically nonlinear theory of static deformation has been developed for sandwich plates and shells with a transversely flexible core and composite facings having low stiffness on transverse shear and transverse compression. This theory is based on a refined Timoshenko shear model taking into account the transverse compression to describe the mechanics of facings. For a transversally flexible core, simplified three-dimensional equations of elasticity theory have been used. These equations allow integration along the transverse coordinate with the introduction into consideration of two two-dimensional unknown functions in the role of which constant in thickness transverse tangential stresses are used. Based on the generalized Lagrange variational principle, two-dimensional geometrically nonlinear equations of equilibrium as well as coupling conditions of facings with a core via tangential displacements are constructed to describe the static deformation process with high rates of variability of stress–strain state parameters. Based on them,

---

V. N. Paimushin (✉) · R. K. Gazizullin · N. V. Polyakova · M. A. Shishov
Kazan National Research Technical University named after A.N. Tupolev,
10 K.Marks str., Kazan 420111, Russian Federation
e-mail: vpajmushin@mail.ru

R. K. Gazizullin
e-mail: gazizullin.rk@yandex.ru

N. V. Polyakova
e-mail: chydo.n@mail.ru

M. A. Shishov
e-mail: maxim.shishov@mail.ru

V. N. Paimushin · M. A. Shishov
Kazan (Volga region) Federal University, 18 Kremlyovskaya str.,
Kazan 420008, Russian Federation

H. Altenbach et al. (eds.), *Multiscale Solid Mechanics*,
Advanced Structured Materials 141,
https://doi.org/10.1007/978-3-030-54928-2_29

an approximate analytical solution of the linearized problem of possible buckling modes has been found for a sandwich beam of a symmetrical in thickness structure under four-point bending.

Keywords Sandwich shells · Composite facings · Transversely soft core · Refined equations · Test specimens under four-point bending · Buckling modes

29.1 Introduction

Application of multilayer structures in different industries has led to intensive research of methods for their calculation. As a result, a separate direction has developed in the mechanics of thin-walled structures related to the study of the problems of mechanics of multilayer (in particular, sandwich) rods, plates, and shells. To date, a huge number of studies [3, 5, 8, 10, 11, 13–18, 27, 28, 44, 46] devoted to both theoretical and experimental research have been published in this area. In particular, a review of such studies is contained in [4, 31].

The problems associated with the identification and classification of possible buckling modes as well as the construction of appropriate mathematical models and resolving equations for their description are one of the main areas of scientific research in the macromechanics of laminated structures. In the scientific literature on the mechanics of such structures, there is generally accepted classification of buckling problems in which the skew-symmetric (in-phase) and symmetric (anti-phase) buckling modes were distinguished as the main ones. Local buckling modes also stood out as a separate group. Such modes are associated with bulging of facings within the cells of discrete structure core (honeycomb, ribbed, etc.) or elements of such core. Large numbers of researchers analyzed the possibility of not taking into account the strain of the transverse compression of core, transverse shear in facings, the momentum of the subcritical state, and other factors in the formulation of buckling problems within the framework of the indicated restrictions on buckling modes. However, in all these studies, the classical statement of buckling problems prevailed, within the framework of which the studied refinements were introduced in the equations when describing only the perturbed state, and the unperturbed equilibrium state of the packet was assumed to be undeformed and momentless. But, as is well known, one of the main advantages of multilayer (in particular, sandwich) structures is their optimality in bending. Therefore, they are used where it is impossible to avoid the momentum of a subcritical stress–strain state. As a rule, moment zones in shells arise near supporting attachment, in areas of rapid changes in the geometric parameters of the shell, etc. In cases of a substantially momentary state of the laminated package, it is possible to realize mixed buckling modes in areas where the unperturbed stress–strain state of facings differs significantly from each other. In the general case, they are characterized by different buckling modes of facings and the greatest amplitudes of buckling in places of predominantly momentary subcritical state. However, the use of the momentless assumption of the subcritical stress–strain state of the laminated package, which has become traditional in the theory of stability of laminated plates and shells, has led researchers to formulate a number of incorrect conclusions and basic provisions regarding both the classification of buckling modes and the construction of the corresponding stability equations. This statement follows from an analysis of the results of [37, 38], devoted to the formulation and solution of the buckling problems of sandwich plates under transverse and longitudinal–transverse bends. These studies, in turn, allowed to give an updated classification of buckling modes and build refined stability equations [36]. In this study, in addition to the well studied in the literature in-phase and anti-phase modes, it is a mixed buckling mode of facings has been also included. The critical loads corresponding to this buckling mode in the case of real conditions of fixing and loading of the facings turn out to be significantly lower (by 2–3 times) than the critical loads corresponding to in-phase or anti-phase buckling forms. To determine them, it is necessary to use stability equations, in which, along with transverse shear, the transverse compression of the core is taken into account when the moment work of the facings and the moment character of the subcritical stress–strain state of the laminated package as a whole are taken into account [36–38].

Coverage of the main results obtained in the field of mechanics of laminated plates and shells until 1996 was given in a review [31]. The classification of buckling modes presented in it contains in-phase and anti-phase modes that have been studied in detail by many authors, local mixed mode, and also the mode of shear corrugation in the core. The latter is similar to the shear form noted in the article [36] in the facings, but it is realized in the core as a result of the predominant formation of transverse shear deformations. This buckling mode, as well as the mode of a shear in the facings noted above, has not been theoretically studied in the scientific literature for a long time. Although judging by the experimental data obtained in different years by many researchers, it can be implemented in real structures in combination with other modes.

In a series of studies [20, 22], it was found that the equations [36], which are used to study mixed buckling modes and based on the "broken line" model [14, 15] taking into account transverse compression (within the framework of this model, the mechanics of deformation of facings were described by the classical Kirchhoff–Love hypotheses, and for the core a linear approximation of the tangential displacements over the thickness was chosen), are extremely simplified and only the main features of facing buckling by mixed modes can be identified.

In this regard, for the sandwich [20, 22] and multilayer [21, 32] shells with transversally flexible core, a refined theory of subcritical deformation and a linearized theory of stability were constructed. In this theory, in contrast to [36] a large indicator of the variability of transverse shear stresses in the core (taking place in the zones of local buckling of facings) is allowed as well as a much more accurate description of the stress–strain state compared to [36] due to the introduction of only two additional required functions for each core layer. The accuracy of the equations of this theory, as shown by subsequent studies [12] on establishing the limits of their applicability, practically approaches the accuracy of the linearized equations of the three-dimensional theory of elasticity in determining such integral characteristics as the crippling load or the frequency of free vibrations of sandwich and multilayer structures with transversally flexible core.

It is known that composite materials are most rationally used in sandwich structures. Facings of such structures are made of composite materials, which, as a rule, have a small transverse shear stiffness. If compression stresses are formed in them during loading, then one of the mechanisms of their destruction is the buckling by the transverse shear mode. This buckling mode has been studied in the mechanics of composites in various approximations in a large number of studies [1, 2, 6, 7, 19, 23, 24, 26, 29, 30, 40, 41, 43, 45, 47]. In accordance with the results of these studies, in order to identify the described buckling modes (their implementation is possible only in sandwich structures with facings of fibrous composites), the construction of the corresponding theory requires taking into account transverse shear strains in the facings of sandwich structures.

29.2 Refined Model of Deformation of Sandwich Shells with Transversally Flexible Core

Consider a sandwich shell with rigid composite external layers (facings) and a low-rigid middle layer (core). We denote the numbers of the lower and upper facings by an index $k = 1, 2$. Indicating numbers k, we put in parentheses. For the parametrization base, we take the middle surface σ of the core. We consider that surface σ is given by the equation $\mathbf{r} = \mathbf{r}\left(x^1, x^2\right) = \mathbf{r}\left(x^i\right)$. We attribute the space of this layer to a semi-orthogonal coordinate system x^i, z normally connected with the surface. By $\mathbf{r}_i = \partial \mathbf{r}/\partial x^i$, $\mathbf{m}$, we denote the basic basis vectors on σ and the unit normal vector, respectively. In what follows, we assume that a change in the metric in the direction of the z axis can be neglected. Thus, we identify the basis vectors for each facings with the basis vectors $\mathbf{r}_i$ onσ. In addition, we introduce the following notation: $z_{(k)}, z$ are coordinates along the normal $\mathbf{m}$ to the median surfaces $\sigma_{(k)}$ and σ of each facings and the core; $t_{(k)}, h$ are the corresponding thickness of the layers, at that $-t_{(k)}/2 \le z_{(k)} \le t_{(k)}/2 \quad (k = 1, 2)$ and $-h/2 \le z \le h/2$; $H = t_{(1)} + t_{(2)} + h$ is total shell thickness; $a_{ij} = \mathbf{r}_i\mathbf{r}_j$, $b_{ij} = -\mathbf{r}_i\mathbf{m}_j$ are covariant components of the first and second metric tensors on σ, which, by virtue of the above assumptions, are unchanged when passing from layer to layer.

We assume that the shell is thin and non-depressed. Its boundary section is a ruled surface formed by the movement of the vector $\mathbf{m}$ along some smooth curve $C \in \sigma$. By $\mathbf{n}$ and $\boldsymbol{\tau}$, we denote the unit normal and tangent vectors on C that make up the right-handed trihedron with vector $\mathbf{m}$ at each point of C.

The variants of the theory of multilayer plates and shells known in the literature mainly differ from each other by models of the core adopted in them. To date, their detailed analysis has been carried out by many authors, in particular, Mushtari [27, 28] and Bolotin [5]. Comprehensive reviews of deformation models of sandwich and multilayer structural elements are contained in [4, 31]. In accordance with the results of [5], a core is called transversely flexible when the density of potential strain energy U in it is calculated by the formula

$$\Pi^{(3)} = \frac{1}{2}\iint\limits_{\sigma}\int\limits_{-h/2}^{h/2}\left(2\sigma^{i3}\varepsilon_{i3}^{z} + \sigma^{33}\varepsilon_{33}^{z}\right)\mathrm{d}z\mathrm{d}\sigma. \tag{29.1}$$

This is equivalent to the adoption of assumptions $\sigma^{ij} \approx 0$. By virtue of these assumptions for cores, the equilibrium equations can be written in the form

$$\frac{\partial\sigma^{i3}}{\partial z} = 0,\ \frac{\partial\sigma^{33}}{\partial z} + \nabla_i\sigma^{i3} = 0, \tag{29.2}$$

from which the equalities $\sigma^{i3} = q^i\left(x^1, x^2\right)$ and the equation

$$\frac{\partial\sigma^{33}}{\partial z} + \nabla_i q^i = 0 \tag{29.3}$$

follows.

If we accept a refined kinematic model Timoshenko to describe the deformation mechanics of facings, then in case of small and medium bending of the shell the displacement vectors and components of tangential deformation tensor in facings are determined by the formulas [9, 42]

$$\mathbf{U}^{z_{(k)}} = u_i^{(k)}\mathbf{r}^i + w^{(k)}\mathbf{m} + z_{(k)}\left(\gamma_i^{(k)}\mathbf{r}^i + \varphi^{(k)}\mathbf{m}\right), \tag{29.4}$$

$$\varepsilon_{ij}^{z_{(k)}} = \varepsilon_{ij}^{(k)} + z_{(k)}\chi_{ij}^{(k)},\ \varepsilon_{33}^{z_{(k)}} = \varepsilon_{33}^{(k)} = \varphi^{(k)},\ 2\varepsilon_{i3}^{z_{(k)}} = 2\varepsilon_{i3}^{(k)} + z_{(k)}\nabla_i\varphi^{(k)}. \tag{29.5}$$

Here $\varepsilon_{ij}^{(k)}$, $\chi_{ij}^{(k)}$, and $2\varepsilon_{i3}^{(k)}$ are covariant components of tangential and bending deformation tensors, as well as transverse shear vectors at the level $z_{(k)} = 0$. To determine them, it follows the expressions

$$\begin{gathered} 2\varepsilon_{ij}^{(k)} = e_{ij}^{(k)} + e_{ji}^{(k)} + \omega_i^{(k)}\omega_j^{(k)},\ e_{ij}^{(k)} = \nabla_i u_j^{(k)} - b_{ij}w^{(k)}, \\ \omega_i^{(k)} = \nabla_i w^{(k)} + u_j^{(k)}b_i^j,\ 2\chi_{ij}^{(k)} = \nabla_i\gamma_j^{(k)} + \nabla_j\gamma_i^{(k)} - b_{ij}\varphi^{(k)}, \\ 2\varepsilon_{i3}^{(k)} = \gamma_i^{(k)} + \omega_i^{(k)}. \end{gathered} \tag{29.6}$$

With the accepted degree of accuracy $\sigma^{ij} \approx 0$ and $\delta_i^j - zb_i^j \approx \delta_i^j$ for the stress σ^{33} in the linear approximation, the following elasticity relation can be written in the linearized approximation

$$\sigma^{33} = E_3\varepsilon_{33}^{z} = E_3^0\frac{\partial U_3}{\partial z}. \tag{29.7}$$

After substituting (29.7) into equations (29.3) and integrating over z, we obtain a following expression

$$U_3 = W_1 + zW_2 - \frac{z^2}{2E_3^0}\nabla_i q^i, \tag{29.8}$$

where W_1 and W_2 are integration functions depending on coordinates x^1, x^2 and time t. Let us determine them from following coupling conditions of the core with the facings

$$U_3\left(z=-h/2\right) = w^{(1)} + \frac{t_{(1)}}{2}\varphi^{(1)},\ \ U_3\left(z=h/2\right) = w^{(2)} - \frac{t_{(2)}}{2}\varphi^{(2)}.$$

Expression (29.8) for the deflection in the core can be reduced to the form

$$\begin{aligned} U_3 = {} & \frac{w^{(1)}+w^{(2)}}{2} + z\frac{w^{(2)}-w^{(1)}}{h} + \frac{1}{2E_3^0}\left(\frac{h^2}{4}-z^2\right)\nabla_i q^i \\ & + \left(\frac{1}{4}-\frac{z}{2h}\right)t_{(1)}\varphi^{(1)} - \left(\frac{1}{4}+\frac{z}{2h}\right)t_{(2)}\varphi^{(2)};\quad k=1,2. \end{aligned} \tag{29.9}$$

To establish the law of change in the tangential components displacements over z in the core, we refer to the elasticity relations for σ^{i3}. These relations in the linear approximation within the accepted degree of accuracy can be represented in the following approximate form (A^{is} is the divalent tensor of shear elastic constants)

$$\sigma^{i3} = q^i = 2A^{is}\varepsilon_{s3}^z = A^{is}\left[\frac{\partial U_s}{\partial z} + \nabla_s U_3\right]. \tag{29.10}$$

These expressions are valid for both small and medium bends. Substituting relations (29.9) into relations (29.10), we obtain the differential equation

$$\begin{aligned} \frac{\partial U_i}{\partial z} = {} & d_{is}q^s - \frac{w_i^{(1)}+w_i^{(2)}}{2} - \frac{z}{h}\left(w_i^{(2)}-w_i^{(1)}\right) \\ & + \frac{4z^2-h^2}{8E_3^0}\nabla_i\nabla_s q^s - \left(\frac{1}{4}-\frac{z}{2h}\right)t_{(1)}\varphi_i^{(1)} + \left(\frac{1}{4}+\frac{z}{2h}\right)t_{(2)}\varphi_i^{(2)}. \end{aligned} \tag{29.11}$$

Here we introduced the notation for the partial derivatives as $w_i^{(k)} = \partial w^{(k)}/\partial x^i$ and $\varphi_i^{(k)} = \partial\varphi^{(k)}/\partial x^i$; through d_{is}, we denote the divalent tensor of the compliance of the core on the transverse shear. Integrating (29.11) over z, we obtain

$$U_i = u_i + zd_{is}q^s - z\frac{w_i^{(1)} + w_i^{(2)}}{2} - \frac{z^2}{2h}\left(w_i^{(2)} - w_i^{(1)}\right)$$
$$+\left(\frac{4}{3}z^3 - h^2 z\right)\frac{\nabla_i\nabla_s q^s}{8E_3^0} - \left(\frac{z}{4} - \frac{z^2}{4h}\right)t_{(1)}\varphi_i^{(1)} + \left(\frac{z}{4} + \frac{z^2}{4h}\right)t_{(2)}\varphi_i^{(2)}, \tag{29.12}$$

where u_i are unknown two-dimensional functions. To determine them, the obtained relations (29.12) must comply with coupling conditions via tangential displacements

$$u_i^{(1)} + \frac{t_{(1)}}{2}\gamma_i^{(1)} = U_i\left(-\frac{h}{2}, x^1, x^2, t\right), \quad u_i^{(2)} - \frac{t_{(2)}}{2}\gamma_i^{(2)} = U_i\left(\frac{h}{2}, x^1, x^2, t\right). \tag{29.13}$$

By substituting (29.12) in (29.13), it follows the equations

$$\mu_i = u_i^{(1)} - u_i^{(2)} - \frac{h}{2}\left(w_i^{(1)} + w_i^{(2)}\right) + \frac{1}{2}\left(t_{(1)}\gamma_i^{(1)} + t_{(2)}\gamma_i^{(2)}\right)$$
$$+ hd_{is}q^s - \frac{h^3}{12E_3^0}\nabla_i\nabla_j q^j - \frac{h}{4}\left(t_{(1)}\varphi_i^{(1)} - t_{(2)}\varphi_i^{(2)}\right) = 0. \tag{29.14}$$

They will be used below in deriving the equilibrium equations and kinematic coupling conditions of facings with the core via tangential displacements.

29.3 Equilibrium Equations and Kinematic Coupling Conditions of Facings with a Core via Tangential Displacements

In accordance with the core model under consideration, we assume that only facings of the sandwich shell are loaded with external forces. We introduce into consideration the vectors of the given forces and moments

$$\mathbf{\Phi}^{(k)} = \Phi_n^{(k)}\mathbf{n} + \Phi_{n\tau}^{(k)}\tau + \Phi_m^{(k)}\mathbf{m}, \quad \mathbf{L}^{(k)} = L_n^{(k)}\mathbf{n} + L_{n\tau}^{(k)}\tau + L_m^{(k)}\mathbf{m}; \quad k = 1, 2,$$

applied to the boundary lines $C_{(k)}$ of the median surfaces $\sigma_{(k)}$ of the facings. We also introduce the vectors of external surface forces and moments

$$\mathbf{X}_{(k)} = X_{(k)}^i\mathbf{r}_i + X_{(k)}^3\mathbf{m}, \quad \mathbf{M}_{(k)} = M_{(k)}^i\mathbf{r}_i + M_{(k)}^3\mathbf{m},$$

applied at points of surface $\sigma_{(k)}$. The variation of work of these forces at the corresponding displacements (29.4) can be determined by the relation

$$\delta A = \sum_{k=1}^{2} \left\{ \int_C \left[\Phi_n^{(k)} \delta u_n^{(k)} + \Phi_{n\tau}^{(k)} \delta u_\tau^{(k)} + \Phi_m^{(k)} \delta w^{(k)} + L_n^{(k)} \delta \gamma_n^{(k)} \right. \right.$$
$$+ L_{n\tau}^{(k)} \delta \gamma_\tau^{(k)} + L_m^{(k)} \delta \varphi^{(k)} \Big] \mathrm{d}s + \iint_\sigma \left[X_{(k)}^i \delta u_i^{(k)} + X_{(k)}^3 \delta w^{(k)} + M_{(k)}^i \delta \gamma_i^{(k)} \right.$$
$$\left. \left. + M_{(k)}^3 \delta \varphi^{(k)} \right] \mathrm{d}\sigma \right\}. \tag{29.15}$$

The variation of potential stain energy of the shell can be calculated by the formula

$$\delta \Pi = \iint_\sigma \int_{h/2}^{h/2} \left(2\sigma^{i3} \delta \varepsilon_{i3}^z + \sigma^{33} \delta \varepsilon_{33}^z \right) \mathrm{d}\sigma \mathrm{d}z$$
$$+ \sum_{k=1}^{2} \iint_\sigma \int_{-t_{(k)}/2}^{t_{(k)}/2} \left(\sigma_{(k)}^{ij} \delta \varepsilon_{ij}^{z(k)} + 2\sigma_{(k)}^{i3} \delta \varepsilon_{i3}^{z(k)} + \sigma_{(k)}^{33} \delta \varepsilon_{33}^{(k)} \right) \mathrm{d}\sigma \mathrm{d}z_{(k)}$$
$$= \sum_{k=1}^{2} \iint_\sigma \left(T_{(k)}^{ij} \delta \varepsilon_{ij}^{(k)} + M_{(k)}^{ij} \delta \chi_{ij}^{(k)} + 2T_{(k)}^{i3} \delta \varepsilon_{i3}^{(k)} + T_{(k)}^{33} \delta \varphi^{(k)} + M_{(k)}^{i3} \delta \varphi_i^{(k)} \right) \mathrm{d}\sigma$$
$$+ \iint_\sigma \left[c_{is} q^s \delta q^i + \sigma_0^{33} \left(\delta w^{(2)} - \delta w^{(1)} - \frac{t_{(1)}}{2} \delta \varphi^{(1)} - \frac{t_{(2)}}{2} \delta \varphi^{(2)} \right) \right.$$
$$\left. + \frac{h^3}{12 E_3^0} \nabla_s q^s \nabla_i \delta q^i \right] \mathrm{d}\sigma, \tag{29.16}$$

obtained using following transformations of the constructed relations (29.7), (29.9)

$$2\sigma^{i3} \delta \varepsilon_{i3}^z = d_{is} q^s \delta q^i, \ \sigma^{33} \delta \varepsilon_{33}^z = E_3^0 \varepsilon_{33}^z \delta \varepsilon_{33}^z$$

and introducing following notation

$$\gamma_n^{(k)} = \gamma_i^{(k)} n^i, \ \gamma_\tau^{(k)} = \gamma_i^{(k)} \tau^i, \ u_n^{(k)} = u_i^{(k)} n^i, \ u_\tau^{(k)} = u_i^{(k)} \tau^i,$$
$$T_{(k)}^{\alpha\beta} = \int_{-h_{(k)}/2}^{h_{(k)}/2} \sigma_{(k)}^{\alpha\beta} \mathrm{d}z_{(k)},$$
$$M_{(k)}^{i\alpha} = \int_{-h_{(k)}/2}^{h_{(k)}/2} \sigma_{(k)}^{i\alpha} z_{(k)} \mathrm{d}z_{(k)}, \quad i = 1, 2; \ \alpha = 1, 2, 3; \ \beta = 1, 2, 3,$$
$$c_{is} = h d_{is}, \quad \sigma_0^{33} = \frac{E_3^0}{h} \left[w^{(2)} - w^{(1)} - \frac{1}{2} \left(t_{(1)} \varphi^{(1)} + t_{(2)} \varphi^{(2)} \right) \right],$$

where $n^i = \mathbf{n}\mathbf{r}^i and \tau^i = \boldsymbol{\tau}\mathbf{r}^i$ are the contravariant components of the vectors $\mathbf{n},\ \boldsymbol{\tau}$ with respect to the basis vectors $\mathbf{r}_i$; ds is an element of the length of the contour line C.

If the kinematic relations (29.4), (29.12) are not submitted to kinematic coupling conditions of facings with the core (29.13), then in accordance with [33] to derive the necessary set of basic equations describing the static equilibrium of the shell a generalized variational equation of the following form should be drawn up

$$\delta I = \delta I_q + \delta A - \delta \Pi = 0. \tag{29.17}$$

Here

$$\begin{gathered}\delta I_q = \delta \iint\limits_{\sigma} q^i \left[u_i^{(1)} - u_i^{(2)} + \frac{t_{(1)}\gamma_i^{(1)}}{2} + \frac{t_{(2)}\gamma_i^{(2)}}{2} \right. \\ \left. + U_i\,(z = h/2\,) - U_i\,(z = -h/2\,) \right] \mathrm{d}\sigma. \end{gathered} \tag{29.18}$$

The difference in (29.18) for U_i is established by integrating (1.26) over z from $-h/2$ to $h/2$

$$\begin{gathered} U_i\,(z = h/2\,) - U_i\,(z = -h/2\,) = c_{is}q^s - \frac{h}{2}\left(w_i^{(1)} + w_i^{(2)}\right) \\ - \frac{h^3}{12E_3^0}\nabla_i\nabla_s q^s - \frac{h}{4}\left(t_{(1)}\varphi_i^{(1)} - t_{(2)}\varphi_i^{(2)}\right). \end{gathered} \tag{29.19}$$

After substitution of expressions (29.15), (29.16), (29.18) into equation (29.17) and implementation of traditional transformations taking into account (29.19), we obtain

$$\begin{aligned} \delta I = & \int\limits_C \left\{ \sum_{k=1}^{2} \left[\left(\Phi_n^{(k)} - T_n^{(k)}\right)\delta u_n^{(k)} + \left(\Phi_{n\tau}^{(k)} - T_{n\tau}^{(k)}\right)\delta u_\tau^{(k)} \right.\right. \\ & + \left(\Phi_m^{(k)} - \tilde{S}_{(k)}^i n_i\right)\delta w^{(k)} + \left(L_n^{(k)} - M_n^{(k)}\right)\delta\gamma_n^{(k)} + \left(L_{n\tau}^{(k)} - M_{n\tau}^{(k)}\right)\delta\gamma_\tau^{(k)} \\ & \left.\left. + \left(L_m^{(k)} - \tilde{M}_{(k)}^{i3} n_i\right)\delta\varphi^{(k)}\right] + \frac{h^3}{12E_3^0} q^i n_i \nabla_s \delta q^s \right\} \mathrm{d}s \\ & - \iint\limits_{\sigma} \left\{ \sum_{k=1}^{2} \left[\left(\nabla_i T_{(k)}^{ij} - S_{(k)}^i b_i^j + \tilde{X}_{(k)}^j\right)\delta u_j^{(k)} + \left(\nabla_i \tilde{S}_{(k)}^i + T_{(k)}^{ij} b_{ij} \right.\right.\right. \\ & \left. + \tilde{X}_{(k)}^3\right)\delta w^{(k)} + \left(\nabla_i M_{(k)}^{ij} - T_{(k)}^{j3} + \tilde{M}_{(k)}^j\right)\delta\gamma_j^{(k)} \\ & \left.\left. + \left(\nabla_i \tilde{M}_{(k)}^{i3} + M_{(k)}^{ij} b_{ij} + \tilde{M}_{(k)}^3 - T_{(k)}^{33}\right)\delta\varphi^{(k)}\right] + \mu_i \delta q^i \right\} \mathrm{d}\sigma = 0. \end{aligned} \tag{29.20}$$

Here the following designations for stresses and moments are introduced $\left(\sigma_{(1)} = -\sigma_{(2)} = 1\right)$

$$
\begin{gathered}
T_n^{(k)} = T^{ij}_{(k)} n_i n_j, \;\; T_{n\tau}^{(k)} = T^{ij}_{(k)} n_i \tau_j, \;\; M_n^{(k)} = M^{ij}_{(k)} n_i n_j, \;\; M_{n\tau}^{(k)} = M^{ij}_{(k)} n_i \tau_j, \\
\tilde{X}^i_{(k)} = X^i_{(k)} + \delta_{(k)} q^i, \quad \tilde{M}^i_{(k)} = M^i_{(k)} + \frac{t_{(k)}}{2} q^i, \\
\tilde{X}^3_{(k)} = X^3_{(k)} + \delta_{(k)} \sigma_0^{33}, \quad \tilde{M}^3_{(k)} = M^3_{(k)} + \frac{t_{(k)}}{2} \sigma_0^{33}, \\
\tilde{S}^i_{(k)} = S^i_{(k)} + \frac{h}{2} q^i, \;\; S^i_{(k)} = T^{i3}_{(k)} + T^{ij}_{(k)} \omega_j^{(k)}, \;\; \tilde{M}^{i3}_{(k)} = M^{i3}_{(k)} + \delta_{(k)} \frac{h t_{(k)}}{4} q^i.
\end{gathered}
$$

Due to the arbitrary of variation of displacements $\delta u_i^{(k)}$, $\delta w^{(k)}$, rotation angles $\delta \gamma_i^{(k)}$, functions of compression $\varphi^{(k)}$, and transverse shear stresses δq^i, from the variational Eq. (29.20) twelve differential equations of the equilibrium of the facings follow

$$
\begin{gathered}
f^i_{(k)} = \nabla_j T^{ij}_{(k)} - S^j_{(k)} b^i_j + \tilde{X}^i_{(k)} = 0, \;\; f^3_{(k)} = \nabla_i \tilde{S}^i_{(k)} + T^{ij}_{(k)} b_{ij} + \tilde{X}^3_{(k)} = 0, \\
f^{j+3}_{(k)} = \nabla_i M^{ij}_{(k)} - T^{j3}_{(k)} + \tilde{M}^j_{(k)} = 0, \\
f^6_{(k)} = \nabla_i \tilde{M}^{i3}_{(k)} + M^{ij}_{(k)} b_{ij} + \tilde{M}^3_{(k)} - T^{33}_{(k)} = 0
\end{gathered} \tag{29.21}
$$

as well as two differential equations of the form

$$
\mu_i = 0, \tag{29.22}
$$

representing the kinematic coupling conditions of facings with the core via tangential displacements. For the obtained equations on the contour C, the following boundary conditions are formulated

$$
\begin{gathered}
T_n^{(k)} = \Phi_n^{(k)} \text{ at } \delta u_n^{(k)} \neq 0, \quad T_{n\tau}^{(k)} = \Phi_{n\tau}^{(k)} \text{ at } \delta u_\tau^{(k)} \neq 0, \\
\tilde{S}^i_{(k)} n_i = {}_m^{(k)} \text{ at } \delta w^{(k)} \neq 0, \quad M_n^{(k)} = L_n^{(k)} \text{ at } \delta \gamma_n^{(k)} \neq 0, \\
M_{n\tau}^{(k)} = L_{n\tau}^{(k)} \text{ at } \delta \gamma_\tau^{(k)} \neq 0, \quad \tilde{M}^{i3}_{(k)} n_i = L_m^{(k)} \text{ at } \delta \varphi^{(k)} \neq 0 \quad (k = 1, 2),
\end{gathered} \tag{29.23}
$$

$$
q^i n_i = 0 \text{ at } \delta \nabla_s q^s \neq 0. \tag{29.24}
$$

29.4 Linearized Problem of Possible Buckling Modes of a Sandwich Test Specimen Under Four-Point Bending

The experimental determination of the mechanical properties of composite materials is an extremely important and integral part in the technological process of designing and manufacturing of structures made of such materials. The most difficult for implementation is compression tests of fiber composites. Even despite the relatively small length of the working part of the test samples, in accordance with existing standards, their compression tests are usually accompanied by longitudinal–transverse bending.

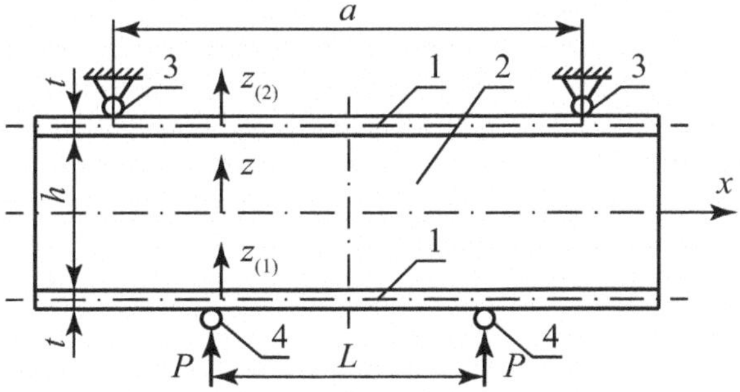

Fig. 29.1 Testing schemes for sandwich specimens under four-point bending (1—facings, 2—core, 3—support rod, 4—loading rod)

Such bending of test samples, even being very small, significantly affects the test results. This effect is especially significant during prolonged exposure of specimens under load.

One of the typical testing schemes for specimens under four-point bending is shown in Fig. 29.1. Such testing is one of the typical experimental studies of sandwich structures with rigid facings made of fibrous composites and a transversely flexible core.

As a rule, these tests are carried out to determine limit of compressive stress formed in the upper facing in the area of the length L of the working length a of the test specimen. Reaching this limit, destruction occurs in the upper facing. It can be assumed that one of the causes of such destruction is the buckling of the compressed facings by one of the possible mode, the classification of which is given in [34–36]. As it will be shown below, the use of the equations constructed above allows one to identify bending, bending shear, and purely shear modes of the upper facing of the test specimen.

Let us consider a sandwich rod of a symmetric structure. The facings of this rod have a thickness t (Fig. 29.1) and are made of a fiber composite. Fiber composite is characterized by thickness-averaged elastic moduli E_1, E_3 in the directions of length and thickness, respectively, by Poisson's ratios ν_{13}, $\nu_{31} = E_3\nu_{13}/E_1$, and transverse shear modulus $G_{13} << E_1$. Moreover, in accordance with previously obtained results [25, 39], it is believed that $G_{13} = G_{13}\left(\varepsilon_{13}^{z}\right)$. We assume that the core is characterized by elastic modulus E_3^0 in the direction of thickness and transverse shear modulus G_{13}^0 in the plane xoz.

In the considered rod under four-point bending (Fig. 29.1) in the case $t << h$ and $E_1^0 \sim E_2^0 << E_1$ over a section of length L in the upper facings membrane, compressive forces are formed, whereas tensile forces are formed in the lower layer. These forces can be determined with a high degree of accuracy by the following formula ($\delta_{(1)} = -\delta_{(2)} = 1$)

$$T_{(k)}^{0\,11} = -\delta_{(k)}T_0, \quad T_0 = \frac{(a-L)}{2\,(h+t/2\,)}P. \tag{29.25}$$

When forces (29.25) are formed in facings over a section of length L, the rod will be in pure bending conditions. Under this condition, a buckling by one of the mode is possible in the upper facing. A study of a mixed bending buckling modes of sandwich rods under pure bending was carried out in [22]. In these studies, refined equations [20] were used to describe the perturbed equilibrium state of a sandwich rod. These equations were based on the description of the stain mechanics of facings by the classical Kirchhoff–Love model.

In contrast to [20], to describe the perturbed stress–strain state of facings, we will further use the refined kinematic model of S.P. Timoshenko (29.4). According to it, for displacements $U^{(k)}$ in the direction of the axis x and deflections $W^{(k)}$ in the direction of the axes $z_{(k)}$, the following representations from (29.4) are taken

$$U^{(k)} = u^{(k)} + z_{(k)}\gamma^{(k)}, \quad W^{(k)} = w^{(k)} + z_{(k)}\phi^{(k)}. \tag{29.26}$$

Here, $u^{(k)}$ and $w^{(k)}$ are the displacements in the directions of the axes x and $z_{(k)}$ of points of the median planes of the facings, $\gamma^{(k)}$ and $\phi^{(k)}$are the angles of the cross sections rotation and the transverse compression function. To describe the stain–stress state of a transversally flexible core in accordance with the above results, it suffices to introduce the function of transverse shear stresses $q = q(x)$. These stresses are constant in the direction of the axis z. Based on (29.21), this allows to compose a system of eight linearized equations of neutral equilibrium of the form

$$\begin{gathered} T^{11}_{(k),x} + \delta_{(k)}q = 0, \quad \left(T^{13}_{(k)} - \delta_{(k)}T_0 w^{(k)}_{,x} + \frac{h}{2}q\right)_{,x} + \delta_{(k)}\sigma^0_{33} = 0, \\ M^{11}_{(k),x} - T^{13}_{(k)} + \frac{t}{2}q = 0, \quad \left(M^{13}_{(k)} + \delta_{(k)}\frac{ht}{4}q\right)_{,x} - T^{33}_{(k)} + \frac{t}{2}\sigma^0_{33} = 0, \end{gathered} \tag{29.27}$$

where the internal forces and moments of the perturbed state introduced into consideration through the sought for functions $u^{(k)}$, $w^{(k)}$, $\gamma^{(k)}$ and $\phi^{(k)}$ are expressed by physical dependencies

$$\begin{gathered} T^{11}_{(k)} = B_1\left(u^{(k)}_{,x} + \nu_{31}\varphi^{(k)}\right), \quad T^{33}_{(k)} = B_3\left(\nu_{13}u^{(k)}_{,x} + \varphi^{(k)}\right), \\ T^{13}_{(k)} = B_{13}\left(w^{(k)}_{,x} + \gamma^{(k)}\right), \quad \sigma^0_{33} = \frac{E^0_3}{h}\left[w^{(2)} - w^{(1)} - \frac{t}{2}\left(\varphi^{(1)} + \varphi^{(2)}\right)\right], \\ M^{11}_{(k)} = D_1\gamma^{(k)}_{,x}, \quad M^{13}_{(k)} = D_{13}\varphi^{(k)}_{,x}, \end{gathered} \tag{29.28}$$

$$B_1 = \frac{E_1 t}{1 - \nu_{13}\nu_{31}}, \quad B_3 = \frac{E_3 t}{1 - \nu_{13}\nu_{31}}, \quad B_{13} = G_{13}t, \quad D_1 = \frac{B_1 t^2}{12}, \quad D_{13} = \frac{B_{13}t^2}{12}. \tag{29.29}$$

Within the framework of the model used, the equilibrium equations (29.27) should be supplemented by the following equation

$$u^{(2)} - u^{(1)} - \frac{t}{2}\left(\gamma^{(1)} + \gamma^{(2)}\right) + \frac{h}{2}\left(w^{(1)}_{,x} + w^{(2)}_{,x}\right) + \frac{ht}{4}\left(\varphi^{(1)}_{,x} - \varphi^{(2)}_{,x}\right)$$
$$- \frac{h}{G^0_{13}} q + \frac{h^3}{12E^0_3} q_{,xx} = 0, \tag{29.30}$$

which are a condition for the kinematic coupling conditions of facings with a core via tangential displacements at points of $z_{(k)} = \delta_{(k)} t/2$ and $z = -\delta_{(k)} h/2$ planes, respectively.

We introduce the new unknowns

$$u^{(-)} = u^{(1)} - u^{(2)}, \quad \varphi^{(-)} = \varphi^{(1)} - \varphi^{(2)}, \quad w^{(+)} = w^{(1)} + w^{(2)}, \quad \gamma^{(+)} = \gamma^{(1)} + \gamma^{(2)},$$
$$u^{(+)} = u^{(1)} + u^{(2)}, \quad \varphi^{(+)} = \varphi^{(1)} + \varphi^{(2)}, \quad \gamma^{(-)} = \gamma_{(1)} - \gamma_{(2)}, \quad w^{(-)} = w^{(1)} - w^{(2)}. \tag{29.31}$$

Then, using relations (29.28) and taking into account (29.31), Eqs. (29.27) and (29.30) admit transformations of the form

$$B_1\left(u^{(-)}_{,xx} + \nu_{31}\varphi^{(-)}_{,x}\right) + 2q = 0, \quad D_{13}\varphi^{(-)}_{,xx} - B_3\left(\nu_{13}u^{(-)}_{,x} + \varphi^{(-)}\right) + \frac{ht}{2}q_{,x} = 0,$$
$$D_1\gamma^{(+)}_{,xx} - B_{13}\left(w^{(+)}_{,x} + \gamma^{(+)}\right) + tq = 0,$$
$$- u^{(-)} - \frac{t}{2}\gamma^{(+)} + \frac{h}{2}w^{(+)}_{,x} + \frac{ht}{4}\varphi^{(-)}_{,x} - \frac{h}{G^0_{13}}q + \frac{h^3}{12E^0_3}q_{,xx} = 0,$$
$$B_{13}\left(w^{(+)}_{,xx} + \gamma^{(+)}_{,x}\right) - T_0 w^{(-)}_{,xx} + hq_{,x} = 0,$$
$$B_1\left(u^{(+)}_{,xx} + \nu_{31}\varphi^{(+)}_{,x}\right) = 0, \quad D_{13}\varphi^{(+)}_{,xx} - B_3\left(\nu_{13}u^{(+)}_{,x} + \varphi^{(+)}\right)$$
$$- \frac{tE^0_3}{h}\left(w^{(-)} + \frac{t}{2}\varphi^{(+)}\right) = 0, \quad D_1\gamma^{(-)}_{,xx} - B_{13}\left(w^{(-)}_{,x} + \gamma^{(-)}\right) = 0,$$
$$B_{13}\left(w^{(-)}_{,xx} + \gamma^{(-)}_{,x}\right) - T_0 w^{(+)}_{,xx} - \frac{2E^0_3}{h}\left(w^{(-)} + \frac{t}{2}\varphi^{(+)}_{,x}\right) = 0. \tag{29.32}$$

We note that in the case under consideration, the system of formulated equations (29.32) describes mixed buckling modes of rod's facings. During their implementation, the largest amplitudes of the bulges should be formed at the lower compressed facing.

Solutions of equations (29.32) will be sought in the form

$$u^{(\pm)} = \tilde{u}^{(\pm)}\cos\lambda x, \quad w^{(\pm)} = \tilde{w}^{(\pm)}\sin\lambda x, \quad \gamma^{(\pm)} = \tilde{\gamma}^{(\pm)}\cos\lambda x,$$
$$\varphi^{(\pm)} = \tilde{\varphi}^{(\pm)}\sin\lambda x, \quad q = \tilde{q}\cos\lambda x, \tag{29.33}$$

where $\lambda = n\pi/L$, $n = 1, 2, \ldots$ are the wave numbers. Substituting functions (29.33) in equations (29.32), the dependences

$$\begin{aligned}
\tilde{\varphi}^{(+)} &= -b_{\varphi}\tilde{w}^{(-)},\\
\tilde{\gamma}^{(-)} &= -\frac{B_{13}\lambda D_1}{D_1\lambda^2 + B_{13}}\tilde{w}^{(-)},\quad \tilde{u}^{(+)} = \frac{\nu_{31}}{\lambda}\tilde{\varphi}^{(+)} = -\frac{\nu_{31}b_{\varphi}}{\lambda}\tilde{w}^{(-)},\\
\tilde{q} &= -\frac{c_w}{r_q}\tilde{w}^{(+)},\quad \tilde{\gamma}^{(+)} = \frac{t}{D_1\lambda^2 + B_{13}}\tilde{q} - \frac{B_{13}\lambda}{D_1\lambda^2 + B_{13}}\tilde{w}^{(+)}\\
&= -\frac{1}{\left(D_1\lambda^2 + B_{13}\right)}\left(\frac{tc_w}{r_q} + B_{13}\lambda\right)\tilde{w}^{(+)},\\
\tilde{u}^{(-)} &= -a_q\tilde{q} = \frac{a_q c_w}{r_q}\tilde{w}^{(+)},\quad \tilde{\varphi}^{(-)} = -b_q\tilde{q} = \frac{b_q c_w}{r_q}\tilde{w}^{(+)}
\end{aligned} \tag{29.34}$$

as well as a system of two homogeneous equations of the form

$$a^{(-)}\tilde{w}^{(-)} + T_0\lambda^2\tilde{w}^{(+)} = 0,\quad a^{(+)}\tilde{w}^{(+)} + T_0\lambda^2\tilde{w}^{(-)} = 0 \tag{29.35}$$

are established. In the last expressions, the following notation is introduced

$$\begin{aligned}
b_{\varphi} &= \frac{tE_3^0}{h\left(\tilde{B}_3 + D_{13}\lambda^2 + \frac{t^2E_3^0}{2h}\right)},\quad \tilde{B}_3 = B_3\left(1 - \nu_{13}\nu_{31}\right),\\
a^{(-)} &= \frac{tE_3^0}{h}b_{\varphi} + \frac{B_{13}^2\lambda^2}{D_1\lambda^2 + B_{13}} - B_{13}\lambda^2 - \frac{2E_3^0}{h},\\
b_q &= \frac{htB_1\lambda^2 - 4B_3\nu_{13}}{2\left(B_1\tilde{B}_3\lambda + B_1D_{13}\lambda^3\right)},\quad c_w = -\left(\frac{h\lambda}{2} + \frac{tB_{13}\lambda}{2\left(D_1\lambda^2 + B_{13}\right)}\right),\\
a_q &= \frac{htB_1\nu_{31}\lambda^2 - 4\left(D_{13}\lambda^2 + B_3\right)}{2\lambda^2\left(B_1\tilde{B}_3 + B_1D_{13}\lambda^2\right)},\\
r_q &= a_q - \frac{t^2}{2\left(D_1\lambda^2 + B_{13}\right)} - \frac{htb_q\lambda}{4} - \frac{h}{G_{13}^0} - \frac{h^3\lambda^2}{12E_3^0},\\
a^{(+)} &= -B_{13}\lambda^2 + \frac{B_{13}^2\lambda^2}{D_1\lambda^2 + B_{13}} + \left(h\lambda + \frac{B_{13}\lambda t}{D_1\lambda^2 + B_{13}}\right)\frac{c_w}{r_q}.
\end{aligned} \tag{29.36}$$

The critical value of the force T_0^{cr} is determined from the condition of non-triviality of the solutions of combined Eq. (29.35). It can be found by the following formula

$$T_0^{cr} = \frac{1}{\lambda^2}\sqrt{a^{(+)}a^{(-)}}. \tag{29.37}$$

The relation between the amplitude values of the functions $w^{(+)}$ and $w^{(-)}$ of the following form

$$\tilde{w}^{(+)} = -\frac{T_0^{cr}\lambda^2}{a^{(+)}}\tilde{w}^{(-)} = -\sqrt{\frac{a^{(-)}}{a^{(+)}}}\tilde{w}^{(-)} \tag{29.38}$$

allows to determine $\tilde{w}^{(+)}$ at a given value $\tilde{w}^{(-)}$. The amplitude values of other functions are determined by the dependences (29.34).

The dependences

$$w^{(1)} = \left(w^{(+)} + w^{(-)}\right)/2\,, \quad w^{(2)} = \left(w^{(+)} - w^{(-)}\right)/2$$

follow from (29.31). When they are taken into account, Eq. (29.35) transforms to the form

Table 29.1 Values of critical stresses and parameters characterizing buckling modes for rods of various lengths L

L, mm	σ^*, MPa	n^*	$\tilde{w}^{(2)}/\tilde{w}^{(1)}$
$t = 1.12$ mm; $G_{13} = 459$ MPa; $E_3^0 = 80$ MPa; $G_{13}^0 = 30$ MPa			
50	45	2	−0.918
60	209	3	−0.575
80	209	4	−0.575
100	45	4	−0.918
120	106	5	−0.807
150	45	6	−0.918
$t = 1.12$ mm; $G_{13} = 459$ MPa; $E_3^0 = 150$ MPa; $G_{13}^0 = 75$ MPa			
50	112	4	−0.732
60	149	5	−0.649
80	181	7	−0.576
100	112	8	−0.732
120	149	10	−0.649
150	112	12	−0.732
$t = 3$ mm; $G_{13} = 459$ MPa; $E_3^0 = 80$ MPa; $G_{13}^0 = 30$ MPa			
50	444	5	−0.0523
60	444	6	−0.0523
80	388	1	−0.804
100	175	1	−0.938
120	443	11	−0.059
150	415	2	−0.768
$t = 3$ mm; $G_{13} = 459$ MPa; $E_3^0 = 150$ MPa; $G_{13}^0 = 75$ MPa			
50	371	1	−0.761
60	93.5	1	−0.952
80	433	9	−0.0934
100	371	2	−0.761
120	93.5	2	−0.952
150	371	3	−0.761

$$\left(a^{(+)} + T_0\lambda^2\right)\tilde{w}^{(1)} + \left(a^{(+)} - T_0\lambda^2\right)\tilde{w}^{(2)} = 0,$$
$$\left(a^{(-)} + T_0\lambda^2\right)\tilde{w}^{(1)} - \left(a^{(-)} - T_0\lambda^2\right)\tilde{w}^{(2)} = 0. \tag{29.39}$$

Taking into account (29.37) from (29.39), we can obtain the dependence between the amplitude values of the deflections $\tilde{w}^{(1)}$ and $\tilde{w}^{(2)}$ of the following form

$$\tilde{w}^{(2)} = \frac{a^{(+)} + \sqrt{a^{(+)}a^{(-)}}}{a^{(+)} - \sqrt{a^{(+)}a^{(-)}}}\tilde{w}^{(1)} = \frac{a^{(-)} + \sqrt{a^{(+)}a^{(-)}}}{-a^{(-)} + \sqrt{a^{(+)}a^{(-)}}}\tilde{w}^{(1)}. \tag{29.40}$$

Based on the found solution, we analyzed the feasibility of implementing the investigated buckling modes of a rod with facings made of unidirectional carbon fiber HSE 180 REM and a binder EDT-69NM. Such a composite material has the elastic characteristics [25, 39]

$$E_1 = 133\,\text{GPa},\ E_3 = 5.9\,\text{GPa},\ \nu_{13} = 0.29$$

under tension in the direction of the fibers (E_1, ν_{13}) and across the fibers (E_3, ν_{31}). Under compression in the direction of the fibers, this composite has an elastic modulus also equal to $E_1 = 133$ GPa, and its failure occurs when limit stress $\sigma_c = 459$ MPa is reached. In accordance with the results in [39], stress σ_c is equal to the transverse shear modulus G_{13}^* corresponds to purely shear buckling mode of the composite structure.

The calculations were carried out for a rod with a core thickness of $h = 10$ mm and two thicknesses of the facings: $t = 1.12$ mm and $t = 3$ mm. For each of the calculation options, two cases are considered. These cases differ in the mechanical characteristics of the core. The numerical results obtained are shown in Table 29.1 for various fixed values L, covering both short and long sections of the rod under pure bending. The critical value of stresses in them is indicated through σ^* and can be found by the formula $\sigma^* = T_0^{\text{cr}}/t$.

For a rod with its various parameters, some calculation results are shown in Fig. 29.2–29.5 as relations $\sigma^* = \sigma^*(n)$ (a), $\sigma^* = \sigma^*(G_{13})$ (b), $n^* = n^*(G_{13})$ (c). Furthermore, relations between dimensionless value $\tilde{w}^{(2)}/\tilde{w}^{(1)}$ and the transverse shear modulus G_{13} of the facing material are also shown in Figs. 29.2, 29.3, 29.4 and 29.5d.

29.5 Conclusion

Results of the approximate solution of the linearized problem constructed show that in flat sandwich plates with rigid facings made of fibrous composites and a transversally flexible core only bending and bending shear buckling modes can be revealed in case of four-point bending. As can be seen from relations (a), the implementation of a purely shear buckling modes is possible only at $n \to \infty$. It also can be seen that there

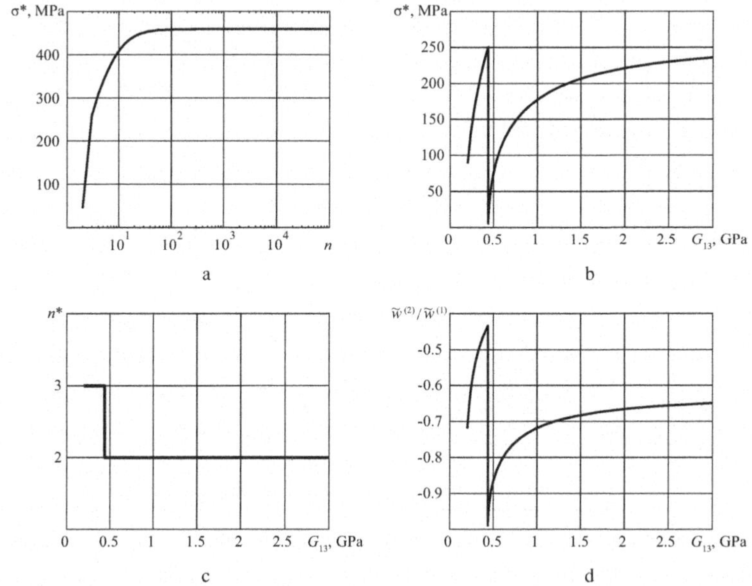

Fig. 29.2 Graphs in case of $t = 1.12$ mm, $E_3^0 = 80$ MPa, $G_{13}^0 = 30$ MPa, $L = 50$ mm

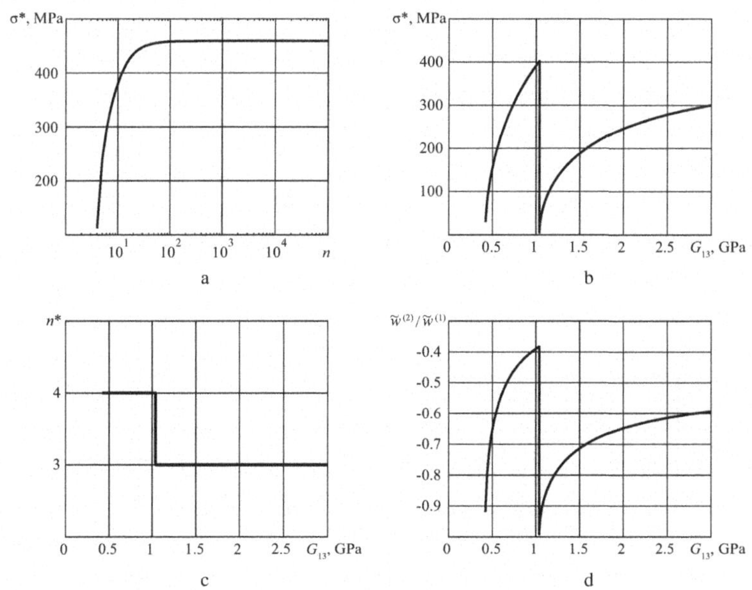

Fig. 29.3 Graphs in case of $t = 1.12$ mm, $E_3^0 = 150$ MPa, $G_{13}^0 = 75$ MPa, $L = 50$ mm

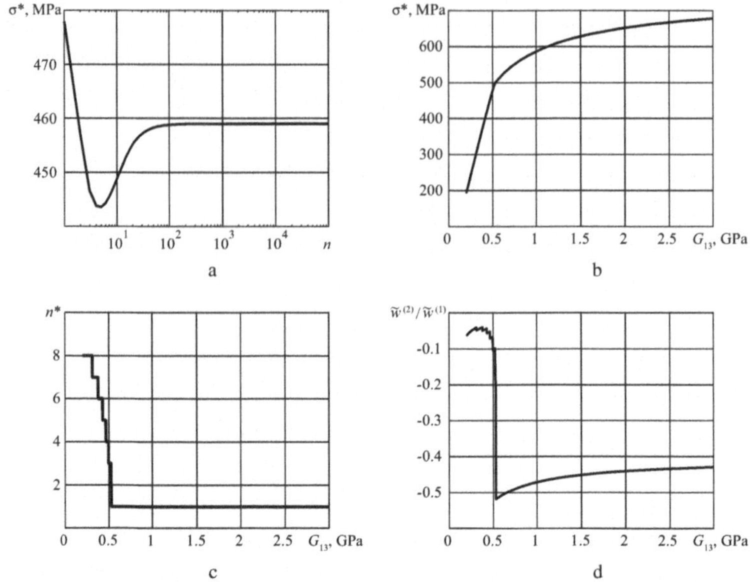

Fig. 29.4 Graphs in case of $t = 3$ mm, $E_3^0 = 80$ MPa, $G_{13}^0 = 30$ MPa, $L = 50$ mm

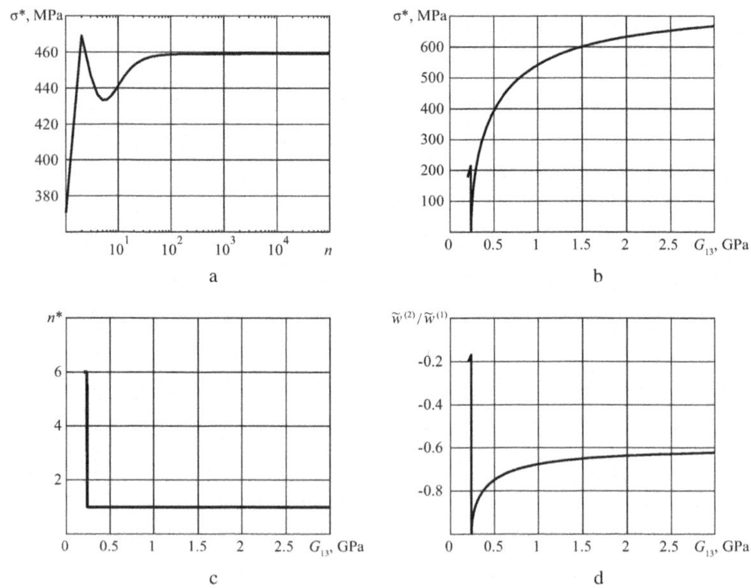

Fig. 29.5 Graphs in case of $t = 3$ mm, $E_3^0 = 150$ MPa, $G_{13}^0 = 75$ MPa, $L = 50$ mm

is such a critical value $G_{13} = G_{13}^*$ where qualitative change in the buckling modes of the facings occurs [see relations (b) and (d)]. We should notice that a deeper and more meaningful study of the problem under consideration, as in [39], is possible only on the basis of a numerical solution of the derived equations, composed taking into account the geometrically and physically nonlinear behavior of the facing materials.

Acknowledgements This work was supported by the Russian Foundation for Basic Research (19-08-00073; Introduction and Sects. 29.2, 29.3) and the Russian Science Foundation (project No. 19-79-10018; Sect. 29.4).

References

1. Adams, D.F., Xie, M.A.: Plasticity model for unidirectional composite materials and its applications in modeling composite testing. Compos. Sci. Technol. **51**(1), 11–21 (1995)
2. Allix, O., Feld, N., Baranger, E., Guimard, J.-M., Ha-Minh, C.: The Compressive Behaviour of Composites Including Fiber kinking: modelling across the scales. Meccanica **49**(11), 2571–2586 (2014)
3. Altenbach, H., Eremeyev, V.A., Naumenko, K.: On the use of the first order shear deformation plate theory for the analysis of three-layer plates with thin soft core layer. Z. Angew. Math. Mech. (2015)
4. Altenbach, J., Altenbach, H., Eremeyev, V.A.: On generalized Cosserat-type theories of plates and shells: a short review and bibliography. Arch. Appl. Mech. **80**(1), 73–92 (2010)
5. Bolotin, V.V., Novichkov, U.N.: Mechanics of multilayer structures. Engineering, Moscow (1980). (in Russian)
6. Budiansky, B., Fleck, N.A.: Compressive failure of fibre composites. J. Mech. Phys. Solids. **41**(1), 183–211 (1993)
7. Davidson, P., Waas, A.M.: Mechanics of kinking in fiber-reinforced composites under compressive loading. Math. Mech. of Solids. **21**(4), 667–684 (2016)
8. Eisentrager, J., Naumenko, K., Altenbach, H., Meenen, J.: A user-defined finite element for laminated glass panels and photovoltaic modules based on a layer-wise theory. Compos. Struct. **133**, 265–277 (2015)
9. Galimov, K.Z.: Fundamentals of the Nonlinear Theory of Thin Shells. Publishing House of Kazan University, Kazan (1975). (in Russian)
10. Galimov, N.K.: On the application of Legendre polynomials to the construction of refined theories of sandwich plates and shells. Res. Theor. Plates Shells **10**, 371–385 (1973). (in Russian)
11. Galimov, N.K.: On the theory of thin depressed shells with a core at finite deflections. Nonlinear Theor. Plates Shells. **61–95** (1963). (in Russian)
12. Galimov, N.K., Ivanov, V.A., Paimushin, V.N.: The problem of stability of the moment equilibrium of sandwich plates and shells and modeling the core in a perturbed state in them. In: Proceedings of the XVII International Conference on Theory of Shells and Plates, pp. 16–23. Publishing house of Kazan University. Kazan (1996) (in Russian)
13. Galimov, N.K., Mushtari, H.M.: On the theory of sandwich plates and shells. Res. Theor. Plates Shells **2**, 61–95 (1964). (in Russian)
14. Grigolyuk, E.I.: Equations of sandwich shells with a lightweight core. Rep. USSR Acad. Sci. OTN. **1**, 77–84 (1957). (in Russian)
15. Grigolyuk, E.I.: Finite deflections of sandwich shells with rigid core. Rep. USSR Acad. Sci. OTN **1**, 26–34 (1958). (in Russian)
16. Grigolyuk, E.I., Chulkov, P.P.: Critical Loads of Cylindrical and Conical Sandwich Shells. West Siberian Book Publishing House, Novosibirsk (1966). (in Russian)

17. Grigolyuk, E.I., Chulkov, P.P.: On the calculation of sandwich plates with rigid core. Rep. USSR Academy Sci. OTN. Mech. Mech. Eng. **1**, 67–74 (1964) (in Russian)
18. Grigolyuk, E.I., Chulkov, P.P.: Stability and vibrations of sandwich shells. Engineering, Moscow (1973). (in Russian)
19. Hapke, J., Gehrig, F., Huber, N., Schulte, K., Lilleodden, E.T.: Compressive failure of UD-CFRP containing void defects: in situ SEM microanalysis. Compos. Sci. Technol. **71**, 1242–1249 (2011)
20. Ivanov, V.A., Paimushin, V.N.: A refined theory of the stability of sandwich structures (nonlinear equations of subcritical equilibrium of shells with a transversally soft core). Izvestiya VUZov Math. **11**, 29–42 (1994). (in Russian)
21. Ivanov, V.A., Paimushin, V.N.: The stability of depressed multilayer shells with transversally soft core. Mech. Comp. Mater. **30**(3), 372–390 (1994). (in Russian)
22. Ivanov, V.A., Paimushin, V.N., Polyakova, T.V.: A refined theory of the stability of sandwich structures (linearized equations of neutral equilibrium and simplest one-dimensional problems). Izvestiya VUZov Math. **3**, 15–24 (1995). (in Russian)
23. Jensen, H.M.: Analysis of Compressive failure of layered materials by kink band broadening. Int. J. Solids Struct. **36**, 3427–3441 (1999)
24. Jumahat, A., Soutis, C., Jones, F.R., Hodzic, A.: Fracture mechanisms and failure analysis of carbon fibre/toughened epoxy composites subjected to compressive loading. Compos. Struct. **92**, 295–305 (2010)
25. Kayumov, R.A., Lukankin, S.A., Paimushin, V.N., Kholmogorov, S.A.: Identification of the mechanical characteristics of fiber-reinforced composites. Uchenye Zapiski Kazanskogo Universiteta Seriya Phys. Math. Sci. **157**(4), 112–132 (2015) (in Russian)
26. Lee, S.H., Yerramalli, C.S., Waas, A.M.: Compressive splitting response of glass reinforced unidirectional composites. Compos. Sci. Technol. **60**(16), 2957–2966 (2000)
27. Mushtari, H.M.: On the applicability of various theories of sandwich plates and shells. Rep. USSR Acad. Sci. OTN. Mech. Mech. Eng. **6**, 163–165 (1960) (in Russian)
28. Mushtari, H.M.: On the general theory of depressed shell with a core. Rep. USSR Academy Sci. OTN. Mech. Mech. Eng. **2**, 24–29 (1961) (in Russian)
29. Naik, N.K., Kumar, R.S.: Compressive strength of unidirectional composites: evaluation and comparison of prediction models. Compos. Struct. **46**, 299–308 (1999)
30. Niu, K., Talreja, R.: Modeling of compressive failure in fiber reinforced composites. Int. J. Solids Struct. **37**(17), 2405–2428 (2000)
31. Noor, A.K., Burton, W.S., Bert, ChW: Computational models for sandwich panels and shells. Appl. Mech. Rev. **49**, 155–199 (1996)
32. Paimushin, V.N.: A refined theory of the stability of multilayer structures (nonlinear equations of subcritical equilibrium of multilayer shells with transversally soft core). In: Scientific works "Fundamental and applied problems of mechanics of deformable media and structures", pp. 44–56. Nizhne Novgorod University Publishing House. Nizhne Novgorod (1993) (in Russian)
33. Paimushin, V.N.: On the variational methods for solving the spatial problems of conjugation of deformable bodies. Rep. USSR Acad. Sci. **273**(5) (1983) (in Russian)
34. Paimushin, V.N.: The theory of stability of sandwich plates and shells (Stages of development, current state and direction of further research). Proc. Russ. Acad. Sci. MMT. **2**, 148–162 (2001). (in Russian)
35. Paimushin, V.N.: The theory of stability of sandwich structural elements. 1. Analysis of the current state and a refined classification of buckling modes. Mech. Compos. Mater. **6**, 230–246 (1999). (in Russian)
36. Paimushin, V.N., Bobrov, S.N.: On the buckling modes of sandwich plates and shells with facings of homogeneous and reinforced materials. Mech. Compos. Mater. **1**, 79–86 (1985). (in Russian)
37. Paimushin, V.N., Galimov, N.K.: Axisymmetric bending and stability of sandwich round plates with lightweight core under combined loading. In: Studies on Linear Problems of the Theory of Plates and Shells, pp. 94–102. Publishing House of Saratov University, Saratov (1974) (in Russian)

38. Paimushin, V.N., Galimov, N.K.: On the stability of sandwich plates with a lightweight core under bending. Proc. theory of shells of Kazan Phys.-Tech. Inst. USSR Acad. Sci. **5**, 35–42 (1974) (in Russian)
39. Paimushin, V.N., Kholmogorov, S.A., Makarov, M.V., Tarlakovskii, D.V., Lukaszewicz, A.: Mechanics of fiber composites: Forms of loss of stability and fracture of test specimens resulting from three-point bending tests. Z. Angew. Math, Mech (2019)
40. Pimenta, S., Gutkin, R., Pinho, S.T., Robinson, P.: A micromechanical model for kink-band formation: Part I Experimental study and numerical modelling. Compos. Sci. Technol. **69**(7–8), 948–955 (2009)
41. Prabhakar, P., Waas, A.M.: Interaction between kinking and splitting it the compressive failure of unidirectional fiber reinforced laminated composites. Compos. Struct. **98**, 85–92 (2013)
42. Rikards, R.B., Teters, G.A.: Stability of Shells made of Composite Materials. Zinatne, Riga (1974). (in Russian)
43. Rosen, B.W.: Mechanics of composite strengthening in fiber composite materials. American Soc. Met. **37–75** (1965)
44. Schulze, S.H., Pander, M., Naumenko, K., Altenbach, H.: Analysis of laminated glass beams for photovoltaic applications. Int. J. Solids Struct. **49**, 2027–2036 (2012)
45. Tarnopolsky, U.M., Rose, A.V.: Features of the calculation of parts made of reinforced plastics. Zinatne, Riga (1969). (in Russian)
46. Weps, M., Naumenko, K., Altenbach, H.: Unsymmetric three-layer laminate with soft core for photovoltaic modules. Compos. Struct. **105**, 332–339 (2013)
47. Wu, L.C., Lo, C.Y., Nakamura, T., Kushner, A.: Identifying failure mechanisms of composite structures under compressive load. Int. J. Solids Struct. **35**(12), 1137–1161 (1998)

Chapter 30
Modeling of the Operation of EHD Devices with the Working Medium of Complex Microstructure

Irina L. Pankratieva and Vitaly A. Polyansky

Abstract The model of a working medium of complex microstructure is used to calculate processes in a channel of one stage of an electrohydrodynamic (EHD) device. The flow of a multicomponent fluid consisting of neutral particles and a small admixture of two kinds of oppositely charged ions through a plane channel with dielectric walls is investigated. Two plane electrodes penetrable by the fluid and exerting no hydrodynamic drag force are placed in the cross sections of the channel. These electrodes and the space between them form a stage of an EHD device applied to increase the pressure in the medium. It is shown that the pressure drop and the total volume charge of the stage almost linearly depend on the applied voltage in a certain range; however, there is a critical value beyond which a further increase in voltage reduces the efficiency of the converter. This is due to the influence of injection of negative ions from the cathode, which grows with increasing field and causes redistribution of the volume charge in the interelectrode space.

Keywords Electrohydrodynamic device · Complex multicomponent fluid · Electric field · Volume electric charge · Pressure of the medium

30.1 Introduction

Electrohydrodynamic (EHD) devices that increase the pressure in continua of complex microstructure are perspective in various technological processes such as cold production [1]. As compared to conventional reciprocating compressors for refrigerators, they have a number of merits of which the most important are the absence of moving mechanical parts and the ability of using refrigerants such as propane/butane

I. L. Pankratieva (✉) · V. A. Polyansky
Institute of Mechanics, Lomonosov Moscow State University,
Michurinsky prosp. 1, Moscow 119192, Russia
e-mail: ilpan@imec.msu.ru

H. Altenbach et al. (eds.), *Multiscale Solid Mechanics*,
Advanced Structured Materials 141,
https://doi.org/10.1007/978-3-030-54928-2_30

gas mixtures, which are environmentally safe and thermodynamically more efficient than halocarbons. The operation of such EHD devices is based on converting the energy of the electric field applied to the electrodes into the internal energy of the fluid in the electrode gap. The modeling of such energy transfer is a rather difficult problem; it is studied in the literature under the assumption that simplifies the structure of the medium, namely in the presence of ions of only one kind (either positive or negative), which are injected in the electrode gap from the surface of one of the electrodes. However, the study of the operation of EHD pumps of another type, which was carried out in [2] in the framework of the model of a medium with a more complex microstructure, showed that the presence of oppositely charged ions in the medium exerts essential influence on the efficiency of energy transform, especially in the case of using non-stationary applied electric fields.

30.2 Model of the Medium

In this paper, the model of the working medium of a complex microstructure [3] is used to calculate the processes in the channel of one stage of an EHD device. The flow of the fluid consisting of neutral particles and a small admixture of two kinds of oppositely charged ions through a plane channel of uniform cross section with dielectric walls is considered. Two plane electrodes penetrable by the fluid and exerting no hydrodynamic drag force are placed in the cross sections of the channel. These electrodes and the space between them form a stage of an EHD device. An electrical potential difference F_w is applied to the electrodes from an external source. It is assumed that under the applied electric field the surface of each electrode may inject (like-charged) ions into the electrode gap. Moreover, at the surface of each electrode, recombination of both kinds of ions is observed. The influent is assumed to be quasineutral and have a given concentration n_0 of charge carriers. The rate of flow of the carrying medium through the stage is also assumed to be constant.

The distributions of the fluid pressure p, the ion concentrations n_m $(m = 1, 2)$, and the electric field strength E over the electrode gap are calculated with the use of the non-stationary system of equations of electrohydrodynamics (30.1)–(30.3), which includes the equation of the bulk motion of the fluid written in hydraulic approximation with regard to the linear (in velocity) friction with the channel walls with the friction coefficient C_x (30.1), the ion continuity equations, which take into account the convective transport of ions with the fluid velocity u, their drift in the electric field with the ion mobility coefficients b_m, diffusion with the diffusion coefficients D_m, and bulk electrochemical processes of dissociation with the dissociation rate w, and recombination with the recombination coefficient α (30.2). Poisson's equation (30.3) for the electric potential F takes into account the volume electric charge q arising in the electrode gap due to the motion of ions and their interaction with the electrode surfaces.

$$\rho \frac{\partial u}{\partial t} = -\frac{\partial p}{\partial x} + qE - C_x u, \quad \frac{\partial u}{\partial x} = 0 \tag{30.1}$$

$$\begin{aligned} \frac{\partial n_1}{\partial t} + \frac{\partial}{\partial x}(n_1 U_1) = w - \alpha n_1 n_2, \\ \frac{\partial n_2}{\partial t} + \frac{\partial}{\partial x}(n_2 U_2) = w - \alpha n_1 n_2 \end{aligned} \tag{30.2}$$

$$n_m U_m = n_m u - D_m \frac{\partial n_m}{\partial x} - \operatorname{sign}(e_m) b_m n_m \frac{\partial F}{\partial x},$$

$$\varepsilon\varepsilon_0 \frac{\partial^2 F}{\partial x^2} = q, \quad E = -\frac{\partial F}{\partial x}, \quad q = \sum_m e_m n_m, \quad m = 1, 2, \tag{30.3}$$

$$e_1 > 0, \quad e_2 < 0$$

Here, x is the space variable along the channel axis, e_m is the ion charge, and $\varepsilon\varepsilon_0$ is the absolute dielectric permittivity of the fluid.

The boundary conditions for the concentrations of ions at the electrodes can be obtained from the condition of balance between the fluxes of the charged particles to the electrodes with due regard to surface electrochemical processes of ionization and recombination [2]:

$$\begin{aligned} n_m U_m &= n_0 u + A_{0m} + B_{0m} E - K_{0m} n_m, m = 1, 2, F = F_w, x = x_0; \\ n_m U_m &= n_0 u - A_{1m} - B_{1m} E + K_m n_m, m = 1, 2, F = 0, x = x_0 + L. \end{aligned} \tag{30.4}$$

Here, x_0 is the coordinate of the anode, L is the distance between the electrodes, and A_{im}, B_{im}, and K_{im} are the parameters of the surface electrochemical processes of ionization and ion recombination. It is assumed that the surface ionization rate depends linearly on the field strength. The ion transport coefficients and the rates of the bulk and surface reactions are assumed to be parameters of the problem given in advance. Note that when EHD flows are investigated in the framework of the model of unipolarly charged liquid, the boundary condition is usually specified by the equilibrium concentration of ions on the surface of the injecting electrode. In the case of non-stationary processes caused, for example, by a pulsating voltage applied to the electrodes, the assumption that the surface electrochemical reaction is in equilibrium may fail to hold owing to an essential change in the distribution of the concentrations of charged particles and the field in the vicinity of the electrode during the pulse. The equilibrium is also violated in the case of a constant applied potential difference when the field strength exceeds the threshold value [4] and the ion injection switches on and in a certain range of parameters the waves of the field strength and of the volume charge density arise in the fluid.

The problem (30.1)–(30.4) is solved numerically by the method of iterations at each time step with a simultaneous integration of all equations involved in the system. The initial state of the ions is the uniform distribution of the quasineutral concentration over the electrode gap. The external electric field applied at the initial instant is also uniform. For a constant flow rate, the pressure drop dP across the stage of the compressor is determined from the relation

$$dP = LC_x u - \left[E^2(L,t) - E^2(x_0,t)\right]. \tag{30.5}$$

The initial pressure drop in the absence of exposure to the electric field corresponds to a given rate of flow of the fluid through the stage. For a constant applied potential difference, the calculations are carried out until the distributions of the parameters become stationary.

30.3 Discussion

Let us briefly present some results of the computational experiments performed to study the efficiency of operation of the stage of the EHD device depending on the values of the model parameters describing the surface and bulk electrochemical processes. We consider a fluid with the characteristic parameters involved in the problem which correspond to a hydrocarbon liquid of complex structure with impurity ions that raise up the initial conductivity $\sigma_0 = 2.5 \times 10^{-11}$ S/m and have the same diffusion coefficient $D_0 = 10^{-9}$ m^2/s.

The pressure drop dP and the total volume electric charge Q of the electrode gap, which are produced at the stage after the stationary state, are established and are presented for different values of applied voltage in the dimensionless form in Fig. 30.1. The distance between the electrodes is $L = 0.005$ m, the characteristic value of the applied field strength is $E_0 = 5 \times 10^6$ V/m. It is assumed in calculations that the electrodes are injecting similarly charged ions and those ions of the opposite charge on their surfaces recombine in accordance with conditions (30.4) with the coefficients $A_m = 0\,(i = 0, 1)$. The value of the coefficient B_{01}, which characterizes the injection of positive ions from the anode, is three times greater than that of the coefficient B_{12} characterizing the injection of negative ions from the cathode.

It is seen that the dependence of the pressure drop and the total volume charge of the electrode gap on the applied voltage F_w in a certain range is close to linear, but as the voltage exceeds some critical value $F_{w\,cr}$, further increase in F_w reduces the efficiency of the operation of the device. This is due to the effect of the injection of negative ions from the cathode, which becomes stronger with an increase in the field strength and causes redistribution of the volume charge over the electrode gap. The non-monotone character of the change in the pressure drop with an increase in voltage is also observed in the case where the injection properties of the electrodes do not depend on the field strength. We note that in the framework of the model of a fluid of simplified microstructure with similarly charged ions, the calculations exhibit a monotone growth of the pressure drop over the entire range of applied potential differences.

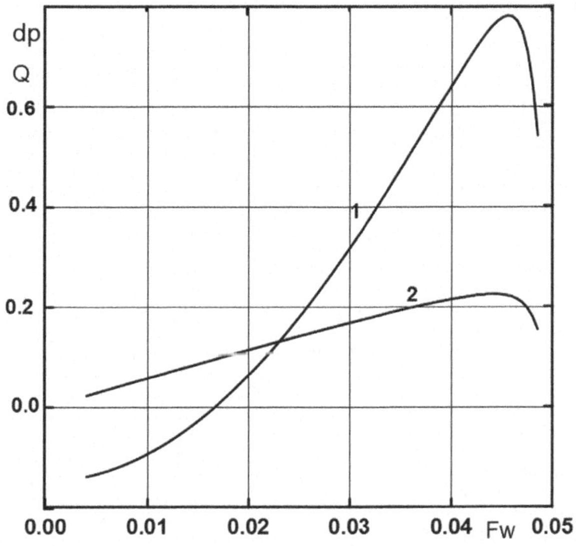

Fig. 30.1 Dependence of the pressure drop dP (curve 1) and the volume charge Q (curve 2) of the electrode gap on the applied voltage

With due account taken of the volume electrochemical processes in the modeling of the operation of the stage, the calculations show that the highest efficiency is observed in the conditions where the process of dissociation of neutral molecules of admixture is frozen and recombination of ions is intensively going on.

30.4 Conclusions

The pressure drop arising in the electrode gap of one stage of an EHD device in the fluid of complex microstructure due to the interaction with the electric field is obtained. The relation between the parameters of the problem and the dependence of the pressure drop on the applied voltage is investigated.

It is established that the non-monotone change of the pressure drop across the stage with an increase in voltage observed in the experiment is determined by the influence of the injection of negative ions from the cathode, which becomes stronger with an increase in the field strength and causes redistribution of the volume charge over the electrode gap.

Acknowledgements This work was in part financially supported by the Russian Foundation for Basic Research, project no. 19-01-00056.

References

1. Bumagin, G.I., Rakhanskii, A.E., Savinov, A.E., Sukhikh, A.A.: Perspectives of application of EHD compressor in refrigerating equipment. Kholodilnaya Tekhnika **2**, 28–30 (1995)
2. Gogosov, V.V., Polyansky, K.V., Polyansky, V.A., Shaposhnikova, G.A., Vartanyan, A.A.: Modeling of nonstationary processes in channels of EHD pump. J. Electrostat. **34**, 245–262 (1995)
3. Pankratieva, I.L., Polyansky, V.A.: Mathematical modeling of electrohydrodynamic phenomena. Intellektualnye Systemy. Teoriya i Prilozheniya **20**(4), 130–147 (2016)
4. Polyansky, V.A., Pankratieva, I.L.: Electric current oscillations in low-conducting liquids. J. Electrostat. **48**, 27–41 (1999)

Chapter 31
Instability of Plastic Deformation in Metals at Low Temperatures

Georgy F. Sarafanov, Fedor G. Sarafanov, and Igor S. Pavlov

Abstract Mathematical models of serrated plastic deformation in metals at ultra-low temperatures are developed. In the framework of the dislocation model, the instability of the plastic flow is shown to be caused by the relay mechanism of overcoming barriers (sub-boundaries, grain boundaries, etc.) by dislocation clusters. Regularities of localization of plastic deformation in alloys are described on the base of the model of interaction of dislocations with point defects. Within the scope of the thermodynamic model, it has been analytically and numerically established that in the range of helium temperatures, a change in the deforming stress, temperature, and plastic strain rate acquires an irregular stochastic nature. The obtained results are in good agreement with experimental data and can be used to construct the theory of plastic deformation of structural metals and alloys in the region of extremely low temperatures.

Keywords Serrated plastic deformation · Low-temperature instability · Dislocations · Plastic strain rate · Slip band

G. F. Sarafanov (✉)
Mechanical Engineering Research Institute of Russian Academy of Sciences, 85 Belinsky Str, Nizhny Novgorod, Russian Federation 603024
e-mail: gf.sarafanov@yandex.ru

F. G. Sarafanov · I. S. Pavlov
National Research Lobachevsky State University of Nizhny Novgorod, Nizhny Novgorod, Russian Federation 603950
e-mail: sfg180@yandex.ru

I. S. Pavlov
e-mail: ispavlov@mail.ru

H. Altenbach et al. (eds.), *Multiscale Solid Mechanics*,
Advanced Structured Materials 141,
https://doi.org/10.1007/978-3-030-54928-2_31

31.1 Introduction

An unstable nature of plastic deformation is currently considered as a universal property of solids that can manifest itself in a wide temperature range [1]. In crystalline materials, it is provided by self-consistent collective motion of dislocations under the action of external and internal stresses [3, 4, 7, 20, 22].

The plastic instability phenomena were proposed in [5, 8] to be classified taking into account the assumed dependence $\tau = \tau(\varepsilon, \dot{\varepsilon})$, according to which, in a linear analysis, the relationship between stress τ, strain ε, and plastic strain rate $\dot{\varepsilon}$ can be represented in the following form:

$$d\tau = h\mathrm{d}\varepsilon + S\mathrm{d}\dot{\varepsilon} \quad (31.1)$$

where $h = (\partial\tau/\partial\varepsilon)_{\dot{\varepsilon}}$ and $S = (\partial\tau/\partial\dot{\varepsilon})_{\varepsilon}$ are the material constants that depend, in a general case, on stresses and a strain rate. Equation (31.1) describes a local stress–strain state, which is identified with global properties, if the deformation of the sample is uniform along its entire length.

According to [5, 8], the plastic instability phenomena are divided into instability of *h*- and *S*-types. For *h*-type instability, sensitivity to changes in the strain rate (the rate sensitivity *S*) is positive, and the rate of strain hardening, *h*, is less than h_c—the instability threshold value that coincides with the deforming stress τ, whereas for *S*-type instability everything is vice versa: $S < 0$ and $h > h_c$.

Conditions for *h*-instability occurrence usually correspond to the late stages of plastic deformation, when increasing of ductility can lead to negative values of *h* [12]. *S*-type instability is most clearly manifested in alloys, when a negative dependence of the drag force of dislocations on their velocity is observed in the field of high temperatures.

At low temperatures, *S*-instability can have a dynamic dislocation nature or can be associated with the thermoplastic instability [21]. For example, at helium temperatures, the plastic deformation instability (serrated deformation) is typical for the field of very low temperatures [20]. Its properties are determined by numerous parameters related to both the deformation conditions (strain rate and temperature) and the properties of the material (lattice type, grain size, etc.) [3, 4, 20]. Various models that enable one describing the instability of deformable metals at low temperatures have been proposed in a large number of works; however, some problems of interpretation and description of these phenomena remain unresolved. Models of the plastic deformation instability in metals at low temperatures, which are provided by both dynamic and thermodynamic mechanisms, are considered in this paper.

31.2 The Dynamic Model

The problem of a theoretical study of unstable modes of plastic deformation can be formulated on the basis of a set of nonlinear evolution equations of hydrodynamic type for the density and the average velocity of continuously distributed dislocations [23, 24]. Evolution equations for density $\rho_a(r, t)$

$$\frac{\partial \rho_a}{\partial t} + \operatorname{div} \rho_a v_a = \Phi_a(\rho_a), \tag{31.2}$$

can be obtained from the compatibility equation [10] and have the form of a continuity equation, where the local interaction of dislocations is neglected ($\Phi_a = 0$). The latter can take place if the process of plastic deformation proceeds in a certain slip band and it is pronounced dynamic in nature. In the case of interest, the dislocation motion equation, which determines $v_a(r, t)$, has the following form

$$m_{ik}^{*} \frac{\mathrm{d}v_k^a}{\mathrm{d}t} = e_{ijm} l_j b_k^a \{\tau_{mk}^{ext} + \tau_{mk}^{\mathrm{int}}(\rho_a) - \tau_{mk}^{f}(C_j, \alpha_j)\} - F_i(v^a). \tag{31.3}$$

The first term in the right-hand side of Eq. (31.3) is the Peach–Köhler force per unit length of the dislocation line, where $\tau_{\mathrm{mk}}^{\mathrm{ext}}$ is the component of the external stress tensor in the slip plane of dislocations; τ_{mk}^{int} is the component of the tensor of long-range internal stresses arising due to a system of distributed dislocation charges, which is functionally dependent on ρ_a; $\tau_{mk}^{\mathrm{f}}(C_j, \alpha_j)$ is the stress of dry friction that is caused by various local obstacles distributed in space with density C_j and possessing force characteristics α_j of locking; m_{ik}^{*} is the tensor of the effective mass of dislocations; $F_i(\mathbf{v}^a)$ is the friction force.

Equations (31.2) and (31.3) are formulated in the approximation when the radius of curvature of the dislocation lines, R_c, significantly exceeds the average distance $\bar{r}$ between the dislocations. In the framework of this approximation, dislocations can be assumed to be generally straightforward.

We will consider a behavior of an ensemble of dislocations in a certain slip band of width w. Let x-axis be selected in the direction of the specified slip system. We shall suppose that the subsystem of positive boundary dislocations ($b \perp l$, where l is the unit vector tangent to the dislocation line), which are characterized by the density $\rho_+(r, t) = \rho(r, t)$ (in equilibrium $\rho_+^0(r, t) = \rho_0$), is excited. Then, the process of plastic deformation in the selected slip band can be described by the following equations:

$$m^*(\frac{\partial v}{\partial t} + v \frac{\partial v}{\partial x}) = b\{\tau_{\mathrm{ext}} - \tau_f(C_j, \alpha_j)\} - F(v), \tag{31.4}$$

$$\frac{\partial \rho}{\partial t} + \frac{\partial}{\partial x}(\rho v) = 0. \tag{31.5}$$

The stress of "dry" friction τ_f can be considered, as a rule, a parameter of the problem, but if we are interested in the case of structural softening of the crystal, then it is necessary to take into account the possible dependence of C_j and α_j on the dislocation density ρ, therefore, in the general case $\tau_f = \tau_f(\rho)$.

The set of Eqs. (31.4) and (31.5) admits a uniform stationary solution

$$\rho = \rho_0, \quad v = v_0, \tag{31.6}$$

in which v_0 is determined from the condition that the right-hand side of Eq. (31.4) vanishes

$$F(v_0) = b\tau_{\text{ext}} - b\tau_f(\rho_0) = b\tau_0, \tag{31.7}$$

and the density ρ_0 is determined from the given conditions of the crystal loading.

Taking into account that Eqs. (31.4) and (31.5) are one-dimensional, we will further assume $\rho = \rho(x, t)$ and $v = v(x, t)$.

Analysis of the stability of the homogeneous stationary state (31.6) of Eqs. (31.4) and (31.5) will be performed under the assumptions made regarding $\tau_{\text{int}}(\rho)$ and $\tau_f(\rho)$. Linearizing the original system and introducing, for convenience, dimensionless variables

$$n(x, t) = \rho(x, t)/\rho_0 - 1, \quad u(x, t) = v(x, t)/v_0 - 1, \tag{31.8}$$

one can obtain the following equations:

$$\frac{\partial u}{\partial t'} + \frac{\partial u}{\partial x'} = an - \gamma u, \tag{31.9}$$

$$\frac{\partial n}{\partial t'} + \frac{\partial n}{\partial x'} + \frac{\partial u}{\partial x'} = 0. \tag{31.10}$$

Here, the dimensionless coordinates $x' = x/t_0 v_0$, $t' = t/t_0$ and parameters $\tau = m^* v_0/bt_0$, $a = -\tau_f'(\rho_0)\rho_0/t_0$, $\gamma = F'(v_0)v_0/bt_0$ are introduced.

Under the assumption $n, u \sim \exp \lambda t' - ikx'$, the dispersion equation

$$\lambda^2 + \lambda(\gamma - 2ik) - ik(\gamma + a) - k^2 = 0$$

yields. Therefore, for the unstable curve $\lambda_+ = \mu + i\omega$ it is possible to find

$$\mu = -\frac{\gamma}{2} + \frac{1}{2\sqrt{2}}\left[\gamma^2 + \sqrt{\gamma^4 + 16k^2a^2}\right]^{1/2}, \tag{31.11}$$

$$\omega = k + \frac{ak}{\gamma + 2\mu}. \tag{31.12}$$

Hence, instability ($\text{Re}\lambda_+ = \mu > 0$) is realized in two cases:

$$\gamma < 0, \tag{31.13}$$

$$a > 0, \quad \gamma > 0. \tag{31.14}$$

The instability of the first type (31.13) is associated with the effect of negative friction $\left(F'(v) < 0\right)$ in the dynamics of dislocations and macroscopically manifests itself as S-instability. The second instability (31.14), which has the nature of h-instability, is caused by decreasing of the barrier drag in the slip band of dislocations, when their density grows $\left(\tau_f'(\rho_0) < 0\right)$.

As it follows from (31.11), instabilities (31.13) and (31.14) are realized in the range of wave numbers $0 < k^2 < \infty$. The presence of harmonics with a zero wavelength in the spectrum of unstable modes means that Eqs. (31.4) and (31.5) do not have stationary inhomogeneous solutions in the class of continuous functions. This difficulty can be overcome by taking into account the dislocation gas viscosity. The introduction of viscosity can be physically explained by the fact that dislocations, when they move, are able to pass into neighboring slip planes, transferring a dislocation pulse there. As it is known, it leads to the internal friction of the gas layers [9].

It is easy to show that taking into account in the right-hand side of Eq. (31.4) the term ηv_{xx} responsible for the viscosity of the dislocation gas (η is the dynamic viscosity coefficient) leads to restriction of the spectrum of unstable modes and, accordingly, to the regularization of the problem (see Fig. 31.1). In this case, expressions (31.11) and (31.12) for the unstable branch are redefined by changing the parameter γ ($\gamma \to \gamma + \nu k^2$, where $\nu = \eta / m^* v_0^2 \tau$ is the dimensionless kinematic viscosity coefficient).

Let us consider a model, where a negative friction instability $\left(F_v'(v) < 0\right)$ is realized. In [2, 13], it is shown that such instability can occur in metals at low temperatures during the interaction of dislocations with point defects.

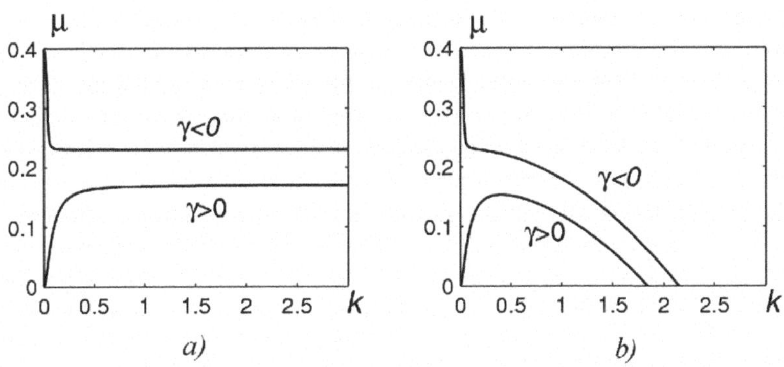

Fig. 31.1 Dependence of the increment of the unstable mode $\mu(k)$ on the wave number k for two types of instabilities without account of viscosity **a** or with account of viscosity **b** in the ensemble of dislocations

In this case, we will neglect the possible dependence of the stress of a dry friction on the density of dislocations. Then, taking into account the term ηv_{xx}, Eq. (31.4) takes the form:

$$m^*(\frac{\partial v}{\partial t} + v\frac{\partial v}{\partial x}) = \eta\frac{\partial^2 v}{\partial x^2} + b\tau - F(v). \tag{31.15}$$

Equation (31.15) does not depend on the density of dislocations; therefore, linearization of Eq. (31.5) with respect to ρ under assumption $|\rho - \rho_0| << \rho_0$ yields

$$\frac{\partial \rho}{\partial t} + \rho_0\frac{\partial}{\partial x}v = 0. \tag{31.16}$$

If there are point defects in crystals that influence the motion of dislocations, they can significantly affect the mechanical characteristics of samples. When dislocations move, their interaction with defects leads to the appearance of dislocation vibrations; therefore, the efficiency of energy dissipation depends on the spectrum form of these vibrations. The motion of a boundary dislocation with a phonon subsystem of a crystal containing randomly distributed point defects (vacancies and interstitial atoms) was studied in [13]. It was shown that in this case, under the certain conditions, the dependence of the drag force of dislocations on the velocity, $F(v)$, can have two extrema. Thus, the function $F(v)$ has an N-shaped dependence, which is given by the expression [13, 14]

$$F(v) = F_d + F_f = \frac{B_d v}{1 + v^2/v_0^2} + Bv, \tag{31.17}$$

where F_d is the drag force due to the interaction of dislocations with point defects, B_d is a quantity proportional to the concentration of point defects [13], F_f is the friction force caused by phonon, magnon, or electronic dissipation mechanisms, and B is the damping constant that substantially depends on temperature T.

It should be noted that the presence of extrema on the curve $F(v)$ can take place only for $B(T) < B_d$ [13]. For typical values of the parameters, including the dislocation density $\rho = 10^{11} m^{-2}$ and the concentration of nonequilibrium point defects (per atomic volume) $n = 10^{-4}$, such values are reached for most crystals at temperatures $T < 25$ K, i.e., in the field of very low temperatures [13, 14].

Further, we shall consider what dynamic effects the dependence of the drag force (31.17) plotted on the phase plane of variables τ and v (Fig. 31.2) leads to.

The straight line $\tau = \tau_0$ can cross the curve $F(v)$ in different ways. We are interested in the case, when the intersection occurs at three points, for example, v_1, v_2 and v_3 ($v_1 < v_2 < v_3$). Then, Eqs. (31.15) and (31.16) have three equilibrium states: two stable ones corresponding to the velocities $v = v_1$ and $v = v_3$, and one unstable state corresponding to the velocity $v = v_2$.

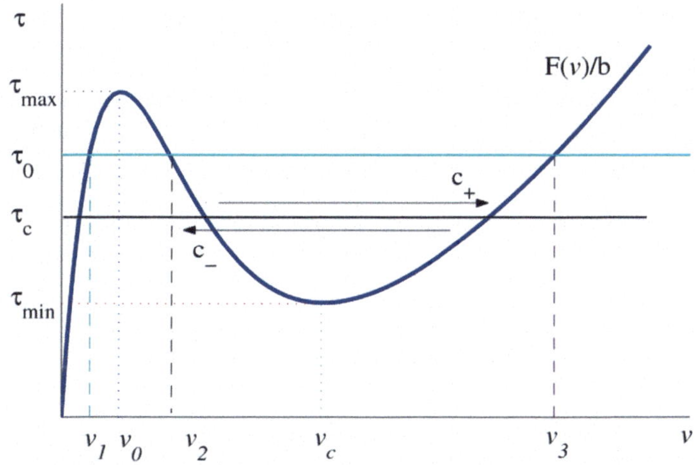

Fig. 31.2 Phase plane for variables τ and v

In order to find stationary solutions of Eqs. (31.15) and (31.16), it is necessary to pass to the traveling variable $\xi = x - ct$ in functions $v = v(\xi)$ and $\rho = \rho(\xi)$. Substitution of the solution in this form into (31.15) and (31.16) and integration of Eq. (31.16) lead to the following equations:

$$m^*(v - c)\frac{\mathrm{d}v}{\mathrm{d}\xi} = +\eta\frac{\mathrm{d}^2v}{\mathrm{d}\xi^2} + b\tau - F(v), \tag{31.18}$$

$$\rho = \frac{v}{c}\rho_0. \tag{31.19}$$

We shall study Eqs. (31.18) and (31.19) on the phase plane (v, W) under the assumption that $W = \mathrm{d}v/\mathrm{d}\xi$. After excluding the variable ρ, one can obtain

$$\eta\frac{\mathrm{d}W}{\mathrm{d}\xi} = -m^*(c - v)W - F(v), \tag{31.20}$$

$$\frac{dv}{d\xi} = W. \tag{31.21}$$

These equations have three fixed points on the phase plane: $(v_1, 0)$, $(v_2, 0)$, and $(v_3, 0)$. The state $(v_2, 0)$ is the focus, and the singular points $(v_1, 0)$ and $(v_3, 0)$ are the saddle points, through which two trajectories pass. The only stable stationary solutions are described by separatrices going from a saddle to a saddle, which correspond to certain values of the wave velocity $c_\pm$.

Equations (31.20) and (31.21) can be reduced to the equation

$$\eta W \frac{\mathrm{d}W}{\mathrm{d}v} + m^*[c - v]W + F(v) = 0, \tag{31.22}$$

with the boundary conditions $W(v_1) = W(v_3) = 0$. In order to obtain results in an analytical form, we shall suppose that $v << c$. Then, in the approximation $F(v) = \kappa(v - v_1)(v - v_2)(v - v_3)$ and under assumption that $W = \delta(v - v_1)(v - v_3)$, one can find

$$c = c_\pm = V_\delta, \tag{31.23}$$

$$\delta = \delta_\pm = \frac{m^*}{4\eta}\left(1 \pm \sqrt{1 + 8\eta\kappa/m^{*2}}\right), \tag{31.24}$$

where $V_\delta = v_2 + (\delta\eta/m^*)(v_1 + v_3 - 2v_2)$.

Since the form of the wave solutions depends on parameter δ, then, as follows from Eqs. (31.23) and (31.24), there are two types of solitary waves corresponding to two values of δ ($\delta_+ > 0$, $\delta_- < 0$) and having different velocities c_+ and c_- of the wave fronts. Let us define profiles of these wave solutions. After integration of $W = \mathrm{d}v/\mathrm{d}\xi = \delta(v - v_1)(v - v_3)$, it is possible to obtain

$$v(x,t) = v_1 + (v_3 - v_1)\left[1 + C_0 \exp\left(z_\delta \frac{x - ct}{\Lambda_\delta}\right)\right]^{-1}, \tag{31.25}$$

where $\Lambda_\delta = |\delta(v_3 - v_1)|^{-1}$ is the characteristic width of the wave front, $z_\delta = \pm$, and C_0 is the integration constant.

From the form of solution (31.25), it follows that a wave of switching from the state v_1 to the state v_3 (a softening wave) corresponds to the positive value $\delta = \delta_+$, and a wave of inverse switching from the state v_3 to the state v_1 (a hardening wave) corresponds to the negative value $\delta = \delta_-$. The width of the front of the softening wave tends to zero ($\eta \to 0$) for $\eta\kappa/m^{*2} << 1$, since in this case $\Lambda_+ \approx 2\eta/m^*(v_3 - v_1)$, whereas $\Lambda_- \approx m^*/\kappa(v_3 - v_1)$ for the strengthening wave. Therefore, in this case $\Lambda_- > \Lambda_+$.

Now, we will consider possible wave patterns arising in the framework of the obtained solutions (31.25). Let at the initial time moment a perturbation of the field of internal stresses τ_{int} arise in a certain area Ω of the slip band. As a result, a velocity perturbation $v(x, 0) \approx v_3$ appears in Ω. Then, from the right boundary of perturbation a softening wave will start to propagate with velocity $c = c_+$, and from the left boundary—a hardening wave with velocity $c = c_-$ (Fig. 31.3).

The softening wave velocity, c_+, is greater than the hardening wave velocity, c_-, provided $v_1 + v_3 - 2v_2 > 0$. In this case, the fronts are scattered with the relative velocity

$$\Delta c = c_+ - c_- = \frac{\sqrt{1 + 8\eta\kappa/m^{*2}}}{2}(v_1 + v_3 - 2v_2). \tag{31.26}$$

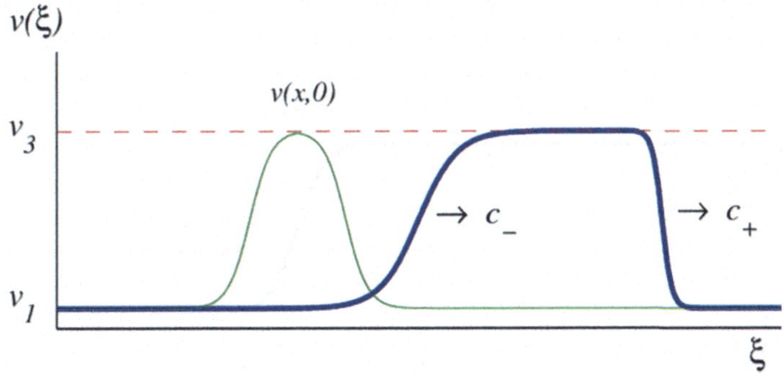

Fig. 31.3 Localized waves of hardening ($c = c_-$) and softening ($c = c_+$), which arise in the field of low temperatures during their formation from the local perturbation $v(x, 0)$

In this case, the propagation velocity of the front of the softening wave, under the abovementioned assumptions ($\eta\kappa/m^{*2} << 1$) and according to Eqs. (31.23) and (31.24), the softening wave front propagation velocity is equal to $c_+ \approx (v_1 + v_3)/2$ and the hardening wave front propagation velocity equals $c_- \approx v_2$.

In the case $v_1 + v_3 - 2v_2 < 0$, a softening wave does not arise, since it is damped by a hardening wave. Thus, there is a certain threshold value of the flow stress $\tau = \tau_c$ corresponding to the condition $\Delta c = 0$, below which ($\tau < \tau_c$) the perturbation is damped and the deformation of the material occurs macroscopically smoothly without the appearance of inhomogeneous wave structures.

For a threshold value of the load ($\tau = \tau_c$), an especial dissipative soliton can propagate, when the leading and trailing edges of the pulse move with the same velocity $c_+ = c_- = v_2$. The width of such a soliton is determined by the conditions of its occurrence.

For the considered initial problem, where the perturbations are given in the form of a pulse, the boundary conditions have the form: $v(\pm\infty, t) = v_1$. However, for example, under active loading of a crystal, when the plastic deformation rate is kept constant, the boundary conditions for this problem should be given in the following form: $v(-\infty, t) = v_3$, $v(\infty, t) = v_1$. Only the softening wave passing through the entire crystal and creating in the low-temperature region a localized zone of plastic strain, i.e., a wave of switching of the plastic deformation rate from the state $\dot{\varepsilon}_1 = b\rho v_2$ to the state $\dot{\varepsilon}_3 = b\rho v = 2b\rho_0 v_3^2/(v_3 + v_1) \approx 2b\rho v_3$ (Fig. 31.4), corresponds to these conditions.

The initial set of equations was considered at a constant load ($\tau = \text{const}$). In fact, it is achieved on a special machine keeping the constant tensile rate of the metal sample and it is described by the equation

$$\frac{\partial \tau}{\partial t} = G^*[\dot{\varepsilon}_0 - \frac{1}{L}\int_0^L \dot{\varepsilon}(x, t)\,\mathrm{d}x]. \tag{31.27}$$

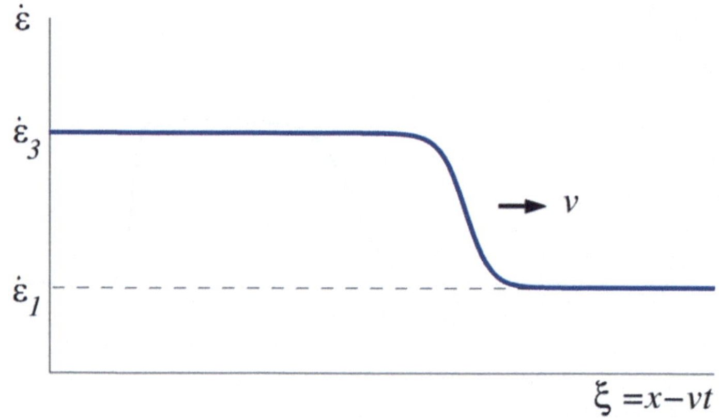

Fig. 31.4 Wave front of the localized plastic strain wave that arises in the low-temperature region

Here, $\dot{\varepsilon}_0$ is the specified rate of plastic strain in the slip strip, $G^* = \kappa h_0/\zeta_1 S$ is the effective modulus of elasticity, κ is the rigidity of the "machine-sample" system, h_0 and S are, accordingly, height and cross section of the sample, L is the length of the plastic strain zone, and ζ_1 is the geometric factor of the unit order.

The plastic strain dislocation velocity containing in (31.27) is defined as a solution of the stationary problem (31.25) for a self-made variable ξ, i.e., $\dot{\varepsilon}(x,t) = \dot{\varepsilon}(x - vt)$.

In the case, when $\tau \sim \tau_c$, the relative acceleration rate Δc of the fronts of randomly generated (at the boundary or in the bulk of the material) and propagating pulses is quite small and does not lead to a noticeable variation of the integral containing in Eq. (31.27). Therefore, a macroscopic change of a load τ does not occur. But if τ slightly exceeds τ_c, then a softening effect takes place, since the deformation zone continuously increases and the integral value grows in (31.27). In accordance with Eq. (31.27), it leads to decreasing of the load τ and, accordingly, to decreasing of Δc up to zero. However, such adjustment occurs with some delay and, as a result, for $\tau \sim \tau_c$ a pulsating regime for the softening zone of a plastic strain and load changes τ are possible.

As it was mentioned above, the original system has two small parameters provided by small quantities m^* and η, which determine the small width of the fronts of the hardening and softening waves in comparison with the parts of the plastic strain zone. Then, the average rate of plastic strain in the entire area $(0, L_p)$ can be approximately expressed by the following way:

$$\bar{\dot{\varepsilon}}(t) = \frac{1}{L}\int_0^L \dot{\varepsilon}(x - ct)\,dx \approx \dot{\varepsilon}_3 l_r + \dot{\varepsilon}_1 (L - l_r)]/L, \tag{31.28}$$

where the width l_r of the softening zone moves with the velocity $\Delta c = \mathrm{d}l_r/\mathrm{d}t$. In the stationary case ($\Delta c = 0$, $\bar{\dot{\varepsilon}}(t) = \dot{\varepsilon}_0$), the plastic deformation zone width l_r can be easily determined from (31.28):

$$l_r = L\frac{\dot{\varepsilon}_0 - \dot{\varepsilon}_1}{\dot{\varepsilon}_3 - \dot{\varepsilon}_1}. \quad (31.29)$$

The plastic zone l_r formed under such conditions can be divided into several zones with the same total "flow rate" as the large zone had. Instead of one large zone of plastic deformation with an increased rate of dislocation motion, several smaller zones can take place. The location and the quantity of such zones are determined by initial conditions. It should be noted that since in the general case the width L of the plastic deformation zone is a quantity, which can be adjusted to the given conditions of a crystal deformation, the considered mechanism can also determine the localization process of slip of dislocations in the direction of their transverse slip; i.e., it can create prerequisites for the deformation regime in the form of narrow lines and slip bands.

It should be noted that, since the considered dislocation motion in the slip band is of an over-barrier type [13, 14], then, according to hypothesis [21], this fact can contribute to the emergence of thermally stimulated avalanches and lead to further instability of plastic deformation—to serrated deformation.

31.3 The Thermodynamic Model

Several hypotheses have been given to explain a physical nature of serrated deformation at low temperatures [19, 21]. At present, a hypothesis explaining the largest number of experimental facts is the hypothesis of the thermal nature of jumps [21]. For example, it was found [17] that, in the range of helium temperatures and plastic strain rates $\dot{\varepsilon} = 10^{-2} - 10^{-5}$, each stress jump was accompanied by an almost delta-shaped temperature growing up to values of the order of 50 K, and the larger the jump amplitude, the higher the temperature. The most strict criterion for thermomechanical instability of deformation at low temperatures was formulated in [19].

Despite the rigor of the criterion obtained in [19], description of serrated deformation and the dynamics of its occurrence in this work and in some other works [16] have not been sufficiently studied. Therefore, in the present work, within the scope of the autowave model, we will consider the instability of plastic flow on the base of the analysis of nonlinear dynamics of serrated deformation.

First, we shall study a model that is often used to establish a criterion for the instability of the plastic deformation during thermally activated sliding of dislocations [15, 19, 25]. Then, for sufficiently thin metal samples ($R << L$, where R is the radius of the cylindrical sample and L is the length of the sample), the processes of deformation and thermal conductivity (inhomogeneous along the axis of the cylinder) can be described by the following equations [15, 25]:

$$c\frac{\partial T}{\partial t} = \kappa\frac{\partial^2 T}{\partial x^2} - \frac{2h}{R}(T - T_0) + \sigma\dot{\varepsilon}. \quad (31.30)$$

$$\frac{\partial \varepsilon(t)}{\partial t} = \dot{\varepsilon}, \qquad \dot{\varepsilon} = \nu \exp[-W/k_B T], \tag{31.31}$$

$$\frac{\partial \sigma}{\partial t} = G^* \left(\dot{\varepsilon}_0 - \frac{1}{L} \int_0^L \dot{\varepsilon}(x, t) \mathrm{d}x \right). \tag{31.32}$$

Here, Eq. (31.30) is a thermal conductivity equation, where T is the temperature of the metal sample, κ is the thermal conductivity coefficient, T_0 is the environment temperature, h is the heat exchange coefficient, σ is the external stress, and c is the heat capacity of the sample, which, like in [19], is considered as a constant value for conveniency of analysis.

Thermal activation mode of plastic strain is characterized by Eq. (31.31), where $\dot{\varepsilon}$ is the local rate of plastic strain in deformation areas, ε is a plastic strain value, W is the activation energy, ν is the pre-exponential multiplier, and k_B is Boltzmann constant.

Equation (31.32) determines the dynamics of the system "a sample—a loading device" at a given constant plastic deformation rate $\dot{\varepsilon}_0$, where $G^* = k_m L/S$ is the effective module of the "sample-machine" system, k_m is the stiffness of the machine, and S is the cross-sectional area of the sample. Low-temperature serrated deformation is similar, in its external manifestations, to mechanical relaxation self-oscillations. A necessary condition for the occurrence of relaxation oscillations in a mechanical system is the presence of a section of negative velocity sensitivity of a friction force. For the plastic deformation case in question, this condition is equivalent to a negative dependence of the deforming stress on the temperature or the strain rate. Experimental studies of speed dependencies of deforming stress in many metals and alloys have shown that the speed sensitivity decreases, when the temperature decreases, and becomes negative at helium temperatures [21].

Equations (31.30)–(31.32) refer to a self-oscillating system, if the function

$$\sigma = \left(\frac{2h}{\nu R}\right)(T - T_0) \exp\left[\frac{W}{kT}\right] \tag{31.33}$$

derived due to equating to zero of the right-hand side of Eq. (31.32) has a descending section; i.e., in a certain temperature interval, the condition $\partial\sigma/\partial T < 0$ is valid and, accordingly, $\partial\sigma/\partial\dot{\varepsilon} = (\partial\sigma/\partial T)(\partial T/\partial\dot{\varepsilon}) < 0$ (as it follows from (31.50), inequality $\partial\dot{\varepsilon}/\partial T > 0$ is always valid).

From the condition $\partial\sigma/\partial T = 0$, it is possible to obtain an equation

$$\left(\frac{T}{T_0}\right)^2 - \frac{W}{kT_0}\left(\frac{T}{T_0}\right) = 0, \tag{31.34}$$

which roots have the form:

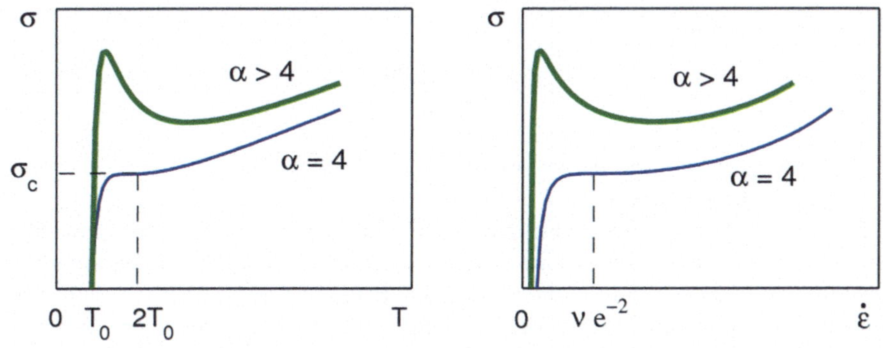

Fig. 31.5 Functions $\sigma = \sigma(T)$ and $\sigma = \sigma(\dot{\varepsilon})$ for various values of the parameter $\alpha = W/k_B T$

$$T_{1,2} = T_0\left[\frac{\alpha}{2} \pm \sqrt{\left(\frac{\alpha}{2} - 1\right)^2 - 1}\right], \tag{31.35}$$

where $\alpha = W/k_B T_0$ is the parameter of a normalized activation energy.

Unifying of roots in (31.35) takes place, when $\alpha = 4$, that corresponds to the inflection point on the curve $\sigma = \sigma(T)$. Thus, for

$$\alpha > \alpha_c = 4 \tag{31.36}$$

there is N-shaped dependence $\sigma(T)$ with the section of the negative dependence of the deforming stress on temperature (Fig. 31.5a). Besides, at the point of inflection the critical stress is related to temperature by the relation

$$\sigma_c = \sigma(2T_0). \tag{31.37}$$

Since the plastic deformation rate $\dot{\varepsilon}$ depends monotonically on the temperature T, then the graph of function $\sigma = \sigma(\dot{\varepsilon})$ also has N shape (Fig. 31.5b).

For further analysis, we write Eqs. (31.30)–(31.32) in a dimensionless form by introducing dimensionless variables

$$u = \frac{\dot{\varepsilon}}{\nu}, \qquad \theta = \frac{T}{T_0}, \qquad \tau = \frac{\sigma}{G^*} \tag{31.38}$$

and time $t' = t\nu$. As a result, these equations yield

$$\frac{d\theta}{dt'} = K\frac{\partial^2\theta}{\partial x'^2} + \mu u\tau - \beta(\theta - 1), \tag{31.39}$$

$$\frac{d\varepsilon}{dt'} = u, \quad u = \exp\left[-\frac{\alpha}{\theta}\right], \tag{31.40}$$

$$\frac{d\tau}{dt'} = u_0 - \int_0^1 u dx'. \qquad (31.41)$$

Here, the dimensionless parameters $\mu = G^*/cT_0$, $\beta = 2h/c\nu R$, $u_0 = \hat{\varepsilon}_0/\nu$, and $K = \kappa/cL^2\nu$ are introduced.

It is possible to show that the instability condition can be reduced to the inequality $\alpha > 4$ and it is a necessary condition, but not sufficient. A sufficient condition for instability is location of $\theta = \alpha/\ln(\nu/\dot{\varepsilon}_0)$ in the range of values, where $\partial\sigma/\partial\theta < 0$, or location of the specified strain rate $\hat{\varepsilon}_0$ in the range of values, where $\partial\sigma/\partial\dot{\varepsilon} < 0$.

Let us analyze the stability of a stationary homogeneous solution of Eqs. (31.39)–(31.41):

$$u = u_0, \quad \theta_0 = \alpha/\ln u_o, \quad \tau = \left(\frac{\beta}{\mu}\right)(\theta_0 - 1)/u_0 \qquad (31.42)$$

After linearization of Eqs. (31.39)–(31.41) in the neighborhood of u_0 and θ_0 and under assumption $u, \theta, \tau \propto \exp(\lambda t + ikx)$, it is possible to obtain the following dispersion equation

$$\lambda^2 + (k^2 K + \beta - \mu\tau_0 u_{\theta'})\lambda + \mu u_{\theta'}\delta(k) = 0, \qquad (31.43)$$

where $\delta(k)$ is the Dirac delta function and $u_{\theta'} = \alpha u_0/\theta_0^2$. In Eqs. (31.39)–(31.41), instability takes place, if $\mathrm{Re}\{\lambda(k)\} > 0$ for at least one of the roots of the dispersion equation. This inequality is valid for

$$\tau > \tau_c = \frac{\beta\theta_0^2}{\mu\alpha u_0}\left(1 + \frac{k^2 K}{\beta}\right) \qquad (31.44)$$

Now let us consider the properties of the solutions of Eqs. (31.39)–(31.41). Numerical investigations show that the plastic flow regime is monotonic, when $\alpha < 4$, but for $\alpha > 4$ there appear uniform regular fluctuations of load, temperature, and strain (Fig. 31.6) that correspond to the limit cycle on the phase plane $\sigma - T$ (Fig. 31.6a). Moreover, if a sawtooth dependence on time takes place for the deforming stress (Fig. 31.6b), then there are regular bursts for the temperature (Fig. 31.6c) and a stepwise dependence for deformation (Fig. 31.6d).

Here, two remarks should be noticed. First, since the oscillations of the system variables are homogeneous ($k = 0$), then condition (31.44) takes on the form

$$\tau > \tau_c = \frac{\beta\theta_0^2}{\mu\alpha u_0}. \qquad (31.45)$$

In this case, at the inflection point $\tau = \tau_c$ the equality $\theta_0 = 2$ is achieved, and accordingly, at this point $\tau_0 = \beta/\mu u_0$. Then, condition (31.45) can also be written

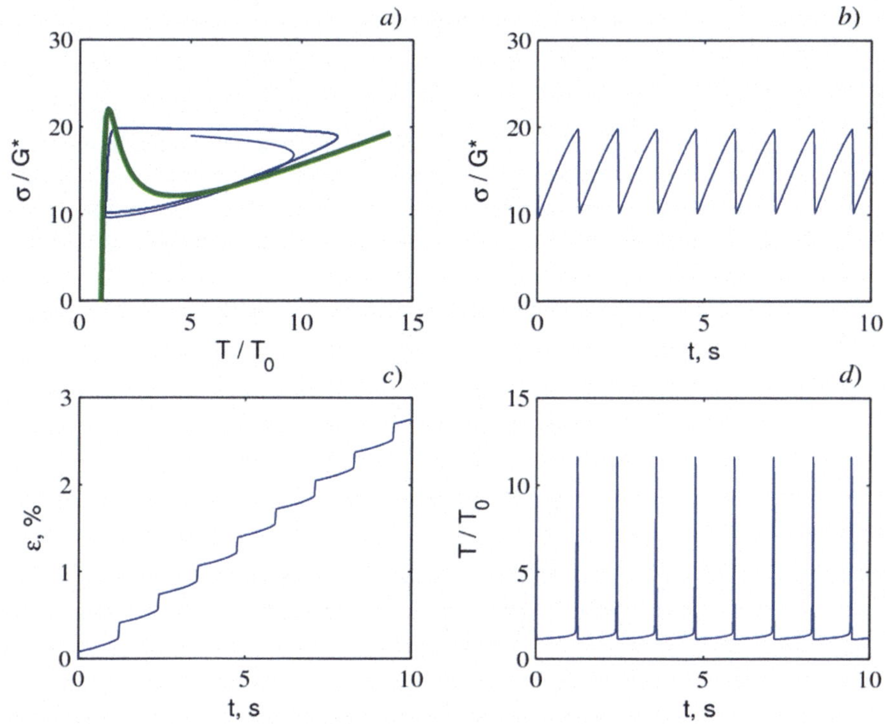

Fig. 31.6 Phase portrait **a** of Eqs. (31.39)–(31.41) for $\alpha = 5.8$, when the isocline $T = T(\dot{\varepsilon}_0)$ is located on the descending section of isocline branch $\sigma = \sigma(T)$ (it corresponds to Eq. 31.33) and the dependences of applied stress σ, temperature T, and deformation ε on time (**b**, **c**, **d**). The graphs have been plotted for the following parameter values: $\mu = 0.5$, $\beta = 1$, $u_0 = 0.015$, $K = 0.01$

in the form

$$\alpha = W/k_B T > 4. \tag{31.46}$$

Inequality (31.46) coincides with the earlier obtained condition (31.36). It follows from (31.46) that the thermomechanical instability is possible only if the average temperature of a sample is less than

$$T < W/4k_B \approx 15^0 K. \tag{31.47}$$

Secondly, it is easy to show that the instability condition (31.45) for the initial dimensional variables takes the form

$$\sigma > \sigma_c = \frac{2hk_B T_0^2}{R\dot{\varepsilon}_0 W}. \tag{31.48}$$

Inequality (31.48) coincides with the instability condition obtained in [19] under the isothermal boundary conditions.

However, the solutions obtained in the framework of the model (31.30)–(31.32) are not completely in agreement with the experimental data, as, in reality, load oscillations are usually not regular, but quasi-stochastic and heterogeneous. In order to describe this behavior, it is necessary to take into account the effect of elastic correlation of neighboring deformation areas (e.g., sliding strips) [11]. As a result of this effect, elastic perturbations $\sigma_i \sim \partial^2_{xx}\varepsilon$ arise and create heterogeneities of stresses along the sample axis.

Then, the original set of equations in the correlation of deformation zones takes the form

$$c\frac{\partial T}{\partial t} = \kappa\frac{\partial^2 T}{\partial x^2} - \frac{2h}{R}(T - T_0) + (\sigma + \sigma_i)\dot{\varepsilon}. \tag{31.49}$$

$$\frac{\partial \varepsilon(t)}{\partial t} = \dot{\varepsilon}, \quad \dot{\varepsilon} = \nu \exp[-W/k_B T], \tag{31.50}$$

$$\frac{\partial \sigma}{\partial t} = G^*\left(\dot{\varepsilon}_0 - \frac{1}{L}\int_0^L \dot{\varepsilon}(x,t)\mathrm{d}x\right). \tag{31.51}$$

$$\frac{\partial \sigma_i}{\partial t} = -\frac{\sigma_i}{t_a} + \gamma_1\frac{\partial^2 \dot{\varepsilon}}{\partial x^2}, \tag{31.52}$$

where parameter $\gamma_1 \approx Gd^2$ is a measure of elastic correlation of neighboring deformation zones, d is their characteristic width, and t_a is the characteristic relaxation time of elastic perturbations.

Numerical investigation of Eqs. (31.49)–(31.52) was carried out after they had been dimensionless. Analysis of solutions shows that irregular dynamics of the system is mainly controlled by the parameter $S = Gd^2/G^*L^2$ of elastic correlation that occurs, when its threshold value is exceeded ($S > S_c = 0.045$). The length of the sample was selected equal $L = 3$ cm and the width of the deformation zones $d = 10\mu$m.

In the case of small values of S, a regular mode is realized in the system similarly to the scenario shown in Fig. 31.6. When the elastic correlation parameter ($S > S_c = 0.045$) grows, the mode of change of load, temperature, and plastic deformation rate becomes irregular (Fig. 31.7).

This regime of a plastic strain and temperature variations can be most clearly observed on the phase portrait of variables σ and T (Fig. 31.7a), where the phase paths are unstable in a limited region. At the same time, the value of load drops is clearly correlated with the value of temperature bursts in deformation zones that was often observed experimentally [22]. Irregular behavior of variables σ, T, and $\dot{\varepsilon}$ presents in the phase space of these variables an attractive set of paths—*strange attractor* [18], on which all the paths belonging to it are unstable (Fig. 31.8).

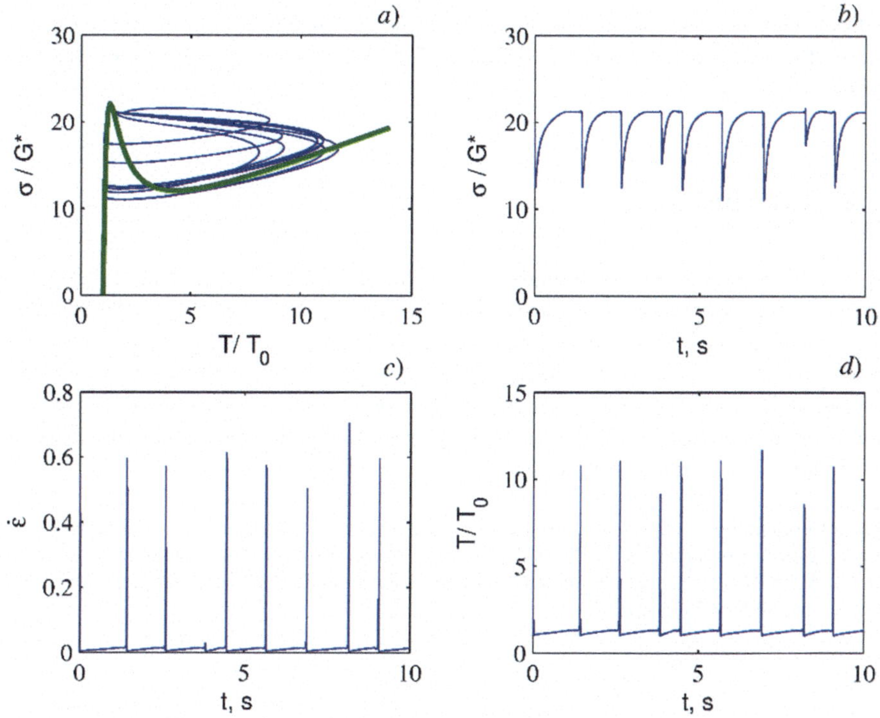

Fig. 31.7 Phase portrait of Eqs. (31.49)–(31.52) for $\alpha = 5.8$ on the plane of variables $\sigma - T$ (**a**) in dependence of applied stress σ, strain rate $\dot{\varepsilon}$, and temperature T on time t (**b**, **c**, **d**). Graphs are plotted for $\mu = 0.5$, $\beta = 1$, $u_0 = 0.015$, $K = 0.01$, $S = 0.05$, and $\gamma = \nu/t_a = 0.01$.

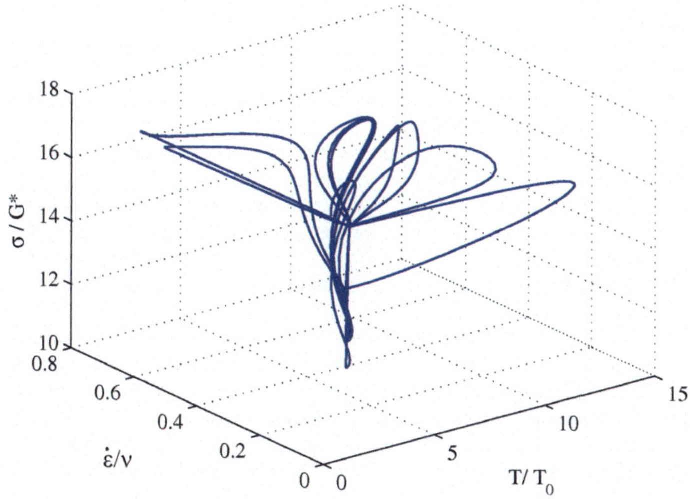

Fig. 31.8 Phase space of the system described by Eqs. (31.49)–(31.52), in which the phase paths behave as a strange attractor

Such a behavior is typical for stochastic systems [6]. With increase of the plastic accommodation parameter ($\gamma > 0.01$), the oscillatory process becomes regular again.

31.4 Conclusion

Thus, analytical and numerical analysis of the model with a heat-activated regime of the plastic flow showed that the plastic deformation mode can become unstable under certain conditions. Taking into account elastic correlation of neighboring deformation zones, there arise elastic perturbations $\sigma_i \sim \partial_{xx}^2 \varepsilon$ leading to irregular dynamics of the system. In this case, the mode of change of load, temperature, and plastic deformation rate becomes stochastic and presents in the phase space of these variables an attractive set of paths—the strange attractor.

Acknowledgements The research was carried out within the Russian state task for fundamental scientific research for 2019-2022 (the topic No. 0035-2019-0026, the state registration No. 01201458049) and under the financial support of the Russian Foundation for Basic Research (projects No 18-02-00444-a).

References

1. Argon, A.S.: Strain avalanches in plasticity. Philos. Magazine. **93**, 3795–3808 (2013)
2. Ashby, M.F.: The deformation of plastically non-homogeneous materials. Philos. Magazine. **21**, 399–424 (1970)
3. Basinski, Z.S.: The instability of plastic flow of metals at very low temperatures. Proc. Roy. Soc. **240A**, 2299–2309 (1957)
4. Dimiduk, D.M., Woodward, C., LeSar, R., Uchic, M.D.: Scale-free intermittent flow in crystal plasticity. Science **312**, 1188–1190 (2006)
5. Estrin,Y., Kubin, L.P.: Micro- and macroscopic aspects of unstable plastic flow. Phase Transform. 185–202, London (1986)
6. Haken, H.: Pattern formation and chaos in synergetic system. Phys. Scr. **9**, 111–118 (1985)
7. Kubin, L.P., Estrin, Y.: Thermal effects in low-temperature deformation. Cryst. Res. Technol. **19**, 853–862 (1984)
8. Kubin, L.P., Estrin, Y.: Strain nonunifomities and plastic instabilities. Rev. Phys. Appl. **28**, 573–583 (1988)
9. Landau, L.D., Lifshitz, E.M.: Fluid Mechanics. Pergamon Press, London (1960)
10. Landau, L.D., Lifshitz, E.M.: Theory of Elasticity. Pergamon Press, London (1981)
11. Lebyodkin, M.A., Dunin-Barkovskii, L.R., Bobrov, V.S., Grпger, V.: Statistical aspects of low temperature discontinuous deformation. Scripta Metall. et Mater. **33**, 773–780 (1995)
12. Luft, A., Richter, J., Schlaubitz, K.: Work softening and microstructural instability of predeformed molybdenum single crystals. Mater. Sci. Eng. **20**, 113–122 (1975)
13. Malashenko, V.V.: Influence of collective effects on the dynamic behavior of a single edge dislocation in a crystal with point defects. Phys. Solid State **49**, 82–86 (2007)
14. Malashenko, V.V.: Dynamic instability of dislocation motion at high-strain-rate deformation of crystals with high concentration of point defects. Phys. Solid State **57**, 2461–2463 (2015)

15. Malygin, G.A.: The low-temperature instability of plastic flow. Phys. Stat. Sol. (b) **61**, 445–448 (1974)
16. Malygin, G.A.: Analysis of discontinuous-deformation parameters of metals at low temperatures. Phys. Metals Metallography. **81**, 235–245 (1996)
17. Ogata, T., Ishikava, K., Hiraga, K., Nagai, K., Yuri, T.: Temperature rise during the tensile test in superfluid helium. Cryogenics **25**, 444–446 (1985)
18. Ott, E.: Strange attractors and chaotic motions of dynamical systems. Rev. Mod. Phys. **83**, 655–671 (1981)
19. Petukhov, B.V., Estrin, Y.Z.: Criterion of instability of a thermally activated plastic deformation. Solid State Phys. (A) **17**, 2041–2044 (1975)
20. Pustovalov, V.V.: Effect of superconducting transition on the low-temperature jumplike deformation of metals and alloys. Low Temp. Phys. **26**, 375–390 (2000)
21. Pustovalov, V.V.: Serrated Deformation of Metals and Alloys at Low Temperatures. Low Temperature Physics. **34**, 683–723 (2008)
22. Rybin, V.V.: Regularities of mesostructures development in metals in the course of plastic deformation. Probl. Mater. Sci. **33**, 9–28 (2003)
23. Sarafanov, G.F.: On the theory of dislocation patterning: evolution of an ensemble of screw and edge dislocations. Phys. Met. Metall. **85**, 276–281 (1998)
24. Sarafanov, G.F., Maksimov, I.L.: Dislocation pattern formation in the course of plastic deformation. Met. Mater. **5**, 323–327 (1999)
25. Zaiser, M.: Spatio—temporal aspects of low—temperature thermomechanical instabilities: A model based dislocation dynamics. Appl. Phys. A **57**, 143–151 (1993)

Chapter 32
Modeling of Attenuation Processes in Piezoceramic Bodies

Alexander S. Skaliukh

Abstract A method has been developed for determining the tangents of the loss angles of polycrystalline ferroelectric materials, based on the hysteretic attenuation model proposed by E. S. Sorokin and Ya. G. Panovko. A three-dimensional polarization model was constructed, which allows one to find the dielectric and deformation hysteresis loops at arbitrary electrical and mechanical loads. Investigations were carried out and the functional dependences of physical modules on the residual polarization, and deformation parameters were established. An approximation of the hysteresis dependences by analytical functions of a special form is proposed for small amplitudes of external impacts. Constitutive relations are obtained, similar to the Maxwell model, from which in the case of harmonic oscillations it is possible to calculate the tangents of the loss angles through the area of the hysteresis loops.

Keywords Tangents of the loss angles · Polycrystalline ferroelectric materials · Polarization model · Constitutive relations · Residual parameters · Approximation · Hysteresis loops · Strain tensor · Polarization vector

32.1 Introduction

The basis for modeling the oscillations of piezoelectric transducers is harmonic and modal analysis. The frequency-dependent characteristics represent the interest and mainly have the main goal of the study. The most preferred and informative may be the amplitude–frequency characteristics (AFC). However, a simplified approach related to neglecting the attenuation phenomena can lead to a discrepancy between the calculated and experimental data. In this case, it is necessary to correct mathematical models, in particular, to take into account more subtle factors and phenomena that affect their work. For example, if attenuation mechanisms are ignored in the calculation model, then at the resonant frequencies the frequency response increases

---

A. S. Skaliukh (✉)
Southern Federal University, Milchakova str. 8A, Rostov-on-Don, Russia
e-mail: a.s.skaliukh@gmail.com

H. Altenbach et al. (eds.), *Multiscale Solid Mechanics*,
Advanced Structured Materials 141,
https://doi.org/10.1007/978-3-030-54928-2_32

unlimitedly, while experiments show that there is no such increase. Therefore, vibration damping mechanisms are an important and interesting problem for research and present the subject of additional research.

The attenuation mechanisms themselves have not yet been fully studied, and the proposed mathematical models sometimes look artificially introduced or even unfounded. This work does not pretend to be a complete study, but presents an interesting approach that takes into account hysteretic attenuation, i.e., attenuation arising from small fluctuations due to changes in irreversible parameters. It was found previously that physical modules depend on residual parameters that change, subjected to hysteretic dependencies. However, the presence of hysteresis leads to the dissipation of energy, and therefore, attenuation occurs, which is proportional to the area of the hysteresis loop. Thus, the mathematical modeling of the attenuation processes is inextricably linked with the modeling of irreversible processes polarization—depolarization in the three-dimensional case. And although a harmonic vibrations are caused by small external electric and mechanical fields, the physical characteristics of the material, which include elastic, piezoelectric and dielectric quantities, being functions of residual parameters, albeit slightly, but also change in a hysteretic manner. To describe this phenomenon, comprehensive studies were carried out in three main directions. Firstly, a model was constructed to determine the residual parameters, which allows finding the deformation and dielectric hysteresis loops due to superimposed electric fields and mechanical stresses [1]. Secondly, the functional dependences were obtained for physical modules on residual parameters [2, 3], such as the polarization vector and strain tensor. Thirdly, approximations of nonlinear functions were used as the approximations of a special form [4], which made it possible to obtain constitutive relations of the type of operators of the Maxwell model, with subsequent determination of the loss tangents. It is interesting to note that the tangents of the loss angles are also found in [5], but in a completely different way.

32.2 Polarization Model of Polycrystalline Ferroelectrics

The first part of the study, i.e., the polarization–depolarization model, is mainly described in [1]. As in many models describing irreversible processes, the parameters of polarization and deformation of a representative volume are divided into the induced (reversible) and residual (irreversible) parts: $\mathbf{P} = \mathbf{P}_e + \mathbf{P}_0,\ \boldsymbol{\varepsilon} = \boldsymbol{\varepsilon}_e + \boldsymbol{\varepsilon}_0$. The representative volume contains many domains with vectors of spontaneous polarization $\mathbf{p}_s$, and, accordingly, with spontaneous deformations $\boldsymbol{\varepsilon}_s$. Spontaneous polarization vectors and spontaneous strain tensors included in a representative volume depend on the angles $\phi,\ \psi$ of the local coordinate system of the unit sphere. Electric and mechanical fields can deform the walls of domains, or they can rotate these vectors. The result of the deformation of the walls of the domains is the appearance $\mathbf{P}_e,\ \boldsymbol{\varepsilon}_e$, and the result of the rotation is $\mathbf{P}_0,\ \boldsymbol{\varepsilon}_0$, and

$$\mathbf{P}_0 = \frac{1}{N}\sum_{k=1}^{N} (\mathbf{p}_s)_k, \quad \boldsymbol{\varepsilon}_0 = \frac{1}{N}\sum_{k=1}^{N} (\boldsymbol{\varepsilon}_s)_k.$$

The main idea of the polarization model is to represent the general process in the form of two components parts, one of which describes the rotations of the spontaneous polarization vectors, and the second takes into account the energy costs of such rotations. According to the first part, turns can be described by utmost dependencies

$$\boldsymbol{\varepsilon}_\infty = \frac{\int_0^{2\pi} \mathrm{d}\phi \int_0^{\pi} e^u \boldsymbol{\varepsilon}_s \sin\psi\, \mathrm{d}\psi}{\int_0^{2\pi} \mathrm{d}\phi \int_0^{\pi} e^u \sin\psi\, \mathrm{d}\psi}, \qquad P_\infty = \frac{\int_0^{2\pi} \mathrm{d}\phi \int_0^{\pi} e^u \mathbf{p}_s \sin\psi\, \mathrm{d}\psi}{\int_0^{2\pi} \mathrm{d}\phi \int_0^{\pi} e^u \sin\psi\, \mathrm{d}\psi}.$$

According to the second part, an energy balance is compiled, in which the work of external fields in the real process of polarization and deformation is equated to the energy costs associated with the destruction of the mechanisms for fixing the domain walls, plus the work of the fields to rotate the spontaneous polarization vectors. The result is a system of equations in differentials (32.1)

$$\begin{aligned} \mathrm{d}\mathbf{P}_0 &= (\mathbf{P}_\infty - \mathbf{P}_0)\frac{|\mathrm{d}\mathbf{E}_{\mathrm{ef}}|}{n_E}, \quad \mathbf{P}_0 = \mathbf{P}_0^*, \quad \mathbf{E} = \mathbf{E}^*, \\ \mathrm{d}\boldsymbol{\varepsilon}_0 &= (\boldsymbol{\varepsilon}_\infty - \boldsymbol{\varepsilon}_0)\frac{|\mathrm{d}(\boldsymbol{\sigma}_{\mathrm{ef}})_{\mathrm{I}} - \mathrm{d}(\boldsymbol{\sigma}_{\mathrm{ef}})_{\mathrm{III}}|}{n_\sigma}, \quad \boldsymbol{\varepsilon}_0 = \boldsymbol{\varepsilon}_0^*, \quad \boldsymbol{\sigma} = \boldsymbol{\sigma}^*, \end{aligned} \tag{32.1}$$

from which increments of irreversible parameters are found. The parameters included in (32.1) are defined as follows:

$$\begin{aligned} &u = \frac{-\mathbf{E}_{\mathrm{ef}} \cdot \mathbf{P}_S/p_S}{a} - \frac{\boldsymbol{\sigma}_{\mathrm{ef}} : \boldsymbol{\varepsilon}_S/\varepsilon_S}{b}, \quad a = \frac{kT}{p_S \Omega_{\mathrm{cr}}}, \quad b = \frac{kT}{\varepsilon_S \Omega_{\mathrm{cr}}}, \\ &\boldsymbol{\sigma}_{\mathrm{ef}} = \boldsymbol{\sigma} + \beta : \boldsymbol{\varepsilon}_0, \quad \mathbf{E}_{\mathrm{ef}} = \boldsymbol{\sigma} + \alpha : \mathbf{P}_0, \end{aligned} \tag{32.2}$$

where $\mathbf{p}_S, \boldsymbol{\varepsilon}_S, k, T, \Omega_{\mathrm{cr}}$—the spontaneous polarization vector, spontaneous polarization tensor, Boltzmann constant, absolute temperature and atomic cell volume, respectively; $\mathrm{d}(\boldsymbol{\sigma}_{\mathrm{ef}})_{\mathrm{I}}, \mathrm{d}(\boldsymbol{\sigma}_{\mathrm{ef}})_{\mathrm{III}}$—the largest and smallest principal values of the tensor of increments of effective stresses. The main parameters of the model are $\alpha, \beta, a, b, n_E, n_\sigma, p_s, \varepsilon_s$. For each type of ceramic, these numbers are selected from the condition that the calculated and experimental hysteresis curves coincide, just as in the linear Hooke model where Young's modulus and Poisson's ratio are determined experimentally. Once found, these characteristics become known numbers and do not change during oscillations.

The proposed model is noteworthy in that there is no need to store information about the position of each vector of spontaneous polarization, but it is enough to operate only with integral characteristics: $\mathbf{P}_0, \ \boldsymbol{\varepsilon}_0$ which, moreover, form the physical properties of the material.

32.3 Dependence of Physical Modules on Residual Parameters

The second part of the study is based on the results of [3], in which, on the basis of the first and second principles of thermodynamics, the constitutive relations for the induced (reversible) components of the strain tensor and the polarization vector (32.3) are obtained:

$$\begin{aligned} \boldsymbol{\varepsilon}_e &= \mathbf{S}(\boldsymbol{\varepsilon}_0) : \boldsymbol{\sigma} + \mathbf{d}^T(\mathbf{P}_0) \cdot \mathbf{E}, \\ \mathbf{D}_e &= \mathbf{d}(\mathbf{P}_0) : \boldsymbol{\sigma} + \boldsymbol{\epsilon}(\boldsymbol{\varepsilon}_0) \cdot \mathbf{E}, \end{aligned} \tag{32.3}$$

and linear dependences of elastic compliance, piezoelectric modules and permittivities on residual parameters were established (32.4):

$$\begin{aligned} \mathbf{S}(\boldsymbol{\varepsilon}_0) &= \mathbf{S}_0 + \mathbf{K} : \boldsymbol{\varepsilon}_0, \\ \mathbf{d}(\mathbf{P}_0) &= \mathbf{N}_1 \cdot \mathbf{P}_0, \\ \boldsymbol{\epsilon}(\boldsymbol{\varepsilon}_0) &= \boldsymbol{\epsilon}_0 + \mathbf{M} : \boldsymbol{\varepsilon}_0, \end{aligned} \tag{32.4}$$

where $\mathbf{S}_0, \boldsymbol{\epsilon}_0$—elastic compliance and permittivity of a thermally depolarized body; $\mathbf{K},\ \mathbf{N}_1,\ \mathbf{M}$—some constant tensors, the components of which will be discussed below. In the future, it is convenient to use the Voigt matrix representations, for which it is necessary to introduce vectors

$$\begin{aligned} &\hat{\sigma} = \{\hat{\sigma}_1\ \hat{\sigma}_2\ \ldots\ \hat{\sigma}_6\}^T \equiv \{\sigma_{11}\ \sigma_{22}\ \ldots\ \sigma_{12}\}^T; \quad \mathbf{E} = \{E_1\ E_2\ E_3\}^T; \\ &\hat{\varepsilon}_e = \{\hat{\varepsilon}_1\ \hat{\varepsilon}_2\ \ldots\ \hat{\varepsilon}_6\}^T \equiv \{\varepsilon_{e\,11}\ \varepsilon_{e\,22}\ \ldots\ \varepsilon_{e\,12}\}^T; \quad \mathbf{D}_e = \{D_{e1}\ D_{e2}\ D_{e3}\}^T. \end{aligned}$$

Then, in the system of axes, when Oz coincides with the direction of the residual polarization vector instead of (32.3), we have the matrix relations:

$$\begin{aligned} \boldsymbol{\varepsilon}_e &= \hat{\mathbf{S}}(\boldsymbol{\varepsilon}_0) \cdot \boldsymbol{\sigma} + \hat{\mathbf{d}}^T(\mathbf{P}_0) \cdot \mathbf{E}, \\ \mathbf{D}_e &= \hat{\mathbf{d}}(\mathbf{P}_0) \cdot \boldsymbol{\sigma} + \hat{\epsilon}(\boldsymbol{\varepsilon}_0) \cdot \mathbf{E}, \end{aligned}$$

in which, $\hat{\mathbf{S}}(\boldsymbol{\varepsilon}_0), \hat{\mathbf{d}}(\mathbf{P}_0), \hat{\epsilon}(\boldsymbol{\varepsilon}_0)$—matrixes of elastic, piezoelectric and dielectric characteristics:

$$\hat{\mathbf{S}}(\boldsymbol{\varepsilon}_0) = \begin{pmatrix} \hat{S}_{11} & \hat{S}_{12} & \hat{S}_{13} & 0 & 0 & 0 \\ \hat{S}_{12} & \hat{S}_{11} & \hat{S}_{13} & 0 & 0 & 0 \\ \hat{S}_{13} & \hat{S}_{13} & \hat{S}_{33} & 0 & 0 & 0 \\ 0 & 0 & 0 & \hat{S}_{44} & 0 & 0 \\ 0 & 0 & 0 & 0 & \hat{S}_{44} & 0 \\ 0 & 0 & 0 & 0 & 0 & 2(\hat{S}_{11} - \hat{S}_{12}) \end{pmatrix} \quad (32.5)$$

$$\hat{\mathbf{d}}(\mathbf{P}_0) = \begin{pmatrix} 0 & 0 & 0 & 0 & \hat{d}_{15} & 0 \\ 0 & 0 & 0 & \hat{d}_{15} & 0 & 0 \\ \hat{d}_{31} & \hat{d}_{31} & \hat{d}_{33} & 0 & 0 & 0 \end{pmatrix}, \quad \hat{\epsilon}(\boldsymbol{\varepsilon}_0) = \begin{pmatrix} \hat{\epsilon}_{11} & 0 & 0 \\ 0 & \hat{\epsilon}_{11} & 0 \\ 0 & 0 & \hat{\epsilon}_{33} \end{pmatrix}.$$

Matrix components can be represented as

$$\hat{S}_{\alpha\beta} = \hat{S}_{0\,\alpha\beta} + k_{\varepsilon\,\alpha\beta}\,(\varepsilon_0)_{III}, \quad \alpha, \beta = 1, \dots, 6$$

$$\hat{d}_{mn} = k_{p\,mn} P_0, \quad \hat{\epsilon}_{mn} = \hat{\epsilon}_{omn} + k_{\varepsilon\,mn}\,(\varepsilon_0)_{III}, \quad m, n = 1, \dots, 3$$

where $\hat{S}_{0\,\alpha\beta}$, $\hat{\epsilon}_{omn}$—the components of the corresponding matrices of elastic compliance and permittivity in an unpolarized state. P_0, $(\varepsilon_0)_{III}$—the modulus of the residual polarization vector and the largest principal value of the residual strain tensor. Odds included here

$$k_{\varepsilon\,\alpha\beta} = \frac{\hat{S}^{\max}_{\alpha\beta} - \hat{S}_{0\,\alpha\beta}}{\varepsilon_{\text{sat}}}, \qquad k_{p\,mn} = \frac{\hat{d}^{\max}_{mn}}{p_{\text{sat}}}, \qquad k_{\varepsilon\,mn} = \frac{\hat{\epsilon}^{\max}_{mn} - \hat{\epsilon}_{o\,mn}}{\varepsilon_{\text{sat}}}$$

are determined through the maximum possible values achieved in the state of saturation, and the values in the unpolarized state.

When considering harmonic oscillations, we assume that the external parameters, which include the electric field and mechanical stresses, change according to the harmonic law in time: $\boldsymbol{\sigma} = \tilde{\sigma}\cos\omega t$, $\mathbf{E} = \tilde{\mathbf{E}}\cos\omega t$. For large amplitudes $\tilde{\sigma}$, $\tilde{\mathbf{E}}$, we will have large loops of dielectric and deformation hysteresis, and for small values of them, small loops. Note that both large and small loops are an informative base for further research. To begin with, we consider the large hysteresis loops shown in Figs. 32.1 and 32.2.

Circles show possible cases of residual polarization and residual deformation when removing external loads. (In real polarization processes, it makes no sense to change the field cyclically: just sufficient set the maximum value, take a shutter speed and turn off the field). When operating piezoelectric ceramic transducers, the external electric field and mechanical stresses are low intensity; i.e., their amplitudes under harmonic modes of oscillations do not go beyond the circles noted above. However, even for such regimes, irreversible repolarization processes take place. In addition, this leads to the situations depicted in Figs. 32.3 and 32.4, where, for example, the upper point of the dielectric hysteresis loop is selected, and the right point of the deformation hysteresis loop.

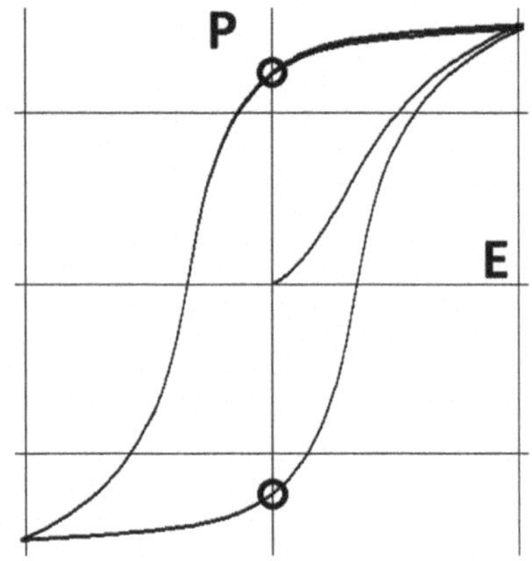

Fig. 32.1 Dielectric loop

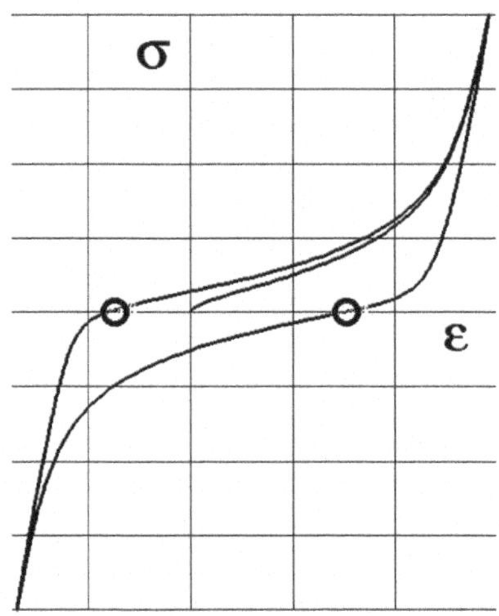

Fig. 32.2 Strain loop

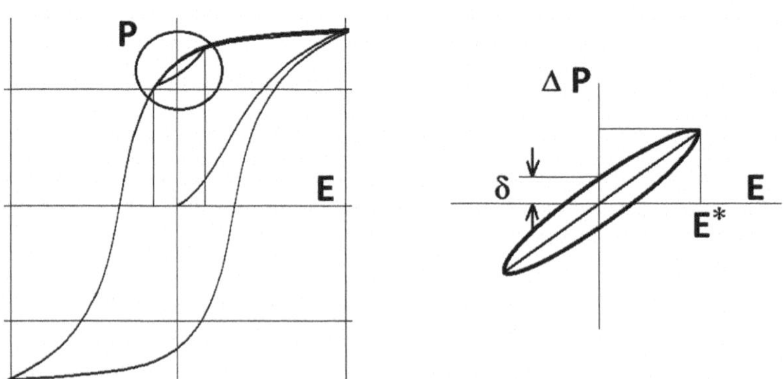

Fig. 32.3 Small dielectric hysteresis loop in harmonic oscillations

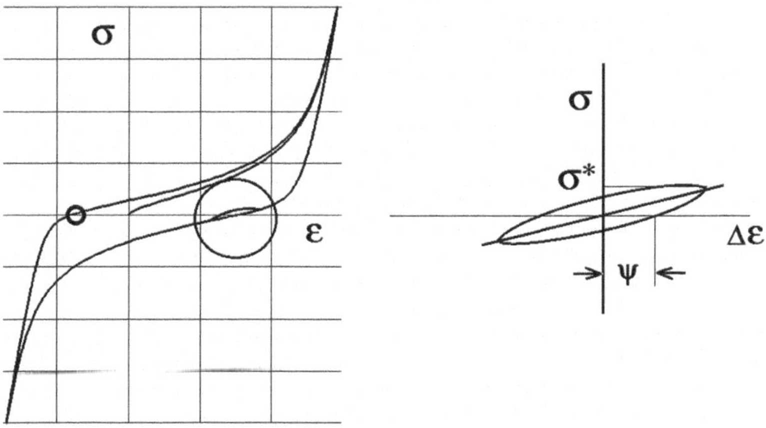

Fig. 32.4 Small loop of deformation hysteresis in harmonic oscillations

In future, we will be interested in the case of small amplitudes $\tilde{\sigma}$, $\tilde{\mathbf{E}}$. It is easy to see that even for the case of small loops, the residual parameters P_0, $(\varepsilon_0)_{III}$, which are in the coefficients $k_{\varepsilon\,\alpha\beta}$, $k_{p\;mn}$, $k_{\varepsilon\;mn}$ of constitutive relations will change. Thus, the physical characteristics of the material will also change. It is known that an electric field generates both dielectric and deformation loops of the butterfly type, and mechanical stresses generate deformation hysteresis loops and depolarize ceramics. In other words, in the three-dimensional case, each of the components of the electric field vector and the stress tensor can generate the corresponding components of the residual parameters, i.e.,

$$P_{0m} = P_{0m}(E_1, E_2, E_3, \sigma_{11}, \sigma_{22}, \ldots, \sigma_{12}),$$
$$(\varepsilon_0)_{III} = (\varepsilon_0)_{III}(E_1, E_2, E_3, \sigma_{11}, \sigma_{22}, \ldots, \sigma_{12}).$$

32.4 Approximations of Nonlinear Functions

From the matrix relations (32.3) based on (32.4), (32.5), the constitutive relations can be represented as:

$$\varepsilon_\alpha = (\hat{S}_{0\alpha\beta} + k^{\varepsilon}_{\alpha\beta}(\varepsilon_0)_{III})\sigma_\beta + k^{p}_{m\alpha} P_0 E_m,$$
$$D_n = k^{p}_{n\beta} P_0 \sigma_\beta + (\hat{\epsilon}_{onm} + k^{\varepsilon}_{nm}(\varepsilon_0)_{III}) E_m.$$

Let the components of the electric field and mechanical stresses change in time according to the harmonic law: $\sigma_\alpha = \sigma^*_\alpha \cos\omega t$, $\quad E^*_n \cos\omega t$. Consider how the equations change, if small hysteresis loops occur. Due to fluctuations of small amplitude, the residual parameters get some increments: $(\varepsilon_0)_{III} \rightarrow (\varepsilon_0)_{III} + \Delta(\varepsilon_0)_{III}$, $P_0 \rightarrow P_0 + \Delta P_0$, which allows us to rewrite the equations in the form:

$$\varepsilon_\alpha = (\hat{S}_{0\alpha\beta} + k^{\varepsilon}_{\alpha\beta}(\varepsilon_0)_{III})\sigma_\beta + k^{p}_{m\alpha} P_0 E_m + k^{\varepsilon}_{\alpha\beta}\Delta(\varepsilon_0)_{III}\sigma_\beta + k^{p}_{m\alpha}\Delta P_0 E_m,$$
$$D_n = k^{p}_{n\beta} P_0 \sigma_\beta + (\hat{\epsilon}_{onm} + k^{\varepsilon}_{nm}(\varepsilon_0)_{III})E_m + k^{p}_{n\beta}\Delta P_0 \sigma_\beta + k^{\varepsilon}_{nm}\Delta(\varepsilon_0)_{III} E_m.$$

It is easy to see that there remained a group of terms with instantaneous modules and a second group appeared with increments of residual parameters. Following the ideas of [2], we approximate these terms with the following functions:

$$k^{\varepsilon}_{\alpha\beta}\Delta(\varepsilon_0)_{III}\sigma_\beta = \psi_{\alpha\beta}\sqrt{1 - \left(\frac{\sigma_\beta}{\sigma^*_\beta}\right)^2}\,\sigma^*_\beta;$$
$$k^{p}_{m\alpha}\Delta P_0 E_m = \delta_{m\alpha}\sqrt{1 - \left(\frac{E_m}{E^*_m}\right)^2}\,E^*_m;$$
$$k^{\varepsilon}_{nm}\Delta(\varepsilon_0)_{III} E_m = \chi_{mn}\sqrt{1 - \left(\frac{E_m}{E^*_m}\right)^2}\,E^*_m.$$

Due to the harmonic process, we have:

$$k^{\varepsilon}_{\alpha\beta}\Delta(\varepsilon_0)_{III}\sigma_\beta = \psi_{\alpha\beta}\sigma^*_\beta \sin\omega t = \psi_{\alpha\beta}\dot{\sigma}_\beta;$$
$$k^{p}_{m\alpha}\Delta P_0 E_m = \delta_{m\alpha} E^*_m \sin\omega t = \delta_{m\alpha}\dot{E}_m;$$
$$k^{\varepsilon}_{nm}\Delta(\varepsilon_0)_{III} E_m = \chi_{mn} E^*_m \sin\omega t = \chi_{mn}\dot{E}_m;$$

After that, the constitutive relations can be represented in the form similar to the Maxwell model:

$$\varepsilon_\alpha = (\hat{S}_{0\alpha\beta} + k^{\varepsilon}_{\alpha\beta}(\varepsilon_0)_{III})\sigma_\beta + \psi_{\alpha\beta}\dot{\sigma}_\beta + k^{p}_{m\alpha} P_0 E_m + \delta_{m\alpha}\dot{E}_m,$$
$$D_n = k^{p}_{n\beta} P_0 \sigma_\beta + \delta_{m\beta}\dot{\sigma}_\beta + (\hat{\epsilon}_{onm} + k^{\varepsilon}_{nm}(\varepsilon_0)_{III})E_m + \chi_{mn}\dot{E}_m.$$

and get the loss tangents:

$$tg\gamma_{\alpha\beta} = \frac{\psi_{\alpha\beta}\omega}{\hat{S}_{0\alpha\beta} + k^{\varepsilon}_{\alpha\beta}\varepsilon_{03}};$$
$$tg\gamma_{m\alpha} = \frac{\delta_{m\alpha}\omega}{k^{p}_{m\alpha}P_0};$$
$$tg\gamma_{nm} = \frac{\chi_{mn}\omega}{\hat{\epsilon}_{onm} + k^{\epsilon}_{nm}\varepsilon_{03}}.$$

32.5 Conclusions

The developed polarization model allows one to find small loops of dielectric and deformation hysteresis (1) and (2). The constitutive relations for the induced components (3) and the functional dependence of the physical modules on the residual parameters (4) allow us to identify additional terms due to the increment of the residual parameters. The approximation of these functions by small hysteresis loops makes it possible to find the loss tangent for ceramic media through the parameters of small hysteresis loops, which are previously found after a series of numerical experiments.

Acknowledgements This work was supported of the Ministry of Science and Higher Education of the Russian Federation, Project No. 9.1001.2017/4.6.

References

1. Belokon, A.V., Skaliukh, A.S.: Mathematical modeling of irreversible processes of polarization. Moscow, Russia, FIZMATLIT (2010)
2. Skaliukh, A.S.: On functions of limitary polarization used in models of polarization of ferroelectrics. In: Materials of the II International Scientific Conference "Autumn Mathematical Readings in Adygea", pp. 218–222. ASU Publishing House, Maykop (2017)
3. Skaliukh, A.S.: Functional dependence of the physical characteristics on irreversible parameters under electromechanical action on the ferroelectric ceramics. Vestnik Tomskogo gosudarstvennogo universiteta. Matematika i mekhanika [Tomsk State University Journal of Mathematics and Mechanics] **58**, 128–143 (2019)
4. Khromov, V.G., Khromov, E.V., Khromov, O.V.: Methods and problematic issues of applying the hysteresis model of internal friction of a material in the theory of oscillations of machines and structures. Collection of scientific papers SNUYAE and P, Applied Physics and Mathematics, pp. 271–284 (2010)
5. Uchino, K.: Loss mechanisms in piezoelectrics: how to measure different losses separately. IEEE Trans Ferroelectr Freq Control **48**(1), 307–321 (2001)

Chapter 33
Modeling of Pentamode Metamaterials Under Dynamic Loading

Vladimir V. Skripnyak and Vladimir A. Skripnyak

Abstract The field of metamaterials has grown considerably in the last few decades due to the advances in new manufacturing technologies. Metamaterials currently are of interest for a wide variety of applications including damping systems. This work is aimed to evaluate dissipative effect of pentamode metamaterials subjected to dynamic loading. The results of numerical modeling of the mechanical behavior of pentamode metamaterials from alpha-titanium alloys were received and compared with available experimental data. The model of inelastic deformation and ductile damage criterion was used to describe the ductility of the framework of metamaterials in a wide range of strain rates, temperature, and stress triaxiality. A methodology to analyze the energy dissipation due to inelastic deformation of metamaterials at high strain rates was presented. It was shown that the values of the energy dissipation coefficient during uniaxial dynamic compression of the pentamode metamaterial are 1.5 times higher than for the bulk alloy counterpart.

Keywords Metamaterials · Alpha-titanium alloys · Energy dissipation · Inelastic response

33.1 Introduction

Devices and aerospace technical objects are often subjected to intensive dynamic impacts during operation. Development of damping materials and technologies of their manufacturing is one of important problem for modern designs. Various composite and porous materials are widely used in modern aerospace objects.

V. V. Skripnyak (✉) · V. A. Skripnyak
National Research Tomsk State University, Tomsk, Russia
e-mail: skrp2012@yandex.ru

V. A. Skripnyak
e-mail: skrp2006@yandex.ru

H. Altenbach et al. (eds.), *Multiscale Solid Mechanics*,
Advanced Structured Materials 141,
https://doi.org/10.1007/978-3-030-54928-2_33

In the works of Bragov et al., it was shown that in the case of pulsed mechanical loadings, including shock impacts, porous, and frame-reinforced polymer, composite materials exhibit higher dissipative properties compared to structural alloys [2–4].

Kadic et al. and Zadpoor in the review articles noted that dissipative effects play a major role in the observed non-proportional stress behavior in the loading path under cyclic loading of cellular metamaterials [11, 20]. Mohsenizadeh et al. showed that lightweight metamaterials have higher dissipative properties compared to other existing materials, such as open-cell metal foams [14].

The concept of mechanical metamaterials has been implemented in recent years in additive manufacturing technologies, which has enabled to produce materials with complex spatial micro/nano-architecture.

Extreme metamaterials produced using honeycomb grids have anisotropic deformation resistance. Milton and Cherkaev showed that specific combinations may be used to manufacture extreme materials [13]. Volumetric cells of extreme materials can be combined to produce materials with the required elasticity tensor. Metamaterials with two, three, four, or five small eigenvalues of the effective elastic moduli tensor components are called bimode, trimode, quadramode, and pentamode metamaterials, respectively.

It was shown by Kadic and Bückmann that experimental samples of pentamode metamaterials may exhibit the Poisson coefficient close to -1 (dilatation metamaterial) [5, 10]. The pentamodal structure of the material, proposed by Milton and Cherkaev [12, 13] and subsequently studied by Kadic [10], Bückmann and his colleagues [5, 6], consists of elements representing the connection of two truncated cones with a fixed length. Ideally, the structural elements are connected to each other in the region of their vertices and form a structure that resembles a face-centered cubic crystalline structure. The work of Mejica and Lantad [15] substantiates the possibility of creating pentamode metamaterials with structures of other Bravais lattices.

Hedayati et al. [8] showed the substantial availability to produce pentamode metamaterials based on a titanium alloy by selective laser melting. It was shown an availability to additively manufacture pentamode materials with a normalized mass density ρ/ρ_s from 0.24 to 4.24 $\pm$ 0.02 (ρ_s is the mass density of the material of the frame elements) from the Ti–6Al–4V alloy powders.

A feature of the lattice structures of pentamode materials is that the stress arising in the elements of the lattice structure is transmitted through the vertices of the cones. The structure of the unit cell of a pentamode metamaterial is shown in Fig. 33.1a. The parameters of the periodic structure and the geometric parameters of the elements can be set in accordance with the capabilities of modern 3D printing technologies. An increase in the diameter of the cones has a minimal effect on the rigidity of the structure of pentamode metamaterials, but affects the mass density. As a result, the properties of extreme pentamode metamaterials differ from the properties of most porous materials and lattice structures, for which the dependence of the elastic modulus on the mass density exhibit a power-law behavior.

Due to this feature of the mechanical properties of pentamode materials, they can be used to control the dynamics of stress waves and acoustic waves [7].

Fig. 33.1 **a** Model of pentamode material volume, **b** pentamode cell element

Of interest is the study on dissipative properties of pentamode metamaterials subjected to pulse mechanical loadings. Pentamode metamaterials based on light alloys can be used as mechanical dampers placed between support plates.

33.2 Simulation the Response of Metamaterial to Pulsed Load

In this work, we performed numerical studies of the mechanical response of a titanium alloy pentamode metamaterial layer with a relative mass density $\rho/\rho_s = 3.15\%$ under initial temperatures of 300 and 900 K and a pulse loading of 100 m/s. To study the dynamic response of pentamode metamaterials with a skeleton structure based on the alpha-titanium alloy Ti-5Al-2.5Sn (Grade 6), a numerical simulation was carried out using finite elements technique. The model metamaterial volume had effective dimensions of 10 mm × 4.8 mm × 5 mm, and frame elements parameters were $D = 0.4$ mm, $d = 0.11$ mm, $h = 1.6$ mm. The explicit finite element LS-DYNA solver of WB ANSYS was used to solve the boundary problem.

The average grain sizes and the relative pore volume of the framework elements material are considered as structural factors affecting the mechanical properties of the framework elements and the metamaterial as a whole.

To account for the indicated structural factors, constitutive relations have been constructed on the base of micromechanical models of physical mechanisms of inelastic deformation (thermally activated and mechanically activated dislocation mechanisms of plasticity, twinning, the interaction of dislocations with dispersed particles, grain boundaries, etc.). The model of mechanical behavior of the pentamode metamaterial includes the system of mass, momentum, and energy conservation equations, kinematic relations, elastic-viscoplastic constitutive equations of the framework elements material, equations of the kinetics of the damage nucleation and growth, and the fracture criterion [1, 9, 16–18].

The Gurson–Tvergaard condition was used as a yield criterion of the framework elements material [18]:

$$\left(\sigma_{eq}^2/\sigma_{ys}^2\right) + 2q_1 f^* \cosh\left(-q_2\, p/2\sigma_{ys}\right) - 1 - q_3\left(f^*\right)^2 = 0, \tag{33.1}$$

where σ_{ys} is the yield stress, $p = -\sigma_{kk}/3$ is the pressure, $\sigma_{eq} = \sqrt{(3/2) S_{ij}\, S_{ij}}$, $S_{ij} = \sigma_{ij} + p\,\delta_{ij}$ are components of the deviatoric stress tensor, q_1, q_2 *and* q_3 are model parameters, and f is the damage parameter of the material.

To describe the change in the stress of plastic flow in alpha-phase titanium alloys with a hexagonal close-packed crystal lattice, the following ratio was used [1]:

$$\begin{aligned}\sigma_{ys} &= \sigma_{s0} + C_2\left[\exp\left(-C_3 T + C_4 T \ln(\dot{\varepsilon}_{eq}/\dot{\varepsilon}_{eq0}\right)\right]\\ &\times \left\{C_5(\varepsilon_{eq}^p)^n + C_6\right\}\left[B_1 + B_2 T + B_3 T^2\right]\end{aligned} \tag{33.2}$$

where σ_{ys} is the yield stress, σ_{s0}, C_2, C_3, C_4, C_5, C_6, n, B_1, B_2, B_3 are coefficients of constitutive equation, $T$is the temperature, $\varepsilon_{eq}^p = \sqrt{(2/3)\varepsilon_{ij}^p \varepsilon_{ij}^p}$ is the equivalent plastic strain, $\dot{\varepsilon}_{eq}$ is the equivalent strain rate, $\dot{\varepsilon}_{eq\,0} = 1.0\ \mathrm{ms}^{-1}$.

To describe the evolution of damage (growth and nucleation of discontinuities) and fracture of the material of frame elements of pentamode metamaterials, the Needleman model was used [18].

The evolution of the damage parameter is described by the kinetic equations:

$$\begin{aligned}\dot{f} &= \dot{f}_{nucl} + \dot{f}_{growth},\\ \dot{f}_{nucl} &= \left(f_N/s_N\right)\exp\left\{-0.5\left[\left(\varepsilon_{eq}^p - \varepsilon_N\right)/s_N\right]^2\right\},\\ \dot{f}_{growth} &= (1-f)\dot{\varepsilon}_{kk}^p,\end{aligned} \tag{33.3}$$

where ε_N and s_N are the parameters that determine strain at which the void coalescence initiates and the standard deviation in the strain distribution corresponding to the evolution of the void system.

Void nucleation during the plastic deformation is determined by the f_N parameter.

$$\begin{aligned}f^* &= f \quad \text{if } f \le f_c;\\ f^* &= f_c + (\bar{f}_F - f_c)/(f_F - f_c) \quad \text{if } f > f_c,\end{aligned} \tag{33.4}$$

where $\bar{f}_F = \left(q_1 + \sqrt{q_1^2 - q_3}\Big/ q_3\right)$, q_1, q_2 and q_3 are model parameters.

The temperature increase in the plastic strain zone at high-speed deformation was calculated in the adiabatic approximation:

$$T = T_0 + \left(0.9/\rho_s C_p\right)\int_0^{\varepsilon_{eq}^p} \sigma_{eq}\, d\varepsilon_{eq}^p, \tag{33.5}$$

where T_0 is the initial temperature in the framework elements material, $\rho_s = 4.4 \times 10^{-6}$ kg/mm^3 is the mass density of the titanium alloy Ti−5Al−2.5Sn, $C_p = 458 + 0.35\,T - 1.929 \times 10^{-4}\,T^2 + 8.758 \times 10^{-8}\,T^3$ J/kg K is the specific heat capacity of the titanium alloy Ti−5Al−2.5Sn.

To estimate the normalized Young's modulus E/E_s for a metamaterial with a diamond-like lattice structure in [8, 21], the following relation was used:

$$E/E_s = [\sqrt{6}\pi(3/4)^2\ (d/2h)^4]/[1 + 1.5(d/2h)^2], \tag{33.6}$$

where E_s is the Young's modulus of the cell elements material and d and h are minimal diameter and length of cell elements, respectively.

To estimate the yield strength of a metamaterial, an analytical formula was used [21]:

$$\sigma_y/\sigma_{ys} = [9\pi/4\sqrt{6}](d/2h)^3, \tag{33.7}$$

where σ_y and σ_{ys} are the effective yield stress of metamaterial and the yield stress of the cell elements material, respectively.

The damping properties characterizing the dissipation of mechanical energy during loading of the pentamode metamaterial with the Ti–5Al–2.5Sn titanium alloy skeleton subjected to high-speed deformation were evaluated by the amount of dissipated energy and the energy dissipation coefficient [19]:

$$\Delta W(t) = W(t) - W_{\text{int}}(t),\ \lambda(t) = \int_0^t \frac{\Delta W(t)}{W(t)} \mathrm{d}t, \tag{33.8}$$

where ΔW is the increment of dissipated energy, W is the work of loading force on displacement, W_{int} is the internal energy of the metamaterial volume, and λ is the energy dissipation coefficient.

The specific energy W supplied to the model volume of the metamaterial under pulse compression was calculated by:

$$W = \frac{1}{m_0} \int_0^t F(t)u(t)\mathrm{d}t, \tag{33.9}$$

where $u(t)$ is the displacement of the movable support of the metamaterial, $F(t)$ is calculated resistant force in the movable support plane, $m_0 = \rho_0 V_0$ is the initial mass of the deformable metamaterial volume, ρ_0 is the initial effective mass density of the metamaterial, and V_0 is the initial effective volume of the metamaterial.

The value of the specific internal energy of the framework structure was calculated by the formula:

$$W_{\text{int}}(t) = \frac{1}{m_0} \int\limits_{v} \left(\int\limits_{0}^{\varepsilon_{ij}^{e}} \sigma_{ij}\, d\varepsilon_{ij}^{e} \right) \mathrm{d}v, \tag{33.10}$$

where W_{int} is the internal energy of deformed framework elements material, V is the total volume of framework elements, σ_{ij} is the stress tensor components, calculated in the framework elements material, and $\mathrm{d}\varepsilon_{ij}^{e}$ is increments of the elastic stain tensor components in the framework elements material.

The initial conditions assumed a constant initial temperature and the absence of stresses and strains in the framework elements.

The boundary conditions were set in accordance with the loading conditions of the model volume of the metamaterial. The model volume of the pentamode metamaterial was placed between the upper fixed plane and the lower movable plane. The upper boundary of the frame support was considered rigid and immovable. Uniaxial compression of the metamaterial volume was realized when the movable plane was displaced at a constant velocity.

The following values of the model coefficients were used in the calculations (33.1)–(33.5): $\sigma_{s0} = 0.25$ GPa, $C_2 = 0.8$ GPa, $C_3 = 0.0043$, C_4, C_5, C_6, $n = 0.25$, $B_1 = 1.0$, $B_2 = 0.000466$, $B_3 = 2.43 \times 10^{-7}$, $q_1 = 1.0$, $q_2 = 1.3$, $q_3 = 1.0$, $\varepsilon_N = 0.25$, $F_N = 0.04$, $S_N = 0.1$, $f_F = 0.26$, $f_c = 0.117$. The grid convergence of the calculation results within a few percent was provided by the choice of the discretization step of the grid model.

33.3 Results and Discussion

The results of simulation indicate the localization of inelastic strains in the region of the joint of the framework elements. The initial normalized mass density of the model volume of the pentamode metamaterial ρ/ρ_s was 3.145%. Figure 33.2a shows

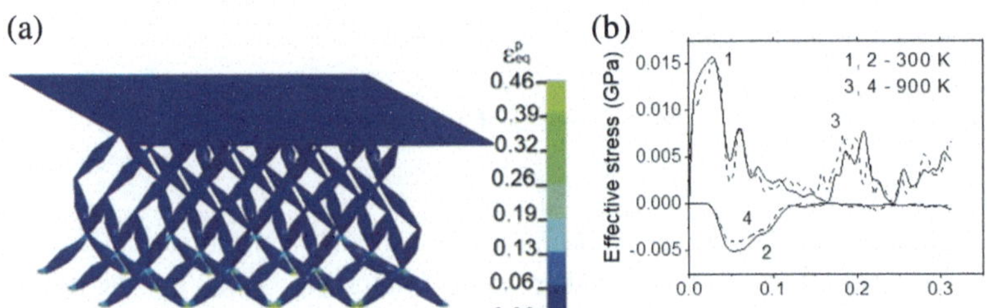

Fig. 33.2 **a** Calculated equivalent plastic strain at $t = 0$. **b** Reaction force under vertical uniaxial compression at the movable plane (curves 1 and 3) and reaction force at the fixed plane (curves 2 and 4) versus time. Curves 1 and 3 correspond to the compression of the pentamode metamaterial volume at an initial temperature of 300 K, and curves 2 and 4 correspond to compression at an initial temperature of 900 K

the calculated equivalent plastic strain in the framework elements of the metamaterial compressed by 39%. Figure 33.2b shows the effective engineering stress versus engineering strain under axial compression of the metamaterial model volume. Curves 1 and 3 correspond to the force acting on the movable support surface, and curves 2 and 4 to the immovable one. Curves 1 and 2 were obtained at the initial temperature of 300 K, and curves 3 and 4 at the initial temperature of 900 K.

The calculated effective stress versus effective engineering strain diagrams are in qualitative agreement with the experimental stress-strain diagrams of titanium pentamode material created using selective laser melting technology [8]. The delay in the arrival of a stress pulse to the fixed support surface is determined by the propagation velocity of elastic waves along the metamaterial framework. The initial stage of deformation was characterized by close to linear dependence of the force on the relative value of compression. This has enabled to identify it as quasi-elastic reaction and evaluate the effective values of elastic modules (Young's modulus, Poisson's ratio, and shear modulus).

The value of $\langle E\rangle/E_s$ obtained by simulation was 2.9×10^{-5} at $E_s = 107$ GPa, and the value of the normalized effective yield stress σ_y/σ_{ys} was estimated as 0.00087 at $\sigma_{ys} = 0.95$ GPa and $T = 300$ K. The data obtained in the calculations are consistent with the experimental data ($\langle E\rangle/E_s = 2.74 \times 10^{-5}$ and $\sigma_y/\sigma_{ys} = 0.00024$ at $\rho/\rho_s = 3.15\%$ for pentamode material obtained by selective laser melting of a Ti–6Al–4V powder using high-energy density laser beam [8]). When comparing the data, the difference between normalization parameters from [8] and simulation was taken into account, such as the values ($E_s = 113$ GPa, $\sigma_{ys} = 0.9$ GPa, $\rho_s = 4.43 \times 10^3$ kg/m^3) of Ti–6Al–4V were slightly differ from the corresponding parameters for the Ti–5Al–2.5Sn alloy.

The theoretical estimation of $\langle E\rangle/E_s$ by Eq. (33.6) and σ_y/σ_{ys} by Eq. (33.7) were of 0.8×10^{-5} and 1.17×10^{-4} (for the parameter $d/2h = 0.034375$ of pentamode metamaterial), respectively.

The metamaterial deforms inelastically until the stability of the frame structure elements is lost as strains within the framework reach critical values. Further deformation has been accompanied by wedging of framework fragments, which result in a temporary increase in the resistance to compression. The energy dissipation occurs as a result of the localization of plastic strain; damage and fracture of the frame elements (see Fig. 33.2b).

The triaxiality stress state parameter $\eta = -p/\sigma_{eq}$ may change significantly during inelastic deformation.

As a result of the work dissipation on the plastic deformation, the temperature rises in local zones of the framework elements. Figure 33.3c shows the temperature in elements of pentamode metamaterial under uniaxial compression at vertical velocity of 100 m/s.

The evolution of the framework structure is the cause of different values of the dissipated work and coefficient of energy dissipation. Figure 33.4a shows calculated values of the specific mechanical energy W over time (curves 1 and 3) and the specific internal energy W_{int} (curves 2 and 4) over time. Curves 5 and 6 indicate the change in W and W_{int} during high-speed deformation of the bulk titanium alloy Ti–5Al–2.5Sn

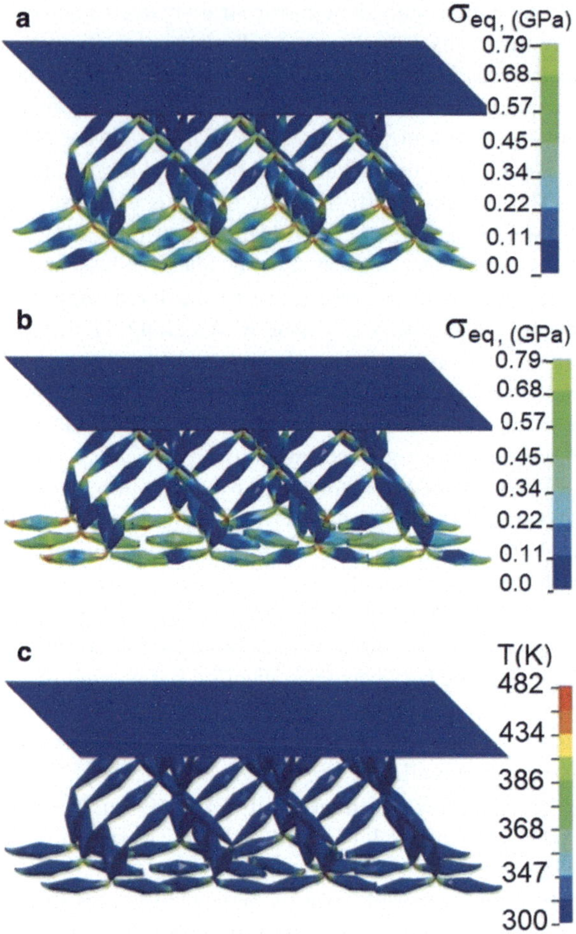

Fig. 33.3 **a** Equivalent stresses in the elements of the frame of the titanium metamaterial at the time 0.01 ms, **b** at the time of 0.01045 ms under uniaxial compression of the pentamode metamaterial volume at 100 m/s at the initial temperature of 300 K. **c** Temperature in frame elements at the time of 0.0104 ms

at a strain rate of 100 s^{-1}. Figure 33.4b shows the coefficient of energy dissipation λ versus the effective deformation $\langle \varepsilon_{yy} \rangle$ at initial temperature of 300 K (curves 1 and 2) and 900 K (curve 3).

As the initial temperature increases from 300 to 900 K, the predicted value of the energy dissipation coefficient λ decreases from 2 to 1.5 times with increasing compression strain of the metamaterial. The decrease in the coefficient of energy dissipation is associated with the decrease in stresses acting in the framework elements at higher temperature. A growth in the energy dissipation coefficient λ along with compression degree shows nonlinear behavior.

The λ increments are predominantly caused by the dissipation of the mechanical work due to plastic deformation in the metamaterial elements. Additional increments of the energy dissipation are associated with the nucleation and growth of damage in

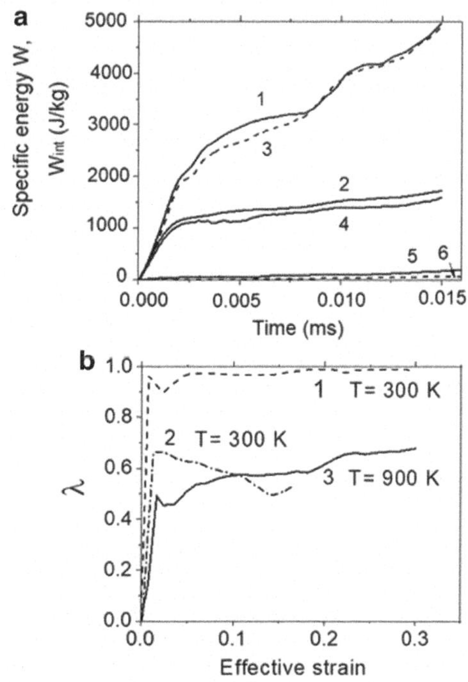

Fig. 33.4 **a** Specific internal energy of deformed framework elements under compression at the velocity of 100 m/s. Curves 1, 2 correspond to $W(t)$, and curves 3, 4 correspond to $W_{int}(t)$, respectively. **b** The coefficient of energy dissipation λ versus effective strain $\langle \varepsilon_{yy} \rangle$ at the initial temperature of 300 K (curves 1 and 2) and 900 K (curve 3)

the material of the framework elements. A slight increase in the energy dissipation coefficient λ takes place due to these increments.

When bulk alloys are deformed, there are no analogs to the physical mechanisms of wedging and compaction of fragmented elements of the metamaterial framework.

The higher values of the energy dissipation coefficient in the case of pentamode metamaterials as compared with the value for the bulk alloy counterpart are due to the significant difference between the specific energy W supplied to the volume of metamaterial and bulk alloy. Thus, pentamode metamaterials based on titanium alloys manufactured by selective laser melting are perspective materials for development structures of pulsed mechanical dampers.

33.4 Conclusion

A computational model describing mechanical response of titanium pentamode metamaterials to pulse loading has been developed. The numerical simulation of the mechanical response of a pentamode metamaterial from the Ti–5Al–2.5Sn titanium alloy was carried out during dynamic compression at 100 m/s. Calculated normalized Young modulus $\langle E \rangle / E_s$ and value of the normalized effective yield stress σ_y / σ_{ys} agree with experimental data obtained by Hedayati and coworkers [9].

A methodology to analyze the energy dissipation due to inelastic deformation of structured materials at high strain rates was presented for metamaterials. The values of the energy dissipation coefficient were determined for uniaxial compression of the pentamode metamaterial with the relative mass density of 3.145% at strain rates of ~20.8×10^3 s^{-1} and initial temperatures of 300 and 900 K. The values of the energy dissipation coefficient during uniaxial dynamic compression of the pentamode metamaterial are 1.5 times higher than for the bulk alloy counterpart. The energy dissipation coefficient under uniaxial compression decreases by a factor of 1.5–2 with an increase in the initial temperature from 300 to 900 K.

Acknowledgements This work was supported by the Russian Science Foundation (RSF), grant No. 16-19-10264. The authors are grateful for the support of this research.

References

1. Armstrong, R.W., Zerilli, F.J.: High rate straining of tantalum and copper. J. Phys. D Appl. Phys. **43**, 492002 (2010)
2. Bragov, A.M., Igumnov, L.A., Konstantinov, A.Y., Lomunov, A.K., Rusin, E.: Effects of high strain rate and self-heating on plastic deformation of metal materials under fast compression loading. J. Dyn. Behav. Mater. **5**, 309–319 (2019)
3. Bragov, A.M., Igumnov, L.A., Konstantinov, A.Y., Lomunov, A.K.: Investigation of the mechanical properties of organoplastic under shock wave loading conditions. J. Phys. Conf. Ser. **991**, 012013 (2018)
4. Bragov, A.M., Konstantinov, A.Y., Lomunov, A.K., Sadyrin, F., Sergeichev, I.V., Kruzka, L.: Dynamic compressibility of high-porosity dampers of thermal and shock loadings: modeling and experiment. Int. J. Mod. Phys. B **22**(9), 1183–1188 (2008)
5. Bückmann, T., Schittny, R., Thiel, M., Kadic, M., Milton, G.W., Wegener, M.: On three-dimensional dilational elastic metamaterials. New J. Phys. **16**, 033032 (2014)
6. Bückmann, T., Stenger, N., Kadic, M., Kaschke, J., Frölich, A., Kennerknecht, T., Eberl, C., Thiel, M., Wegener, M.: Tailored 3D mechanical metamaterials made by dip-in direct-laser-writing optical lithography. Adv. Mater. **24**, 2710–2714 (2012)
7. Del Vescovo, D., Giorgio, I.: Dynamic problems for metamaterials: review of existing models and ideas for further research. Inter. J. Eng. Sci. **80**, 153–172 (2014)
8. Hedayati, R., Leeflang, A.M., Zadpoor, A.A.: Additively manufactured metallic pentamode meta-materials. Appl. Phys. Lett. **110**, 091905 (2017)
9. Herzig, N., Meyer, L.W., Musch, D., Halle, T., Skripnyak, V.A., Skripnyak, E.G., Razorenov, S.V., Krüger, L. The mechanical behaviour of ultrafine grained titanium alloys at high strain rates. Paper presented at 3rd international conference on high speed forming, pp. 141–150. Dortmund, Germany, 11–12 March 2008
10. Kadic, M., Bückmann, T., Stenger, N., Thiel, M., Wegener, M.: On the practicability of pentamode mechanical metamaterials. Appl. Phys. Lett. **100**, 191901 (2012)
11. Kadic, M., Milton, G.W., Hecke, M.L., Van Wegener, M.: 3D metamaterials. Nat. Rev. Phys. **1**, 198–210 (2019)
12. Milton, G.W.: The Theory of Composites. Cambridge University Press, Cambridge, UK (2002)
13. Milton, G.W., Cherkaev, A.V.: Which elasticity tensors are realizable? J. Eng. Mater. Technol. **117**, 483–493 (1995)
14. Mohsenizadeh, M., Gasbarri, F., Munther, M., Beheshti, A., Davami, K.: Additively-manufactured lightweight metamaterials for energy absorption. Mater. Des. **139**, 521–530 (2018)

15. Mejica, F.M., Lantada, A.D.: Comparative study of potential pentamodal metamaterials inspired by Bravais lattices. Smart Mater. Struct. **22**, 115013 (2013)
16. Skripnyak, V.V., Kozulyn, A.A., Skripnyak, V.A.: The influence of stress triaxiality on ductility of α titanium alloy in a wide range of strain rates. Mater. Phys. Mech. **42**, 415–422 (2019)
17. Skripnyak, V.V., Skripnyak, V.A., Skripnyak, E. G. Ductility of titanium alloys in a wide range of strain rates. In: Silva Gomes J.F., and Meguid S.A. (eds.) Paper presented at 6th international conference integrity-reliability-failure. Lisbon/Portugal, 22–26 July 2018. Publ. INEGI/FEUP (2018); Paper REF: 7127
18. Tvergaard, V.: Behavior of porous ductile solids at low stress triaxiality in different modes of deformation. Inter. J. Sol. Struct. **60–61**, 28–34 (2015)
19. Wu, C.-Y., Li, L.-Y., Thornton, C.: Energy dissipation during normal impact of elastic and elastic–plastic spheres. Int. J. Impact Eng **2**(1–4), 593–604 (2005)
20. Zadpoor, A.A.: Mechanical meta-materials. Mater. Horiz. **3**, 371–381 (2016)
21. Zadpoor, A.A., Hedayati, R.: Analytical relationships for prediction of the mechanical properties of additively manufactured porous biomaterials. J. Biomed. Mater. Res., Part A **104A**, 3164–3174 (2016)

Chapter 34
Deformation of Inhomogeneous Elastic Strip

Alexander O. Vatulyan, Julia A. Morozova, and Dmitrii K. Plotnikov

Abstract In the framework of the plane problem of linear elasticity, equilibrium of a vertically inhomogeneous elastic isotropic strip under load with arbitrary laws of heterogeneity is considered. Using Fourier transform, canonical system of differential equations for displacements and stress components is formulated. Problems on the action of forces (normal and tangent) on the upper boundary of the strip are considered. Computational scheme for constructing transfer functions depending on the parameter of the Fourier transform connecting the transformants of displacements and loads for an inhomogeneous strip is presented. Computational scheme is based on the shooting method. The symbols of the kernels of the integral equations arising in the contact problems for the strip are determined through the transfer functions. Based on the solution of auxiliary Cauchy problems and asymptotic analysis, the properties of the transfer functions are analyzed, and behavior at zero and infinity are investigated. Comparison with known expressions for the case of a homogeneous strip is carried out. A comparative analysis of the transfer functions for various laws of heterogeneity, both continuous and containing discontinuities of the first kind, is performed. An asymptotic analysis of the transfer functions for small and large values of the Fourier transform parameter is carried out. Displacements of the upper boundary of the strip for various laws of heterogeneity are built. Comparison of the obtained solutions with solutions based on an approximate model of an inhomogeneous elastic strip is carried out.

A. O. Vatulyan (✉) · J. A. Morozova
Institute of Mathematics, Mechanics and Computer Science named after I.I. Vorovich, Southern Federal University, Rostov-on-Don, Russia
e-mail: vatulyan@math.rsu.ru

J. A. Morozova
e-mail: ulia_morozova_95@mail.ru

D. K. Plotnikov
Southern Mathematical Institute of the Vladikavkaz Scientific Center of the Russian Academy of Sciences, Vladikavkaz, Russia
e-mail: dustheap@mail.ru

H. Altenbach et al. (eds.), *Multiscale Solid Mechanics*,
Advanced Structured Materials 141,
https://doi.org/10.1007/978-3-030-54928-2_34

Keywords Coating · Functionally graded material · Inhomogeneous strip · Asymptotic analysis

34.1 Introduction

Currently, an increasing number of composite structures, coatings, structural elements made of composite materials are being introduced into engineering practice. Modeling of objects with significantly inhomogeneous properties requires studying the deformation of inhomogeneous layered structures under static and dynamic loads. One of the most popular and rapidly developing areas in the design of heterogeneous objects is the production of functionally graded materials (FGM). FGM is a medium whose properties change with coordinates. This is used for various purposes, for example, to reduce the level of stress inside the structural element.

One of the most promising fields of application of FGM is the manufacture of gradient coatings for various purposes. The application of coatings made from FGM is widely used in the manufacture of various elements of engineering structures, cutting tools, medical devices, and in the components of engines in aviation and space technology. Continuous change of properties with the coating depth leads to a decrease in stress concentration, which in turn reduces the probability of cracking and other defects.

Monographs [7, 8] are devoted to the study of deformation for inhomogeneous elastic bodies and construction of solutions for new problems. In [7], based on linear elasticity, problems for structural elements whose mechanical characteristics are continuous coordinate functions are studied.

Among the problems of heterogeneous elasticity theory, when modeling bases and coatings, an important place is occupied by problems for the strip. The main tool in the study of such problems is the integral Fourier transform allowing to reduce the boundary value problems of linear elasticity to systems of ordinary linear differential equations. Important objects in the study of such problems are transfer functions connecting transformants of displacements and loads. Note that for an inhomogeneous strip, these functions cannot be found explicitly but are constructed either numerically or based on approximate models.

Problems for a homogeneous strip (including contact problems) were investigated in [6, 11]. A detailed review of works devoted to the solution of mixed problems for the strip can be found in [12].

The most effective analytical method for solving such problems for the strip are asymptotic methods, in particular, the so-called large lambda method and the "small lambda" method, which were first applied in the study of the problem for an axisymmetric stamp and an elastic layer of relatively large thickness [20] and in the problem of pressing a stamp into an elastic layer of relatively small thickness [3]. The results of the further development of these works are summarized in the monograph [21]. The symbol of the Fourier kernel of the integral equation for contact problem in these works is constructed explicitly and is a meromorphic function in the complex plane that has a countable set of zeros and poles.

In monograph [2], asymptotic analysis of the problem for a homogeneous elastic strip with small thickness is carried out. Based on this analysis, several approximate models of contact interaction for bodies with thin coatings and interlayers are constructed.

The monograph [4] is devoted to the development of numerical analytical methods for solving static contact problems for continuously inhomogeneous media. The kernels of the integral operators for the contact problem are constructed in two steps. At the first step, a boundary value problem for a differential operator with a parameter (Fourier transform parameter) is solved; at the second step, using fractional rational and power functions, an approximation of the function obtained numerically is constructed. In [13, 19], this approach has been expanded for multilayer structures with inhomogeneous coating.

It is also possible to construct numerical solutions for problems on the deformation of an inhomogeneous strip based on the finite element method.

Another way to study the deformation of inhomogeneous foundations is to construct approximate models of an elastic body based on some hypotheses about the structure of displacement fields. In monographs [17, 18], models of elastic bases, including layered ones, are presented. An approximate approach is based on the representation of the desired functions (for example, deflection in the plate bending problem) as the multiplication of two functions, one of which is a given function of one variable, and the other is an unknown function of the second variable. Further, using the principle of possible displacements, a linear system of ordinary differential equations is formulated. Based on the variational approach, the deformation of plates and shells was studied, stability issues were investigated, models of beams, plates and shells on an elastic base were constructed.

In [10] the basics for calculation of foundations on an elastic base, based on a method that generalizes, the Winkler model and introduces two bed coefficients is stated.

Vatulyan and Plotnikov [14, 15], Vatulyan et al. [15] are devoted to the development of models of deformation for an inhomogeneous elastic strip. Models allowing to consider arbitrary laws of heterogeneity, including discontinuous, are presented. Transfer functions are constructed approximately in the form of fractional rational functions of the Fourier transform parameter.

In this paper, a series of problems on the equilibrium of an inhomogeneous elastic strip whose elastic moduli are arbitrary positive bounded functions of the vertical coordinate (smooth or discontinuous) is presented.

34.2 Formulation of the Problem

Consider the plane deformation of an inhomogeneous isotropic elastic strip with thickness h rigidly coupled to a non-deformable base. A rectangular coordinate system (x_1, x_3) is connected to the strip, the origin of the coordinate axes is located on the lower border of the strip. It is assumed that the Lame coefficients of the strip are arbitrary positive functions of the vertical coordinate x_3: $\lambda = \lambda(x_3),\ \mu = \mu(x_3)$. Assuming that $u_1 = u_1(x_1, x_3),\ u_2 = 0,\ u_3 = u_3(x_1, x_3)$, a statement of the boundary value problem is presented.

Equilibrium equations of a plane problem:

$$\sigma_{11,1} + \sigma_{13,3} = 0,\ \sigma_{31,1} + \sigma_{33,3} = 0,$$

Hooke's law for an isotropic body:

$$\begin{aligned}\sigma_{11} &= (\lambda + 2\mu)u_{1,1} + \lambda u_{3,3},\\ \sigma_{33} &= \lambda u_{1,1} + (\lambda + 2\mu)u_{3,3},\\ \sigma_{13} &= \sigma_{31} = \mu\left(u_{1,3} + u_{3,1}\right),\end{aligned}$$

where σ_{ij} are the stress tensor components.

The boundary conditions of the problem have the form

$$x_3 = 0 : u_1 = u_3 = 0;\ x_3 = h : \sigma_{13} = p_1(x_1),\ \sigma_{33} = p_3(x_1).$$

34.3 The Formulation of the Canonical System. Construction of Transfer Functions

Dimensionless parameters are introduced as follows

$$\begin{aligned}&x_k = \xi_k h,\ u_k = v_k h,\ \sigma_{kl} = \mu_0 t_{kl},\\ &\lambda(x_3) = \mu_0 f_1(\xi_3),\ \mu(x_3) = \mu_0 f_2(\xi_3),\ p_k = q\mu_0 h,\end{aligned}$$

where μ_0 is characteristic value of the shear modulus, for example, its average value in the interval $[0, h]$.

Fourier transform to a variable ξ_1 is applied and the following canonical system of differential equations for a vector $X_k(\beta,\ \xi_3),\ k = 1, \ldots, 4$ is formulated.

$$\begin{aligned}X_1'(\beta, \xi_3) &= \frac{1}{f_2(\xi_3)} X_3(\beta, \xi_3) - \beta X_2(\beta, \xi_3),\\ X_2'(\beta, \xi_3) &= \frac{1}{f_1(\xi_3) + 2f_2(\xi_3)}(X_4(\beta, \xi_3) + f_1(\xi_3)\beta X_1(\beta, \xi_3)),\\ X_3'(\beta, \xi_3) &= \frac{4f_2(\xi_3)(f_1(\xi_3) + f_2(\xi_3))}{f_1(\xi_3) + 2f_2(\xi_3)}\beta^2 X_1(\beta, \xi_3)\end{aligned}$$

$$
-\frac{f_1(\xi_3)}{f_1(\xi_3)+2f_2(\xi_3)}\beta X_4(\beta,\xi_3),
$$

$$
X_4'(\beta,\xi_3)=\beta X_3(\beta,\xi_3), \tag{34.1}
$$

where $X_1 = iU_1$, $X_2 = U_3$, $X_3 = i\Sigma_{13}$, $X_4 = \Sigma_{33}$, U_1, U_3, Σ_{13}, Σ_{33} are Fourier transforms of the displacement vector components and the corresponding components of the stress tensor.

System (34.1) has variable coefficients and its solution cannot be found explicitly for arbitrary laws of change in material characteristics. Note that system (34.1) can be used to analyze the initial problem not only in the case of continuous laws of heterogeneity, but also for discontinuous ones, which is an undoubted advantage, in contrast to the approaches implemented in [4, 19], where smoothness of functions characterizing heterogeneity is required.

Due to the linearity of the original boundary value problem, for convenience, two auxiliary problems will be considered: loading by normal force and loading by tangent force.

34.3.1 Problem 1

The first boundary value problem for the canonical system (34.1) corresponds to the action of a normal force $q_1(\xi_1) = 0$, $q_3(\xi_1) = \delta(\xi_1)$ in the center of the upper border. For the components of the vector, the boundary conditions are as follows

$$
X_1(\beta,0)=0,\ X_2(\beta,0)=0,\ X_3(\beta,1)=0,\ X_4(\beta,1)=1. \tag{34.2}
$$

When solving the canonical system of Eqs. (34.1), the shooting method is used. Two auxiliary Cauchy problems have the following initial conditions

$$
X_1^{(1)}(\beta,0)=0,\ X_2^{(1)}(\beta,0)=0,\ X_3^{(1)}(\beta,0)=1,\ X_4^{(1)}(\beta,0)=0, \tag{34.3}
$$

$$
X_1^{(2)}(\beta,0)=0,\ X_2^{(2)}(\beta,0)=0,\ X_3^{(2)}(\beta,0)=0,\ X_4^{(2)}(\beta,0)=1\,. \tag{34.4}
$$

Then, the desired solution of problem (34.1), (34.2) can be found in the form of a linear combination of auxiliary problems solutions as $X_k = C_1X_k^{(1)} + C_2X_k^{(2)}$, $k = 1, .., 4$, where C_1 and C_2 are unknown constants, $X_k^{(1)}$ and $X_k^{(2)}$ are solutions of Cauchy problems (34.1), (34.3) and (34.1), (34.4). Constants C_1 and C_2 is obtained from boundary conditions (34.2), which give a linear system

$$
\begin{aligned}
C_1X_3^{(1)}(\beta,1)+C_2X_3^{(2)}(\beta,1)=0,\\
C_1X_4^{(1)}(\beta,1)+C_2X_4^{(2)}(\beta,1)=1.
\end{aligned}
$$

Solution of this system gives a solution of problem 1.

An important role in the analysis of displacements of the upper boundary of the strip is played by the transfer functions $K_{13}(\beta)$ and $K_{33}(\beta)$, introduced as follows

$$X_1(\beta, 1) = K_{13}(\beta),\ X_2(\beta, 1) = K_{33}(\beta)\,.$$

It is easy to show changing in the system (34.1) β to $-\beta$, that functions $X_k(\beta, \xi_3)$ satisfying conditions (34.2) possess the following symmetry

$$X_1(-\beta, \xi_3) = -X_1(\beta, \xi_3),\ X_2(-\beta, \xi_3) = X_2(\beta, \xi_3),$$
$$X_3(-\beta, \xi_3) = -X_3(\beta, \xi_3),\ X_4(-\beta, \xi_3) = X_4(\beta, \xi_3).$$

From here, one can establish the parity and oddness properties of the transfer functions

$$K_{13}(-\beta) = -K_{13}(\beta), K_{33}(-\beta) = K_{33}(\beta). \tag{34.5}$$

34.3.2 Problem 2

In the second boundary value problem, the boundary conditions for the canonical system (34.1) have the form

$$X_1(\beta,\ 0) = 0, X_2(\beta,\ 0) = 0, X_3(\beta,\ 1) = 1, X_4(\beta,\ 1) = 0. \tag{34.6}$$

These conditions correspond to the action of tangential force $q_1(\xi_1) = \delta(\xi_1), q_3(\xi_1) = 0$ on the upper border of the strip.

To solve problem (34.1), (34.6), two auxiliary Cauchy problems with initial conditions (34.3), (34.4) are also used. The further solution is constructed similarly to the solution of the first problem, while the linear system has the form

$$\begin{aligned} C_1 X_3^{(1)}(\beta, 1) + C_2 X_3^{(2)}(\beta, 1) = 1,\\ C_1 X_4^{(1)}(\beta, 1) + C_2 X_4^{(2)}(\beta, 1) = 0. \end{aligned} \tag{34.7}$$

Determination of unknown constants from (34.7) gives a solution to problem 2.

Similarly to Sect. 34.3.1, it is easy to show that the transfer functions $K_{11}(\beta)$ and $K_{31}(\beta)$ defined by the expressions

$$X_1(\beta,\ 1) = K_{11}(\beta),\ X_2(\beta,\ 1) = K_{31}(\beta)$$

possess the following parity and oddity properties

$$K_{11}(-\beta) = K_{11}(\beta),\ K_{31}(-\beta) = -K_{31}(\beta). \tag{34.8}$$

34.4 Asymptotic Analysis

In this section, an asymptotic analysis of the solutions of the boundary value problems for large and small values of the Fourier transform parameter β is carried out and some analytic estimates for arbitrary laws of non-uniformity is obtained.

Assuming that the parameter β is small, solutions of boundary value problems 1 and 2 are presented in the form of the following regular asymptotic expansions [16]

$$X_j = X_{j0} + \beta X_{j1} + \beta^2 X_{j2} + \cdots + \beta^n X_{jn} \tag{34.9}$$

Problem 1. After substituting expression (34.9) into the original system (34.1) and equating the coefficients with the same powers of β, boundary value problems for determining the coefficients in the expansion (34.9) is obtained. For the problem with normal force, coefficients in (34.9) are found in the form

$$X_{10} = 0,\ X_{20} = \int\limits_0^{\xi_3} \frac{d\tau}{f_1(\tau) + 2f_2(\tau)},\ X_{20} = \int\limits_0^{\xi_3} \frac{d\tau}{f_1(\tau) + 2f_2(\tau)},$$
$$X_{30} = 0,\ X_{40} = 1. \tag{34.10}$$

$$X_{11} = \int\limits_0^{\xi_3} \left(\frac{X_{31}(\tau)}{f_2(\tau)} - X_{20}(\tau) \right) d\tau,\ X_{21} = 0,$$

$$X_{31} = -\int\limits_1^{\xi_3} \left(\frac{f_1(\tau)}{f_1(\tau) + 2f_2(\tau)} \right) d\tau,\ X_{41} = 0.$$

For β^n, when $n \geq 2$ solutions of the corresponding boundary value problems are determined by recurrence formulas as follows

$$X_{1n}(\xi_3) = \int\limits_0^{\xi_3} \left(\frac{X_{3n}(\tau)}{f_2(\tau)} - X_{2(n-1)}(\tau) \right) d\tau,$$

$$X_{2n}(\xi_3) = \int\limits_0^{\xi_3} \frac{1}{f_1(\tau) + 2f_2(\tau)} (X_{4n}(\tau) + f_1(\tau) X_{1(n-1)}(\tau)) d\tau,$$

$$X_{3n}(\xi_3) = \int\limits_1^{\xi_3} \left(\frac{4f_2(\tau) \cdot (f_1(\tau) + f_2(\tau))}{f_1(\tau) + 2f_2(\tau)} X_{1(n-2)}(\tau) \right.$$
$$\left. - \frac{f_1(\tau)}{f_1(\tau) + 2f_2(\tau)} X_{4(n-1)}(\tau) \right) d\tau,$$

$$X_{4n}(\xi_3) = \int\limits_1^{\xi_3} X_{3(n-1)}(\tau)\mathrm{d}\tau. \tag{34.11}$$

Problem 2: For the problem of the action of the tangent force, the expansion coefficients (34.9) are represented by the formulas

$$X_{10} = \int\limits_0^{\xi_3} \frac{1}{f_2(\tau)}\mathrm{d}\tau,\ X_{20} = 0,\ X_{30} = 1,\ X_{40} = 0. \tag{34.12}$$

$$X_{11} = 0, X_{21} = \int\limits_0^{\xi_3} \left(\frac{X_{41}(\tau) + f_1(\tau)X_{10}(\tau)}{f_1(\tau) + 2f_2(\tau)}\right)\mathrm{d}\tau, X_{31} = 0, X_{41} = \xi_3 - 1.$$

Expressions for coefficients X_{jn} when $n \geq 2$ are similar to the formulas (34.11).

The result of the comparison of the transfer functions $K_{11}(\beta)$, $K_{13}(\beta)$ and $K_{33}(\beta)$, constructed using asymptotic analysis with preservation of the expansion coefficients up to β^2 with transfer functions constructed numerically for small values β, showed a coincidence of results when $\beta < 0,2$ for $K_{11}(\beta)$, $K_{13}(\beta)$, when $\beta < 0,3$ for $K_{33}(\beta)$.

Further, the behavior of the transfer functions when $|\beta| \to \infty$ is investigated. It leads to the analysis of a boundary value problem with a small parameter for the highest derivative. The solution to the problem is constructed in the form

$$X_j(\beta, \xi_3) = e^{S(\beta,\xi_3)} \sum_{k=0}^{\infty} \beta^{-k} Y_{jk}(\xi_3). \tag{34.13}$$

Substitution of representation (34.13) in (34.1) leads to a recurrent system of equations with respect to $Y_{jk}(\xi_3)$; moreover, $S\prime(\beta, \xi_3)$ is an eigenvalue of the coefficient matrix of the canonical system (34.1), whose spectrum consists of two values β and $-\beta$ with algebraic multiplicity equals 2. The multiplicity of the eigenvalues of the coefficient matrix for the system of differential equations leads to a significant complication of constructing a solution of the form (13) in the framework of well-known approaches.

Analysis of the known solution of the boundary value problem for large β in the case of a uniform strip showed that asymptotic behavior of transfer functions in the inhomogeneous case for $\beta > 0$ can be built using a boundary layer solution in neighborhood of $\xi_3 = 1$ in the form

$$\begin{aligned} X_k(\beta, \xi_3) &= Y_k(\beta, \xi_3)\exp(\beta(\xi_3 - 1)), \\ Y_k(\beta, \xi_3) &= Y_{k0}(\xi_3)\beta + Y_{k1}(\xi_3) + Y_{k2}(\xi_3)\beta^{-1} + o(\beta^{-1}). \end{aligned} \tag{34.14}$$

Substitution of the representation (14) into the canonical system (34.1) allows to express vector Y components in terms of two independent functions Y_{10} and Y_{11} as follows

$$\begin{aligned}
Y_1 &= Y_{10} + Y_{11}\beta^{-1} + o(\beta^{-1}),\\
Y_2 &= Y_{10} + \left(Y_{11} - \tfrac{2f_2'}{f_1+f_2}Y_{10} - \tfrac{f_1+3f_2}{f_1+f_2}Y_{10}'\right)\beta^{-1} + o(\beta^{-1}),\\
Y_3 &= 2f_2Y_{10}\beta + 2f_2Y_{11} - \tfrac{f_2}{f_1+f_2}(2f_2Y_{10})' + Y_{31}\beta^{-1} + o(\beta^{-1}),\\
Y_4 &= 2f_2Y_{10}\beta + 2f_2Y_{11} - \tfrac{f_1+2f_2}{f_1+f_2}(2f_2Y_{10})' + Y_{41}\beta^{-1} + o(\beta^{-1}).
\end{aligned}$$

Functions Y_{31} and Y_{41} not given due to bulky. Satisfying the boundary conditions (34.3) and (34.4) in problems 1 and 2, respectively, the leading terms when $\beta \to +\infty$ are obtained in the form

$$\begin{aligned}
&K_{13}(\beta) = c_1\beta^{-1} + o(\beta^{-1}),\ K_{31}(\beta) = -c_1\beta^{-1} + o(\beta^{-1}),\\
&K_{11}(\beta) = c_2\beta^{-1} + o(\beta^{-1}),\ K_{33}(\beta) = c_2\beta^{-1} + o(\beta^{-1}),\\
&c_1 = -\tfrac{1}{2(f_1+f_2)}\Big|_{\xi_1=1},\ c_2 = \tfrac{f_1+2f_2}{2f_2(f_1+f_2)}\Big|_{\xi_1=1}.
\end{aligned} \tag{34.15}$$

The obtained asymptotic expressions of the transfer functions are consistent with the known asymptotics of known solutions for the case of a homogeneous strip [1].

34.5 Transfer Function Properties

The results of analytical constructions and computational experiments for problems 1 and 2 revealed some properties of the transfer functions $K_{11}(\beta)$, $K_{13}(\beta)$, $K_{31}(\beta)$ and $K_{33}(\beta)$.

1. Using the reciprocity theorem [9], it can be shown that equality $K_{31}(\beta) = -K_{13}(\beta)$ holds for any laws of strip heterogeneity;
2. From (10) and (12), it is easy to find the values of the functions $K_{11}(\beta)$ and $K_{33}(\beta)$ when $\beta = 0$

$$K_{11}(0) = \int_0^1 \frac{d\xi_3}{f_2(\xi_3)},\ K_{33}(0) = \int_0^1 \frac{d\xi_3}{f_1(\xi_3) + 2f_2(\xi_3)};$$

3. Functions $K_{ij}(\beta)$ are decreasing when $|\beta| \to \infty$, and taking into account the properties (34.5), (34.8) have the following character of decrease

$$K_{11}(\beta) = c_2|\beta|^{-1} + o(\beta^{-1}),\ K_{13}(\beta) = c_1\beta^{-1} + o(\beta^{-1}),$$

$$K_{33}(\beta) = c_2|\beta|^{-1} + o(\beta^{-1}).$$

To verify the correctness of the calculations and the operability of the proposed scheme based on the shooting method, a comparison with the case of a uniform strip, for which there are explicit representations of the transfer functions [1, 21], is made. A comparative analysis showed that the transfer functions calculated using the algorithm constructed in Sect. 34.3 coincide with the exact values of the transfer functions accurate to 8 decimal places. Since system (34.1) at large values is a system with a small parameter at the highest derivative, the algorithm used provides a fairly accurate calculation of the transfer functions for the values $\beta < 600$.

34.6 Numerical Experiments

Below are some results of computational experiments for various laws of heterogeneity. The following laws of variation of the elastic properties of the strip are considered

1. monotonically increasing $f_1(\xi_3) = \frac{3}{2} + \xi_3,\ f_2(\xi_3) = \frac{1}{2} + \xi_3$
2. monotonically decreasing $f_1(\xi_3) = \frac{5}{2} - \xi_3,\ f_2(\xi_3) = \frac{3}{2} - \xi_3$
3. non-monotonous $f_1(\xi_3) = \frac{11}{6} + 2\xi_3^2 - \xi_3,\ f_2(\xi_3) = \frac{5}{6} + 2\xi_3^2 - \xi_3$
4. piecewise constant $f_1(\xi_3) = \begin{cases} \frac{40}{17}, & 0 \le \xi_3 \le 0,7 \\ \frac{20}{17}, & 0,7 < \xi_3 \le 1 \end{cases}$, $f_2(\xi_3) = \begin{cases} \frac{20}{17}, & 0 \le \xi_3 \le 0,7 \\ \frac{10}{17}, & 0,7 < \xi_3 \le 1 \end{cases}$

Note that the laws of heterogeneity introduced above have the same average values on the interval [0, 1]: $\int_0^1 f_1(\xi_3)d\xi_3 = 2,\ \int_0^1 f_2(\xi_3)d\xi_3 = 1$.

In Figs. 34.1 and 34.2 are graphs of the transfer functions $K_{11}(\beta)$, $K_{13}(\beta)$ and $K_{33}(\beta)$, numerically constructed for the laws of heterogeneity 1–4. Solid, dashed, dash-dot and dotted lines represent graphs for laws 1–4, respectively. Note that the

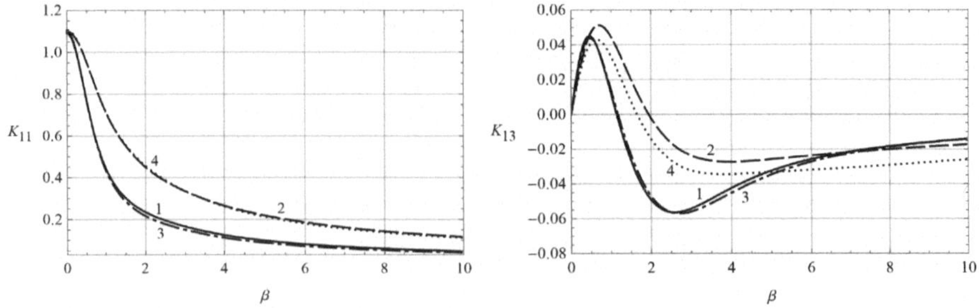

Fig. 34.1 Transfer functions $K_{11}(\beta)$, $K_{13}(\beta)$ for the laws of heterogeneity 1–4

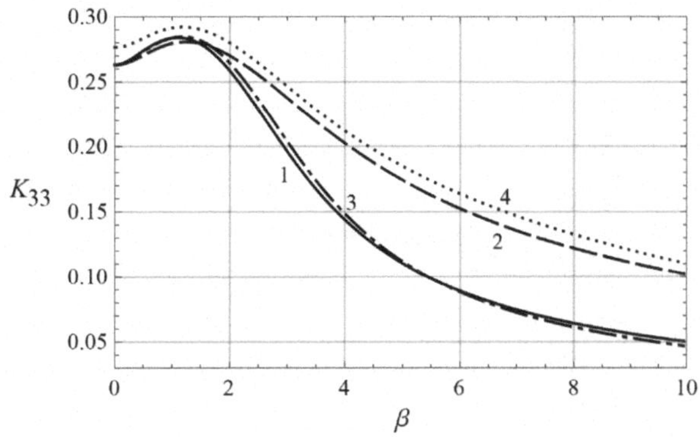

Fig. 34.2 Transfer function $K_{33}(\beta)$ for the laws of heterogeneity 1–4

transfer function graphs can be divided into two groups. The graphs turned out to be close for the laws of heterogeneity 1, 3 and 1, 4, respectively; one of the possible reasons for this circumstance is the proximity of their values at the upper boundary of the strip.

34.7 Calculation of Displacements of the Upper Border of the Strip

The displacement of the upper boundary of the strip under the action of normal (problem 1) or tangential (problem 2) force are found using the inverse Fourier transform, formulas (34.16) and (34.17), respectively.

$$v_1(\xi_1, 1) = \frac{1}{2\pi i} \int\limits_{-\infty}^{\infty} K_{13}(\beta) e^{-i\beta\xi_1} d\beta, \; v_3(\xi_1, 1) = \frac{1}{2\pi} \int\limits_{-\infty}^{\infty} K_{33}(\beta) e^{-i\beta\xi_1} d\beta. \tag{34.16}$$

$$v_1(\xi_1, 1) = \frac{1}{2\pi} \int\limits_{-\infty}^{\infty} K_{11}(\beta) e^{-i\beta\xi_1} d\beta, \; v_3(\xi_1, 1) = -\frac{1}{2\pi i} \int\limits_{-\infty}^{\infty} K_{13}(\beta) e^{-i\beta\xi_1} d\beta. \tag{34.17}$$

The integrals in (16) and (17) are calculated by direct numerical integration. Integration is done over the final segment $[-R, \; R]$, which is divided into N sections $[a_k, b_k]$. In each of the sections, the integral is calculated using the Filon method [5] for oscillating functions.

In Figs. 34.3 and 34.4 graphs of displacements of the upper boundary of the strip v_1, v_3 for problems 1 and 2 are presented. The laws of heterogeneity with larger

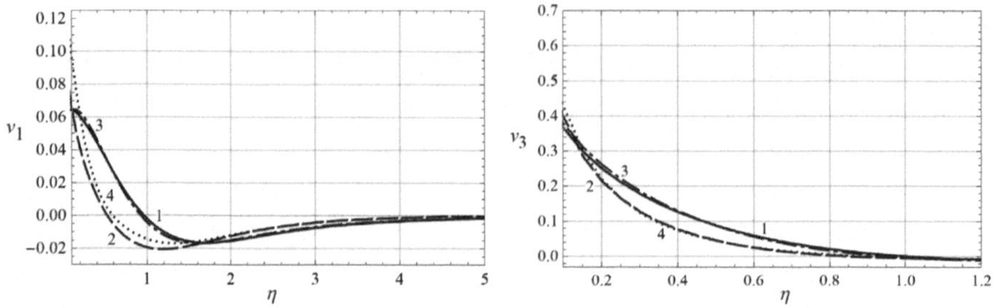

Fig. 34.3 Displacements of the upper boundary of the strip in problem 1 for the laws of heterogeneity 1–4

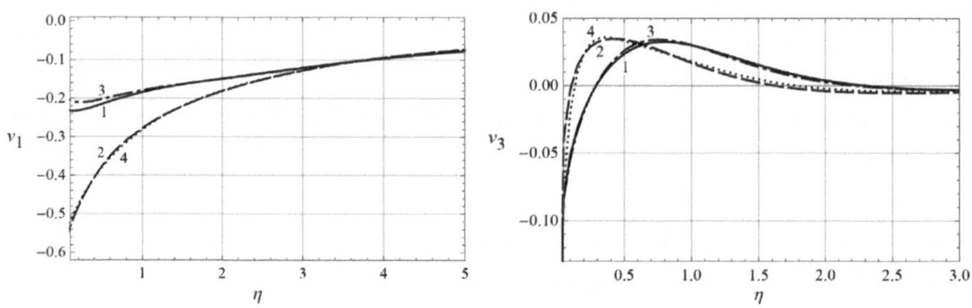

Fig. 34.4 Displacements of the upper boundary of the strip in problem 2 for the laws of heterogeneity 1–4

values of the elastic moduli at the upper boundary of the strip give larger values of displacements in the direction of the applied load.

In [14], an approximate model of deformation for an inhomogeneous elastic strip is presented. The model is based on hypotheses about the nature of the components of the displacement field in the form

$$\begin{gathered} u_1 = \psi_1(x_3)u(x_1), \quad u_3 = \psi_3(x_3)w(x_1), \\ \psi_1 = g_1(x_3)/g_1(h), \quad \psi_3 = g_3(x_3)/g_3(h), \\ g_1 = \int\limits_0^{x_3} \frac{1}{\mu}\mathrm{d}x_3, \quad g_3 = \int\limits_0^{x_3} \frac{1}{\lambda + 2\mu}\mathrm{d}x_3 \end{gathered} \tag{34.18}$$

Below is a comparison of a solution constructed using a computational scheme with an approximate solution based on a model (34.18). Figure 34.5 shows the displacements of the upper boundary of the strip in problem 1 for the law of heterogeneity 1, the solid curve denotes the solution constructed by the formulas (34.17), and the dashed curve is a solution within the model (18). There is some discrepancy in the results in the near zone; when moving away from it, the results are close.

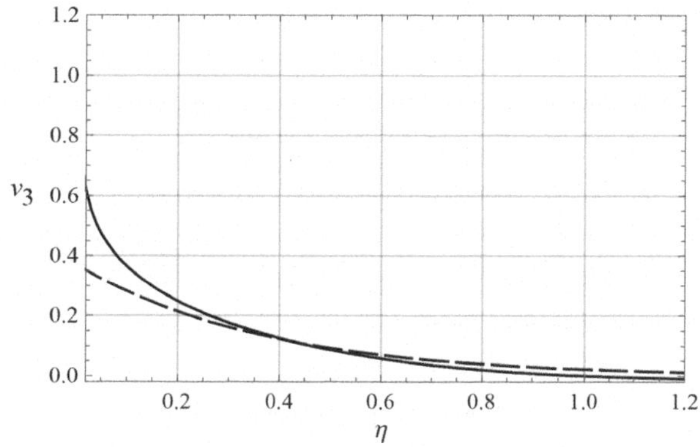

Fig. 34.5 Vertical displacement of the upper boundary of the strip in problem 1 for the law 1

For other laws of changing the type of heterogeneity, the results are similar. Note that horizontal displacements are much smaller than vertical ones, and the discrepancy between the results is very significant.

34.8 Conclusion

Study of the equilibrium of an elastic strip with inhomogeneous material properties in thickness under the action of concentrated forces is carried out. A canonical first-order system of differential equations with variable coefficients with respect to Fourier transforms is formulated. A computational scheme based on the shooting method has been developed, and program for solving auxiliary Cauchy problems for the constructed system of differential equations with parameter is developed. Based on the developed scheme, transfer functions connecting transformants of the displacements of upper boundary of the strip with the transformants of the applied loads are constructed. Comparative analysis of transfer functions for various laws of strip heterogeneity is carried out.

Based on the asymptotic analysis of the boundary value problem, approximate expressions of transfer functions for small and large values of the Fourier transform parameter are found.

The displacements of the upper border of the strip for various methods of concentrated loading are determined. The influence of the laws of strip heterogeneity on the solution of the problem is analyzed. Comparison with applied theory is carried out.

The research had been supported by the Russian Science Foundation: project No. 18-11-00069.

References

1. Aleksandrov, V.M., Chebakov, M.I.: Introduction to the mechanics of contact interactions. CVVR, Rostov-on-Don (2007). [In Russian]
2. Aleksandrov, V.M., Mkhitaryan, S.M.: Contact problems for bodies with thin coatings and interlayers. Nauka, Moscow (1983). [In Russian]
3. Aleksandrov, V.M., Vorovich, I.I.: On the action of a stamp on an elastic layer of finite thickness. J. Appl. Math. Mech. **24**, 323–333 (1960). [in Russian]
4. Ayzikovich, S.M., Aleksandrov, V.M., Belokon, A.V., Krenev, L.I., Trubchik, I.S.: Contact problems of the theory of elasticity for inhomogeneous media. FIZMATLIT, Moscow (2006). [In Russian]
5. Bakhvalov, N.S.: Numerical methods (analysis, algebra, ordinary differential equations). Nauka, Moscow (1975). [In Russian]
6. Filon, L.N.G.: In the approximate solution for the bending of a beam of rectangular cross-section under any system of load, with special reference to points of concentrated or discon-tinuous loading. Phil. Trans. Roy. Soc. Lond. **201**, 491–496 (1903)
7. Kolchin, G.B.: Calculation of structural elements from elastic heterogeneous materials. Kartya Moldovenyaske, Kishinev (1971). [In Russian]
8. Lomakin, V.A.: Elasticity theory of inhomogeneous bodies. MSU Publ, Moscow (1976). [In Russian]
9. Novatskiy, V.: Theory of elasticity. Mir, Moscow (1975). [In Russian]
10. Pasternak, P.L.: Fundamentals of a new method for calculating foundations on an elastic foundation using two bed coefficients. Stroyizdat, Moscow (1954). [In Russian]
11. Sneddon, I.: Fourier transform. Foreign literature, Moscow (1995). [In Russian]
12. Uflyand, Y.S.: Integral transformations in problems of the theory of elasticity. Nauka, Leningrad (1968). [In Russian]
13. Vasiliev, A.S., Volkov, S.S., Aizikovich, S.M.: Approximated analytical solution of contact problem on indentation of elastic half-space with coating reinforced with inhomogeneous in-terlayer. Mater. Phys. Mech. **35**, 175–180 (2018)
14. Vatulyan, A.O., Plotnikov, D.K.: A model of indentation for a functionally graded strip. Dokl. Phys. **64**, 173–175 (2019)
15. Vatulyan A.O., Plotnikov D.K., Poddubny A.A.: On some models of indentation for functionally graded coatings. Izv. Saratov Univ. (N.S.), Ser. Math. Mech. Inform. **18**, 421–432 (2018). [In Russian]
16. Vazov, V.: Asymptotic expansions of solutions of ordinary differential equations. Mir, Moscow (1968). [In Russian]
17. Vlasov, V.Z.: Selected works. Thin-walled spatial systems, vol. 3. Nauka, Moscow (1964). [In Russian]
18. Vlasov, V.Z., Leont'yev, N.N.: Beams, plates and shells on an elastic foundation. Nauka, Moscow (1964). [In Russian]
19. Volkov, S.S., Vasil'yev, A.S., Ayzikovich, S.M., Seleznev, N.M., Leont'yeva, A.V.: Stress-strain state of an elastic soft functionally graded coating with the introduction of a spherical indenter. PNRPU Mech. Bull. **4**, 20–34 (2016). [in Russian]
20. Vorovich, I.I., Ustinov, YuA: On the pressure of a stamp on a layer of finite thickness. J. Appl. Math. Mech. **23**, 445–455 (1959). [In Russian]
21. Vorovich, I.I., Aleksandrov, V.M., Babeshko, V.A.: Nonclassical mixed problems of elasticity theory. Nauka, Moscow (1974). [In Russian]

Chapter 35
Electroelastic Deformation of a Piezoelectric FGM Coated Half-Plane Caused by an Electrostatic Potential Difference Applied to the Strip Electrode on the Surface

Sergey S. Volkov, Andrey S. Vasiliev, Sergei M. Aizikovich, and Alexandr N. Litvinenko

Abstract Solution of a new contact problem is constructed on determining the displacements and electric induction of a piezoelectric half-plane with a piezoelectric FGM coating during electroelastic deformation caused by the electrostatic potential difference applied to the strip electrode on the coating surface and to the substrate. The coating and substrate are transverselly isotropic, the axis of isotropy coincides with the axis of polarization and is normal to the surface of the coating. Electroelastic properties of the coating vary with depth according to arbitrary laws independent of each other. Using integral Fourier transforms, the problem reduced to solution of a dual integral equation, which is solved using the bilateral asymptotic method. Approximated analytical expressions are obtained for electric induction, vertical and horizontal displacements under the electrode. These expressions are asymptotically exact for large and small values of the relative thickness of the coating. Specific non-uniform distribution of the displacements is obtained which is not observed for non-coated materials.

Keywords Piezoelectricity · Elasticity · Coatings · Functionally graded materials · Plane deformation · Electrode · Contact mechanics

---

S. S. Volkov · A. S. Vasiliev (✉) · S. M. Aizikovich
Don State Technical University, 1 Gagarin Sq., Rostov-on-Don 344000, Russia
e-mail: andre.vasiliev@gmail.com

S. S. Volkov
e-mail: fenix_rsu@mail.ru

S. M. Aizikovich
e-mail: saizikovich@gmail.com

A. N. Litvinenko
South Federal University, 105/42 Bolshaya Sadovaya str., Rostov-on-Don 344006, Russia
e-mail: litva@sfedu.ru

H. Altenbach et al. (eds.), *Multiscale Solid Mechanics*,
Advanced Structured Materials 141,
https://doi.org/10.1007/978-3-030-54928-2_35

35.1 Introduction

Piezomaterials are a widespread class of materials that can convert mechanical energy into electrical energy and vice versa. They find application in engineering and microelectronics, where they are used as sensors, electromechanical transducers, etc. [1–4]. In this regard, interest in mathematical modeling of structural elements from piezoelectric materials has increased. Two-dimensional contact problems were solved in an exact analytical form: indentation by conducting and insulating punches [5, 6], sliding contact [7], frictional sliding contact [8] and indentation of materials with general case of symmetry of class 6 [9]. Exact solution for frictional sliding contact for a magnetoelectroelastic half-plane was obtained in [10, 11]. General correspondence principle between elastic and piezoelectric problems was established for transversely isotropic homogeneous materials, in such a way that the knowledge of an elastic solution yields fully coupled electroelastic fields for the corresponding piezoelectric problem [12]. Application of these results for the interpretation of piezoresponse force microscopy and scanning probe microscopy is discussed in [13, 14].

Functionally graded materials (FGMs) are the composite materials with properties varying continuously in one direction. FGMs are widely used as a thermal barrier coatings [15], to improve the performance and lifetime of piezoelectric materials, to reduce stress concentration, improve residual stress distribution, and reduce delamination [16], to improve biocompatibility [17] and implant life [18], etc. Therefore, a huge amount of research is devoted to modeling of FGMs [19–24].

Much attention is paid to study of various types of two-dimensional plane and axisymmetric contact of piezoelectric functionally graded coatings in last decade: frictionless indentation [25, 26], sliding frictional contact [27], fretting sliding contact [28, 29], torsion [30], etc. Most of the results are obtained only for an exponential variation of properties inside the coating. It should be also mentioned a large number of papers where the contact of homogeneous piezoelectric coatings was studied [31–34].

Present paper continues the investigation of contact mechanics of piezoelectric FGM coatings with arbitrary independent variation of all properties in depth of the coating [35, 36]. New contact problem on electroelastic deformation of a piezoelectric FGM coating caused by an electrostatic potential difference applied through a strip electrode on a surface is considered. The solution is obtained in an approximated analytical form using the original semi-analytical method developed by the authors earlier [37]. This method was effectively used before for solution of various contact problems for FGM coatings [38, 39] and FGM interlayers [40]. Analytical expressions are obtained for the distribution of the electric induction, shear and normal displacements on the surface. Specific non-uniform distribution of the displacements is obtained which is not observed for uncoated materials.

35.2 Contact Problem Statement

Let us consider a piezoelectric electroelastic half-plane with a functionally graded coating of thickness H. The coating and the substrate materials are piezoelectric transversely isotropic. Let us use the Cartesian coordinate system (x, z), the z-axis coincides with the axis of preliminary polarization of piezoelectric material. Elastic moduli $c_{11}, c_{12}, c_{13}, c_{33}, c_{44}$, piezoelectric moduli e_{31}, e_{15}, e_{33} and dielectric permeabilities $\varepsilon_{11}, \varepsilon_{33}$ in the coating are described by continuously differentiable functions of depth coordinate—$c_{kj}^{(c)}(z)$, $e_{kj}^{(c)}(z)$, $\varepsilon_{kj}^{(c)}(z)$, and are constant in the substrate—$c_{kj}^{(s)}$, $e_{kj}^{(s)}$, $\varepsilon_{kj}^{(s)}$. Hereinafter, indices (c) and (s) denote the characteristics corresponding to the coating and substrate, respectively.

In the framework of linear electroelasticity, taking into account the piezoelectric effect, the constitutive equations are:

$$\frac{\partial \sigma_x}{\partial x} + \frac{\partial \tau_{xz}}{\partial z} = 0; \ \frac{\partial \sigma_z}{\partial z} + \frac{\partial \tau_{xz}}{\partial x} = 0; \ \frac{\partial D_x}{\partial x} + \frac{\partial D_z}{\partial z} = 0 \tag{35.1}$$

$$\sigma_x = c_{11}\frac{\partial u}{\partial x} + c_{13}\frac{\partial w}{\partial z} + e_{31}\frac{\partial \varphi}{\partial z} \tag{35.2}$$

$$\sigma_z = c_{13}\frac{\partial u}{\partial x} + c_{33}\frac{\partial w}{\partial z} + e_{33}\frac{\partial \varphi}{\partial z} \tag{35.3}$$

$$\tau_{xz} = c_{44}\left(\frac{\partial u}{\partial z} + \frac{\partial w}{\partial x}\right) + e_{15}\frac{\partial \varphi}{\partial x} \tag{35.4}$$

$$D_x = e_{15}\left(\frac{\partial u}{\partial z} + \frac{\partial w}{\partial x}\right) - \varepsilon_{11}\frac{\partial \varphi}{\partial x} \tag{35.5}$$

$$D_z = e_{31}\frac{\partial u}{\partial x} + e_{33}\frac{\partial w}{\partial z} - \varepsilon_{33}\frac{\partial \varphi}{\partial z} \tag{35.6}$$

Perfect bonding is assumed on the coating-substrate interface:

$$z = -H: \quad w^{(c)} = w^{(s)}, \ u^{(c)} = u^{(s)}, \ \varphi^{(c)} = \varphi^{(s)} \tag{35.7}$$

$$z = -H: \quad \sigma_z^{(c)} = \sigma_z^{(s)}, \ \tau_{xz}^{(c)} = \tau_{xz}^{(s)}, \ D_z^{(c)} = D_z^{(s)} \tag{35.8}$$

Here, u, w are horizontal and vertical displacements; σ_x, σ_z τ_{xz} are the components of the stress tensor; φ is the electrostatic potential; D_x, D_z are components of the vector of electric induction.

A strip electrode of width $2a$ is located on the surface of the coating. The electrostatic potential difference φ_0 is applied to the electrode and the infinitely distant boundary of the substrate. Outside the electrode, the coating surface is insulated and free from mechanical stresses:

$$z = 0 : \tau_{xz} = 0, \sigma_z = 0, \quad \begin{cases} \varphi = -\varphi_0, & |x| \leq a \\ D_z = 0, & |x| > a \end{cases} \tag{35.9}$$

$$z \to \infty : \quad \varphi(x, z), u(x, z), w(x, z) \to 0 \tag{35.10}$$

The paper addresses to the determination of displacements under the electrode and electric induction. Let us notate distribution of electric induction on the surface:

$$D_z|_{z=0} = -g_a(x), \quad |x| \leq a \tag{35.11}$$

35.3 Dual Integral Equation and Its Solution

Fourier transforms of displacements U, W and electric potential Φ on the surface can be written as [41]:

$$\begin{gathered} U(\alpha, 0) = -2i\frac{L_{13}(\alpha)}{E_{13}^{(c)}|\alpha|}G_a(\alpha), \quad W(\alpha, 0) = -2\frac{L_{23}(\alpha)}{E_{23}^{(c)}|\alpha|}G_a(\alpha) \\ \Phi(\alpha, 0) = -2\frac{L_{33}(\alpha)}{E_{33}^{(c)}|\alpha|}G_a(\alpha) \end{gathered} \tag{35.12}$$

Here, $G_a(\alpha)$ is the Fourier transform of electric induction on the surface, $L_{ij}(\alpha)$ are the compliance functions. For them $\lim_{\alpha\to 0}L_{ij}(\alpha) = E_{ij}^{(c)}/E_{ij}^{(s)}$ is fulfilled, where $E_{ij}^{(c)}, E_{ij}^{(s)}$ are effective electroelastic moduli describing a bulk piezoelectric material with properties corresponding to the coating surface ($z = 0$)—$E_{ij}^{(c)}$ or to the substrate ($z < -H$)—$E_{ij}^{(s)}$.

The calculation of the compliance functions reduces to solution of a boundary value problem for a system of ordinary differential equations with variable coefficients and is described in details earlier [42] for the more general case of the electromagnetoelastic half-plane. It is also shown that the functions L_{23} and L_{33} are even, while L_{13} is odd. The calculation of effective electroelastic moduli is described in [5, 43].

Let us introduce the dimensionless variables:

$$\lambda = \frac{H}{a}, \ x' = \frac{x}{a}, \ \alpha' = \alpha a, \ g(x') = g_a(x'a), \ L'_{kj}(\alpha) = L_{kj}\left(\frac{\alpha}{H}\right) \tag{35.13}$$

Applying the integral transformations to (35.9) and taking into account the introduced notations, omitting the primes, we obtain a dual integral equation:

$$\begin{cases} \int\limits_0^\infty G(\alpha)\,\frac{L_{33}(\lambda\alpha)}{\alpha}\cos(\alpha x)\,\mathrm{d}\alpha = \pi E_{33}^{(c)}\frac{\varphi_0}{2a}, & |x| \le 1, \\ \int\limits_0^\infty G(\alpha)\cos(\alpha x)\,\mathrm{d}\alpha = 0, & |x| > 1. \end{cases} \tag{35.14}$$

Here, $G(\alpha)$ is the Fourier transform of the function $g(x)$.

Using the idea of [37], we approximate the compliance functions by the following expressions:

$$L_{ij}(\lambda\alpha) \approx L_{ij}^{+}(\lambda\alpha), \quad L_{ij}^{\pm}(\lambda\alpha) = \prod_{n=1}^{N_{ij}} \frac{\left(\lambda^2\alpha^2 \pm A_{ijn}^2\right)}{\left(\lambda^2\alpha^2 \pm B_{ijn}^2\right)} \tag{35.15}$$

Parameters (A_{ijn},B_{ijn}), n=1,2,…,N_{ij} are selected so that $A_{ijk} \neq B_{lmn}$, $A_{ijk} \neq A_{ijn}$, $B_{ijk} \neq B_{ijn}$ and $L_{ij}(0)=L_{ij}^{+}(0)$.

The solution of the dual integral Eq. (35.14) has the following form [44]:

$$G(\alpha) = \frac{Q}{a}\left(J_0(\alpha) - \sum_{i=1}^{N_{33}} \frac{C_i\alpha J_1(\alpha)}{\alpha^2 + A_{33i}^2\lambda^{-2}}\left(\frac{A_{33i}I_0(A_{33i}\lambda^{-1})}{\lambda I_1(A_{33i}\lambda^{-1})} - \frac{\alpha J_0(\alpha)}{J_1(\alpha)}\right)\right) \tag{35.16}$$

Here, Q is a total electric charge on the electrode, constants C_i are determined from the following system of linear algebraic equations:

$$\sum_{i=1}^{N_{33}} \frac{C_i}{A_{33i}^2 - B_{33n}^2}\left(A_{33i}\frac{K_1\left(B_{33n}\lambda^{-1}\right)I_0\left(A_{33i}\lambda^{-1}\right)}{K_0\left(B_{33n}\lambda^{-1}\right)I_1\left(A_{33i}\lambda^{-1}\right)} + B_{33n}\right) = \frac{1}{B_{33n}} \tag{35.17}$$

Here, $n = 1, \ldots, N_{33}$; I_0, I_1 and K_0, K_1 are the modified Bessel functions of the first and second kind, respectively.

After inverting the Fourier transform in (35.16), we obtain:

$$g(x) = \frac{Q}{\pi a}\left(\frac{1}{\sqrt{1-x^2}} + \sum_{i=1}^{N_{33}} C_i\left(\frac{1}{\sqrt{1-x^2}} - \frac{A_{33i}}{\lambda}g_{\mathrm{coat}}\left(x, \frac{A_{33i}}{\lambda}\right)\right)\right) \tag{35.18}$$

$$g_{\mathrm{coat}}(x, A) = \frac{I_0(A)}{I_1(A)}\int\limits_x^1 \frac{t\cosh(A(t-x))}{\sqrt{1-t^2}}dt - \int\limits_x^1 \frac{\sinh(A(t-x))}{\sqrt{1-t^2}}dt \tag{35.19}$$

Expression (35.18) is asymptotically exact for $\lambda \to 0$ and $\lambda \to \infty$. The error for an arbitrary value of λ was studied earlier for similar problems [45, 46] and it was established that it is a quantity of the same order of magnitude as the error of approximation of the compliance function L_{33}.

35.4 Displacements of the Surface

Inverting the Fourier transform in (35.12) and taking into account the properties of the evenness and oddness of the compliance functions, we rewrite the displacements and electrostatic potential on the surface in the form of some quadratures:

$$
\begin{aligned}
u(x,0) &\approx -\frac{2Q}{\pi E_{13}^{(c)}}\int_0^\infty \frac{G(\alpha)}{\alpha} L_{13}^{+}(\lambda\alpha)\sin(\alpha x)\mathrm{d}\alpha \\
w(x,0) &\approx -\frac{2Q}{\pi E_{23}^{(c)}}\int_0^\infty \frac{G(\alpha)}{\alpha} L_{23}^{+}(\lambda\alpha)\cos(\alpha x)\mathrm{d}\alpha \\
\varphi(x,0) &\approx -\frac{2Q}{\pi E_{33}^{(c)}}\int_0^\infty \frac{G(\alpha)}{\alpha} L_{33}^{+}(\lambda\alpha)\cos(\alpha x)\mathrm{d}\alpha
\end{aligned}
\tag{35.20}
$$

We introduce dimensionless expressions:

$$
u_s(x) = \frac{E_{13}^{(s)}}{Q}u(x,0), \quad w_s(x) = \frac{E_{23}^{(s)}}{Q}w(x,0), \ \varphi_s(x) = \frac{E_{33}^{(s)}}{Q}\varphi(x,0), \ \beta_{ij} = \frac{E_{ij}^{(s)}}{E_{ij}^{(c)}}
\tag{35.21}
$$

Calculating the quadratures in (35.20), we obtain the expressions for the displacements and electric potential under the electrode:

$$
\varphi_s(x) = -d_1 + \frac{2\beta_{33}}{\pi}\lambda\sum_{n=1}^{N_{33}} \frac{K_0(B_{33n}\lambda^{-1})F_{33n}}{K_1(B_{33n}\lambda^{-1})B_{33n}}, \quad |x| \le 1
\tag{35.22}
$$

$$
u_s(x) = -\frac{2}{\pi}\arcsin(x) + \frac{2\beta_{13}}{\pi}u_{\mathrm{coat}}(x,\lambda), \ |x| \le 1
\tag{35.23}
$$

$$
w_s(x) = -d_1 + \frac{2\beta_{23}}{\pi}w_{\mathrm{coat}}(x,\lambda), \ |x| \le 1
\tag{35.24}
$$

$$
\begin{aligned}
u_{\text{coat}}(x,\lambda) = \sum_{m=1}^{N_{33}} C_m L_{13}^{-}(A_{33m}) \left(Z_1\left(x, \tfrac{A_{33m}}{\lambda}\right) - \tfrac{I_0\left(A_{33m}\lambda^{-1}\right)}{I_1\left(A_{33m}\lambda^{-1}\right)} Z_2\left(x, \tfrac{A_{33m}}{\lambda}\right)\right) \\
+ \sum_{n=1}^{N_{13}} \left[\tfrac{\pi}{2} \left(F_{13n} I_0\left(\tfrac{B_{13n}}{\lambda}\right) - \tfrac{P_{13n}}{B_{13n}} I_1\left(\tfrac{B_{13n}}{\lambda}\right)\right) e^{-\frac{B_{13n}}{\lambda}x} + \right. \\
\left. + \tfrac{P_{13n}}{B_{13n}} Z_2\left(x, \tfrac{B_{13n}}{\lambda}\right) - F_{13n} Z_1\left(x, \tfrac{B_{13n}}{\lambda}\right)\right],
\end{aligned}
\tag{35.25}
$$

$$
Z_1(x,A) = \int_x^1 \frac{\cosh(A(t-x))}{\sqrt{1-t^2}} \mathrm{d}t, \quad Z_2(x,A) = \int_x^1 \frac{t\sinh(A(t-x))}{\sqrt{1-t^2}} \mathrm{d}t \tag{35.26}
$$

$$
\begin{aligned}
w_{\text{coat}}(x,\lambda) = \sum_{m=1}^{N_{33}} C_m L_{23}^{-}(A_{33m}) \left[-K_0\left(\tfrac{A_{33m}}{\lambda}\right) \cosh\left(\tfrac{A_{33m}}{\lambda}x\right) + \right. \\
\left. + \tfrac{I_0\left(A_{33m}\lambda^{-1}\right)}{I_1\left(A_{33m}\lambda^{-1}\right)} \left(\tfrac{\lambda}{A_{33m}} - K_1\left(\tfrac{A_{33m}}{\lambda}\right) \cosh\left(\tfrac{A_{33m}}{\lambda}x\right)\right)\right] + \\
+ \sum_{n=1}^{N_{23}} \left(\left(F_{23n} K_0\left(\tfrac{B_{23n}}{\lambda}\right) + \tfrac{P_{23n}}{B_{23n}} K_1\left(\tfrac{B_{23n}}{\lambda}\right)\right) \cosh\left(\tfrac{B_{23n}}{\lambda}x\right) - \tfrac{\lambda P_{23n}}{B_{23n}^2}\right)
\end{aligned}
\tag{35.27}
$$

$$
d_1 = \frac{2}{\pi} \int_0^\infty \frac{J_0(\alpha)}{\alpha} \mathrm{d}\alpha = \infty \tag{35.28}
$$

$$
\begin{aligned}
F_{ijn} &= \Pi_{ijn} \left(\tfrac{1}{B_{ijn}^2} + \sum_{m=1}^{N_{33}} \tfrac{C_m}{\left(B_{ijn}^2 - A_{33m}^2\right)}\right), \\
P_{ijn} &= \Pi_{ijn} \sum_{m=1}^{N_{33}} \tfrac{C_m A_{33m}}{\left(B_{ijn}^2 - A_{33m}^2\right)} \tfrac{I_0\left(A_{33m}\lambda^{-1}\right)}{I_1\left(A_{33m}\lambda^{-1}\right)}, \\
\Pi_{ijn} &= \left(A_{ijn}^2 - B_{ijn}^2\right) \prod_{m=1, m\neq n}^{N_{ij}} \tfrac{A_{ijm}^2 - B_{ijn}^2}{B_{ijm}^2 - B_{ijn}^2}
\end{aligned}
\tag{35.29}
$$

Note that for a homogeneous half-plane without a coating, dimensionless displacements and electric potential are determined similarly to the plane elastic contact problem on indentation by a rigid punch and have the following form [48]:

$$
|x| \leq 1 : u_s^{\text{hom}}(x) = -\frac{2}{\pi} \arcsin(x) \quad w_s^{\text{hom}}(x) = -d_1, \; \varphi_s^{\text{hom}}(x) = -d_1 \tag{35.30}
$$

It is important to note that, as in the case of a homogeneous half-plane [47], displacements and electrostatic potential are determined up to a some constant. When determining horizontal displacements, similarly to [48], it is assumed that the center of the electrode does not shift horizontally (deformation occurs symmetrically).

Expression (35.22) is a constant value with respect to the coordinate, which corresponds to the given boundary conditions. Taking into account (35.21) and (35.9), it can be seen that expression (35.22) defines the relationship between the electrostatic potential difference applied to the piezoelectric material and the total electric charge on the electrode. This ratio depends on the dimensionless coating thickness λ and is determined to within an infinite constant d_1.

35.5 Numerical Results and Discussion

Let the substrate be made of PZT-5H piezoceramics and have the properties [49]: $\varepsilon_{11} = 15.04 \times 10^{-9}\ \frac{C}{V\times m}$, $\varepsilon_{33} = 13 \times 10^{-9}\ \frac{C}{V\cdot m}$, $c_{11} = 126$ GPa, $c_{13} = 84.1$ GPa, $c_{33} = 117$ GPa, $c_{44} = 23$ GPa, $e_{15} = 17\ \frac{C}{m^2}$, $e_{31} = -6.5\ \frac{C}{m^2}$, $e_{33} = 23.3\ \frac{C}{m^2}$. For numerical illustrations for simplicity, let us consider only a homogeneous coating, whose properties are two times greater than that of the substrate: $c_{kj}^{(c)}(z) = \text{const} = 2c_{kj}^{(s)}$, $e_{kj}^{(c)}(z) = \text{const} = 2e_{kj}^{(s)}$, $\varepsilon_{kj}^{(c)}(z) = \text{const} = 2\varepsilon_{kj}^{(s)}$.

Let us introduce a relative value characterizing redistribution of horizontal displacements in comparison with a homogeneous half-plane with properties that coincide with the substrate and for the same value of the total electric charge (Q=Q_{hom}):

$$u_s^{\text{rel}}(x) = \frac{u_s(x)}{u_s^{\text{hom}}(x)} \tag{35.31}$$

Figures 35.1 and 35.2 show graphs of the quantities $u_s(x)$ and $u_s^{\text{rel}}(x)$ under the electrode.

It is clear that if the distribution of $u_s^{\text{rel}}(x)$ is close to constant, then the presence of a coating has an insignificant effect on the shear displacements, since in this case, they can be described by a homogeneous half-plane with some modified effective elastic modulus, taking into account the properties of the coating and substrate. Figure 35.2 shows that for small and medium values of λ, the distribution of $u_s^{\text{rel}}(x)$ is different from constant, which indicates that the tangential displacements for the coated half-plane are significantly different from those for the homogeneous half-plane. The most

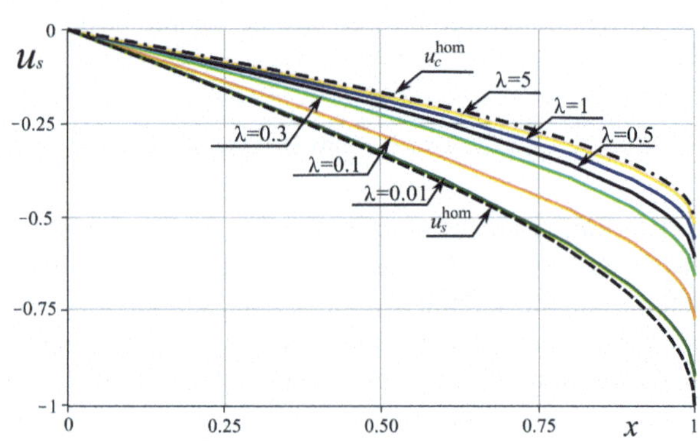

Fig. 35.1 Horizontal displacements $u_s(x)$ for various values of the relative coating thickness λ

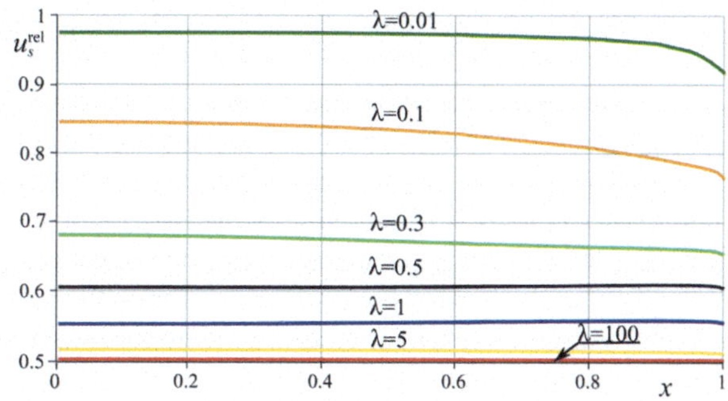

Fig. 35.2 Relative horizontal displacements $u_s^{rel}(x)$ for various values of the relative coating thickness λ

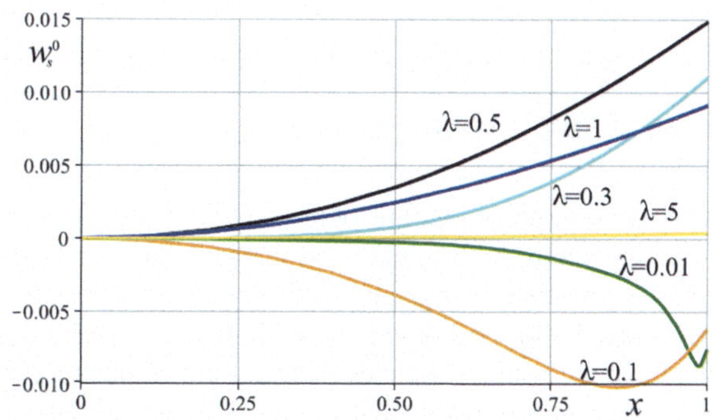

Fig. 35.3 Vertical displacements relative to the center point of the electrode at various values of the relative coating thickness λ

clearly redistribution of displacements is noticeable near the edge of the electrode. For large and very small values of the parameter λ, the shear displacements reach the limiting values corresponding to the values for the half-plane with the properties of the coating—$u_s^{rel}(x)$ (everywhere close to 0.5) and the substrate—$u_s^{rel}(x)$ (close to 1 everywhere except for the vicinity of the electrode edge), respectively.

Since vertical displacements are determined to within a constant, it is convenient to analyze the difference of displacements under the electrode with the value in the center:

$$w_s^0(x) = w_s(x) - w_s(0) \tag{35.32}$$

Figure 35.3 illustrates the distribution of vertical displacements for a coating. Unlike a homogeneous half-plane without a coating the vertical displacements under the electrode are not constant. The distribution of vertical displacements varies greatly depending on the thickness of the coating. For coatings with a large relative thickness, the displacements at the edge exceed the displacements of the central point (the edges of the electrode bend upward) and increase monotonically with distance from the center of the electrode. As the coating $w_s^0(x)$ thickness increases, it tends to zero uniformly. The maximum displacements in magnitude are observed at $\lambda \approx 0.5$.

With decreasing the coating thickness, the displacement distribution becomes nonmonotonic (decreases in the vicinity of the center and increases when approaching the edge of the electrode), this is noticeable at $\lambda < 0.4$ for a homogeneous coating considered. In a certain narrow range of λ, the graphs become alternating, i.e., both larger and smaller values are observed than at the central point, for example, at $\lambda = 0.3$. Starting from some small value of λ (0.2 for the considered homogeneous coating), the value of $w_s^0(x)$ becomes negative in the entire range under the electrode, i.e., the displacement of the central point is maximum. In this case, a point of minimum displacements is observed near the edge, and as it approaches the edge of the electrode, the value $w_s^0(x)$ increases (the edge of the electrode bends locally upward). With a further decreasing the coating thickness, the value of $w_s^0(x)$ becomes close to zero everywhere, except for a narrow neighborhood near the edge of the electrode, where a similar nonmonotonic behavior is observed, however, this region becomes smaller by decreasing λ.

35.6 Conclusion

Analytical expressions are obtained for the distribution of surface displacements under an electrode located on the surface of a piezoelectric FGM coating. It is shown that the presence of a coating fundamentally changes the displacements under the electrode. Unlike a homogeneous half-plane without a coating, for which normal displacements under the electrode are constant, the presence of even a thin coating leads to a non-uniform distribution of vertical displacements over the electrode, which leads to bending of the electrode. The most redistribution of both vertical and horizontal displacements is noticeable for thin coatings (the half-width of the electrode is 3–10 times greater than the thickness of the coating) and near the edge of the electrode.

Acknowledgements This research was supported by the Russian Science Foundation (RSF) through grant no. 19-19-00444.

References

1. Fetisov, Y.K., Srinivasan, G.: Electric field tuning characteristics of a ferrite-piezoelectric microwave resonator. Appl. Phys. Lett. **88**(14), 143503 (2006)
2. Zhai, J., Dong, S., Xing, Z., Li, J., Viehland, D.: Geomagnetic sensor based on giant magnetoelectric effect. Appl. Phys. Lett. **91**(12), 123513 (2007)
3. Wang, M.L., Wang, G.: Electromagnetic sensors for assessing and monitoring civil infrastructure. In: Sensor Technologies for Civil Infrastructures, vol. 1, pp. 238–264. Woodhead Publishing Series in Electronic and Optical Materials, Elsevier Science (2014)
4. Soloviev, A.N., Oganesyan, P.A., Lupeiko, T.G., Kirillova, E.V.: Modeling of non-uniform polarization for multi-layered piezoelectric transducer for energy harvesting devices. In: Springer Proceedings in Physics, vol. 175, pp. 651–658 (2016)
5. Giannakopoulos, A.E., Suresh, S.: Theory of indentation of piezoelectric materials. Acta Mater. **4**(7), 2153–2164 (1999)
6. Chen, W.Q., Ding, H.: Indentation of a transversely isotropic piezoelectric half-space by a rigid sphere. Acta Mech. Solida Sinica **12**, 114–120 (1999)

7. Zhou, Y.T., Lee, K.Y.: Exact solutions of a new, 2D frictionless contact model for orthotropic piezoelectric materials indented by a rigid sliding punch. Philos. Mag. **92**(15), 1937–1965 (2012)
8. Makagon, A., Kachanov, M., Karapetian, E., Kalinin, S.V.: Piezoelectric indentation of a flat circular punch accompanied by frictional sliding and applications to scanning probe microscopy. Int. J. Eng. Sci. **47**(2), 221–239 (2009)
9. Berndt, E.A., Sevostianov, I.: Action of a smooth flat charged punch on the piezoelectric half-space possessing symmetry of class 6. Int. J. Eng. Sci. **103**, 77–96 (2016)
10. Elloumi, R., Guler, M.A., Kallel-Kamoun, I., El-Borgi, S.: Closed-form solutions of the frictional sliding contact problem for a magneto-electro-elastic half-plane indented by a rigid conducting punch. Int. J. Solids Struct. **50**(24), 3778–3792 (2013)
11. Elloumi, R., Kallel-Kamoun, I., El-Borgi, S., Guler, M.A.: On the frictional sliding contact problem between a rigid circular conducting punch and a magneto-electro-elastic half-plane. Int. J. Mech. Sci. **87**, 1–1 (2014)
12. Karapetian, E., Kachanov, M., Sevostianov, I.: The principle of correspondence between elastic and piezoelectric problems. Arch. Appl. Mech. **72**, 564–587 (2002)
13. Kalinin, S.V., Karapetian, E., Kachanov, M.: Nanoelectromechanics of piezoresponse force microscopy Engineering. Phys. Rev. B Condens. Matter Mater. Phys. **70**(18), 184101 (2004)
14. Makagon, A., Kachanov, M., Kalinin, S.V., Karapetian, E.: Indentation of spherical and conical punches into piezoelectric half-space with frictional sliding: applications to scanning probe microscopy. Phys. Rev. B **76**(6), 1–1 (2007)
15. Koizumi, M.: FGM activities in Japan. Compos. Part B Eng. **28**(1–2), 1–4 (1997)
16. Müller, E., Drašar, Č., Schilz, J., Kaysser, W.A.: Functionally graded materials for sensor and energy applications. Mater. Sci. Eng. A **362**(1–2), 17–39 (2003)
17. Matuła, I., Dercz, G., Barczyk, J.: Titanium/Zirconium functionally graded materials with porosity gradients for potential biomedical applications. Mater. Sci. Technol. **36**(9), 972–977 (2020)
18. Watari, F., Yokoyama, A., Omori, M., Hirai, T., Kondo, H., Uo, M., Kawasaki, T.: Biocompatibility of materials and development to functionally graded implant for bio-medical application. Compos. Sci. Technol. **64**(6), 893–908 (2004)
19. Altenbach, H., Eremeyev, V.A.: Direct approach-based analysis of plates composed of functionally graded materials. Arch. Appl. Mech. **78**(10), 775–794 (2008)
20. Tokovyy, Y., Ma, C.C.: Analytical solutions to the axisymmetric elasticity and thermoelasticity problems for an arbitrarily inhomogeneous layer. Int. J. Eng. Sci. **92**, 1–17 (2015)
21. Petrova, V., Schmauder, S.A.: Theoretical model for the study of thermal fracture of functionally graded thermal barrier coatings with a system of edge and internal cracks. Theor. Appl. Fract. Mech. **108**, 102605 (2020)
22. Volkov, S.S., Litvinenko, A.N., Aizikovich, S.M., Wang, Y.C., Vasiliev, A.S.: Axisymmetric bending of a circular plate with stiff edge on a soft FGM layer. Struct. Eng. Mech. **59**(2), 227–241 (2016)
23. Zelentsov, V.B., Lapina, P.A., Mitrin, B.I., Eremeyev, V.A.: Characterization of the functionally graded shear modulus of a half-space. Mathematics **8**(4), 640 (2016)
24. Zelentsov, V.B., Lapina, P.A., Mitrin, B.I., Kudish, I.I.: An antiplane deformation of a functionally graded half-space. In: Continuum Mechanics and Thermodynamics, pp. 1–12 (2019)
25. Su, J., Ke, L.L., Wang, Y.S.: Axisymmetric frictionless contact of a functionally graded piezoelectric layered half-space under a conducting punch. Int. J. Solids Struct. **90**, 45–49 (2016)
26. Ma, J., Ke, L.L., Wang, Y.S.: Frictionless contact of a functionally graded magneto-electro-elastic layered half-plane under a conducting punch. Int. J. Solids Struct. **51**, 2791–2806 (2016)
27. Su, J., Ke, L.L., El-Borgi, S., Xiang, Y., Wang, Y.S.: An effective method for the sliding frictional contact of a conducting cylindrical punch on FGPMs. Int. J. Solids Struct. **141**, 127–136 (2018)
28. Ma, J., El-Borgi, S., Ke, L.L., Wang, Y.S.: Frictional contact problem between a functionally graded magnetoelectroelastic layer and a rigid conducting flat punch with frictional heat generation. J. Therm. Stresses **39**, 245–277 (2018)

29. Su, J., Ke, L.L., Wang, Y.S.: Fretting contact of a functionally graded piezoelectric layered half-plane under a conducting punch. Smart Mater. Struct. **25**(2), 025014 (2016)
30. Su, J., Ke, L.L., Wang, Y.S., Xiang, Y.: Axisymmetric torsional fretting contact between a spherical punch and an FGPM coating. Appl. Math. Modell. **52**, 576–589 (2017)
31. Çömez, İ., Güler, M.A., El-Borgi, S.: Continuous and discontinuous contact problems of a homogeneous piezoelectric layer pressed by a conducting rigid flat punch. Acta Mech. **231**(3), 957–976 (2020)
32. Wang, J.H., Chen, C.Q., Lu, T.J.: Indentation responses of piezoelectric films. J. Mech. Phys. Solids **56**(12), 3331–3351 (2008)
33. Lapina, P.A., Mitrin, B.I., Kuznetsova, T.A., Lapitskaya, V.A.: Dynamics of thermoelastic frictional wear of a coating with piezoelectric interlayer. MATEC Web of Conferences, vol. 226, p. 03031 (2018)
34. Wu, Y.F., Yu, H.Y., Chen, W.Q.: Indentation responses of piezoelectric layered half-space. Smart Mater. Struct. **22**(1), 015007 (2013)
35. Vasiliev, A.S.: Penetration of a spherical conductive punch into a piezoelectric half-space with a functionally graded coating. Int. J. Eng. Sci. **142**, 230–241 (2019)
36. Volkov, S.S., Vasiliev, A.S., Aizikovich, S.M., Mitrin, B.I.: Axisymmetric indentation of an electroelastic piezoelectric half-space with functionally graded piezoelectric coating by a circular punch. Acta Mech. **230**(4), 1289–1302 (2019)
37. Vasiliev, A.S., Volkov, S.S., Aizikovich, S.M.: Approximated analytical solution of a problem on indentation of an electro-elastic half-space with inhomogeneous coating by a conductive punch. Doklady Phys. **63**(1), 18–22 (2018)
38. Vasiliev, A.S., Swain, M.V., Aizikovich, S.M., Sadyrin, E.V.: Torsion of a circular punch attached to an elastic half-space with a coating with periodically depth-varying elastic properties. Arch. Appl. Mech. **86**(7), 1247–1254 (2016)
39. Kudish, I.I., Volkov, S.S., Vasiliev, A.S., Aizikovich, S.M.: Lubricated point heavily loaded contacts of functionally graded materials. Part 2. Lubricated contacts. Math. Mech. Solids **23**(7), 1081–1103 (2016)
40. Vasiliev, A.S., Volkov, S.S., Aizikovich, S.M.: Approximated analytical solution of contact problem on indentation of elastic half-space with coating reinforced with inhomogeneous interlayer. Mater. Phys. Mech. **35**(1), 175–180 (2018)
41. Vasiliev, A.S., Volkov, S.S., Aizikovich, S.M.: Normal point force and point electric charge in a piezoelectric transversely isotropic functionally graded half-space. Acta Mech. **227**(1), 263–273 (2018)
42. Vasiliev, A.S.: Compliance functions of electromagnetoelastic piezoelectric and piezomagnetic half-plane and half-space with functionally graded or layered coatings. Vestnik Samarskogo Gosudarstvennogo Tekhnicheskogo Universiteta, Seriya Fiziko-Matematicheskie Nauki **23**(3), 475–496 (2019)
43. Melkumyan, S.A., Ulitko, A.F.: Axissymmetric contact problem of electroelasticity for a half-space. Sov. Appl. Mech. **23**(9), 836–843 (1987)
44. Vasiliev, A.S., Volkov, S.S., Aizikovich, S.M., Mitrin, B.I.: Plane contact problem on indentation of a flat punch into a transversely-isotropic half-plane with functionally graded transversely-isotropic coating. Zeitschrift fur Angewandte Mathematik und Physik **68**(1), 4 (2017)
45. Sadyrin, E.V., Vasiliev, A.S., Volkov, S.S., Mitrin, B.I., Aizikovich, S.M.: Simplified analytical solution of the contact problem on indentation of a coated half-space by a spherical punch. WIT Trans. Eng. Sci. **122**, 209–221 (2019)
46. Vasiliev, A.S., Sadyrin, E.V., Volkov, S.S., Kislyakov, E.A., Sevostianov, I.: Construction of the simplified analytical solution of the flat punch indentation contact problem. AIP Conf. Proc. **2188**(1), 040017 (2019)
47. Sneddon, I.N.: Fourier Transforms. McGraw-Hill, New York (1951)
48. Johnson, K.: Contact Mechanics. Cambridge University Press, Cambridge (1985)
49. Ueda, S.: Functionally graded piezoelectric strip with a penny-shaped crack under electromechanical loadings. Eur. J. Mech. A/Solids **27**(1), 50–60 (2008)

Chapter 36
Unsteady Elastic–Diffusion Vibrations of a Simply Supported Euler–Bernoulli Beam Under the Distributed Transverse Load

Andrei V. Zemskov, Anatoly S. Okonechnikov, and Dmitry V. Tarlakovskii

Abstract The unsteady vibrations problem of a simply supported Euler–Bernoulli beam under the distributed transverse load is considered. For the mathematical problem formulation, we use the system of a beam deflections equations with inner diffusion processes. The system is obtained using the d'Alembert variational principle from a generalized elastic–diffusion problem with the nonzero diffusion fluxes relaxation. To solve the system, the Green's function method is used. To find the Green's functions, the Laplace integral transform and Fourier series expansion are used. The Laplace transform inversion is done using residues and operational calculus tables. Calculation examples are considered for a rectangular dural beam. The beam deflections and components concentration increments are calculated in the alloy under distributed transverse load. The influence of mass transfer and diffusion flows relaxation on the displacement field inside the beam is analyzed.

Keywords Elastic–diffusion · Coupled problem · Unsteady problem · Green's function · Integral transformation · Multicomponent continuum · Euler–Bernoulli beam

36.1 Introduction

The interaction of diffusion and stresses is the subject of long-standing experimental and theoretical studies. The first experimental confirmations of the interaction between deformations and diffusion were obtained in the 1930s by Gorskij [14].

A. V. Zemskov (✉) · D. V. Tarlakovskii
Moscow Aviation Institute, Institute of Mechanics Lomonosov Moscow State University, Moscow, Russia
e-mail: azemskov1975@mail.ru

D. V. Tarlakovskii
e-mail: tdvhome@mail.ru

A. S. Okonechnikov
Moscow Aviation Institute, Moscow, Russia
e-mail: leon_lionheart@mail.ru

H. Altenbach et al. (eds.), *Multiscale Solid Mechanics*,
Advanced Structured Materials 141,
https://doi.org/10.1007/978-3-030-54928-2_36

Further, experimental research in this area was restarted in the middle of the 20th century. One of the first theoretical research on the influence of the generated stresses on the diffusion kinetics of binary solid mixtures is [17].

In the second half of the twentieth century, the theory of mass transfer in multi-component medium under extremely general assumptions regarding the rheological properties of the medium was developed in [12, 18, 19]. At present, the problem of the mechanodiffusion processes analysis in continuum also remains relevant. In this area, the range of problems under study is expanding due to the inclusion of temperature effects [1, 6, 8, 15, 20] and electromagnetic disturbances [2, 7, 10, 16, 24]. In addition, refined mass transfer models are used that take into account the finite propagation speed of diffusion flows. These models play an important role in the quick unsteady processes calculations. The listed publications consider both stationary [6, 8, 10, 16] and unsteady [1, 2, 20, 24]-related problems for a layer or half-space, mainly in the Cartesian coordinate system.

Since real bodies have a finite size, then the problems of mechanodiffusion for beams, plates and shells are of the greatest practical interest, as modern works devoted to modeling the interaction of mechanical and diffusion fields in these bodies are [3–5, 9, 21]. Here, the effect of diffusion processes on the permissible load of a thin transversally isotropic shell [21], contact interaction of a rod with an elastic half-space [5, 9] and mechanodiffusion processes in plates [3, 4] is considered too. All these problems are solved in the stationary formulation. The formulation of problems on unsteady elastic–diffusion vibrations of beams and plates and methods for their solution is not available in currently known publications.

In this article, unsteady mechanodiffusion vibrations of an Euler–Bernoulli beam are considered. The elastic–diffusion vibrations mathematical model was obtained on the variational principles basis that is described in [22, 23]. An analytical method is proposed for solving this problem, based on using of the Laplace integral transform and Fourier series expansion in eigenfunctions of the elastic–diffusion problem.

36.2 Problem Formulation

We consider the beam under a distributed transverse load action. The material of the beam is a multicomponent medium. Diffusion flows in the medium arise under the action of internal forces. In this case, mass transfer, which causes volumetric deformations, affects the displacement field inside the beam. For the mathematical problem formulation, the elastic–diffusion vibrations model of an isotropic Euler–Bernoulli beam is used (infinite speed of propagation is assumed). It is obtained by using variational principles from the elastic–diffusion continuum model in a rectangular Cartesian coordinate system [22, 23] (Fig. 36.1):

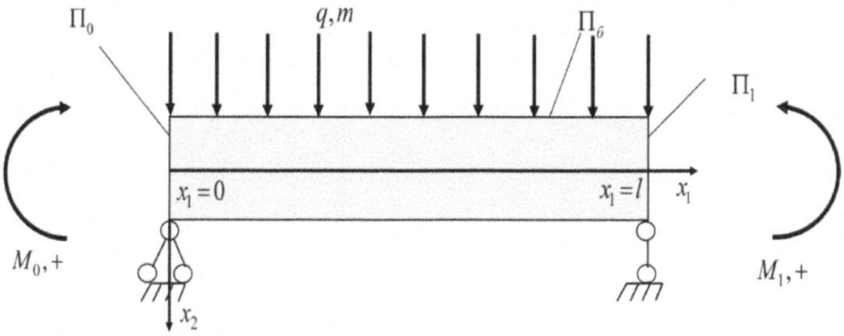

Fig. 36.1 Figure to the problem formulation. Forces and bending moments acting upon the beam

$$\ddot{u}'' - \eta\ddot{u} = u^{IV} + \sum_{j=1}^{N}\alpha_j H_j'' - F_1,\ F_1 = \frac{q+m'}{J_3},\ \eta = \frac{F}{J_3},$$
$$\dot{H}_q + \tau_q\ddot{H}_q = D_q H_q'' + \Lambda_q u^{IV} + F_{q+1},\ F_{q+1} = \frac{z^{(q)}}{J_3}; \tag{36.1}$$

$$\left.\left(u'' + \sum_{j=1}^{N}\alpha_j H_j\right)\right|_{x=0} = 0,\ u|_{x=0} = 0,\ H_q\big|_{x=0} = 0,$$
$$\left.\left(u'' + \sum_{j=1}^{N}\alpha_j H_j\right)\right|_{x=1} = 0,\ u|_{x=1} = 0,\ H_q\big|_{x=1} = 0. \tag{36.2}$$

Here, the dots denote time derivatives, and the dashes are the spatial derivatives in the coordinate $x = x_1$. All quantities in (36.1) and (36.2) are dimensionless. For them, the following notation is used

$$x_i = \frac{x_i^*}{L},\ u = \frac{u^*}{L},\ \tau = \frac{ct}{L},\ \lambda = \frac{\lambda^*}{\lambda^* + 2\mu^*},\ \mu = \frac{\mu^*}{\lambda^* + 2\mu^*},\ C^2 = \frac{\lambda^* + 2\mu^*}{\rho},$$
$$\alpha_q = \frac{\alpha^{*(q)}}{\lambda^* + 2\mu^*},\ D_q = \frac{D^{*(q)}}{CL},\ \Lambda_q = \frac{m^{(q)}D^{*(q)}\alpha^{*(q)}n_0^{(q)}}{\rho RT_0 CL},\ \tau_q = \frac{C\tau^{(q)}}{L},$$
$$q = \frac{\rho L q^*}{\lambda^* + 2\mu^*},\ m = \frac{\rho m^*}{\lambda^* + 2\mu^*},\ z^{(q)} = \frac{L z^{(q)*}}{C},\ F = \frac{F^*}{L^2},\ J_3 = \frac{J_3^*}{L^4}, \tag{36.3}$$

where t is a time; x_i^* are rectangular Cartesian coordinates; $u^*\left(x_1^*, t\right)$ are displacement of the beam; L is the beam length; $H_q\left(x_1^*, t\right)$ is the concentration increment of q-th component in the N-component medium; $n_0^{(q)}$ are initial concentrations; λ^* and μ^* are Lame coefficients; ρ is the medium density; $\alpha^{*(q)}$ are coefficients characterizing the medium volumetric changes due to diffusion; $D^{*(q)}$ are the self-diffusion coefficients; R is the universal gas constant; T_0 is the initial temperature; $m^{(q)}$ is the molar mass; $\tau^{(q)}$ is the relaxation time of diffusion perturbations; q^* is a transverse load density;

m^* is a moment density (moment per unit volume); $z^{(q)*}$ is a linear distribution of diffusion volume source; J_3 is the moment of inertia of the beam section relative to the axis Ox_3; F is the cross-sectional area.

36.3 Method of Solution

The solutions of the problem (36.1), (36.2) are represented in the form ($k = \overline{1, N+1}$):

$$\begin{aligned} u(x,\tau) &= \sum_{k=1}^{N+1} \int_0^\tau \int_0^1 G_{1k}(x,\xi,\tau-t) F_k(\xi,t)\, d\xi dt, \\ H_q(x,\tau) &= \sum_{k=1}^{N+1} \int_0^\tau \int_0^1 G_{1k}(x,\xi,\tau-t) F_k(\xi,t)\, d\xi dt. \end{aligned} \tag{36.4}$$

Here G_{mk} are the bulk Green's functions, which satisfy the equations

$$\begin{aligned} \ddot{G}''_{1k} - \eta \ddot{G}_{1k} &= G^{IV}_{1k} + \sum_{j=1}^{N} \alpha_j G''_{j+1,k} - \delta_{1k}\delta(x-\xi)\delta(\tau), \\ \dot{G}_{q+1,k} + \tau_q \ddot{G}_{q+1,k} &= D_q G''_{q+1,k} + \Lambda_q G^{IV}_{1k} + \delta_{q+1,k}\delta(x-\xi)\delta(\tau), \end{aligned}$$

and homogeneous boundary conditions

$$\begin{aligned} \left.\left(G''_{1k} + \sum_{j=1}^{N} \alpha_j G_{j+1,k}\right)\right|_{x=0} &= 0, \quad G_{1k}|_{x=0} = 0, \quad G_{q+1,k}\big|_{x=0} = 0, \\ \left.\left(G''_{1k} + \sum_{j=1}^{N} \alpha_j G_{j+1,k}\right)\right|_{x=1} &= 0, \quad G_{1k}|_{x=1} = 0, \quad G_{q+1,k}\big|_{x=1} = 0. \end{aligned}$$

Applying to system (36.4), the Laplace transform with respect to time and expanding the Green's functions into Fourier series in a variable x, we obtain

$$\begin{aligned} & k_1(\lambda_n, s)\, G^{Ls}_{1k}(\lambda_n, \xi, s) - \lambda_n^2 \sum_{j=1}^{N} \alpha_j G^{Ls}_{j+1,k}(\lambda_n, \xi, s) = 2\delta_{1k} \sin \lambda_n \xi, \\ & -\Lambda_q \lambda_n^4 G^{Ls}_{1k}(\lambda_n, \xi, s) + k_{q+1}(\lambda_n, s)\, G^{Ls}_{q+1,k}(\lambda_n, \xi, s) = 2\delta_{q+1,k} \sin \lambda_n \xi, \\ & k_1(\lambda_n, s) = \left(\lambda_n^2 + \eta\right) s^2 + \lambda_n^4, \quad k_{q+1}(\lambda_n, s) = s + \tau_q s^2 + D_q \lambda_n^2, \end{aligned}$$

where

$$G^L_{mk}(x, \xi, s) = \int_0^\infty G_{mk}(x, \xi, \tau)\, e^{-s\tau} d\tau,$$

$$G_{mk}^{Ls}(\lambda_n,\xi,s)=2\int_0^1 G_{mk}^{L}(x,\xi,s)\sin\lambda_n x dx,\ \ G_{mk}^{L}(x,\xi,s)=\sum_{n=1}^{\infty}G_{mk}^{Ls}(\lambda_n,\xi,s)\sin\lambda_n x.$$

The solution of this system has the following form $\left(q,p=\overline{1,N},\ k=\overline{1,N+1}\right)$:

$$G_{1k}^{Ls}=\frac{2P_{1k}(\lambda_n,s)}{P(\lambda_n,s)}\sin\lambda_n\xi,$$
$$G_{q+1,k}^{Ls}=\frac{2\delta_{q+1,k}}{k_{q+1}(\lambda_n,s)}\sin\lambda\xi+\frac{2P_{q+1,k}(\lambda_n,s)}{Q_q(\lambda_n,s)}\sin\lambda_n\xi. \tag{36.5}$$

Here

$$P(\lambda_n,s)=\left[\left(\lambda_n^2+\eta\right)s^2+\lambda_n^4\right]\Pi(\lambda_n,s)-\lambda_n^6\sum_{j=1}^{N}\alpha_j\Lambda_j\Pi_j(\lambda_n,s),$$
$$Q_q(\lambda_n,s)=k_{q+1}(\lambda_n,s)P(\lambda_n,s); \tag{36.6}$$

$$P_{11}(\lambda_n,s)=\Pi(\lambda_n,s),\ P_{q+1,1}(\lambda_n,s)=\Lambda_q\lambda_n^4\Pi_q(\lambda_n,s),$$
$$P_{1,q+1}(\lambda_n,s)=\lambda_n^2\alpha_q\Pi_q(\lambda_n,s),\ P_{q+1,p+1}(\lambda_n,s)=\lambda_n^6\Lambda_q\alpha_p\Pi_p(\lambda_n,s), \tag{36.7}$$

$$\Pi(\lambda_n,s)=\prod_{j=1}^{N}k_{j+1}(\lambda_n,s),\ \Pi_q(\lambda_n,s)=\prod_{j=1,j\neq q}^{N}k_{j+1}(\lambda_n,s).$$

The polynomial $P(\lambda_n,s)$ has degree $2N+2$ in variable s. Then, with the help of residue and the tables of operational calculus [11], we obtain the originals of the functions G_{mk}^{Ls} in the formulas (36.5):

$$G_{1k}^{S}(\lambda_n,\xi,s)=2\sum_{j=1}^{2N+2}A_{1k}^{(j)}(\lambda_n)e^{s_j(\lambda_n)\tau}\sin\lambda_n\xi,$$
$$G_{q+1,k}^{S}(\lambda_n,\xi,\tau)=2\sum_{l=1}^{2}\left[\frac{\delta_{q+1,k}}{1+2\tau_q\xi_l(\lambda_n)}+A_{q+1,k}^{(2N+2+l)}(\lambda_n)\right]e^{\xi_l(\lambda_n)\tau}\sin\lambda_n\xi$$
$$+2\sum_{j=1}^{2N+2}A_{q+1,k}^{(j)}(\lambda_n)e^{s_j(\lambda_n)\tau}\sin\lambda_n\xi,$$
$$\xi_{1,2}(\lambda_n)=\frac{-1\pm\sqrt{1-4\tau_qD_q\lambda_n^2}}{2\tau_q}, \tag{36.8}$$

where $s_j(\lambda_n)$, $j = \overline{1, 2N+2}$ are zeros of polynomial $P(\lambda_n, s)$, the coefficients $A_{1k}^{(j)}(\lambda_n)$ are determined by the formulas

$$A_{1k}^{(j)}(\lambda_n) = \frac{P_{1k}(\lambda_n, s_j(\lambda_n))}{P'(\lambda_n, s_j(\lambda_n))} \left(j = \overline{1, 2N+2},\ k = \overline{1, N+1},\ q = \overline{1, N}\right),$$
$$A_{q+1,1}^{(j)}(\lambda_n) = \frac{P_{q+1,1}(\lambda_n, s_j(\lambda_n))}{P'(\lambda_n, s_j(\lambda_n))} \left(l = 1, 2,\ m = \overline{2, N+1}\right), \tag{36.9}$$
$$A_{q+1,m}^{(j)}(\lambda_n) = \frac{P_{q+1,k}(\lambda_n, s_r(\lambda_n))}{Q_q'(\lambda_n, s_r(\lambda_n))},\ A_{q+1,m}^{(2N+2+l)} = \frac{P_{q+1,k}(\lambda_n, \xi_l(\lambda_n))}{Q_q'(\lambda_n, \xi_l(\lambda_n))}.$$

36.4 The Transition to Diffusion Perturbations Infinite Propagation Velocities and to Classical Problems of the Theory of Elasticity

Since any problem mathematical formulation is ideal in comparison with real processes, it is always necessary to use some simplifications. So, if the time interval over which the system is considered is much longer than the relaxation time τ_q (the average time the system returns to equilibrium), then the corresponding relaxation effects can be neglected. For this, in the expressions obtained for the Green's functions (36.5) and (36.9), it is necessary to perform passage to the limit $\tau_q \to 0$. The number of zeros of the polynomial $P(\lambda_n, s)$ under condition $\tau_q \to 0$ is reduced from $2N+2$ to $N+2$. For additional zeros of the polynomial $Q_q(\lambda_n, s)$, we get

$$2N+2 \to N+2,$$
$$k_{q+1}(\lambda_n, s) = s + \tau_q s^2 + D_q \lambda_n^2 \to s + D_q \lambda_n^2; \tag{36.10}$$

$$\xi_1(\lambda_n)\} = s_{N+3}(\lambda_n) = \frac{-1 - \sqrt{1 - 4\tau_q D_q \lambda_n^2}}{2\tau_q} \to -\infty\ (\tau_q \to 0), \tag{36.11}$$

$$\xi_2(\lambda_n) = s_{N+4}(\lambda_n) = \frac{-1 + \sqrt{1 - 4\tau_q D_q \lambda_n^2}}{2\tau_q} \to -D_q \lambda_n^2\ (\tau_q \to 0)$$

Then, formulas (36.8) are written as follows:

$$G_{mk}(x, \xi, \tau) = \sum_{n=1}^{\infty} G_{mk}^s(\lambda_n, \xi, \tau) \sin \lambda_n x,$$
$$G_{1k}^s(\lambda_n, \xi, s) = 2\sum_{j=1}^{N+2} A_{1k}^{(j)}(\lambda_n)\, e^{s_j(\lambda_n)\tau} \sin \lambda_n \xi,$$

$$G^{s}_{q+1,k}(\lambda_n,\xi,\tau)=2\sum_{l=1}^{2}\left[\delta_{q+1,k}+A^{(N+2+l)}_{q+1,k}(\lambda_n)\right]e^{\xi_l(\lambda_n)\tau}\sin\lambda_n\xi$$
$$+2\sum_{j=1}^{N+2}A^{(j)}_{q+1,k}(\lambda_n)\,e^{s_j(\lambda_n)\tau}\sin\lambda_n\xi, \tag{36.12}$$

where the coefficients $A^{(j)}_{mk}(\lambda_n)$ are found by formulas (36.9) taking into account the correction equalities (36.10) and (36.11).

Assuming further in equalities (36.6), (36.7), (36.9) and (36.12) that $\alpha_q=0$, we obtain the Green's functions for the classical elastic problem for an Euler–Bernoulli beam

$$G^{L(el)}(x,\xi,s)=\sum_{n=1}^{\infty}G^{Ls}_{11}(\lambda_n,\xi,s)\sin\lambda_n x=-2\sum_{n=1}^{\infty}\frac{\sin\lambda_n\xi\sin\lambda_n x}{\left(\lambda_n^2+\eta\right)s^2+\lambda_n^4}.$$

The original of this expression with help of the tables of operational calculus is sought [11]

$$G^{(el)}(x,\xi,\tau)=-2\sum_{n=1}^{\infty}\frac{\gamma_n\sin\gamma_n\tau\sin\lambda_n\xi\sin\lambda_n x}{\lambda_n^4},\quad \gamma_n=\frac{\lambda_n^2}{\sqrt{\lambda_n^2+\eta}}. \tag{36.13}$$

36.5 Example

For the calculation, we assume that the external perturbations in equations (36.1) are given as follows:

$$F_1(x,\tau)=\frac{q(x,\tau)}{J_3}+\frac{m'(x,\tau)}{J_3}=H(x)H(\tau),$$
$$F_{q+1}(x,\tau)=\frac{z^{(q)}}{J_3}=0. \tag{36.14}$$

Then, according to (36.4), the solution of the elastic–diffusion problem (36.1) and (36.2) has the form

$$u(x,\tau)=\int_0^{\tau}\int_0^1 G_{11}(x,\xi,\tau-t)F_1(\xi,t)\,\mathrm{d}\xi\,\mathrm{d}t$$
$$=2\sum_{n=1}^{\infty}\left[\sum_{j=1}^{2N+2}A^{(j)}_{11}(\lambda_n)\int_0^{\tau}H(t)\,e^{s_j(\lambda_n)(\tau-t)}\mathrm{d}t\int_0^1 H(\xi)\sin\lambda_n\xi\,\mathrm{d}\xi\right]\sin\lambda_n x$$
$$=2\sum_{n=1}^{\infty}\sum_{j=1}^{2N+2}\left[1-(-1)^n\right]\frac{\left(e^{s_j(\lambda_n)\tau}-1\right)A^{(j)}_{11}(\lambda_n)}{\lambda_n s_j(\lambda_n)}\sin\lambda_n x,$$

$$
\begin{aligned}
H_q(x,\tau) &= \int_0^\tau \int_0^1 G_{q+1,1}(x,\xi,\tau-t)\, F_1(\xi,t)\, d\xi\, dt \\
&= 2\sum_{n=1}^{\infty}\left[\sum_{j=1}^{2N+4} A_{q+1,1}^{(j)}(\lambda_n) \int_0^\tau H(t)\, e^{s_j(\lambda_n)(\tau-t)} dt \int_0^1 H(\xi)\sin\lambda_n\xi\, d\xi\right]\sin\lambda_n x \\
&= 2\sum_{n=1}^{\infty}\sum_{j=1}^{2N+4}\left[1-(-1)^n\right]\frac{\left(e^{s_j(\lambda_n)\tau}-1\right)A_{q+1,1}^{(j)}(\lambda_n)}{\lambda_n s_j(\lambda_n)}\sin\lambda_n x.
\end{aligned}
$$

Here $s_{2N+3}(\lambda_n) = \xi_1(\lambda_n)$, $s_{2N+4}(\lambda_n) = \xi_2(\lambda_n)$.

For calculation, we take the duralumin beam of length $L = 10^{-1}$ and rectangular cross section $h \times b = 0.1L \times 0.05L$. We consider duralumin as two-component material ($N = 2$), the physical characteristics of which [13] after applying the procedure of dimensionless (36.3) are as follows:

$$
\lambda = 4.92\times10^{-1},\ \mu = 2.54\times10^{-1},\ \alpha_1 = 1.50\times10^{-4},\ \alpha_2 = 5.92\times10^{-4},
$$
$$
D_1 = 1.27\times10^{-16}, D_2 = 5.02\times10^{-21}, \Lambda_1 = 2.77\times10^{-18}, \Lambda_2 = 5.50\times10^{-23}.
$$

The dimensionless geometric characteristics of the section are as follows:

$$
F = 5.00\times10^{-3},\ J_3 = 4.16\times10^{-6}.
$$

The calculation results are presented in Figs. 36.2, 36.3 and 36.4.

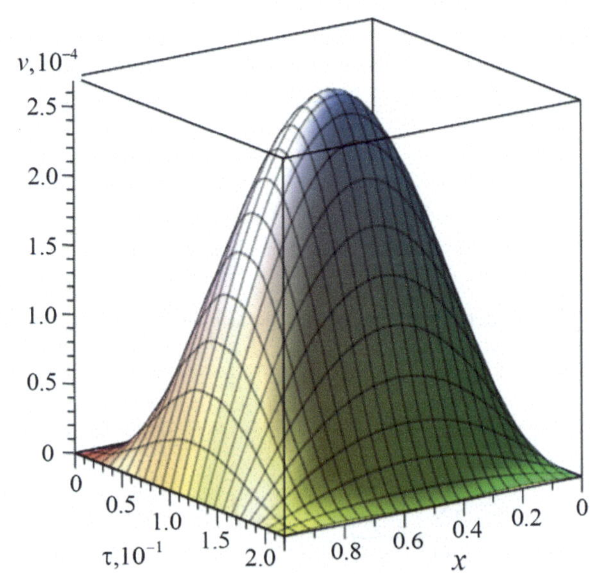

Fig. 36.2 Beam deflections $u(x,\tau)$

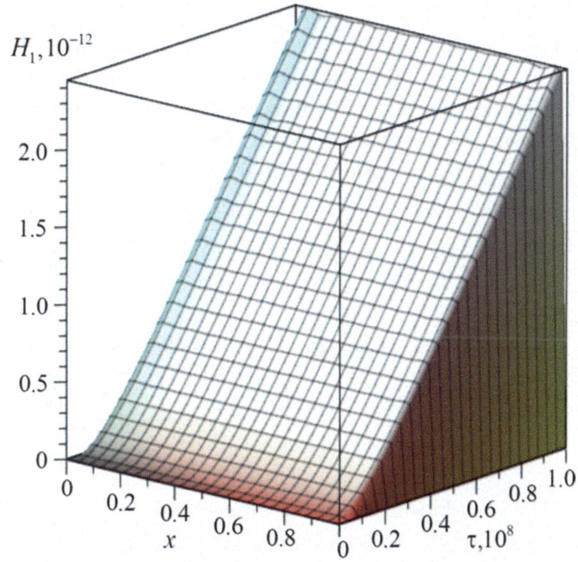

Fig. 36.3 Concentration increment H_1 (x, τ) at a point $x = 0.5$

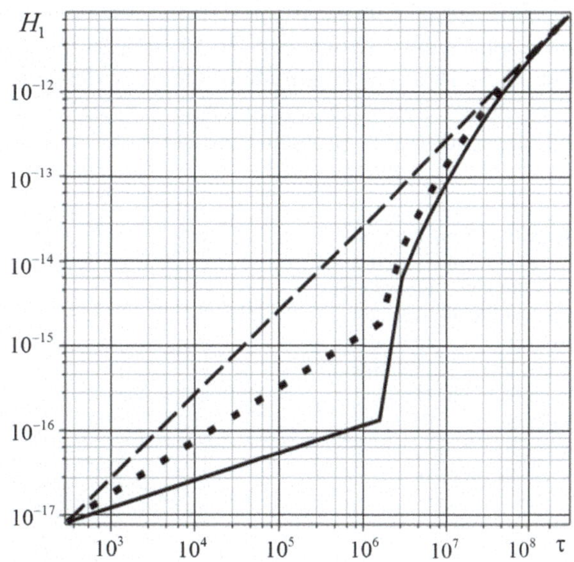

Fig. 36.4 Concentration increment H_1 (x, τ) at a point $x = 0.5$. The solid line is the solution of the problem (36.1) and (36.2) at $\tau^{(q)} = 200\,\text{s}$, the dashed line is the solution of the problem (36.1) and (36.2) at $\tau^{(q)} = 100\,\text{s}$, the hashed line is the solution of the elastic problem

As can be seen from the graphs shown in Fig. 36.4, the relaxation processes of mass transfer affect for a relatively short period of time. Here dimensionless time 10^8 is corresponding to about 3 s.

To assess the effect of mass transfer on the beam stress–strain state, we use the elastic problem solution. Using the Green's functions (36.13) and calculating the corresponding convolution (36.4), we obtain

$$\begin{aligned} u^{(el)}(x,\tau) &= \int_0^{\tau}\int_0^1 G^{(el)}(x,\xi,\tau)\,F_1(\xi,t)\,d\xi dt \\ &= -2\sum_{n=1}^{\infty}\left[\int_0^{\tau} H(t)\sin\left[\gamma_n(\tau-t)\right]dt\int_0^1 H(\xi)\sin\lambda_n\xi d\xi\right]\frac{\gamma_n\sin\lambda_n x}{\lambda_n^4} \\ &= 2\sum_{n=1}^{\infty}\left[1-(-1)^n\right](1-\cos\gamma_n\tau)\,\frac{\sin\lambda_n x}{\lambda_n^5}. \end{aligned}$$

The calculation results for various relaxation times for elastic-diffusive and elastic vibrations of the beam are presented in Figs. 36.5, 36.6 and 36.7.

For the results, presented in Figs. 36.5, 36.6 and 36.7, it can be concluded that the influence of both mass transfer and relaxation of diffusion fluxes on a beam deflections in the case of a load (36.14) appears initially at a very small time interval (Fig. 36.5). Here, the amplitude of the beam deflections can vary by several orders of magnitude. Further, at time intervals $\tau \sim 10^{-1}$ to 10^9, the solutions of the elastic and elastic–diffusion problems are coincided. Starting from the time $\tau \sim 10^9$, a

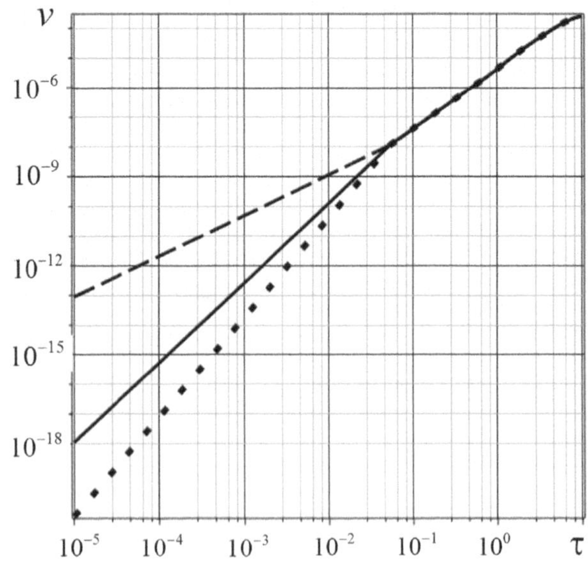

Fig. 36.5 Beam deflections $u(x,\tau)$ at a point $x = 0.5$. The solid line is the solution of the problem (36.1) and (36.2) at $\tau^{(q)} = 200$ s, the dashed line is the solution of the problem (36.1) and (36.2) at $\tau^{(q)} = 100$ s, the hashed line is the solution of the elastic problem

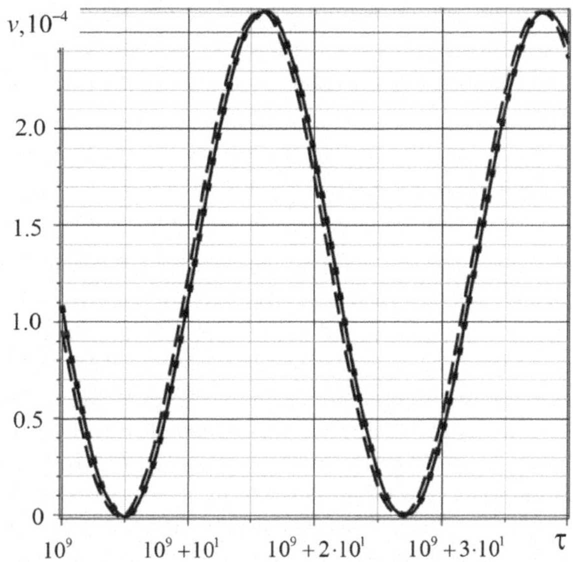

Fig. 36.6 Beam deflections $u\,(x,\,\tau)$ at a point $x = 0.5$. The solid line is the solution of the problem (36.1) and (36.2) at $\tau^{(q)} = 200\,\text{s}$, the dashed line is the solution of the problem (36.1) and (36.2) at $\tau^{(q)} = 100\,\text{s}$, the hashed line is the solution of the elastic problem

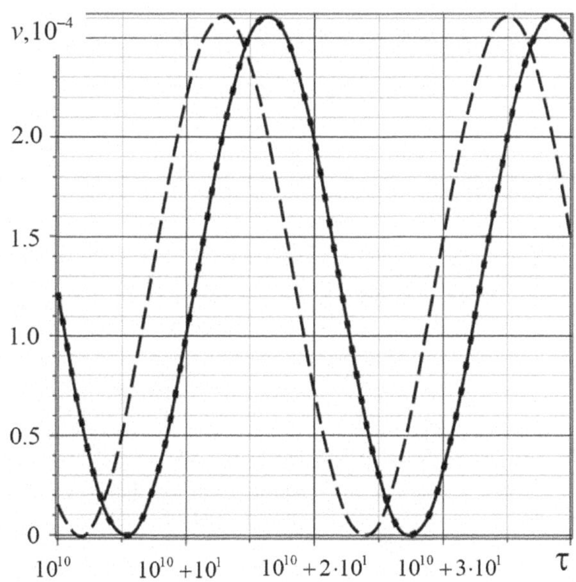

Fig. 36.7 Beam deflections $u\,(x,\,\tau)$ at a point $x = 0.5$. The solid line is the solution of the problem (36.1) and (36.2) at $\tau^{(q)} = 200\ s$, the dashed line is the solution of the problem (36.1) and (36.2) at $\tau^{(q)} = 100\,\text{s}$, the hashed line is the solution of the elastic problem

noticeable phase shift of the elastic–diffusion vibrations with respect to the elastic vibrations begins to be observed (Figs. 36.6 and 36.7). The deflection amplitude of the beam remains constant.

36.6 Conclusions

The unsteady elastic–diffusion vibrations of an Euler–Bernoulli beam are investigated. An algorithm is proposed for constructing bulk Green's functions. The elastic–diffusion vibrations problem of a simply supported beam under the distributed transverse load action is solved. Based on the developed model and the considered example, the interaction of the mechanical and diffusion fields is studied. Calculations using the example for the duralumin alloy show that mass transfer over time begins to affect the displacement field inside the beam, which manifests itself only in the form of a phase delay in the elastic–diffusion vibrations with respect to the elastic vibrations. The relaxation of diffusion fluxes practically does not affect beam deflections, but during relatively small time interval significantly affects the diffusion process.

This work was supported by the Russian Science Foundation (Project No. 20-19-00217).

References

1. Afram, A.Y., Khader, S.E.: 2D problem for a half-space under the theory of fractional thermoelastic diffusion. Am. J. Sci. Ind. Res. **6**(3), 47–57 (2014)
2. Aouadi, M.: Variable electrical and thermal conductivity in the theory of generalized thermoelastic diffusion. Z. Angew. Math. Phys. **57**(2), 350–366 (2005)
3. Aouadi, M.: A Smooth attractor for a nonlinear thermoelastic diffusion thin plate based on Gurtin-Pipkins model. Asymptot. Anal. **95**, 129–160 (2015)
4. Aouadi, M.: On thermoelastic diffusion thin plate theory. Appl. Math. Mech. Engl. Ed. **36**(5), 619–632 (2015)
5. Aouadi, M., Copetti, M.I.M.: Analytical and numerical results for a dynamic contact problem with two stops in thermoelastic diffusion theory. ZAMM Z. Angew. Math. Mech. (2015). https://doi.org/10.1002/zamm.201400285
6. Atwa, S.Y., Egypt, Z.: Generalized thermoelastic diffusion with effect of fractional parameter on plane waves temperature-dependent elastic medium. J. Mater. Chem. Eng. **1**(2), 55–74 (2013)
7. Bachher, M., Sarkar, N.: Fractional order magneto-thermoelasticity in a rotating media with one relaxation time. Math. Models Eng. **2**(1), 57–68 (2016)
8. Belova, I.V., Murch, G.E.: Thermal and diffusion-induced stresses in crystalline solids. J. Appl. Phys. **77**(1), 127–134 (1995)
9. Copetti, M.I.M., Aouadi, M.: A quasi-static contact problem in thermoviscoelastic diffusion theory. Appl. Numer. Math. **109**, 157–183 (2016)
10. Deswal, S., Kalkal, K.: A two-dimensional generalized electro-magneto-thermoviscoelastic problem for a half-space with diffusion. Int. J. Therm. Sci. **50**(5), 749–759 (2011)

11. Ditkin, V.A., Prudnikov, A.P.: Handbook on Operational Calculus. Vysshaya Shkola, Moscow (1965). [In Russian]
12. Eremeev, V.S.: Diffuziya i napryazheniya. Energoatomizdat, Moscow (1985). [In Russian]
13. Grigoriev, I.S., Meylikhov, I.Z. (ed.): Fizicheskiye velichiny. Spravochnik. Energoatomizdat, Moscow (1991) (in Russian)
14. Gorskij, V.S.: Issledovanie uprugogo posledejstviya v splave Si-Au s uporyadochennoj reshetkoj. Zhurnal e'ksperimental'noj i teoreticheskoj fiziki **6**(3), 272–276 (1936) [in Russian]
15. Knyazeva, A.G.: Model of medium with diffusion and internal surfaces and some applied problems. Mater. Phys. Mech. **7**(1), 29–36 (2004)
16. Kumar, R., Chawla, V.: A study of Green's functions for two-dimensional problem in orthotropic magnetothermoelastic media with mass diffusion. Mater. Phys. Mech. **15**, 78–95 (2012)
17. Lyubov, B.Ya., Fastov, N.S.: Vliyanie koncentracionnyx napryazhenij na processy diffuzii v tverdyx rastvorax. Dop. Akad. Nauk USSR **54**(5), 939–941 (1965) (in Russian)
18. Nowacki, W.: Dynamical problems of thermodiffusion in solids. Proc. Vib. Prob. **15**, 105–128 (1974)
19. Pidstryhach, Ya.S.: Differential equations of the problem of thermodiffusion in a solid deformable isotropic body. Dop. Akad. Nauk USSR **2**, 169–172 (1961)
20. Sherief, H.H., El-Maghraby, N.M.: A thick plate problem in the theory of generalized thermoelastic diffusion. Int. J. Thermophys. **30**, 2044–2057 (2009)
21. Shvets, R.N., Flyachok, V.M.: The equations of mechanothermodiffusion of anisotropic shells taking account of transverse strains. Mat. Met. Fiz.-Mekh. Polya **20**, 54–61 (1984)
22. Tarlakovskii, D.V., Zemskov, A.V.: An elastodiffusive orthotropic Euler-Bernoulli beam with considering diffusion flux relaxation. Math. Comput. Appl. **24**(1), 23 (2019). https://doi.org/10.3390/mca24010023
23. Zemskov, A.V., Tarlakovskii, D.V.: Unsteady vibration model of the Euler-Bernoulli beam taking into account diffusion. J. Phys.: Conf. Ser. **1158**, 042043 (2019). https://doi.org/10.1088/1742-6596/1158/4/042043
24. Zhang, J., Li, Y.: A two-dimensional generalized electromagnetothermoelastic diffusion problem for a rotating half-space. Math. Probl. Eng. **2014** (2014). https://doi.org/10.4172/2165-7890.S1-007

The manufacturer's authorised representative in the EU is Springer Nature Customer Service Centre GmbH, Europaplatz 3, 69115 Heidelberg, Germany. If you have any concerns regarding our products, please contact ProductSafety@springernature.com

Printed and bound by CPI Group (UK) Ltd, Croydon, CR0 4YY
30/09/2026
02225506-0015